Organic Chemistry

J. Williams Suggs, Ph.D.
Associate Professor of Chemistry and Biochemistry
Brown University

All inquiries should be addressed to:
Barron's Educational Series, Inc.
250 Wireless Boulevard
Hauppauge, New York 11788
http://www.barronseduc.com

ISBN-13: 978-0-7641-1925-5
ISBN-10: 0-7641-1925-7

Library of Congress Catalog Card No. 2002018597

Library of Congress Cataloging-in-Publication Data

Suggs, J. William.
Organic chemistry / J. William Suggs.
p. cm.—(Barron's college review series)
Includes index.
ISBN 0-7641-1925-7
1. Chemistry, Organic—Outlines, syllabi, etc. I. Title.
II. College review series.
QD256.5 .S84 2002
547′.076—dc21

2002018597

PRINTED IN CANADA

9 8 7 6 5 4

CONTENTS

1
CHEMICAL BONDS AND LEWIS STRUCTURES

ATOMIC STRUCTURE

The Organic Chemist's Periodic Table

Carbon is unusual because it provides strong, stable bonds to almost every atom in the periodic table, including itself. Carbon is the framework atom that forms linkages to build biological molecules, drugs, synthetic polymers like polylethylene, and even molecule-sized objects that resemble bowls, windows, and wires. While professional organic chemists regularly use compounds with carbon bonded to ruthenium, tin, indium, and many other rare elements, introductory organic chemistry focuses on the carbon compounds of hydrogen and on the **second-row** elements of the periodic table—boron, carbon, nitrogen, and oxygen. Other elements also play important roles in organic chemistry: the alkali metals lithium, sodium, and potassium; the halogens fluorine, chlorine, bromine, and iodine; and the **third-row** elements magnesium, aluminum, silicon, phosphorus, and sulfur.

The Structure of the Atom

Throughout this chapter, we will assume that the reader has completed a course in general chemistry, including the material covered in Barron's College Review Series *Chemistry* by Neil Jespersen, so that only a brief overview of atomic structure is given. Each element has an **atomic number**, which is the number of protons in the nucleus or the number of electrons around the nucleus. The atomic number of carbon is 6. The same element can exist as different **isotopes** arising from different numbers of neutrons in the nucleus. Carbon has three important isotopes: ^{12}C, present in approximately 99 percent natural abundance, with six neutrons and six protons in the nucleus; ^{13}C, present in 1.1 percent natural abundance; and radioactive ^{14}C, which has a half-life of 5730 years. The **atomic mass** of carbon, taking the isotopic abundances into account, is 12.011. The atomic mass of the ^{12}C isotope is the reference for all the elements and is defined as 12.0000.

Most of organic chemistry ultimately comes down to the action of the **electromagnetic force**. This is the force that causes like charges (positive-positive or negative-negative) to repel one another and unlike charges (negative-positive) to attract one another. Quantum mechanics was

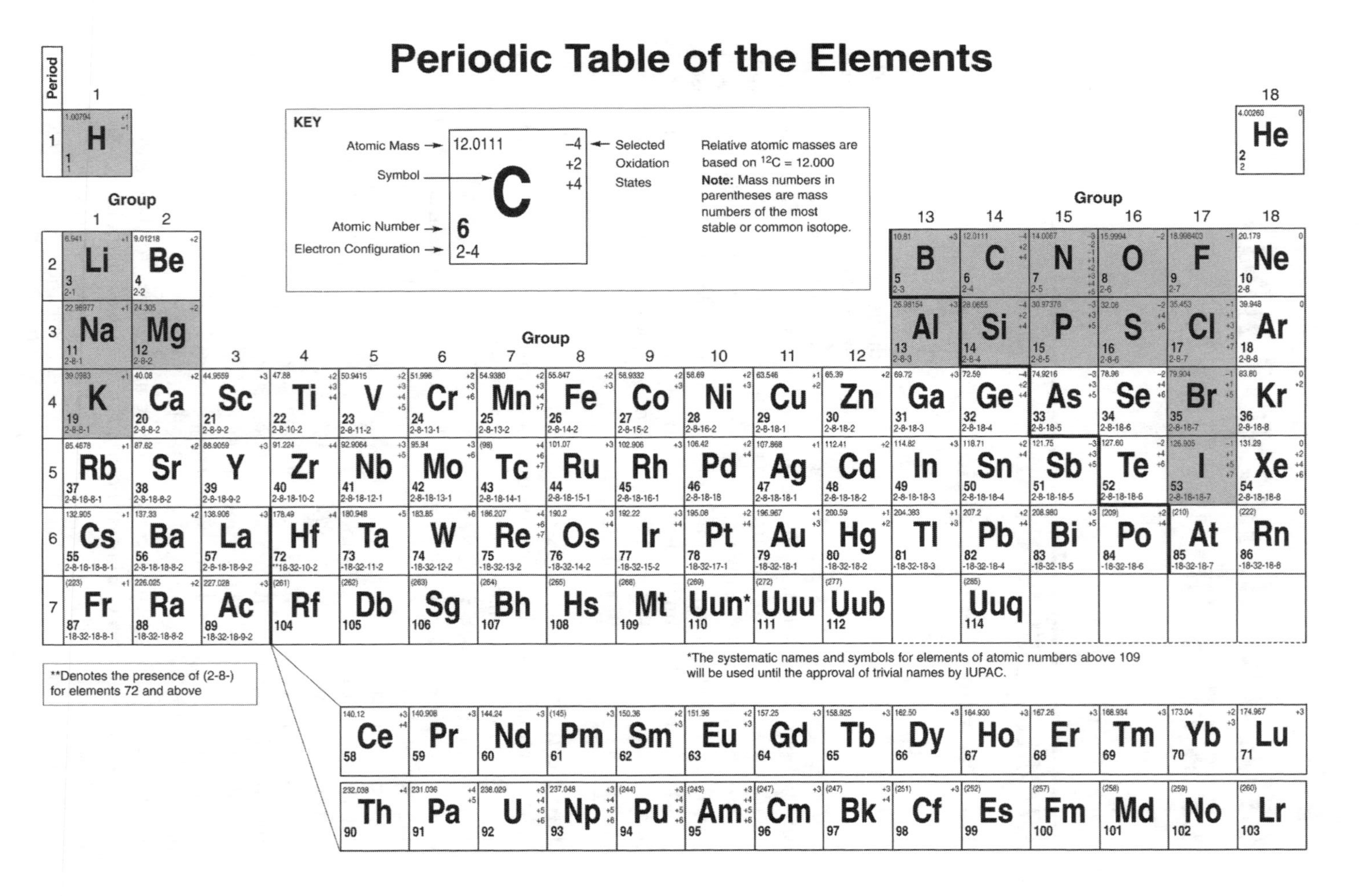
Periodic Table of the Elements
KEY
Atomic Mass → 12.0111
Symbol → C
Atomic Number → 6
Electron Configuration → 2-4
−4 +2 +4 ← Selected Oxidation States
Relative atomic masses are based on 12C = 12.000
Note: Mass numbers in parentheses are mass numbers of the most stable or common isotope.
Period
Group
*The systematic names and symbols for elements of atomic numbers above 109 will be used until the approval of trivial names by IUPAC.
**Denotes the presence of (2-8-) for elements 72 and above

triumphantly able to explain why electrons in atoms do not spiral into the nucleus under the influence of the electromagnetic force. Nevertheless, simple classical physics concepts of the electromagnetic force explain the basics of how atoms and molecules are held together.

Electron Energy Levels in Atoms

Quantum mechanics limits the positions of electrons in atoms to certain quantized positions, termed **principal energy levels**, around the nucleus. The first level holds no more than 2 electrons, the second level holds no more than 8 electrons, and the third level holds up to 18. The maximum number of electrons a level can hold is found using the formula $2n^2$, where n is the **principal quantum number**. Electrons in the first principal energy level are, on average, very close to the positively charged nucleus and are thus in a most stable environment. It takes a great deal of energy to move an electron in the first level to a higher level, and if an electron drops from a higher level into a vacant spot in the first level, energy is released in the form of X rays.

Within each principal energy level are sublevels. These sublevels are called **orbitals**. An orbital is a region of space where there is a certain probability of finding an electron. Each individual orbital can hold no more than two electrons. The first principal energy level contains just one orbital, the ***1s* orbital**. This orbital, like all orbitals in the *s* family, has spherical symmetry. An electron in this orbital in the hydrogen atom has the highest probability of being found 53 pm (10^{-12} m) from the nucleus. However, there is a lesser probability of finding the electron 50 or 60 pm from the nucleus. A diagram of an orbital sometimes tries to represent the electron probability distribution by showing the orbital as a shaded cloud, with the density of the shading meant to represent the probability of finding the electron at a given spot. More often, orbitals are drawn as boundary surfaces where the surface encloses a 90 percent or 95 percent probability of finding the electron within this boundary.

Second-row elements like carbon have four orbitals containing up to eight electrons: a ***2s*** orbital and three ***2p*** orbitals, the $\mathbf{2p_x}$, $\mathbf{2p_y}$, and $\mathbf{2p_z}$ orbitals. Orbitals in the second principal energy level have one **node**, which is a region of space where the probability of finding an electron is zero. For the 2*s* orbital, the node is a spherical shell, while for the *p* orbitals the node is a plane, which includes the nucleus, termed the **nodal plane**. Electrons in *s* orbitals have a greater probability of being found near the nucleus than electrons in *p* orbitals, making 2*s* electrons lower in energy than 2*p* orbitals.

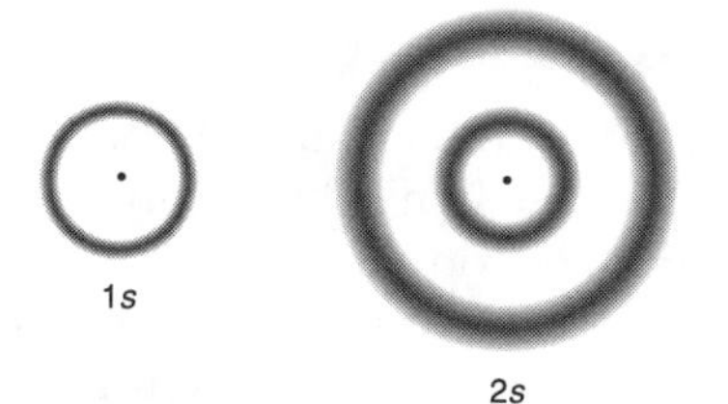

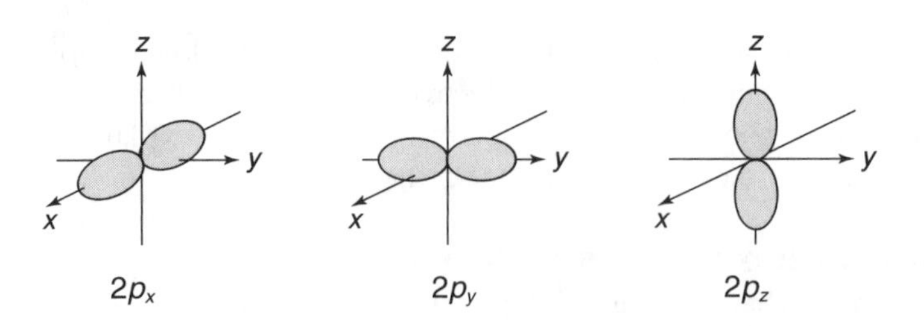

Electrons have a property called **spin** that can take the values +1/2 and −1/2. According to the **Pauli exclusion principle**, two electrons can occupy the same orbital only if they have opposite spins. Electrons with opposite spins are said to be paired and are usually written as arrows: ↑↓.

The ground state distribution of electrons in atoms can be summarized by the **ground state electron configuration**, which lists the orbitals that are filled or partly filled. Helium has the configuration $1s^2$, where the superscript indicates the number of electrons in the orbital. Carbon's electron configuration is $1s^2\,2s^2\,2p_x^{\,1}2p_y^{\,1}$. The three *p* orbitals are equal in energy (the technical term is **degenerate**), and it is arbitrary which *p* orbitals are filled and which are empty. However, **Hund's rule** states that for degenerate orbitals, electrons enter the orbitals singly before they pair up. To give a second example, in fluorine, the nine electrons give the configuration $1s^2\,2s^2\,2p_x^{\,2}\,2p_y^{\,2}\,2p_z^{\,1}$. Only the outermost electrons (for second-row elements, the 2*s* and 2*p* electrons) are involved in bonding. These electrons are the element's **valence electrons**. Carbon has four valence electrons, and fluorine has seven.

Example What is the ground state electron configuration of boron and of sulfur? How many valence electrons does each have?

Solution: Boron is $1s^2\,2s^2\,2p_x^{\,1}$ for a total of three valence electrons. Sulfur is $1s^2 2s^2 2p_x^{\,2} 2p_y^{\,2} 2p_z^{\,2} 3s^2 3p_x^{\,2} 3p_y^{\,1} 3p_z^{\,1}$ for a total of six valence electrons.

COVALENT BONDING AND MULTIPLE BONDS

The Electron Pair Bond

More than 95 years ago the American chemist Gilbert Newton Lewis first proposed that atoms form stable bonds by sharing two electrons. These **covalent** bonds with a shared electron pair continue to be formed by an atom until it attains the closed-shell electron configuration of a noble gas (neon in the case of second-row elements). **Lewis structures** are molecular structures in which valence electrons are represented by dots and bonds in molecules are represented by pairs of dots.

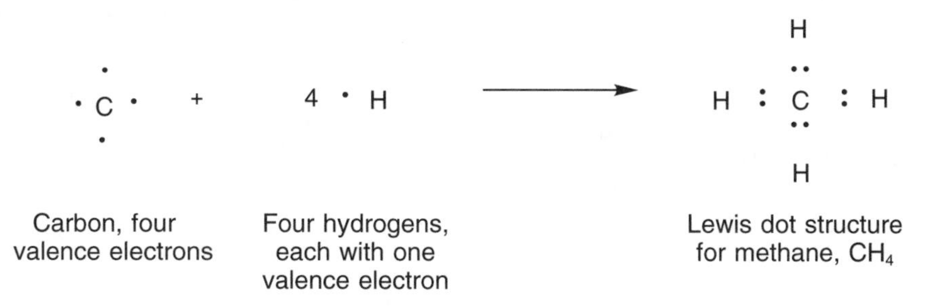

It is easy to make an electrostatic argument for the formation of molecules from atoms. In a molecule, the electron pairs are close to two positively charged nuclei, instead of just one as in the atom.

A number of atoms have some of their electrons paired in their ground state electron configuration: nitrogen and oxygen, for example. These electron pairs are normally maintained when the atoms form compounds and are called **lone pairs** or **nonbonding pairs**.

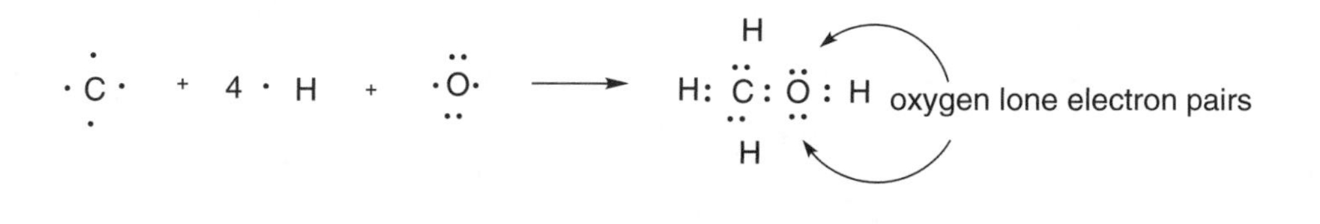

The Octet Rule

When writing Lewis structures, one invariant rule is the **octet rule**, which states that one can place no more than eight electrons around any second-row atom. Later, you will see examples of reactive organic molecules with six or seven electrons around a carbon atom, but never nine or ten electrons. Any molecule with five bonds to carbon or with four bonds to nitrogen plus a lone electron pair on nitrogen is an example of a molecule that violates the octet rule and is an invalid molecular structure. The octet rule does not hold for third-row elements, so one can draw valid structures with five or six bonds to atoms such as phosphorus and sulfur.

Because Lewis dot structures are cumbersome to draw, especially for larger molecules, it is usual to replace the bonding pair of dots with a line, giving H—H in place of H**:**H. The following are examples of structures that obey and structures that violate the octet rule.

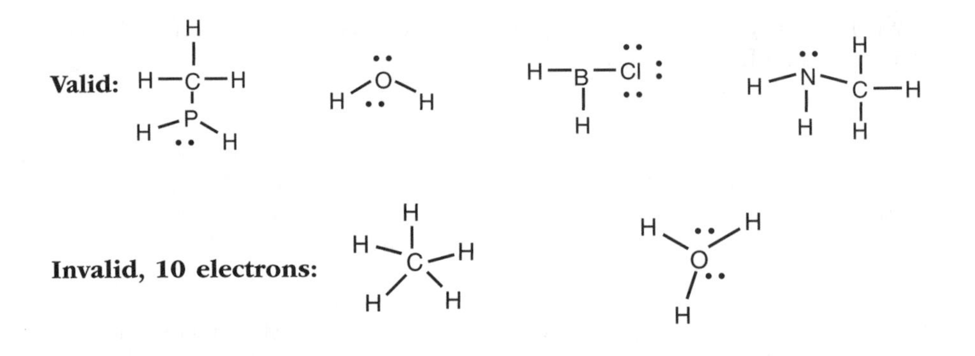

Multiple Bonds

When two atoms share more than one pair of electrons, the result is a **double bond** (two electron pairs) or a **triple bond** (three electron pairs). Second-row elements cannot form quadruple bonds. Double or triple bonds occur when there are not enough atoms to form single bonds between all the bonding partners, as shown below for ethane, C_2H_6, ethene (historical name, ethylene), C_2H_4, and ethyne (historical name, acetylene) C_2H_2.

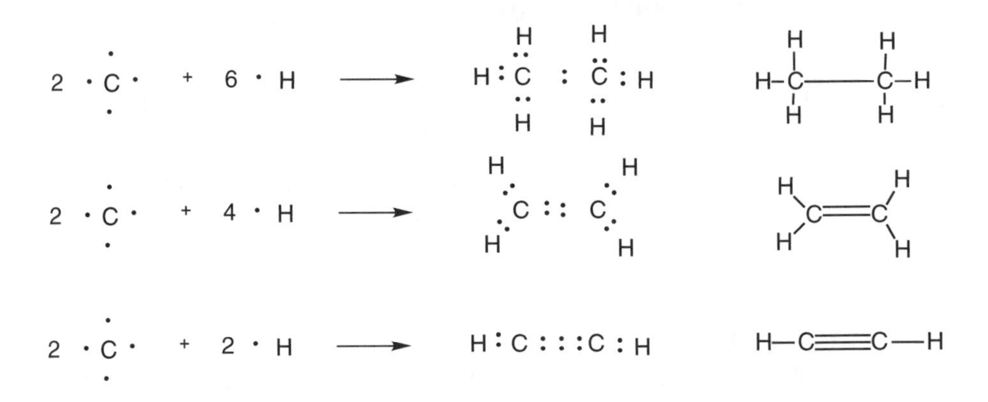

Formal Charges

Some molecules cannot be drawn without the inclusion of charges on individual atoms even when the overall molecule is neutral. These **formal charges** arise when the assigned number of electrons for an atom in a molecule differs from the atom's number of valence electrons. To calculate an atom's formal charge, take its number of valence electrons (its group number in the periodic table—5 for nitrogen, 6 for oxygen, and so on) and subtract the sum of the number of electrons that are lone pairs and one-half the number of electrons in bonding pairs. Since a covalently bound hydrogen does not have a formal charge, only the heavy atoms in a molecule need to be considered. For example, consider the solvent and racing fuel nitromethane:

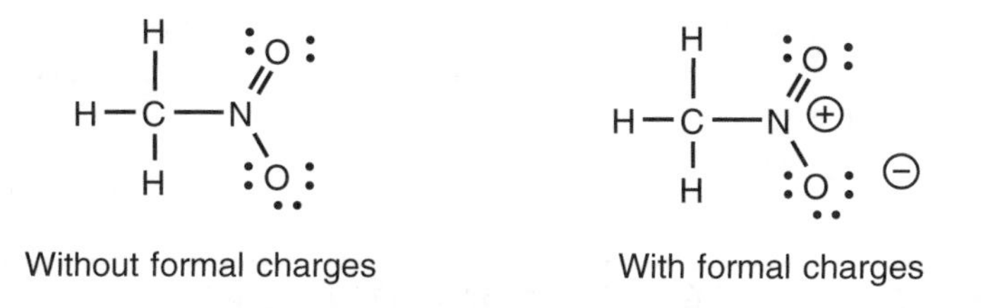

Without formal charges   With formal charges

Carbon has four valence electrons. Here there are four single bonds to carbon, and the formal charge is 4 – 4, or zero. The upper oxygen atom has two lone electron pairs and two bonding pairs, so its formal charge is 6 – [4 + 1/2(4)] = 0. The nitrogen atom has no lone pairs and four bonds. Its formal charge is 5 – 1/2(8) = +1. The lower oxygen atom's formal charge is –1 since it has three lone pairs and one bonding pair of electrons: 6 – [6 + 1/2(2)] = –1. In general, nitrogen atoms with four bonds have a formal charge of +1. Such nitrogens cannot have any lone pairs since that would violate the octet rule. Likewise, oxygen atoms with three lone pairs and a single covalent bond have a formal charge of –1.

EXERCISES

Assign formal charges to the following molecules.

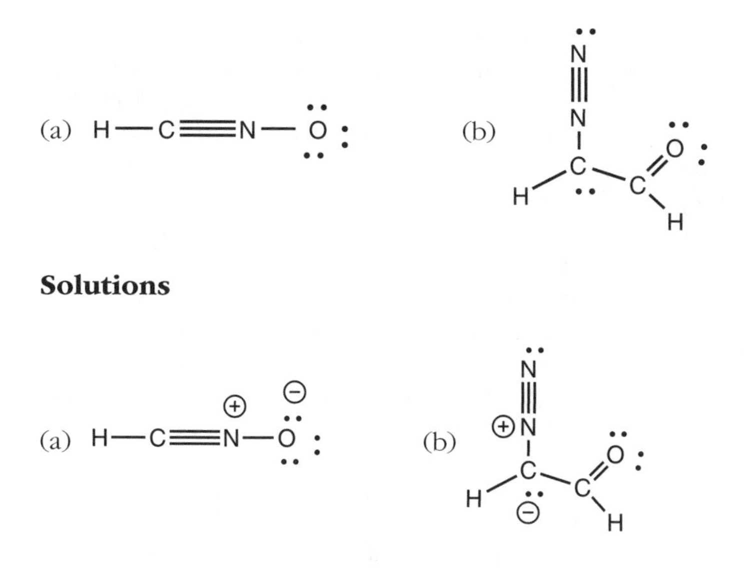

Solutions

The normal valences for carbon are four bonds and no lone electron pairs; for nitrogen, three bonds and one lone pair; and for oxygen, two bonds and two lone pairs. Therefore only the atoms that deviate from this pattern need to be considered. In (a) the nitrogen is 5 – [1/2(8)] = +1 and the O is 6 – [6 + 1/2(2)] = –1. In (b) the nitrogen is 5 – [1/2(8)] = +1 and the adjacent carbon is 4 – [2 + 1/2(6)] = –1.

ORGANIC IONS AND RADICALS

Reactive Intermediates

Carbon compounds with an octet of electrons are usually stable, whether they possess single, double, or triple bonds. However, carbon compounds are also known that have fewer than eight electrons. **Carbocations** are one such class of compounds. These carbon species have three bonds to carbon (or some other atom) and no lone electron pairs, and thus have a charge of +1.

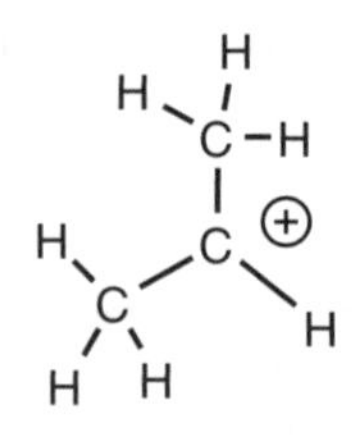

A carbocation

A carbocation (pronounced carbo-cat-ion not carbo-ka-shun) is a positively charged carbon ion, just as Na^+ is a sodium ion. The central carbon has only six electrons, not eight, surrounding it. The octet rule gives an upper limit to the number of electrons around a second-row element, but a fewer number of electrons is possible. Normally, carbocations are unstable species that have lifetimes much shorter than seconds.

A carbocation is an example of an organic **reactive intermediate**, a high-energy species that is generated and converted during the course of many organic reactions. Much of organic chemistry is devoted to understanding the creation, properties, and control of many classes of reactive intermediates.

Organic radicals are a second common class of reactive intermediates. These species also have three bonds to carbon, but in addition there is a single unpaired electron on the central carbon atom. Since three shared bonding pairs plus one unshared electron give an assignment of four electrons to the central carbon of a carbon radical, these species are uncharged. Nevertheless, because the central carbon is surrounded by only seven electrons, it is highly reactive. Since electrons have the property of spin, a single electron is sometimes called an unpaired spin.

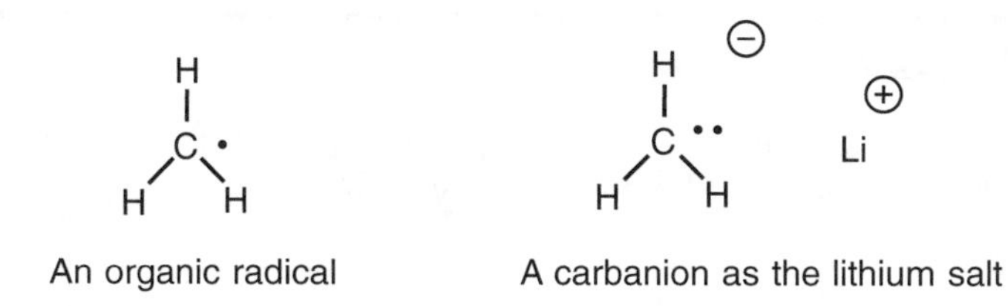

An organic radical A carbanion as the lithium salt

The third common reactive intermediate in organic chemistry is the **carbanion**, which is a carbon anion. These species have three bonds to carbon (which can include single, double, or triple bonds) and a lone electron pair on carbon. Following the formal charge rules, the central carbon is negatively charged. Ions have to have counterions to balance the charges, and in the case of carbanions, we will see that Li^+ and Mg^{2+} are common counterions. While carbanions have an octet of electrons around the central carbon, it is more stabilizing to share an electron pair on carbon (thus forming a bond) than to have the electron pair unshared. Therefore, carbanions are also reactive intermediates.

Organic Compounds with Charged Heteroatoms

In organic chemistry, nonmetallic atoms other than carbon and hydrogen are termed **heteroatoms**. Organic ions are often made that have either positively or negatively charged heteroatoms. As we have seen in our discussion of formal charges, four single bonds to nitrogen should result in a positively charged ion. Likewise, an oxygen or sulfur atom with one single bond and three lone electron pairs is normally an oxygen or sulfur anion. To keep drawings from being overwhelmed by lone pairs, the lone pairs on negatively charged oxygen and sulfur anions are usually not explicitly drawn in. It is up to the student to remember their presence and their number.

Heteroatomions:

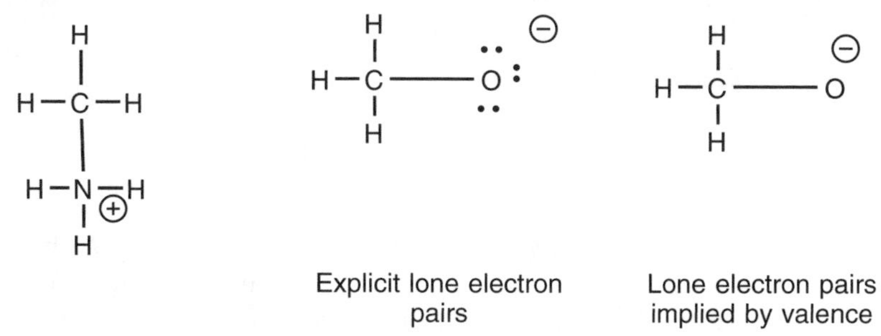

Explicit lone electron pairs Lone electron pairs implied by valence

In subsequent chapters, even neutral heteroatom-containing molecules (such as those with three bonds to nitrogen or two bonds to oxygen) will not have the lone electron pairs explicitly drawn on the molecule. A common mistake beginning students make is to forget about the presence of these lone pairs.

EXERCISES

Draw the following molecules with appropriate charges and show all the lone electron pairs on each heteroatom.

(a)

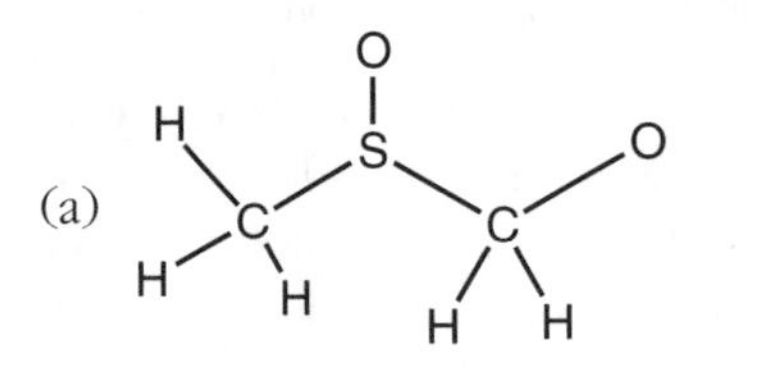

(b)

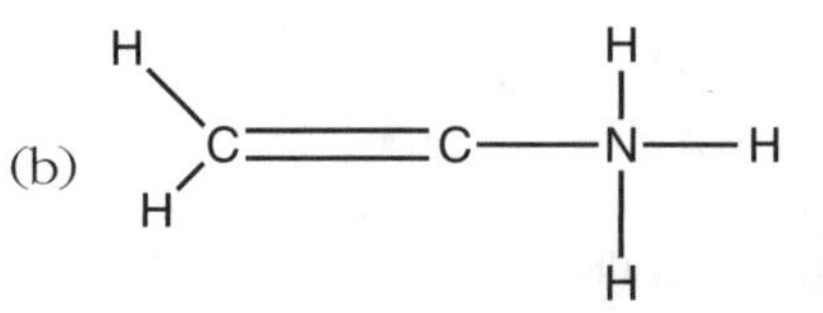

Both a radical and positively charged

Solutions

(a)

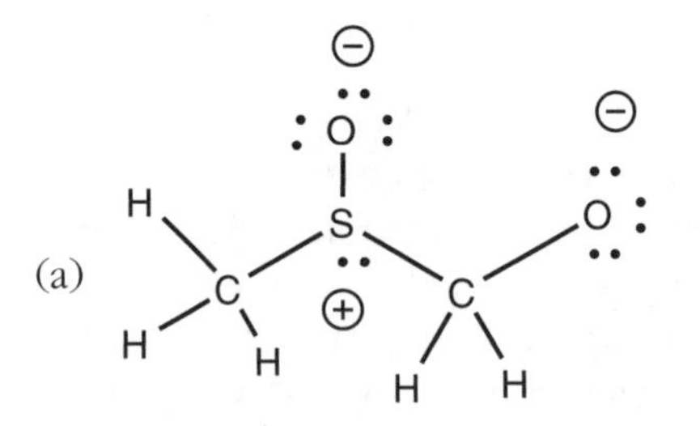

Sulfur is in the same group as oxygen. The bonds to oxygen or sulfur result in a positive formal charge and a lone electron pair is needed for an octet of electrons. A single bond to oxygen normally leads to an oxygen anion with three lone pairs.

(b)

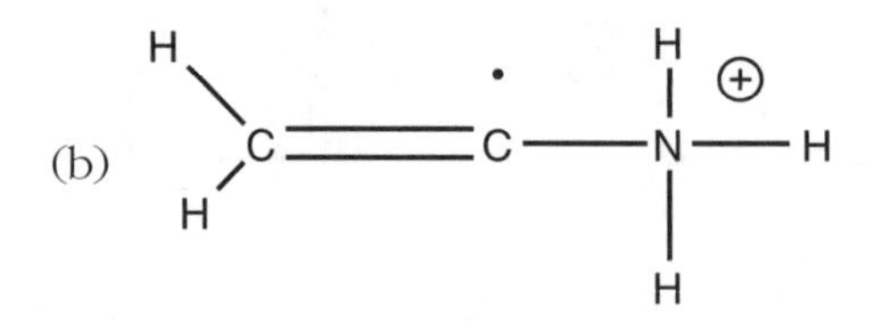

The middle carbon atom, with three bonds, could in principle be a carbocation, radical, or carbanion. Since nitrogen, with four bonds, has a formal charge of +1, the carbon is a radical with one unpaired spin and seven electrons in its valence shell.

ELECTRONEGATIVITY AND BOND POLARITY

Electronegativity

Two extremes exist in chemical bonding. When an electron pair is shared equally between two atoms, the result is a covalent bond. If the electron pair is completely transferred to one atom of a pair, the result is an **ionic bond**, as in NaCl. Normally, however, the distribution of electrons in a chemical bond is asymmetric, with the electrons more concentrated near one atom of the bonding pair. This distribution results in a **polar covalent bond**.

The concept of **electronegativity**, introduced into chemistry by Linus Pauling, is a measure of how strongly different elements attract electrons and thus lead to polar covalent or even ionic bonds with different atoms. Values of electronegativity for common elements are tabulated here.

PAULING ELECTRONEGATIVITIES

H 2.1						
Li 1.0	Be 1.5	B 2.0	C 2.5	N 3.0	O 3.5	F 4.0
Na 0.9	Mg 1.2	Al 1.5	Si 1.8	P 2.1	S 2.5	Cl 3.0
K 0.8	Ca 1.0	Ga 1.8	Ge 2.0	As 2.2	Se 2.5	Br 2.8
						I 2.5

The most electronegative, electron-attracting elements are found in the upper right hand corner of the periodic table. The least electronegative elements are the alkali and alkali earth elements. With its intermediate electronegativity, carbon forms covalent or polar covalent bonds to almost all the elements except the alkali metals. Even C—Li bonds are not purely ionic.

Dipole Moments

The degree of charge asymmetry in a bond can be quantified by calculating the bond's **dipole moment** μ, which is the product of the charge separation e and the bond distance d. Large dipole moments arise when the atoms in a bond have very different electronegativities or when the bond is long.

Dipole moments are measured in units of debyes (D) (1 D = 1 × 10^{-20} electrostatic units per meter). A dipole moment is a vector:

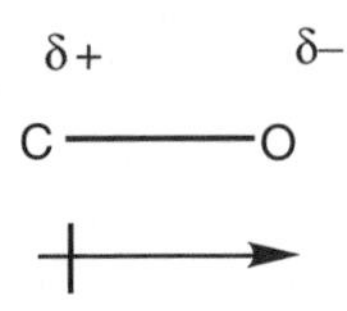

A C—O bond is polar covalent because of the electronegativity difference between the two atoms. The carbon is at the positive end of the dipole (δ+), and the oxygen at the negative end (δ–). The arrow represents the direction of the dipole moment, pointing to the negative end of the dipole.

Some dipole moments of common bonds are listed in the accompanying table. The dipole moment of a molecule is the vector sum of all the molecule's bond dipole moments. Understanding the electron distribution in organic molecules is central to predicting the reactivity of organic molecules. Most reactions are explained by the negative region of one molecule interacting with the positive region of another molecule, and it is electronegativity that largely explains the charge distribution in molecules.

BOND DIPOLE MOMENTS[a]

Bond	Dipole Moment (D)	Bond	Dipole Moment (D)
C—N	1.3	C=O	3.2
C—F	2.2	C—Br	2.2
C—OH	1.5	C—S	1.6
C—C1	2.3	C—I	2.05

[a]Taking C—H to be 0.4 D, δ^{+}C—Hδ^{-}.
From A. J. Gordon and R. A. Ford, *Chemist's Companion*, Wiley, New York, 1972, p. 126.

Hydrogen Bonds

A special type of bonding takes place in organic and inorganic molecules with O—H or N—H bonds. Because of electronegativity differences, the H in either bond is the positive end of a dipole. The lone electron pairs of N and O atoms are the negative ends of dipoles (arising from the positive nucleus and the negative charge of nonbonding electrons). Each dipole attracts the other in a **dipole-dipole interaction**, which in this case is called **hydrogen bonding**:

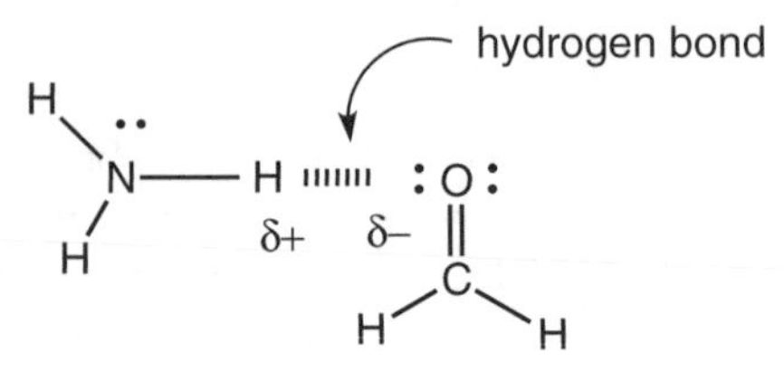

A hydrogen bond is much weaker than a covalent bond, by 1/15 to 1/50, but it is still strong enough to be chemically significant.

REPRESENTING STRUCTURAL FORMULAS

Constitutional Isomers and Condensed Structural Formulas

Isomers are molecules with the same molecular formula that differ in some way. **Constitutional isomers** differ in the way the atoms are connected. Because even a simple molecule like C_5H_{12} can have three constitutional isomers, organic chemists need a convenient notation for representing molecules. The solution that has been adopted uses **condensed structures** that leave out most or all of the lines that indicate bonds and uses subscripts to indicate the number of atoms or groups attached to particular carbons. These condensed structures are not too difficult to figure out as long as one keeps in mind the normal valences of carbon (4), nitrogen (3), oxygen (2), hydrogen (1), and the halogens (1). Shown below are different ways to draw butane, used in disposable lighters, and 2-propanol, sold as rubbing alcohol.

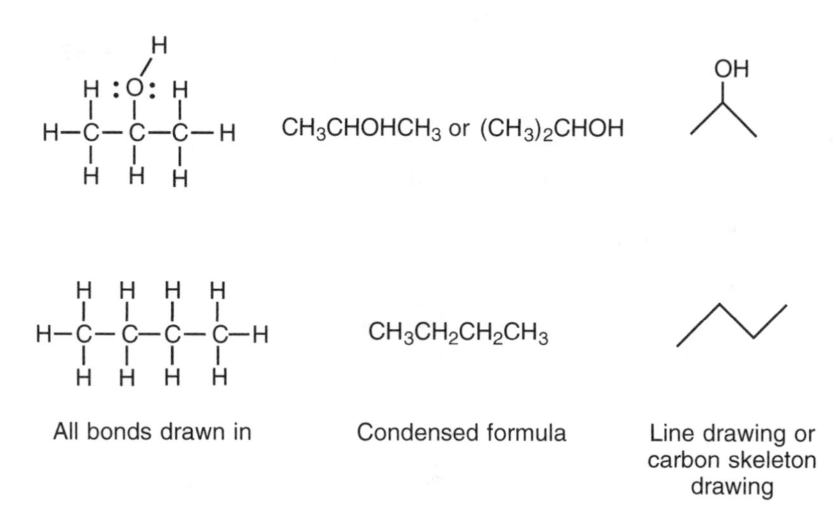

Carbon skeleton drawings are the most commonly used depictions of organic molecules. Each vertex or chain end is a carbon atom, and hydrogens are added to each carbon to bring it up to a valence of four. Heteroatoms always have their hydrogen atoms added explicitly.

One way to reconstruct molecules from condensed structural formulas is to think of each part of the molecule as a piece of a jigsaw puzzle. For example, $CH_3CH_2OCH_3$ is made up of four pieces, two CH_3, one CH_2, and one O.

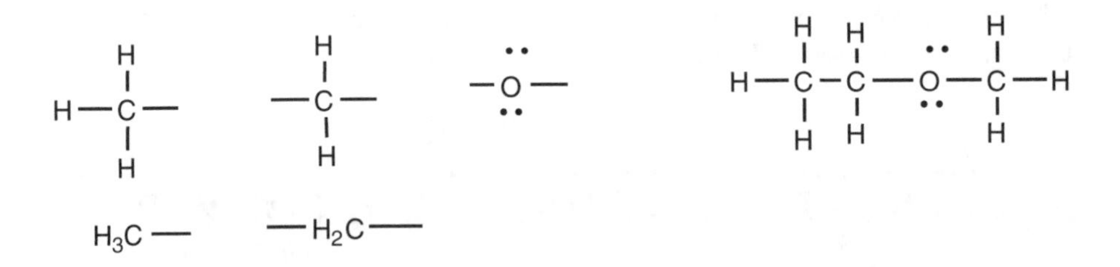

The CH_3 group is able to form one bond, and the CH_2 group and the O can each form two additional bonds. The order of groups in the structural formula equals the order in which the groups and atoms are joined. Just as in putting together a jigsaw puzzle, some of the pieces have to be turned around to fit; so here the H_3C— and the —CH_3 groups are identical. In these drawings, it is just the connectivity, which atoms are joined to which other atoms, that is represented.

EXERCISES

Draw condensed structures and complete the structures for the following carbon skeleton structures.

(a) 

(b) 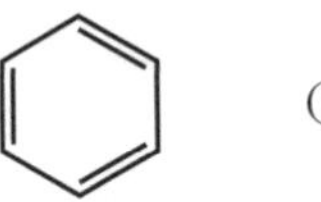
(c) 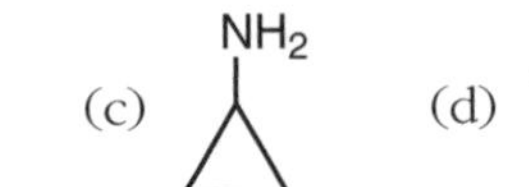

(d) 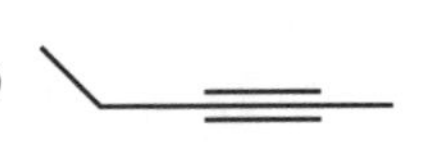

Solutions

Note that in (a) the connectivity is the same for both structures.

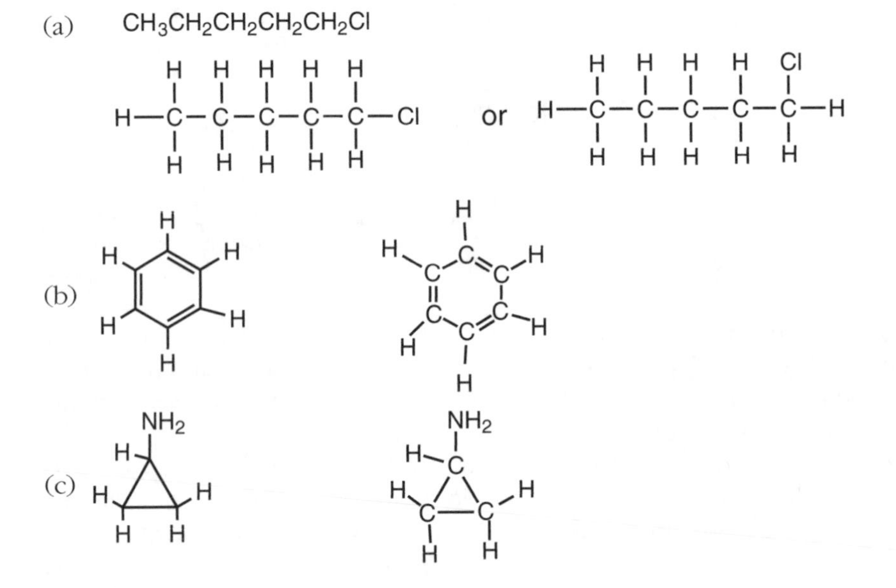

(d) $CH_3CH_2C_2CH_3$

Degree of Unsaturation

For a **hydrocarbon**, a compound containing only carbons and hydrogens, the molecular formula for a compound with no rings or **multiple bonds** (double or triple bonds) in its structure is C_nH_{2n+2}. For example, if you look at the structure of butane shown above, you will see that butane is C_4H_{10}. Thus, butane is a **saturated hydrocarbon**, one with no rings or multiple bonds. The molecule C_4H_8 has two fewer hydrogens than butane, and it is said to be **unsaturated**. There are numerous possible constitutional isomers for the formula C_4H_8, including

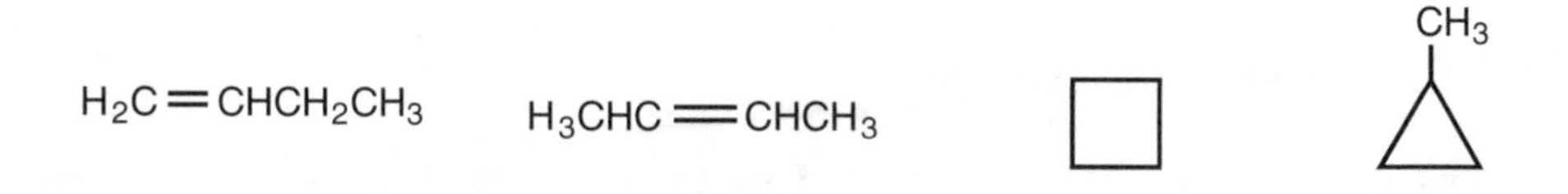

All these structures have either one ring or one double bond. Either a ring or a double bond produces one **degree of unsaturation**. Some organic molecules have many degrees of unsaturation. For example, benzene, C_6H_6, has one ring and three double bonds, for a total of four degrees of unsaturation.

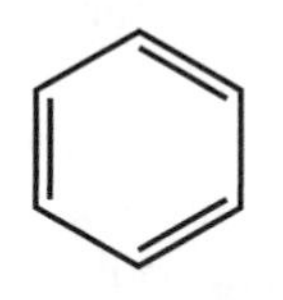

Benzene, C_6H_6

$$\text{Degree of unsaturation} = \frac{(\text{no. of H in saturated hydrocarbon}) - (\text{no. of H in } C_nH_{2n+2})}{2}$$

Using this formula, a saturated hydrocarbon with six carbons has the formula C_6H_{14}. Benzene has $(14 - 6)/2 = 4$ degrees of unsaturation. Without more information, these four degrees of unsaturation could be four rings, four double bonds, two triple bonds (a triple bond is two degrees of unsaturation), or some other combination of rings and multiple bonds that add up to four.

EXERCISES

Indicate how many degrees of unsaturation the following hydrocarbons possess and draw two possible constitutional isomers for each.
(a) C_3H_4 (b) C_5H_{10} (c) C_4H_6

Solutions

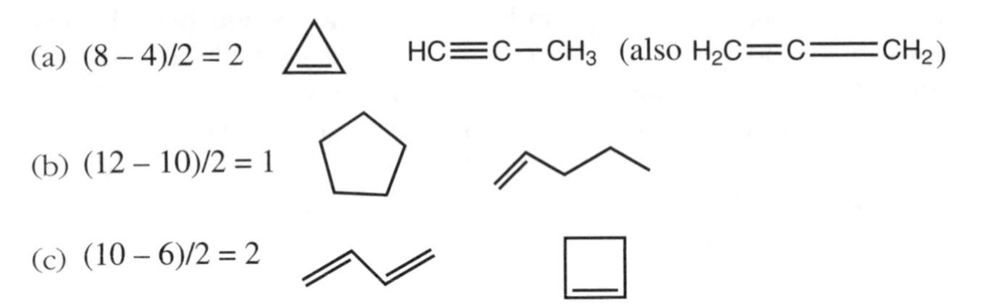

Other answers are also possible.

Unsaturation in Heteroatom-Containing Molecules

In addition to carbon and hydrogen, many organic molecules contain halogens, nitrogen, or oxygen. Since oxygen can form both single and double bonds and nitrogen can form single, double, and triple bonds, unsaturation can arise from multiple bonds to these heteroatoms as well as from multiple bonding among carbon atoms. The molecular formula can be used to determine the degree of unsaturation if heteroatoms are present by converting it to its hydrocarbon equivalent formula. For $C_aH_bN_xO_yX_z$, where X is any one of the halogens F, Cl, Br, or I, (1) add the number of halogens to the number of hydrogens and cross out the halogens, (2) ignore the number of oxygens, and (3) for each N, add one C and one H, crossing out the nitrogens. This results in the hydrocarbon equivalent formula $C_{a+x}H_{b+x+z}$. Finally, use the hydrocarbon formula to determine the degree of unsaturation. For example, C_7H_4NOBr has the equivalent formula C_8H_6. One possible constitutional isomer for this formula that has six degrees of unsaturation is

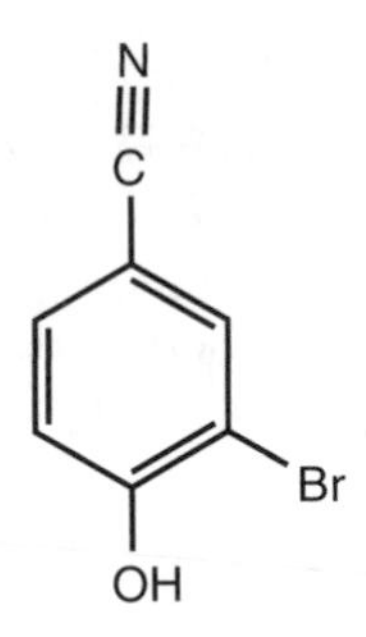

Equivalent Structures

Molecules are three-dimensional objects, and this fact has to be recognized when we draw them. Just as a person can be viewed from the front, the side, or the back, so can a molecule. It is a common mistake to draw the same molecule in two different orientations and think that one has drawn two different molecules. If two molecules have the same kinds of atoms and groups and are connected in exactly the same way, they are the same. For larger molecules it may be hard at first to see if two structures are the same or are constitutional isomers. In such cases it helps to make an inventory of how many CH groups, CH_2 groups, and CH_3 groups there are and what they are bonded to. Shown here are some examples of identical molecules in different orientations.

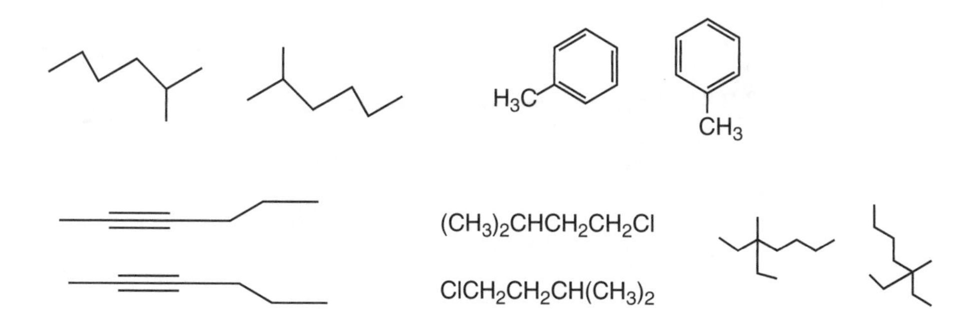

Recognizing equivalent molecules is complicated by the fact that single bonds act like molecular axles. Groups can spin around a single bond, pointing in any arbitrary direction. As in our analogy with a person, someone is the same whether she is standing with her arms outstretched or at her sides. This property of single bonds is discussed in detail in Chapter 6.

EXERCISES

Indicate which of the following molecules are identical and which are constitutional isomers.

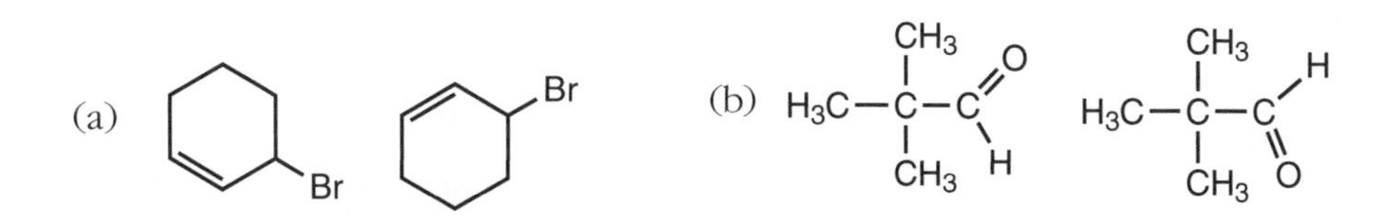

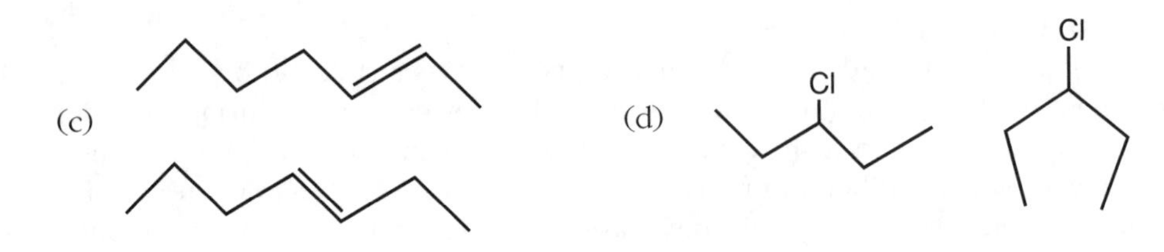

Solutions

The molecules in (a), (b), and (d) are identical. Those in (c) are constitutional isomers.

2
ORBITALS AND RESONANCE

BONDING THEORIES IN ORGANIC CHEMISTRY

Three theories are commonly used by organic chemists to explain bonding. The Lewis electron pair theory was introduced in Chapter 1. The other two theories are the **valence bond (VB) theory** and the **molecular orbital (MO) theory**. All three theories are based on the idea of sharing electron pairs, but the VB and MO theories differ in describing what contains the electron pairs. Valence bond theory places electrons in the **hybrid orbitals** created by mixing *s* and *p* orbitals on an individual atom. Molecular orbital theory uses **molecular orbitals** made up by combining atomic orbitals from some or all of the atoms in the molecule. MO theory is a much better description of reality. It is used to explain and predict molecular properties, but advanced mathematics and complex computer programs are required to make these predictions for most molecules. A beginning organic student can understand the results of MO calculations but cannot be expected to derive them. VB theory provides a qualitative description of bonding and does a good job of explaining molecular shapes.

VALENCE BOND THEORY

Hybrid Orbitals

Atomic carbon has two electrons in the 2*s* orbital and a total of two electrons in the equivalent p_x, p_y, and p_z orbitals. The only way carbon can form four bonds to other atoms is to promote one of the 2*s* electrons into a higher-energy 2*p* orbital. This produces the electronic state $2s^1 2p_x^1 2p_y^1 2p_z^1$. With four half-filled orbitals, carbon can form four new bonds, the maximum allowed by the octet rule. However, an *s* orbital is spherically symmetric and *p* orbitals are mutually perpendicular. This geometry is not what is found for molecules such as CH_4, which has the four C—H bonds directed toward the corners of a tetrahedron (Figure 2.1).

In the 1930s Linus Pauling proposed an explanation for the tetrahedral structure of CH_4 and similar compounds. He suggested that the 2*s* and 2*p* orbitals on carbon mix to form four equivalent orbitals he termed ***sp*^{3} hybrid orbitals**,

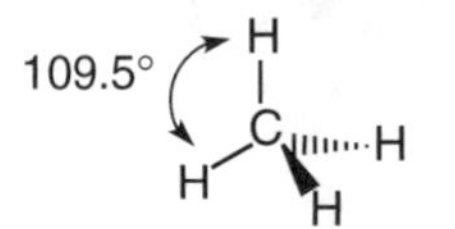

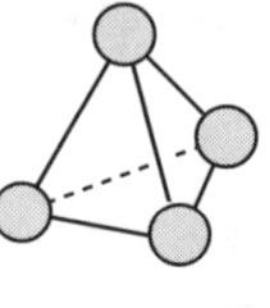

In CH_4 the H atoms are at the vertices of a tetrahedron. Each H is equidistant from the other three.

FIGURE 2.1. *Tetrahedral structure of CH_4. All H—C—H angles are 109.5°.*

which are directed toward the corners of a tetrahedron. The nomenclature sp^3 is used to indicate that one s orbital and three p orbitals are mixed (or **hybridized**). Each sp^3 hybrid orbital is 25 percent s and 75 percent p. Bonds are formed by overlap of the hybrid orbitals on carbon with other orbitals, which can be the hybrid orbitals of other carbons, or atomic orbitals such as the hydrogen 1s orbital.

sp^3 Hybrid Orbitals

The shape of an individual sp^3 orbital resembles a p orbital with one large lobe and one small lobe (the small lobe is sometimes called the tail of the orbital). The **carbon nucleus is located at the intersection of the two lobes**.

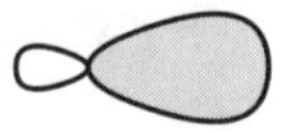

Recall that the shading represents **phases** since orbitals are like three-dimensional waves:

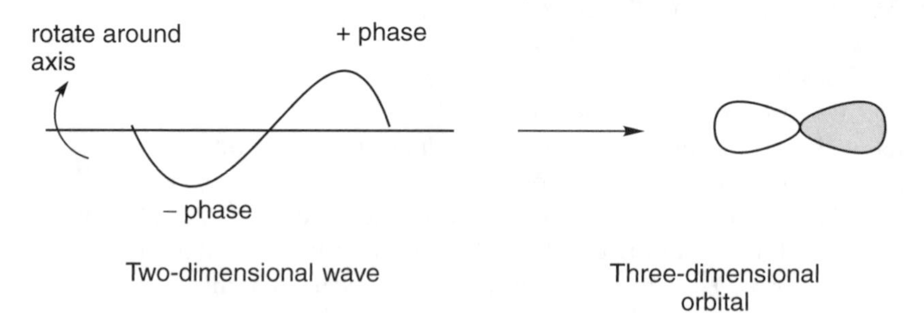

The phases of an orbital on any atom are arbitrary with regard to which lobe is + and which is –. However, according to VB theory, bonds are formed when orbitals overlap. In this case, the phases of the overlapping orbitals should be the same. Such overlap corresponds to two waves adding constructively rather than destructively:

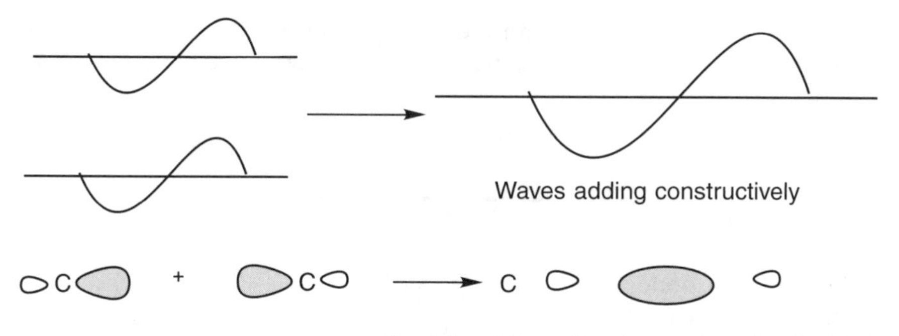

C—C bond formation by overlap of sp^3 orbitals

In the case of methane, the four C—H single bonds arise from an overlap of the four tetrahedral carbon sp^3 orbitals with the four hydrogen 1*s* orbitals:

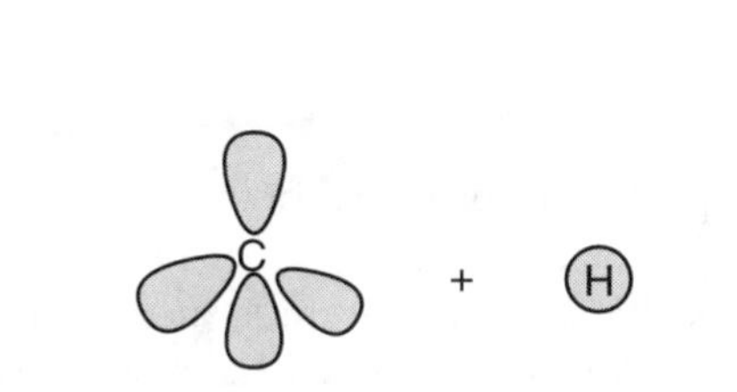

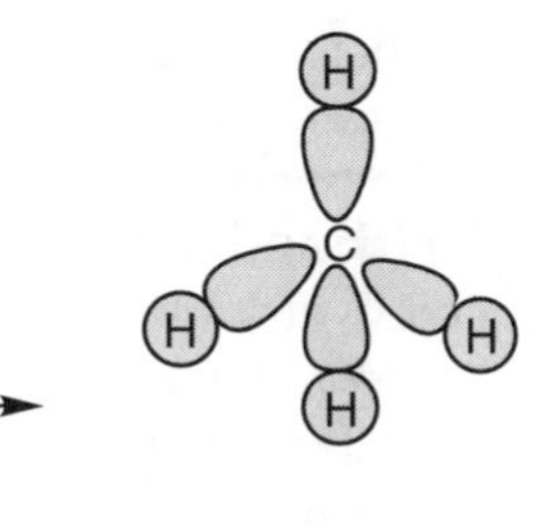

Four sp^3 carbon orbitals (tails omitted for clarity) Hydrogen is orbital

CH_4

For ethane, $H_3C—CH_3$, the central carbon-carbon bond arises from the overlap of two carbon sp^3 orbitals:

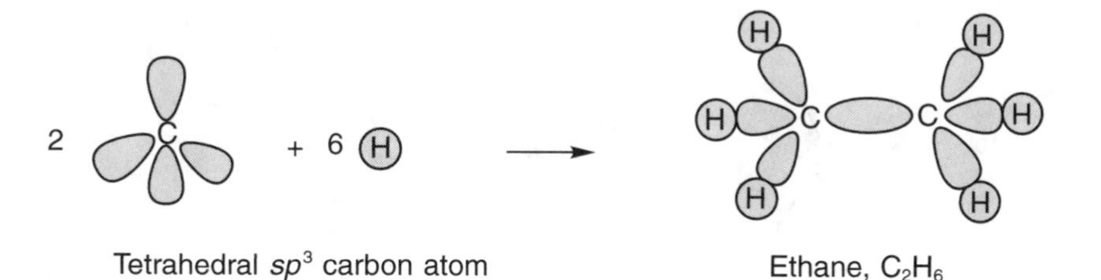

The single bond formed from the overlap of two sp^3 hybridized orbitals, or an sp^3 orbital and a hydrogen 1*s* orbital, is a **sigma (σ)** bond. Sigma bonds are cylindrically symmetric, and all single bonds in organic chemistry are σ bonds. These bonds are relatively strong since the electron pair is located along the axis joining the two positively charged nuclei that defines the bond. One can categorize the bonding in ethane as consisting of six Csp^3—H1*s* σ bonds and one Csp^3—Csp^3 σ bond.

Because of the symmetry properties of a σ bond, there is free rotation around it. In ethane, for example, the $—CH_3$ groups spin like windmills around the σ-bond axis.

EXERCISE

What kinds of bonds make up propane, $CH_3—CH_2—CH_3$, and what are the approximate bond angles?

Solution

Propane consists of eight $Csp^3—H1s$ σ bonds and two $Csp^3—Csp^3$ σ bonds. All the H—C—H, H—C—C, and C—C—C angles have approximately the tetrahedral value of 109.5°. The different sizes of H and C atoms lead to small differences between the experimentally determined bond angles and the theoretical tetrahedral value.

sp^2 Hybridized Carbons

Saturated hydrocarbons are made up entirely of sp^3 carbons and σ bonds. Unsaturated compounds have carbons with a different sort of hybridization. For ethylene, $H_2C{=}CH_2$, each carbon is bonded to only three atoms. Since σ bonds are strong bonds and *s* orbitals allow electrons to be closest to the nucleus, it is reasonable to mix a carbon 2*s* orbital with two carbon 2*p* orbitals to form three equivalent $\boldsymbol{sp^2}$ hybrid σ orbitals. Which *p* orbitals one uses is arbitrary, but if the p_x and p_y orbitals are used, three orbitals in the *xy* plane, making an angle of 120° with one another, result. This is called **trigonal** geometry. Left over is the unhybridized p_z orbital, perpendicular to the three σ orbitals:

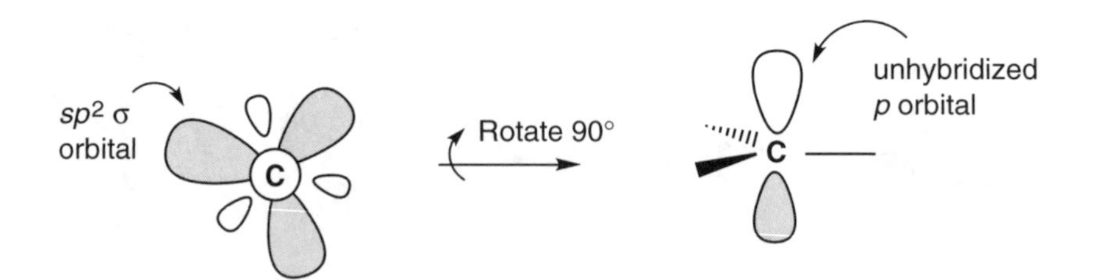

When two sp^2 hybridized carbons come together, they can form a double bond from the overlap of two sp^2 hybridized σ orbitals and the overlap of adjacent unhybridized *p* orbitals. In this model, one bond is a $Csp^2—Csp^2$ σ bond, and the other bond, arising from the overlap of two *p* orbitals, is called a **pi (π)** bond, in this case a $Cp_z—Cp_z$ π bond. In $H_2C{=}CH_2$ all the C and H atoms associated with σ bonds are in the *xy* plane and the π bond's electrons occupy the space above and below the C—C σ bond. Because the *p* orbitals that make up the π bond have a node at the nucleus, the resulting π bond has a **nodal plane**, the *xy* plane containing the carbon nuclei, where no π-bond electrons can be found. Therefore, a π bond is expected to be weaker than a σ bond since the π electrons cannot be in the region of space enjoying the best electrostatic interactions.

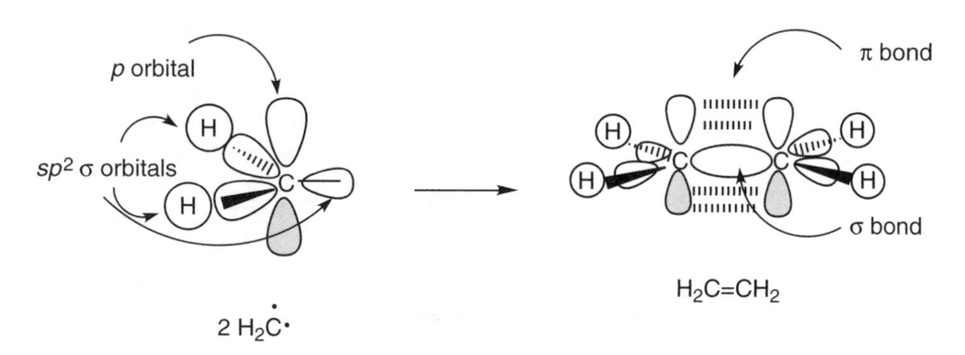

Sigma bonds, being cylindrically symmetric, act as molecular axles allowing free rotation around a σ bond. Pi bonds are different. Good overlap between *p* orbitals occurs only when they are parallel, or nearly parallel, to one another. When they are 90°, or **orthogonal**, to one another, they do not interact. For this reason, rotation around π bonds does not take place. Twisting a π bond 90° causes it to break.

EXERCISE

What kind of bonds make up propene, CH_3—CH=CH_2, and what is the C—C—C bond angle?

Solution

There are two kinds of C—H σ bonds: three Csp^3—H1*s* σ bonds and three Csp^2—H1*s* σ bonds. There are two kinds of C—C σ bonds: a Csp^3—Csp^2 σ bond and a Csp^2—Csp^2 σ bond. In addition, there is one Cp_z—Cp_z π bond. The C—C—C angle is 120°, determined by the sp^2 hybridization of the central C atom.

sp Hybridized Carbons

Carbons bonded to four other atoms are said to have a **coordination number** of 4 and are sp^3 hybridized. Carbon atoms with a coordination number of 3 are usually sp^2 hybridized. Carbons bonded to only two other atoms are normally *sp* hybridized. In this instance the carbon hybridizes one *s* and one *p* orbital to form two *sp* σ orbitals. Two *p* orbitals are left unhybridized. The *sp* orbitals are pointed 180° from one another and are perpendicular to the two *p* orbitals.

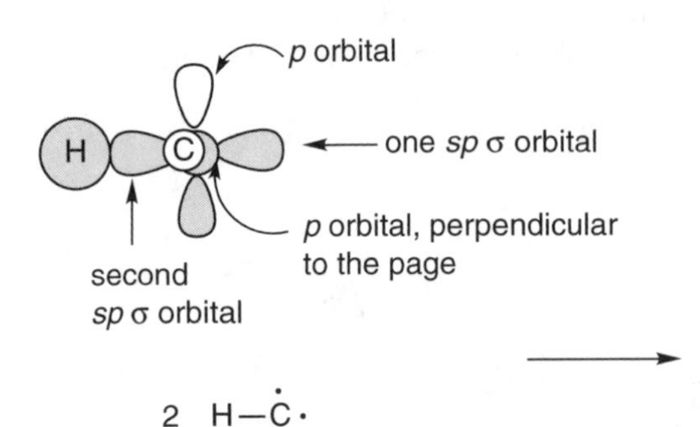

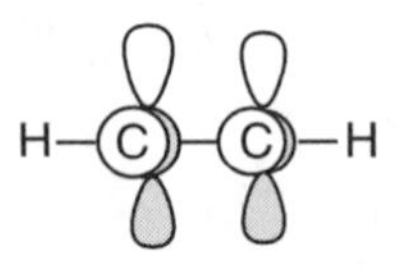

Acetylene, showing only the two perpendicular π bonds for clarity

H—C≡C—H

The small lobe of each *sp* σ orbital is not shown. The *sp* orbital has the largest amount of *s* character, at 50 percent.

EXERCISES

Indicate the hybridization for all the carbons in the following compounds, sketch the location of the *p* orbitals, and indicate the hybridization of all the bonds.

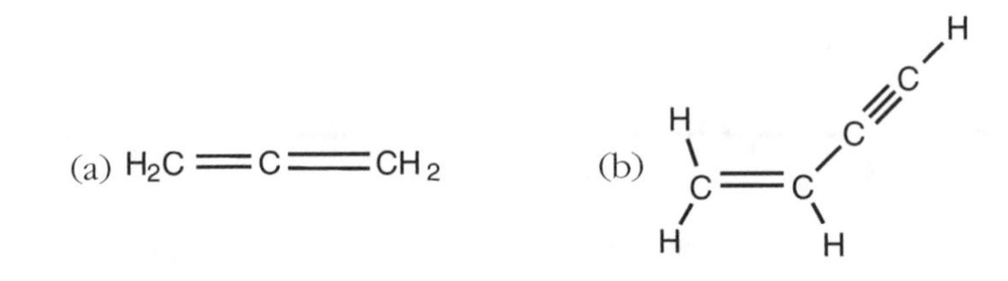

Solutions

(a) The central carbon is *sp* hybridized, and the two terminal carbons are sp^2 hybridized. There are four Csp^2—H1*s* σ bonds, two *Csp*—Csp^2 σ bonds, a Cp_y—Cp_y π bond, and a Cp_z—Cp_z π bond (the directions of the *p* orbitals are arbitrary).

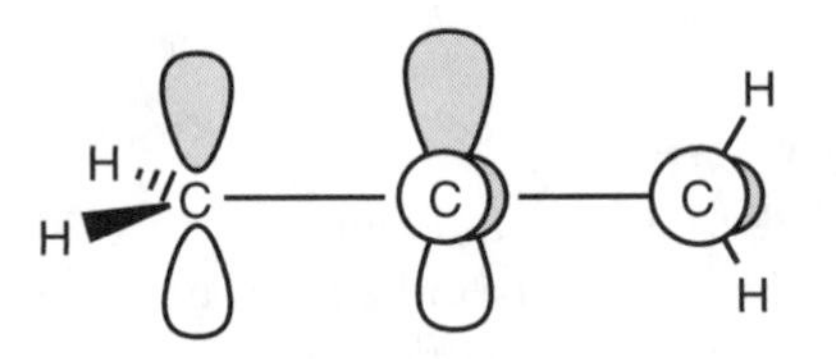

Notice that to allow for overlap of the *p* orbitals, the two CH_2 groups must be rotated 90° relative to one another.

(b) The double-bonded carbons are *sp*^{2} hybridized, and the triple-bonded carbons are *sp* hybridized. There are three C*sp*^{2}—H1*s* σ bonds, one C*sp*—H1*s* σ bond, a C*sp*^{2}—C*sp*^{2} σ bond, a C*sp*—C*sp* σ bond, a C*sp*—C*sp*^{2} σ bond, two Cp_y—Cp_y π bonds, and a Cp_z—Cp_z π bond. (The directions of the *p* orbitals are arbitrary.) There is free rotation around the single bond, but for simplicity all the atoms are shown in one plane.

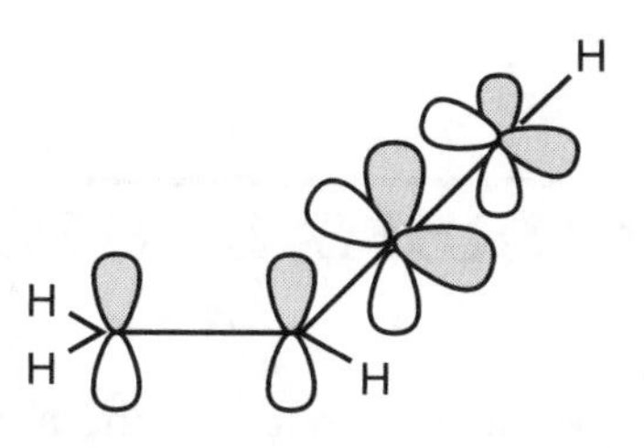

Hybridization of Reactive Intermediates

Because the reactive intermediates introduced in Chapter 1—carbocations, carbon radicals, and carbanions—are so important in understanding organic reactivity, one should know each intermediate's hybridization and structure. For a general carbocation, R_3C^+, the three-coordinate carbon is *sp*^{2} hybridized. All three groups and the central carbon atom are in the same plane, and a *p* orbital empty of electrons is perpendicular to the molecular plane. The central carbon of a carbon radical is *sp*^{2} hybridized also, and thus the structure possesses a trigonal geometry. For carbanions, the electron pair acts like a fourth substituent, giving rise to *sp*^{3} tetrahedral geometry around the negatively charged carbon. Nevertheless, carbanions readily undergo a process called **inversion**. All the groups on one side of the central carbon move to a corresponding position on the other side. The movement of the three R groups is similar to what happens when an umbrella is blown inside out by a strong wind. Since carbanions can invert thousands of times a second (or more), the average structure is planar.

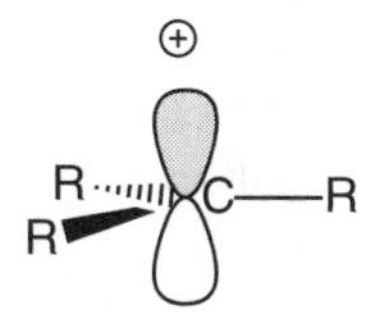

In a carbocation the *p* orbital has no electrons in it.

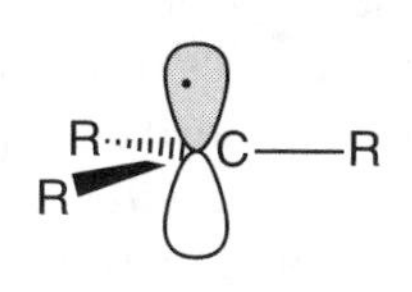

In a carbon radical, only a single electron occupies the *p* orbital.

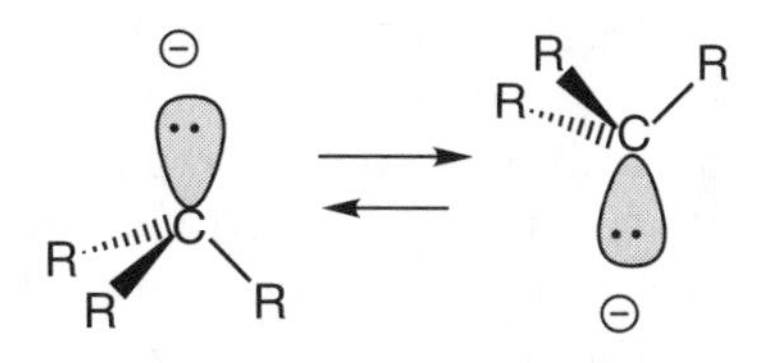

Carbanions are tetrahedral, but they easily invert their shape.

Bond Lengths and Hybridization

Since electrons in *s* orbitals are on average closer to the nucleus than electrons in *p* orbitals, bonds with more *s*-orbital character are shorter than bonds with more *p*-orbital character. Thus, *sp* σ bonds (50 percent *s*) are shorter than sp^2 σ bonds (33 percent *s*), and sp^3 σ bonds (25 percent *s*) are the longest σ bonds. A C—C bond is 1.54 Å, a C=C bond is 1.31 Å, and a C≡C bond is 1.18 Å, while a single bond between two sp^2 carbons, such as in the molecule $H_2C{=}CH{-}CH{=}CH_2$, is 1.48 Å (1 Å = 10^{-10} m).

MOLECULAR ORBITAL THEORY OF BONDING

Molecular orbital (MO) theory is not only the most powerful and quantitative theory of bonding but also the most mathematical and least intuitive. We can provide only a brief description of MO theory. Molecular orbitals are created by adding combinations of a molecule's atomic orbitals. For example, in ethylene, $H_2C{=}CH_2$, together the two carbons and four hydrogens have a total of 12 valence atomic orbitals. These can be added together to form 12 molecular orbitals. Six of them will be **bonding molecular orbitals**. Usually a molecule has as many bonding MOs as it has bonds in its Lewis structure. An MO can hold up to two electrons. However, unlike a valence bond hybridized orbital, an MO is not necessarily localized between two atoms. It can extend over some or all of the atoms of the molecule. A molecule's MOs are ordered from low energy to high energy. For example, in ethylene, the lowest-energy MO allows two electrons to interact with all six of the molecule's nuclei, resulting in a very stabilizing interaction. The highest-energy bonding MO in ethylene looks very much like the π orbital of valence bond theory. The two electrons are localized between the carbon atoms, above and below the molecular plane.

In addition to bonding molecular orbitals, MO theory introduces the idea of **antibonding molecular orbitals**. Molecules normally have equal numbers of bonding and antibonding orbitals. Ethylene, for example, has six filled bonding MOs and six empty antibonding MOs. Antibonding orbitals are higher in energy than bonding MOs. Putting electrons into antibonding MOs destabilizes the molecule (usually more than putting electrons into bonding orbitals stabilizes the molecule). Antibonding orbitals have shapes that result in electron-electron and nucleus-nucleus repulsion and very little electron-nucleus stabilization. Essentially all stable molecules have just enough electrons so that all their bonding MOs are filled and all their antibonding MOs are empty of electrons.

The importance of understanding chemical reactivity in terms of the properties of antibonding MOs was developed by Kenichi Fukui, who was awarded the Nobel Prize in chemistry in 1981. He found that many reactions could be explained by considering just a few molecular orbitals, **frontier orbitals**. These are the **highest occupied molecular orbital** (HOMO) and the **lowest unoccupied molecular orbital** (LUMO). The HOMO is the highest energy bonding orbital, the one whose electrons are most weakly held. The LUMO is the antibonding orbital of lowest energy. Often when a chemical reaction takes place, electrons flow from the highest occupied molecular orbital on one molecule to the lowest unoccupied molecular orbital on the other molecule. In this process, bonds are broken and new bonds are made.

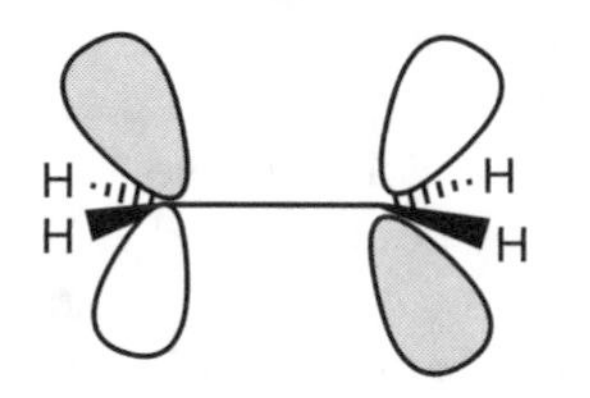

The LUMO of ethylene, an antibonding π^* orbital that is empty

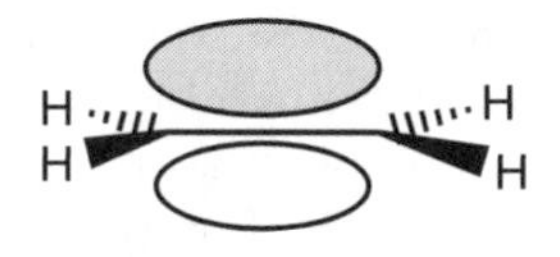

The HOMO of ethylene, a bonding π orbital with two electrons

From the shapes of the MOs shown above, it is seen that the bonding MO has one node—in the molecular plane. The antibonding MO has two nodes—one in the molecular plane and one passing through the center of the C—C bond perpendicular to the surface of the paper. In general, antibonding orbitals have more nodes than bonding orbitals.

RESONANCE STRUCTURES

Not all molecules can be adequately described by a single Lewis structure. For example, consider the reactive intermediate the allyl radical:

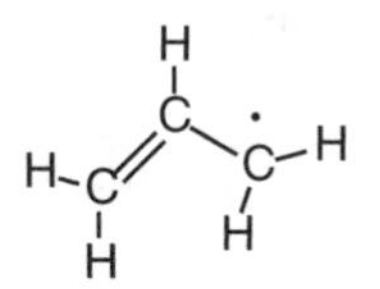

Allyl radical, Lewis structure

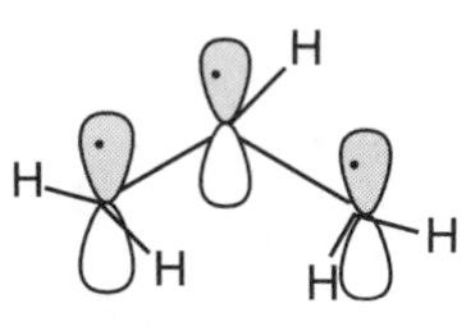

Valence bond structure for the allyl radical

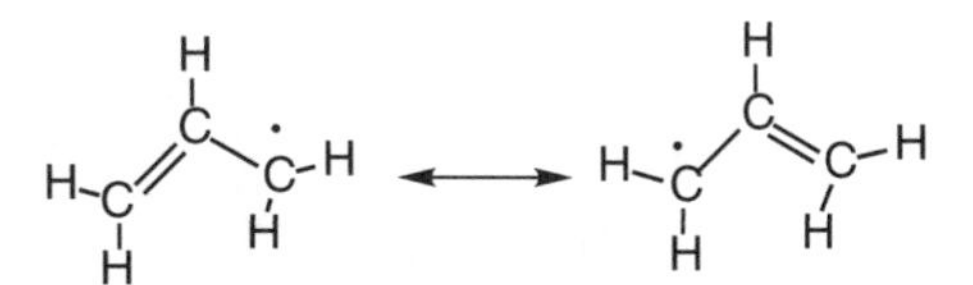

Contributing structures for the resonance hybrid model of the allyl radical's structure. Note the use of a double-headed arrow.

The Lewis structure for the allyl radical shows one C=C double bond and one C—C single bond. However, experimental data show that the two C—C bonds in the allyl radical are of equal length.

The symmetry in the allyl radical is clear from the valence bond depiction, showing three *p* orbitals, each with one electron. Obviously, from the valence bond picture, it is arbitrary which two carbons are double-bonded and which carbon possesses the unpaired electron. In order to accurately represent the bonding in the allyl radical (and in many other structures) using Lewis structures, the concept of **resonance** must be introduced.

Resonance theory states that for some molecules, the electron distribution is a **resonance hybrid** of two or more **contributing structures** or **resonance structures**. In the case of the allyl radical, two equivalent contributing structures are necessary to describe the bonding. Individual contributing structures do not really exist. The molecule does not rapidly change from one structure to the other but is a blend of both structures at once. For example, in the allyl radical, the bonds do not oscillate between a short double bond and a long single bond. The bonds are always equal in length, with a value intermediate between that of a single bond and double bond. What the contributing structures represent is the **delocalization** of electrons among several nuclei.

Notice that contributing structures are always joined by doubled-headed arrows. Whenever you see such arrows in an organic chemistry text, the structures are resonance hybrids.

Rules for Drawing Resonance Structures

Drawing resonance structures is one key to understanding organic chemistry. The following simple rules will make it easier to construct proper resonance structures.

1. *First-row elements must obey the octet rule. They cannot have more than eight electrons in their valence shell although they can have fewer than eight.*

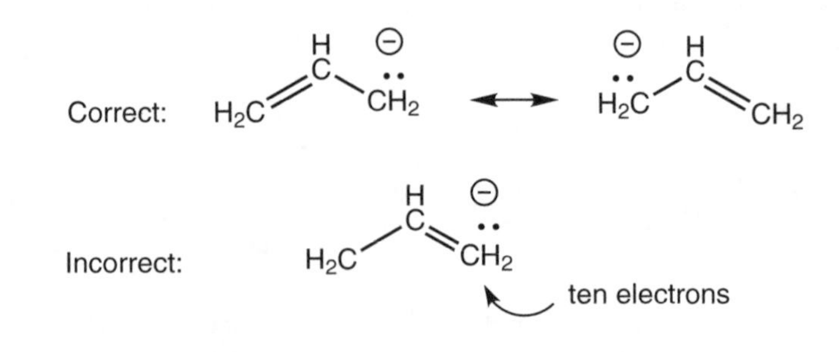

2. *Only electrons are moved in drawing resonance structures, not nuclei. All atoms must be in the same position in all resonance structures.*

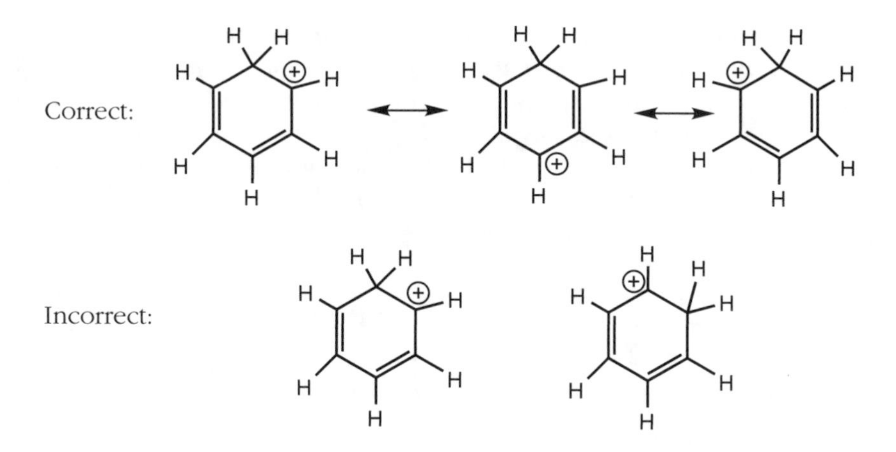

3. *Each resonance structure must have the same net charge as any other.* Note than an uncharged structure and a structure with separated positive and negative charges both have net zero charge.

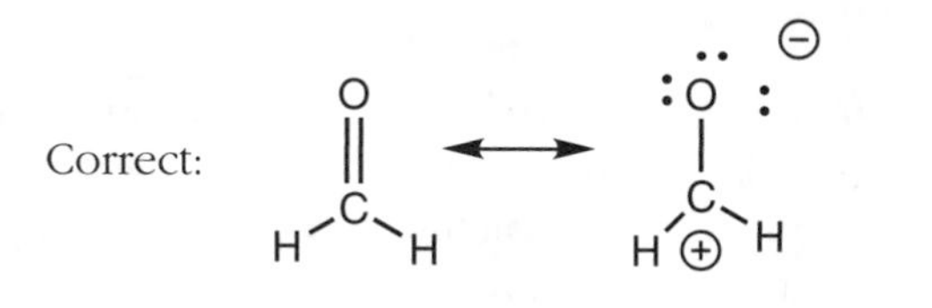

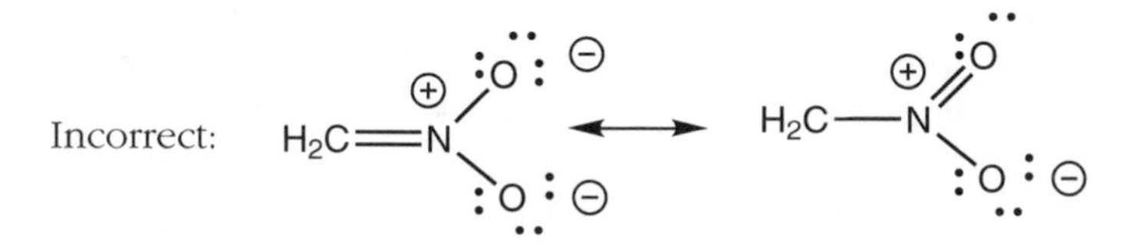

4. *Resonance structures must all have the same number of unpaired electrons and the same total number of electrons.* A subtle mistake is to draw some structures with different numbers of electrons, leaving out a lone pair for example. Such mistakes are difficult to pick up.

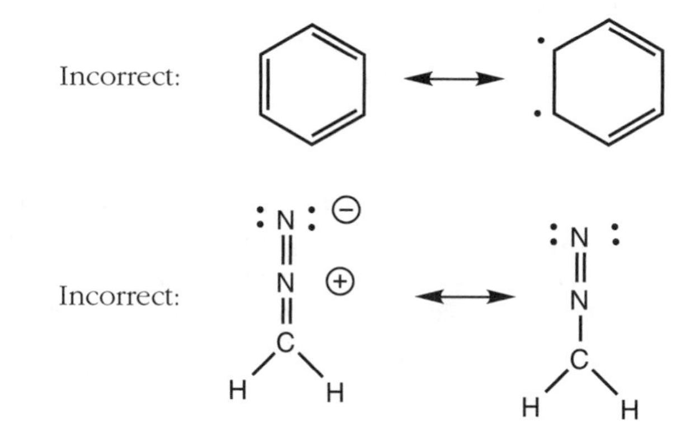

5. *Resonance structures with no separation of charge are more important than structures that have charge separation.* It is always possible to draw resonance structures where a double bond is replaced by adjacent positive and negative charges. Where the two double-bonded atoms are the same, as is the case here for a C=C bond, such a resonance structure is usually unimportant and does not really depict the electron distribution in the molecule.

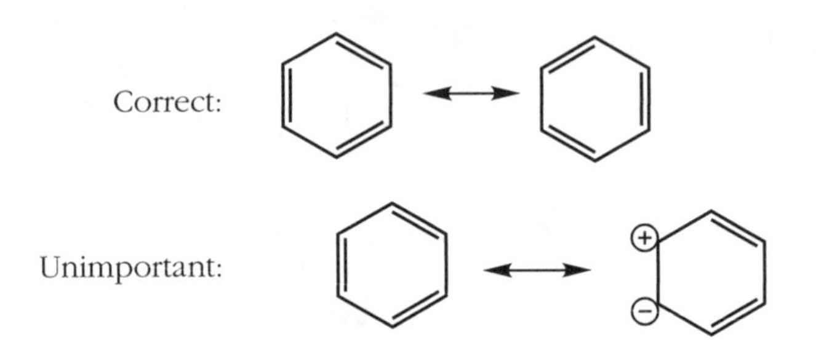

6. *For resonance structures with formal charges, the negative charge should reside on the most electronegative atom, and the positive charge on the most electropositive atom.* Sometimes two contributing structures are of unequal importance. For example, in the ketone structures here, the uncharged contributing structure is more important than the charge-separated contributing structure. Nevertheless, because oxygen is so much more electronegative than carbon, the charge-separated structure is of some importance. Its existence implies that negatively charged reagents tend to react at the C of a C=O double bond, and positively charged reagents at the O of a C=O double bond.

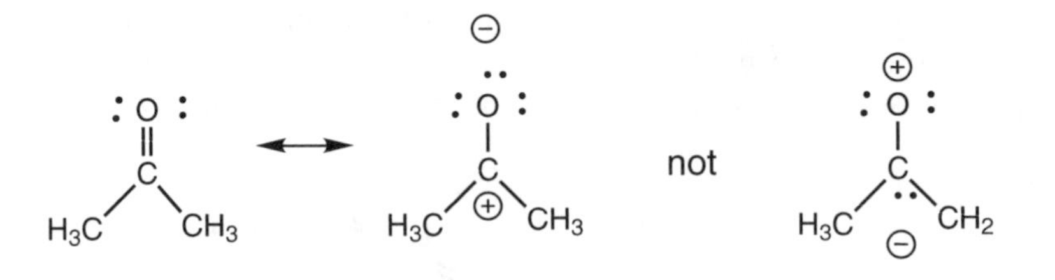

A molecule with several equivalent resonance structures is more stable than a molecule without the electron delocalization implied by resonance. In other words, *resonance stabilizes molecules.* The **resonance stabilization** (the energy between the real molecule with resonance and a hypothetical molecule with no resonance) will be low if the resonance structures differ from one another, as is the case for the ketone group, and will be substantial if all the contributing structures are similar, as is the case for the allyl radical.

Resonance, as it has been presented here, is a phenomenon of electrons in *p* orbitals and lone pair electrons. The electrons in σ orbitals do not participate in resonance. In addition, for resonance interactions to take place, the *p* orbitals must be parallel to one another. *p* orbitals that are perpendicular to one another do not interact, and electrons cannot move between them.

EXERCISES

1. Draw all the important contributing structures for the following molecules.

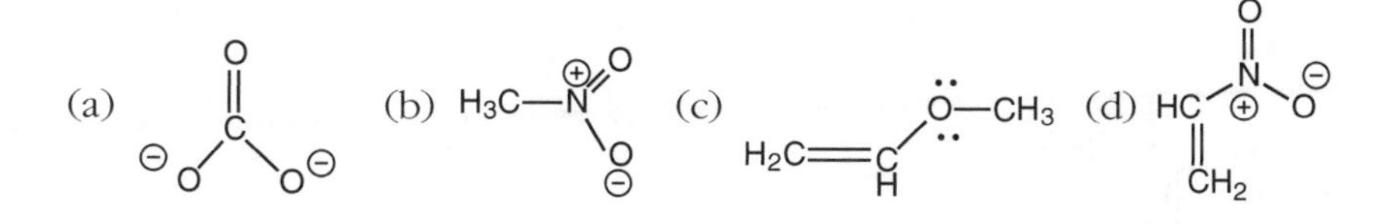

2. Seven acceptable contributing structures can be drawn for the following molecule. Indicate which structures are the most important contributing structures and give a reason for your choice.

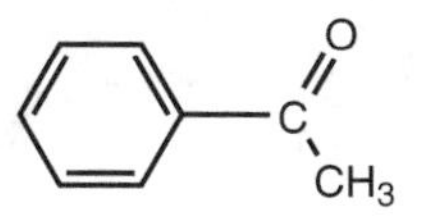

3. The molecule phenanthrene has five acceptable resonance restructures. Based on these structures, which double bond in phenanthrene is most localized and thus most like an ordinary double bond?

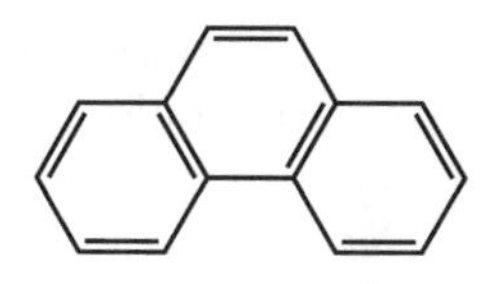

Phenanthrene

Solutions

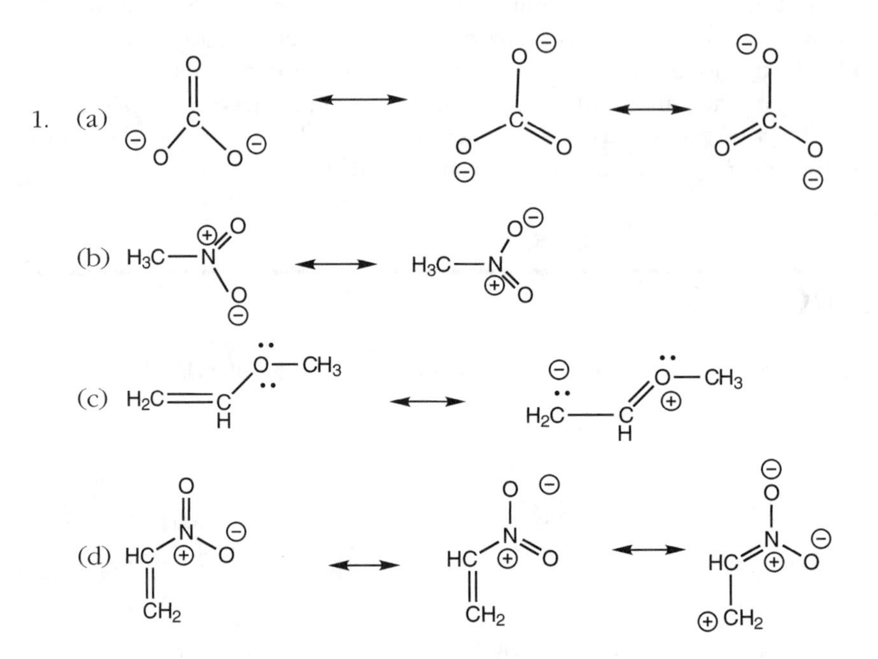

The last structure is the least important because of charge separation.

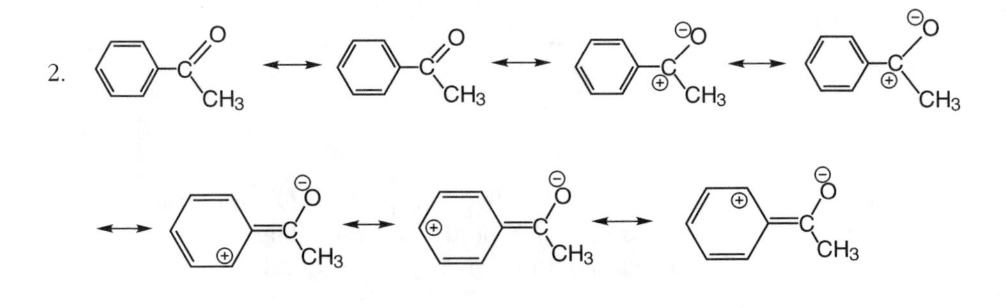

The first two structures are the most important contributing structures because they have no charge separation.

3.

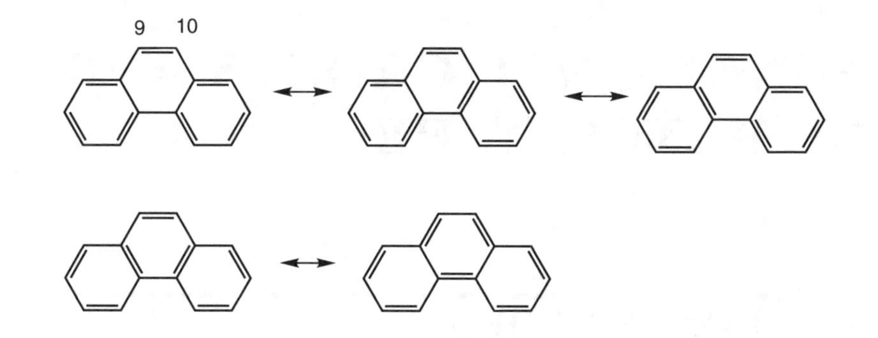

In four out of the five contributing structures the bond labeled 9,10 is a double bond. From simple statistics is has 80 percent double-bond character. Experimentally, it is the shortest carbon-carbon bond in the molecule.

3
FUNCTIONAL GROUPS AND NOMENCLATURE

THE CONCEPT OF A FUNCTIONAL GROUP

The properties and reactivities of organic molecules are largely determined by the kinds of **functional groups** a given molecule possesses. Functional groups are groups of atoms that react in characteristic ways more or less independently of their surroundings. For example, a carboxylic acid functional group consists of a carbon atom double-bonded to an oxygen atom and single-bonded to an OH group.

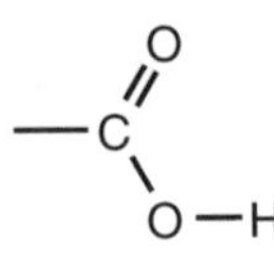

The fourth valence on carbon can be attached anywhere on a carbon framework. Any molecule that has the carboxylic acid functional group is expected to be an acid, approximately as acidic as vinegar. One can make lists of reagents that react with this functional group and what the products of these reactions will be. The functional group concept is the central organizing principle in organic chemistry.

Since we are usually concerned with the customary properties of a functional group and not the specific properties of an individual molecule, we need a symbol to represent a generic carbon framework. The letter R is normally used for this purpose. Thus, RCO_2H represents any molecule with a carboxylic acid functional group.

COMMON ORGANIC FUNCTIONAL GROUPS

Functional groups contain either heteroatoms or carbon-carbon double or triple bonds. Normally, the heteroatoms are halogens, nitrogen, or oxygen. Functional groups containing other atoms, especially sulfur or phosphorus, are also important in organic chemistry but are less common and will be covered later. A list of important functional groups is given in Table 3.1. Notice that

TABLE 3.1. COMMON FUNCTIONAL GROUP STRUCTURES

Structure	Functional Group Name	Example
$C=C$	Alkene	$H_2C=CH-CH_3$
$C\equiv C$	Alkyne	$H-C\equiv C-CH_2-CH_3$
$-C(=O)-O-H$	Carboxylic acid	$(H_3C)_2CH-C(=O)-O-H$
$-C(=O)-O-C(=O)-$	Carboxylic acid anhydride	$H_3C-C(=O)-O-C(=O)-CH_3$
$-C(=O)-O-R$	Carboxylic acid ester	$H_3C-C(=O)-O-CH_2CH_3$
$-C(=O)-Cl$	Carboxylic acid chloride	$H_3C-C(=O)-Cl$
$-C(=O)-N<$	Amide	$CH_3CH_2CH_2-C(=O)-N(H)-CH_3$
$-C\equiv N$	Nitrile	$CH_3CH_2-CH(CH_3)-C\equiv N$
$-C(=O)-H$	Aldehyde	$H_3C-C(=O)-H$
$R-C(=O)-R$	Ketone	$H_3C-C(=O)-CH_3$
$R-OH$	Alcohol	$CH_3CH_2CH_2OH$
$R-N<$	Amine	$H_3C-N(H)-CH_3$
$R-O-R$	Ether	$H_3C-O-CH_2CH_3$
$R-X$, where X = F, Cl, Br, I	Alkyl halide	CH_3CH_2Br

some functional groups have closely related structures. For example, carboxylic acids and carboxylic acid esters differ only in the presence of an OH or OR group. Aldehydes and ketones differ in having one H and one R or two R (carbon) groups attached to the C=O. Any organic molecule can contain one or more than one of these functional groups.

EXERCISES

Each of the following molecules contains two functional groups. Identify them.

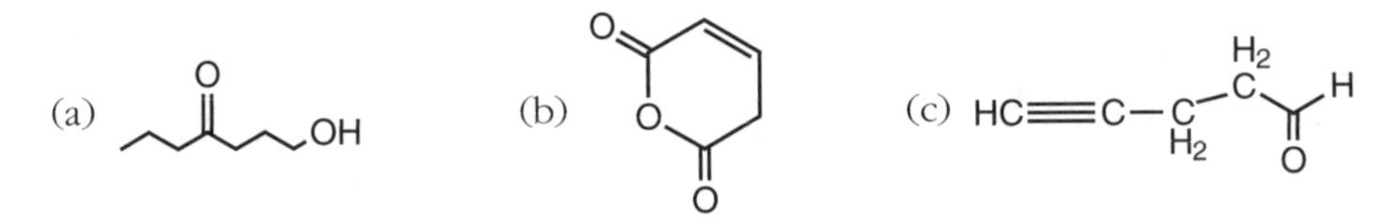

Solutions

(a) Ketone, alcohol; (b) carboxylic acid anhydride, alkene; (c) alkyne, aldehyde.

ORGANIC CHEMISTRY NOMENCLATURE

Nomenclature Problems

There are two kinds of nomenclature problems in organic chemistry. One type involves being given the name of a compound and having to draw its structure. Alternatively, one may be given a structural drawing and be expected to provide a correct name for the molecule. A particular teacher or examination may focus on one or both types of naming exercises. Increasingly, the tendency in organic chemistry is to concentrate on the reactions of organic molecules and depict molecules graphically, while deemphasizing nomenclature. However, the emphasis on nomenclature varies greatly from one course or examination to another. Students should have a clear understanding of the level of competence in nomenclature that the circumstances require. This chapter provides information on naming all the common classes of molecules. *For the beginning student, it may be best to concentrate on the nomenclature of hydrocarbons. As different functional groups and their reactions are covered in detail, the reader can return to the appropriate section of this chapter.*

Common and IUPAC Nomenclature

Some compounds have been known for hundreds or thousands of years. Grain alcohol, alcohol, ethyl alcohol, and ethanol are all names that have been used for CH_3CH_2OH. The names of compounds that have a historical origin are called **common names** or trivial names. However, because an unlimited number of possible organic structures exists, a systematic method of naming organic compounds has been developed by the International Union of Pure and Applied Chemistry called **IUPAC nomenclature** or **systematic nomenclature**. Once the IUPAC rules are understood, any IUPAC name can be used to draw a unique molecular structure.

Nomenclature of Unbranched Alkanes

Saturated hydrocarbons, which were defined in Chapter 1 as having the empirical formula C_nH_{2n+2}, are called **alkanes**. The names of **linear alkanes**, which are straight chains of carbons with no branches, are the basis of all organic nomenclature (Table 3.2).

TABLE 3.2. NOMENCLATURE OF LINEAR ALKANES

Number of Carbons	Molecular Formula	Name	Condensed Structure
1	CH_4	Methane	CH_4
2	C_2H_6	Ethane	CH_3CH_3
3	C_3H_8	Propane	$CH_3CH_2CH_3$
4	C_4H_{10}	Butane	$CH_3(CH_2)_2CH_3$
5	C_5H_{12}	Pentane	$CH_3(CH_2)_3CH_3$
6	C_6H_{14}	Hexane	$CH_3(CH_2)_4CH_3$
7	C_7H_{16}	Heptane	$CH_3(CH_2)_5CH_3$
8	C_8H_{18}	Octane	$CH_3(CH_2)_6CH_3$
9	C_9H_{20}	Nonane	$CH_3(CH_2)_7CH_3$
10	$C_{10}H_{22}$	Decane	$CH_3(CH_2)_8CH_3$
11	$C_{11}H_{24}$	Undecane	$CH_3(CH_2)_9CH_3$
12	$C_{12}H_{26}$	Dodecane	$CH_3(CH_2)_{10}CH_3$
20	$C_{20}H_{42}$	Icosane	$CH_3(CH_2)_{18}CH_3$
30	$C_{30}H_{62}$	Triacontane	$CH_3(CH_2)_{28}CH_3$
100	$C_{100}H_{202}$	Hectane	$CH_3(CH_2)_{98}CH_3$

The names of the first four alkanes are common names. Subsequent alkanes have names derived from Greek words. Linear alkanes make up a **homologous series** of compounds, differing by the successive insertion of a **methylene group (—CH_2—)** into the chain.

Nomenclature of Branched Alkanes

Branched alkanes have hydrocarbon substituents branching off a linear chain. Like all IUPAC names, the systematic names for these compounds have three parts: a number and prefix (designating position and the number of carbon atoms in a branch); a root, also called an infix (designating the number of carbons in the main chain); and a suffix (designating what functional group or functional groups the molecule contains). The name of the simple hydrocarbon 4-methyloctane illustrates these parts. The prefix *4-methyl-* indicates that a methyl group is attached at the fourth carbon. The root *-oct-* designates the root chain octane, an eight-carbon chain. The suffix *-ane* indicates that the molecule is an alkane.

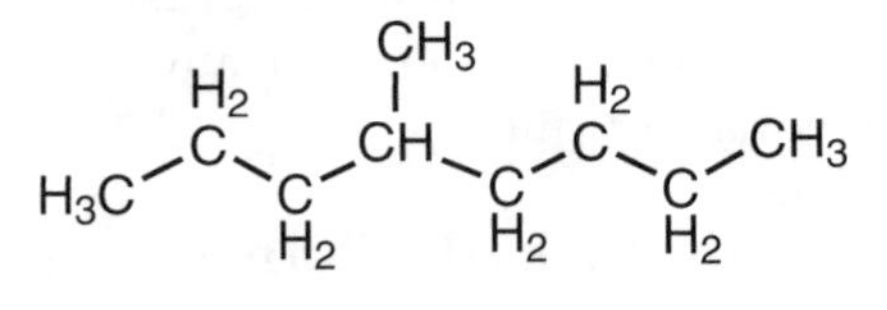

4-Methyloctane

The names of some common groups are given in Figure 3.1.

Linear alkyl groups:

CH_3— CH_3CH_2— $CH_3CH_2CH_2$— $CH_3CH_2CH_2CH_2$—

Methyl Ethyl Propyl Butyl

Branched alkyl groups:

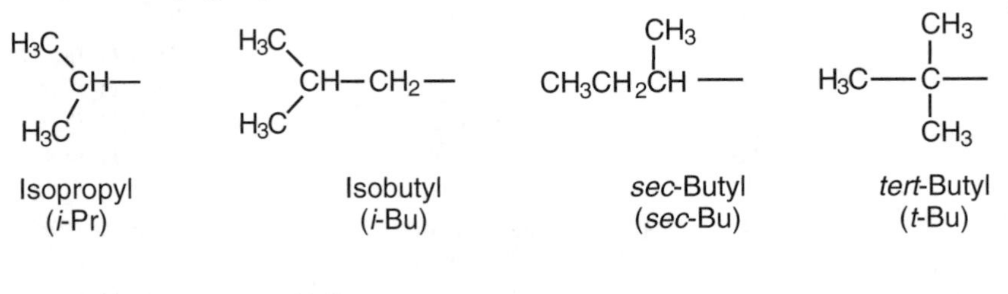

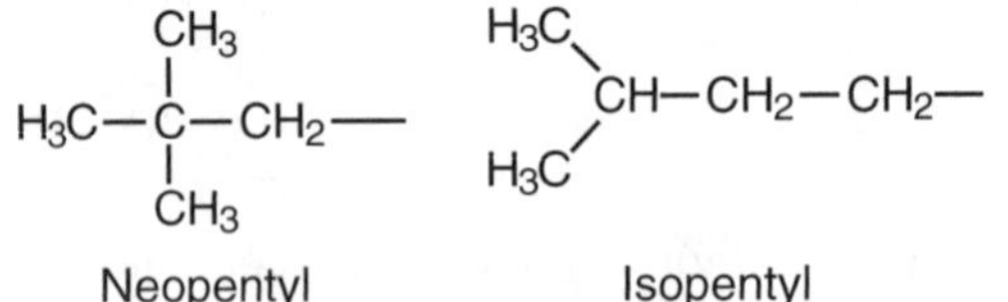

Unsaturated alkyl groups:

$H_2C{=}CH$— $H_2C{=}CH{-}CH_2$—

Phenyl (Ph) Vinyl Allyl

FIGURE 3.1. *Common substituent groups.*

The general steps taken to name any molecule in the IUPAC system are, first, to determine the functional group class to which the molecule belongs. The functional group determines the suffix. For alkanes, the suffix is *-ane*. Second, the principal carbon chain must be identified. For alkanes, this is the longest chain (*but-* for four carbons, *pent-* for five carbons, and so on). Third, determine what groups (such as those listed in Figure 3.1) are attached to the principal chain and their numerical position on the chain. Finally, assemble all the components into a complete name, using alphabetical order for all the group names.

The process of finding the longest chain in a structure and the numerical position of groups on the chain is one of the most difficult parts of nomenclature. In a complex structure, all the paths should be counted to ensure that the longest chain is found. The way a molecule is drawn may obscure the longest chain.

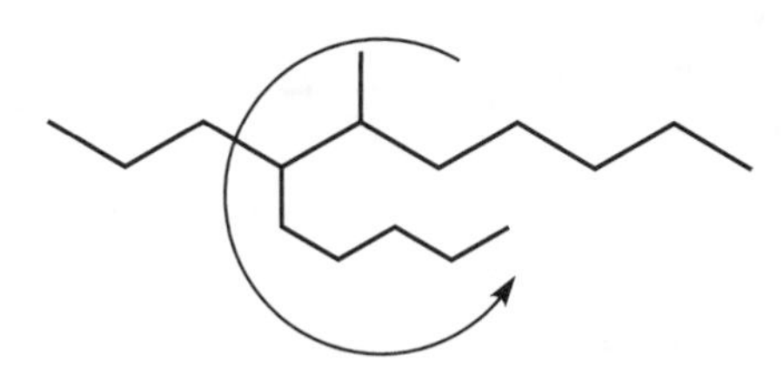

A dodecane derivative

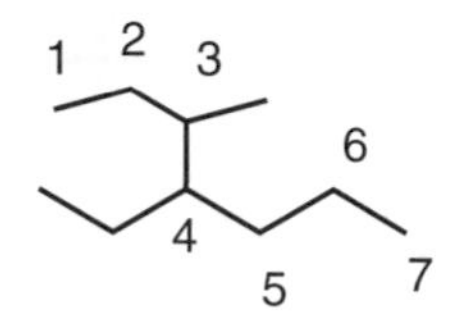

A heptane derivative

The longest chain is numbered from one end to the other, and the direction is chosen to give the lowest possible numbers to the side chains. For two possible directions of chain numbering, when a series of terms is compared term by term, the lowest is the order that contains the lowest number on the occasion of the first difference.

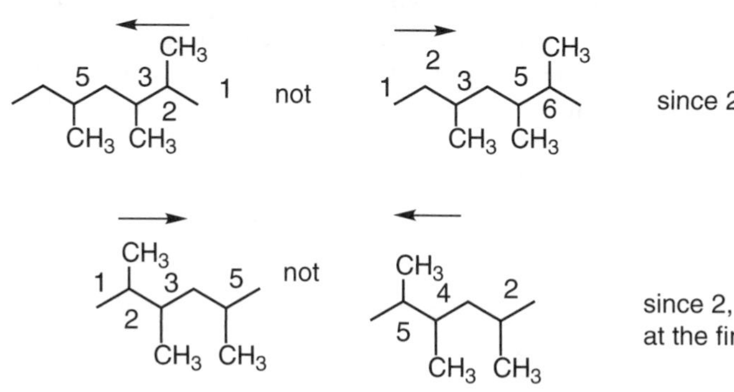

since 2 is lower than 3

since 2, 3, 5 is lower than 2, 4, 5 at the first difference

Figure 3.1 gives the names of the groups most often found branching from a chain. An unbranched alkyl substituent is named by taking its full name, removing the suffix *-ane* and replacing it with the suffix *-yl*. Thus, a six-carbon chain branching off a structure is a hexyl group. Sometimes a linear carbon chain has the prefix *n*- (for normal). Alkyl and *n*-alkyl are the same, so the *n*- prefix is redundant. When there are two or more substituents of the same kind, they are given the prefixes *di-*, *tri-*, *tetra-*, *penta-*, and so on, to indicate their number. Thus, three methyl (CH_3—) groups are indicated as trimethyl, and numerical prefixes start the name, separated by commas, to indicate the positions of the methyl groups on the chain. The two hydrocarbons drawn above are named 2,3,5-trimethylheptane and 2,4,5-trimethylhexane. There are no spaces in the name. When different groups are on a chain, they are arranged in alphabetical order. The numerical prefixes *di-*, *tri-*, *tetra-* (and so on) and *sec-* and *tert-* are ignored in alphabetizing groups. *tert*-Butyl is alphabetized under B and hexamethyl under M, for example. If the chain numbering gives identical substituent numbers in both directions, the group that is alphabetically first is given the lower number.

Examples

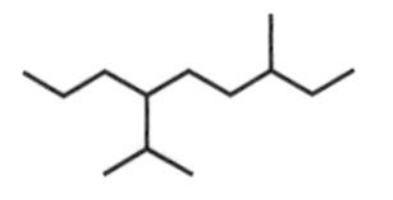

5-Isopropyl-2-methylnonane

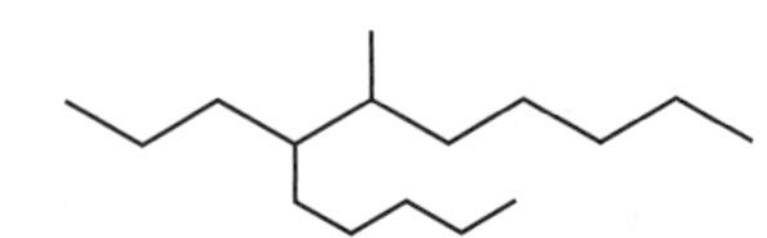

6-Methyl-7-propyldodecane, not 6-propyl-7-methyldecane (M before P)

4-*tert*-butyl-2-methyloctane

Branched groups coming off a primary chain can be named not only as isopropyl, *tert*-butyl, and so on, but systematically as well. In some cases the groups do not have a common name and must be named systematically. The branched group is assumed to be attached to the longest chain at the C-1 of branched group. In naming the whole molecule the branched group's name is placed in parentheses. Outside the parentheses is placed the numerical prefix that indicates the position of the group on the longest chain. The name of the branched group, for purposes of alphabetization, is considered to begin with the first letter of the complete name. Thus, a branched group (1,2-dimethylpentyl) is alphabetized under D.

Examples

6-(2,2-Dimethylpropyl)-
5-(1-methylethyl)dodecane or
5-isopropyl-6-neopentyldodecane

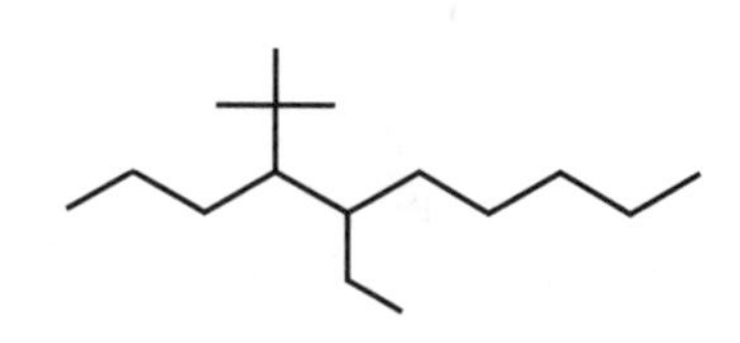

4-(1,1-Dimethylethyl)-
5-ethyldecane or 4-*tert*-butyl-
5-ethyldecane

In rare cases, chains of equal length may be chosen as the longest chain in a molecule. To break the tie, the chain with the greatest number of side chains is chosen as the main chain. Finally, one may have more than one of the same highly branched group on a chain. Then, instead of using *di-*, *tri-*, and so on, the multiplying prefix is derived from the Greek and is *bis-*, *tris-*, *tetrakis-*, *pentakis-*, and so on. Thus, if one had three (1,2-dimethylpentyl) groups on a chain, the numerical prefix would be tris(1,2-dimethylpentyl).

The above are not the complete rules for naming branched hydrocarbons, but they should provide the basis for naming any saturated hydrocarbon an introductory student is likely to encounter. One should also be able to derive structures of hydrocarbons from lists of their names.

EXERCISES

1. Draw structures for the following hydrocarbons.
   (a) 2,2-dimethylhexane
   (b) 4-ethyl-5-methyloctane
   (c) 6-(1-methylbutyl)dodecane

2. Give IUPAC names for the following structures.

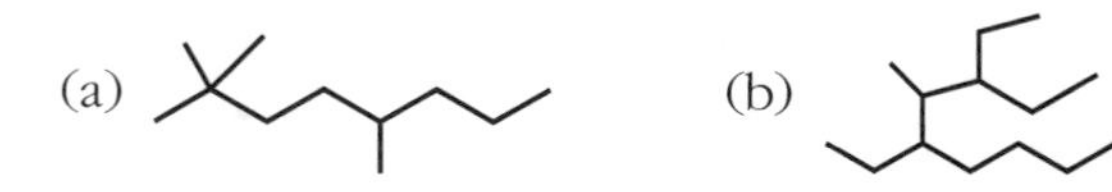

(c) 

Solutions

1. 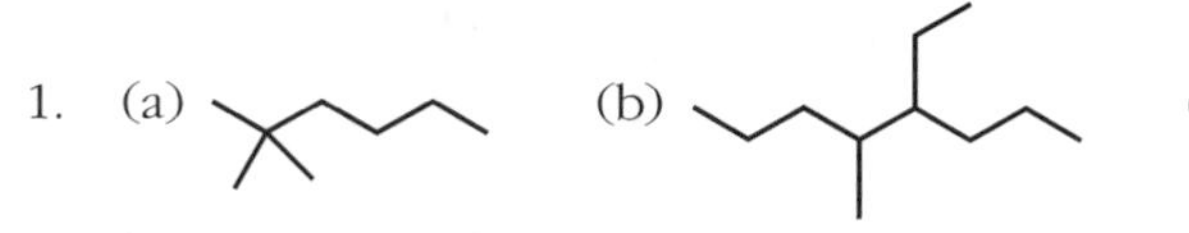

(c) 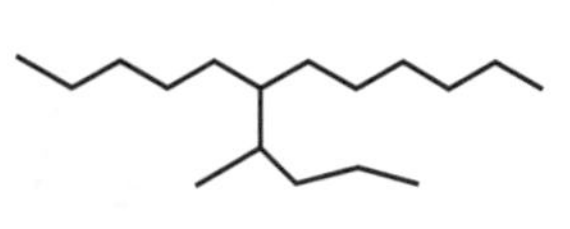

2. (a) 2,2,5-Trimethyloctane;
   (b) 3,5-diethyl-4-methylundecane;
   (c) 2,2,3,3-tetramethylbutane

Note: Make sure that you are comfortable drawing molecules in the carbon skeleton form used above. The molecular formulas for structures 1(a) through (c) are C_8H_{18}, $C_{11}H_{24}$, $C_{17}H_{36}$.

Cyclic Alkanes

Hydrocarbons can form ring structures, resulting in molecules with the formula C_nH_{2n}. They are named using the prefix *cyclo-*:

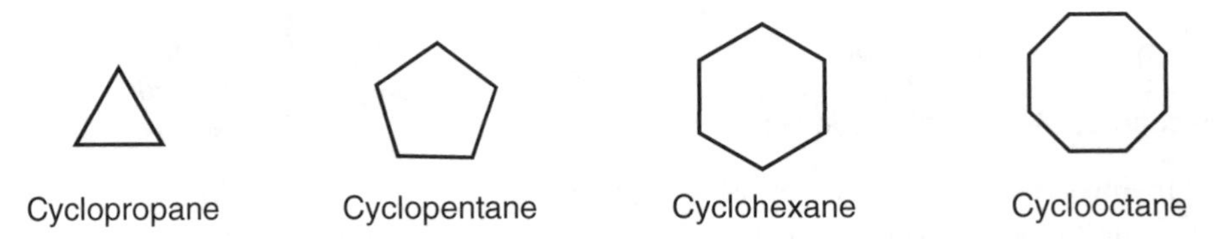

When the ring is attached to a chain with more carbons than are in the ring itself, the ring is named as an alkyl substituent, replacing the *-ane* suffix with *-yl*. 2-Cyclopropylhexane and 3-cyclohexyldecane are examples of this rule. Otherwise, the ring is given the root name. No number is needed if only one group is on the ring. All positions on a ring are initially equivalent. Two substituents are listed in alphabetical order with position 1 for the group listed first. Rings with three or more substituents are numbered in either a clockwise or a counterclockwise direction, according to which sense gives the lowest numbering to the subsequent substituents.

Examples

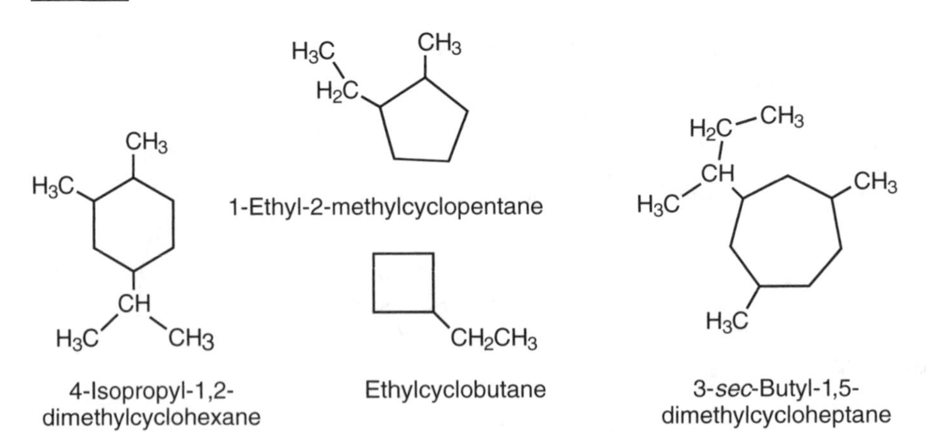

Alkenes and Alkynes

The suffix for compounds with a carbon-carbon double bond is *-ene*, and *-yne* is the suffix for triple bonds. An older name for the alkene functional group is **olefin**. Nomenclature rules for alkenes and alkynes are similar to those used for saturated hydrocarbons except that the longest carbon chain is the one that contains both unsaturated carbons. The chain is numbered in the direction that gives the first unsaturated carbon the lower number. If two or more double or triple bonds are present, the prefixes *di-*, *tri-*, and so on, are used. A triple bond takes precedence over a double bond for the final suffix.

When a double bond is present in a ring, the two unsaturated carbons are always numbered 1 and 2. When one of the unsaturated carbons is bonded to a group, that carbon becomes C-1, even when it results in higher numbers for groups elsewhere on the ring.

Examples

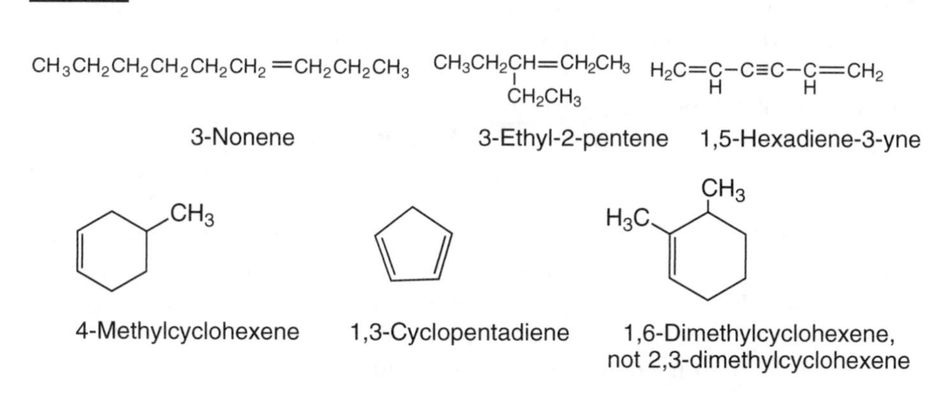

When a branch off a principal chain contains a double or triple bond, the position of the double bond in the branch is indicated by a numerical prefix. The suffix of the group is *-enyl* for a double bond and *-ynyl* for a triple bond. C-1 is the carbon bonded to the main chain. A substituent with the structure $CH_3CH{=}CHCH_2{-}$ is named 2-butenyl.

The most recent revision of IUPAC rules makes a change in the preferred position of the numerical prefix indicating the position of double or triple bonds. The new rules prefer that the numbers be placed immediately before the part of the name to which they relate. Thus, in the examples above, 3-nonene is named non-3-ene and 1,3-cyclopentadiene is cyclopenta-1,3-diene. Students may encounter either style in lists of compounds.

Alkyl Halides

Alkyl halides are organic derivatives of F, Cl, Br, and I. Many simple halides are used in organic chemistry, and their names are based on the organic groups listed in Figure 2.1 followed by chloride, bromide, and so forth. For example, CH_3I is methyl iodide, and $CH_2{=}CHCH_2{-}Br$ is allyl bromide. For more complex structures, the prefix *fluoro-*, *chloro-*, *bromo-*, or *iodo-* is added to the name of the parent compound. When there are several groups on the longest chain, the halogens and alkyl groups are given equivalent priority and the chain is numbered to give the lowest number to the substituents:

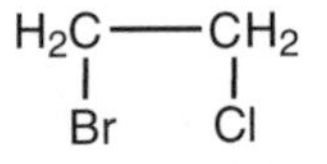

1-Bromo-2-chloroethane

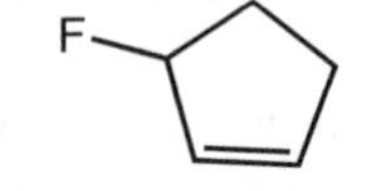

3-Fluorocyclopentene

4-Ethyl-1-iodoheptane

Alcohols

IUPAC rules permit two different strategies for naming alcohols, as well as molecules with other kinds of functional groups. **Functional class nomenclature** involves naming the alkyl group that contains the HO— group (the **hydroxy** substituent) and adding the word *alcohol* as a second word. This is usually used for common alcohols of simple structure; for example, CH_3OH is methyl alcohol, and $(CH_3)_2CHOH$ is isopropyl alcohol.

The other type of nomenclature, **substitutive nomenclature**, adds the suffix *-ol* to the name of the parent hydrocarbon chain with removal of a terminal *-e* before the vowel (i.e., ethanol and not ethaneol). The general procedure of finding the longest chain containing the functional group, numbering the chain from the direction that gives the functional group the smaller number, and adding branching groups in alphabetical order is followed. The HO— group takes precedence over alkyl, halogen, and double or triple bonds in determining the direction of chain numbering.

Examples

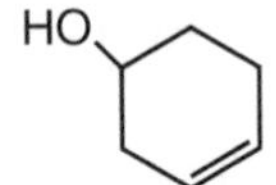

Cyclohex-3-en-1-ol

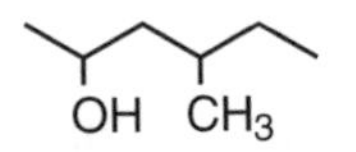

4-Methyl-2-hexanol

Pentane-1,5-diol or 1,5-pentanediol

2-Bromoheptan-3-ol or 2-bromo-3-heptanol

Priority of Functional Groups

When a molecule has more than one functional group, one of them takes priority for the last suffix. The order of functional group priorities is given in Table 3.2 along with the suffixes used to identify each functional group. Prefixes are used to denote the presence of functional groups along a chain when the molecule contains several different functional groups.

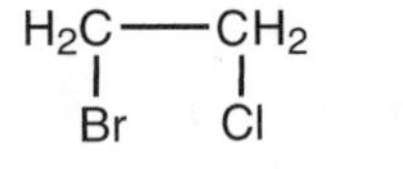

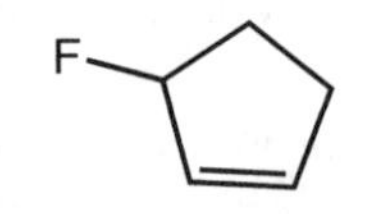

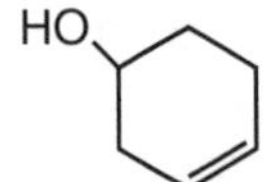

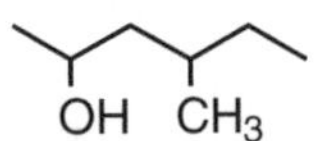

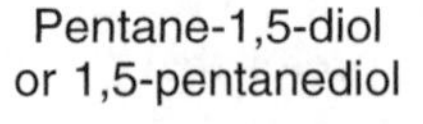

TABLE 3.2. SUFFIXES AND PREFIXES USED IN IUPAC NOMENCLATURE[a]

Class	Prefix	Suffix
Carboxylic acid	*carboxy-*	*-carboxylic acid* or *-oic acid*
Carboxylic anhydride	—	*-anhydride*
Carboxylic ester	(*R*)-*oxocarbonyl-*	(*R*)-*carboxylate* or (*R*)-*oate*
Carboxylic acid chloride	*chlorocarbonyl-*	*-carbonyl chloride* or *-oyl chloride*
Amide	*carbamoyl-*	*-carboxamide* or *-amide*
Nitrile	*cyano-*	*-carbonitrile* or *-nitrile*
Aldehyde	*formyl-* or *oxo-*	*-carbaldehyde* or *-al*
Ketone	*oxo-*	*-one*
Alcohol	*hydroxy-*	*-ol*
Amine	*amino-*	*-amine*
Ether	(*R*)-*oxy*	—

[a]The functional groups are listed in order of decreasing priority.

Ethers

Ethers are normally not very reactive compounds and are usually used as solvents in organic reactions. Simple ethers such as $CH_3CH_2OCH_3$ are usually named by citing the names of the two R groups attached to the oxygen in alphabetical order followed by the class name **ether**, in the above case ethyl methyl ether. Alternatively, ethers are named as **alkoxy** derivatives of alkanes. The suffix *-oxy* is used to replace the *-ane* ending in the RO— substituent. A CH_3O— group is a methoxy group, and a CH_3CH_2O— group is an ethoxy group, for example. A three-membered ring with one oxygen atom in the ring is called **oxirane** (common name, ethylene oxide), and a five membered ring with one oxygen atom is called tetrahydrofuran (THF), a common solvent.

Examples

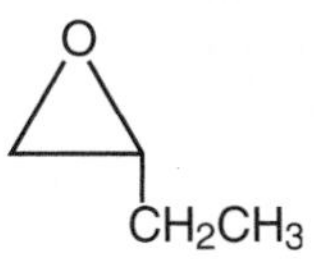

Ethyloxirane

Tetrahydrofuran

3-Ethoxyheptane

Diethylether
(common name, ether)

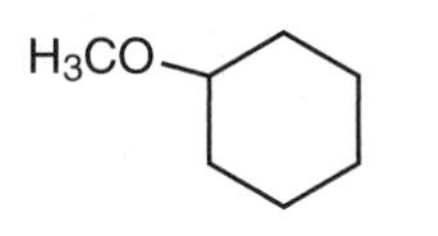

Methoxycyclohexane

Amines

Amines are organic derivatives of nitrogen. One or more of the hydrogens of ammonia are replaced by a carbon group. Simple amines are usually named using the functional class method, where the alkyl groups attached to nitrogen are listed, without spaces, ending with the group name amine. $(CH_3CH_2)_2NCH_3$ is diethylmethylamine. Or amines can be named in the same way one names alcohols. The longest carbon chain containing the nitrogen atom is found and named, the chain direction being chosen so that the nitrogen has the lower number. The suffix *-amine* is added, removing the terminal *-e* of the hydrocarbon name. If there are other, smaller, groups on the nitrogen, they are given the prefix *N*-:

H_2C—N—CH_2CH_3

N-Ethyl-*N*-methyl-3-pentanamine

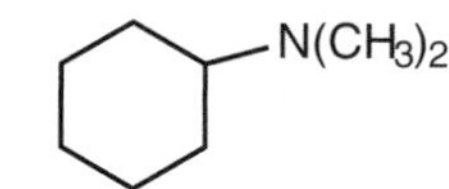

N,*N*-dimethylcyclohexanamine

H_2N

4-Methyl-1-hexanamine

When another functional group takes priority over an amine, the presence of the amine is established using the prefix *amino*-. For example, $NH_2CH_2CH_2CH_2CH_2OH$ is named 4-amino-1-butanol. Finally, an amine can have two or more groups with their own functional groups on the nitrogen. For example, in the molecule $(ClCH_2CH_2)_2NH$ there are two chlorine atoms, one at the 2 position on each ethyl group. Primes are used on the numerical prefixes to show that each group has a chloride: 2,2′-dichlorodiethylamine. The molecule $(Cl_2CHCH_2)_2NH$ is named 2,2,2′,2′-tetrachlorodiethylamine.

EXERCISES

Draw structures based on the following names.

(a) isopropyl methyl ether
(b) *N*-methylhept-2-en-4-amine
(c) 2,2-diethoxypropane
(d) *N*-(2-bromoethyl)propan-1-amine
(e) 4-ethylcyclohexylamine

Solutions

(a) H_3C / H_3C CH—O—CH_3

(b)

(c)

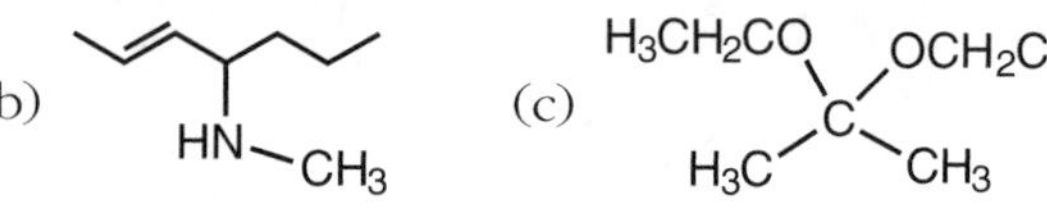

TABLE 3.2. SUFFIXES AND PREFIXES USED IN IUPAC NOMENCLATURE[a]

Class	Prefix	Suffix
Carboxylic acid	*carboxy-*	*-carboxylic acid* or *-oic acid*
Carboxylic anhydride	—	*-anhydride*
Carboxylic ester	*(R)-oxocarbonyl-*	*(R)-carboxylate* or *(R)-oate*
Carboxylic acid chloride	*chlorocarbonyl-*	*-carbonyl chloride* or *-oyl chloride*
Amide	*carbamoyl-*	*-carboxamide* or *-amide*
Nitrile	*cyano-*	*-carbonitrile* or *-nitrile*
Aldehyde	*formyl-* or *oxo-*	*-carbaldehyde* or *-al*
Ketone	*oxo-*	*-one*
Alcohol	*hydroxy-*	*-ol*
Amine	*amino-*	*-amine*
Ether	*(R)-oxy*	—

[a]The functional groups are listed in order of decreasing priority.

Ethers

Ethers are normally not very reactive compounds and are usually used as solvents in organic reactions. Simple ethers such as $CH_3CH_2OCH_3$ are usually named by citing the names of the two R groups attached to the oxygen in alphabetical order followed by the class name **ether**, in the above case ethyl methyl ether. Alternatively, ethers are named as **alkoxy** derivatives of alkanes. The suffix *-oxy* is used to replace the *-ane* ending in the RO— substituent. A CH_3O— group is a methoxy group, and a CH_3CH_2O— group is an ethoxy group, for example. A three-membered ring with one oxygen atom in the ring is called **oxirane** (common name, ethylene oxide), and a five membered ring with one oxygen atom is called tetrahydrofuran (THF), a common solvent.

Examples

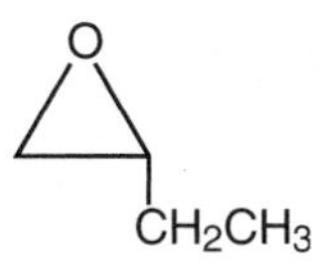

Ethyloxirane

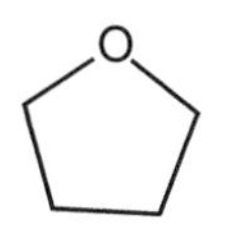

Tetrahydrofuran

3-Ethoxyheptane

Diethylether
(common name, ether)

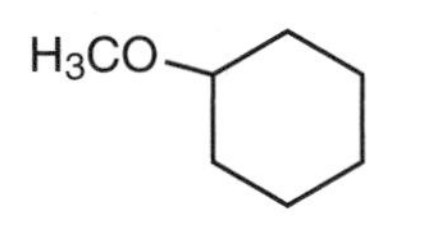

Methoxycyclohexane

Amines

Amines are organic derivatives of nitrogen. One or more of the hydrogens of ammonia are replaced by a carbon group. Simple amines are usually named using the functional class method, where the alkyl groups attached to nitrogen are listed, without spaces, ending with the group name amine. $(CH_3CH_2)_2NCH_3$ is diethylmethylamine. Or amines can be named in the same way one names alcohols. The longest carbon chain containing the nitrogen atom is found and named, the chain direction being chosen so that the nitrogen has the lower number. The suffix *-amine* is added, removing the terminal *-e* of the hydrocarbon name. If there are other, smaller, groups on the nitrogen, they are given the prefix *N*-:

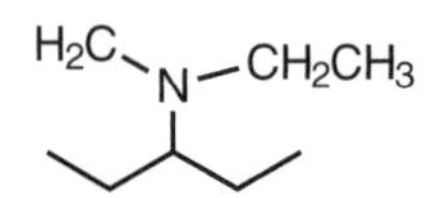

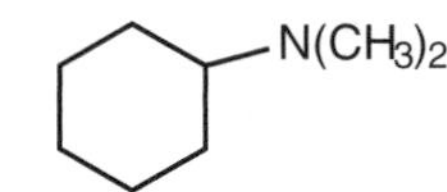

H2N

N-Ethyl-*N*-methyl-3-pentanamine

N,*N*-dimethylcyclohexanamine

4-Methyl-1-hexanamine

When another functional group takes priority over an amine, the presence of the amine is established using the prefix *amino*-. For example, $NH_2CH_2CH_2CH_2CH_2OH$ is named 4-amino-1-butanol. Finally, an amine can have two or more groups with their own functional groups on the nitrogen. For example, in the molecule $(ClCH_2CH_2)_2NH$ there are two chlorine atoms, one at the 2 position on each ethyl group. Primes are used on the numerical prefixes to show that each group has a chloride: 2,2′-dichlorodiethylamine. The molecule $(Cl_2CHCH_2)_2NH$ is named 2,2,2′,2′-tetrachlorodiethylamine.

EXERCISES

Draw structures based on the following names.

(a) isopropyl methyl ether
(b) *N*-methylhept-2-en-4-amine
(c) 2,2-diethoxypropane
(d) *N*-(2-bromoethyl)propan-1-amine
(e) 4-ethylcyclohexylamine

Solutions

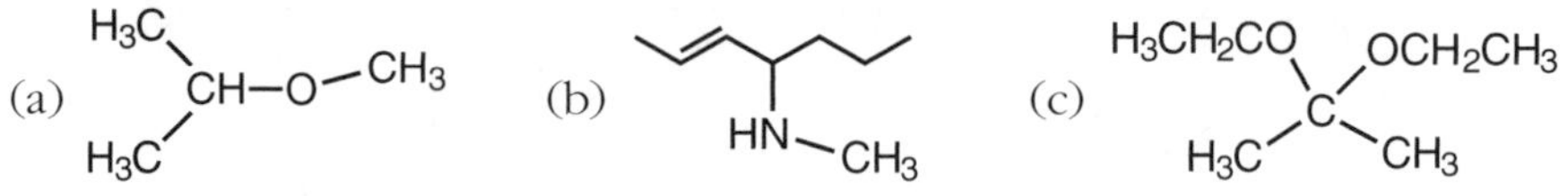

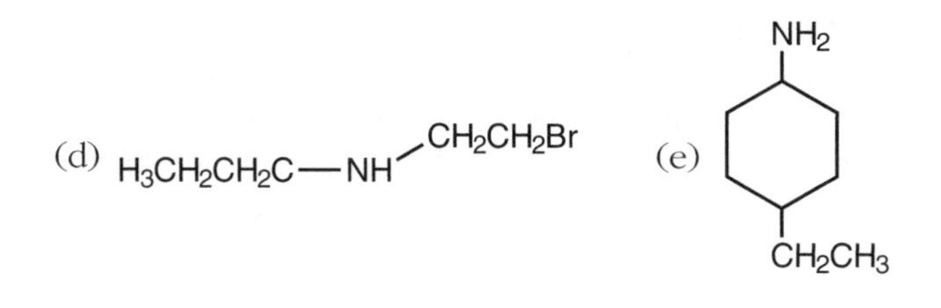

Ketones and Aldehydes

Two of the most important functional groups in organic chemistry are ketones and aldehydes. Much of the second half of a one-year course in organic chemistry is taken up with their chemistry. Because of their structure, aldehydes are always at the end of a chain, and ketones are located in the interior of a carbon chain.

In naming aldehydes, the longest carbon chain that contains the —CHO group is identified, and the name of the alkane chain has its terminal *-e* replaced by the suffix *-al*. The carbon of the aldehyde group is C-1. If the —CHO group, which is called a **formyl group**, is not part of an acyclic hydrocarbon, the suffix used is *-carbaldehyde*. In the case of a cyclopentane attached to a —CHO group, the name of the compound is cyclopentanecarbaldehyde.

Ketones are named by taking the name of the longest alkyl chain that contains the —CO— group (the **carbonyl group**), replacing the terminal *-e* with the suffix *-one*, and numbering the chain in the direction that gives the —CO— carbon the lower number. The functional class nomenclature method is often used for simple ketones. The alkyl groups attached to the carbonyl group are named as separate words followed by the group name **ketone**. A common solvent is ethyl methyl ketone, $CH_3CH_2COCH_3$ (usually sold under its improperly alphabetized name methyl ethyl ketone). Examples of some aldehydes and ketones known by their common names are given here.

Examples

Formaldehyde

Acetaldehyde

Benzaldehyde

Acrolein

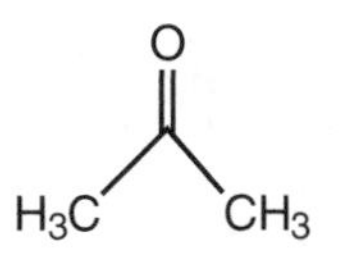

Acetone

Methyl vinyl ketone

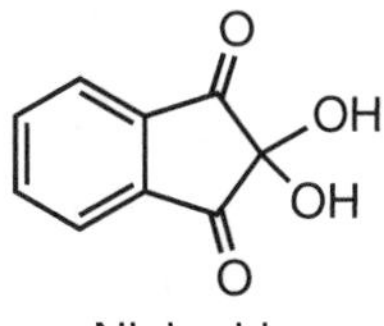

Ninhydrin

When an aldehyde or ketone is present in a compound with a functional group of higher priority, its presence is indicated by the prefix *formyl-* for an aldehyde or *oxo-* for a ketone. Since a carboxylic acid group has higher priority than a ketone group, the derivative of $CH_3CH_2CH_2CH_2COOH$, pentanoic acid, with a ketone group at C-4, $CH_3COCH_2CH_2COOH$, is named 4-oxopentanoic acid.

EXERCISES

1. Draw structures of the following compounds.
   (a) 4,4-dimethyl-2-heptanone (b) 2,4-hexanedione
   (c) 2-ethoxycyclobutanecarbaldehyde

2. Assign IUPAC names to the following compounds.

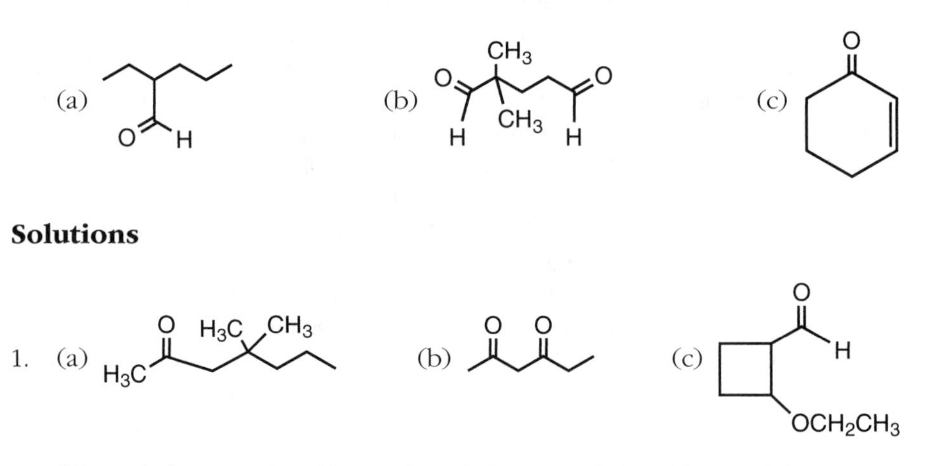

2. (a) 2-ethylpropanal; (b) 2,2-dimethylpentanedial; (c) 2-cyclohexenone

Carboxylic Acids and Their Derivatives

The first six functional groups listed in Table 3.2, from carboxylic acids to nitriles, are all closely related chemically. All have a carbon atom with three bonds to an electronegative atom, either oxygen or nitrogen, and conversions among these groups are easy. The carboxylic acid group has the highest priority among the common functional groups. Where the —COOH substituent conceptually replaces a $—CH_3$ group in an acyclic hydrocarbon, the suffix *-oic acid* is added to the name of the hydrocarbon with loss of the terminal *-e*. For example, the linear six-carbon carboxylic acid is hexanoic acid. When the —COOH group is attached to a ring, the suffix *-carboxylic acid* is added to the name of the parent hydrocarbon. When all carboxylic acid groups cannot be described with a suffix, a carboxylic group is described by the prefix *carboxy-*. Examples are given below, along with some common names of simple carboxylic acids.

Examples

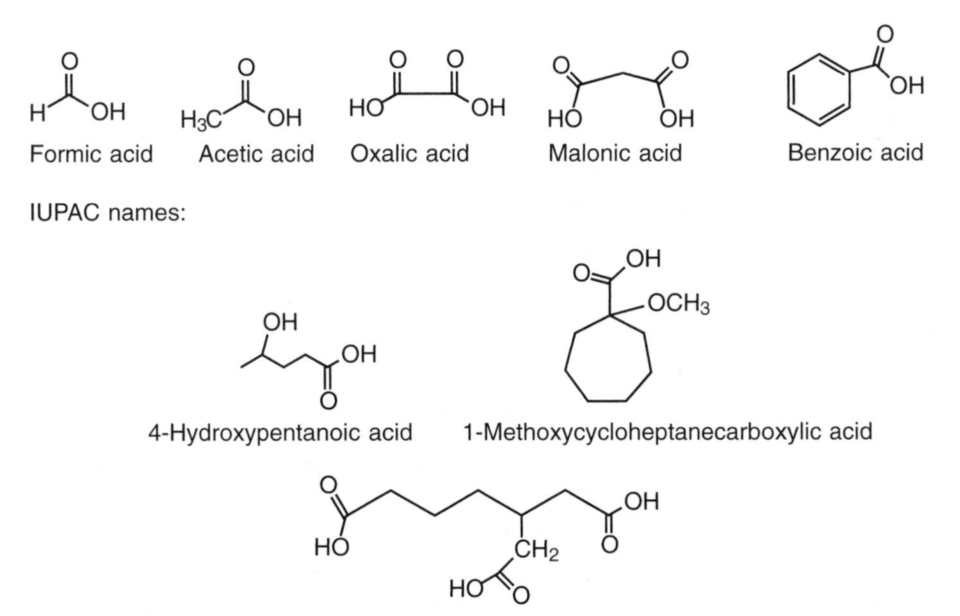

Formic acid Acetic acid Oxalic acid Malonic acid Benzoic acid

4-Hydroxypentanoic acid 1-Methoxycycloheptanecarboxylic acid

3-(Carboxymethyl)heptanedioic acid

Nomenclature of Acyl Derivatives of Carboxylic Acids

The **acyl group** is one of the most important sites of reactivity in any organic molecule that contains it.

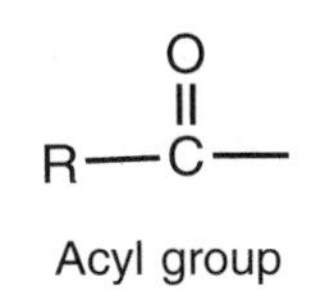

Acyl group

When a heteroatom is bonded to the carbon, the result is a carboxylic acid, a carboxylic acid anhyhdride, a carboxylic acid chloride, a carboxylic ester, or an amide. All these acyl derivatives have names based on the corresponding carboxylic acid.

Carboxylic acid chlorides (commonly called **acid chlorides**) are named by replacing the *-ic acid* suffix of the parent carboxylic acid with the suffix *-yl chloride*. For acids ending in *-carboxylic acid* the suffix *-carbonyl chloride* replaces it. Only rarely are other halogens seen in the place of chloride. Two very commonly used acid chlorides are acetyl chloride, CH_3COCl, and oxalyl chloride, ClCOCOCl.

Carboxylic acid anhydrides, RCO—O—COR, are usually just called **anhydrides**. The word *anhydride* means without water, and these compounds can be made by taking two carboxylic groups and removing one molecule of water. When both R groups are the same, the anhydride is termed symmetric. Symmetric anhydrides are the most common type. They are named by replacing the term *acid* in the parent acid's name by the word *anhydride*.

Examples

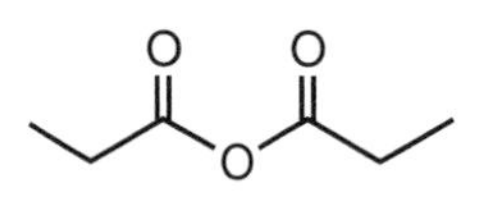

Propionic anhydride

Succinic anhydride

from

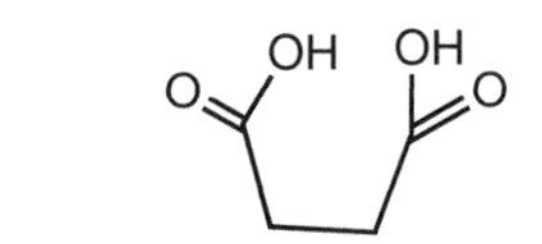

Succinic acid (common name)

Carboxylic acid esters, normally referred to as **esters**, are named as derivatives of both carboxylic acids and alcohols. Historically, esters have been made by combining alcohols and carboxylic acids and removing one molecule of water. In an ester, RCO—OR′, the RCO— portion is derived from a carboxylic acid and the OR′ part is derived from an alcohol. Ester names are composed of two words. The first word denotes the alcohol part, and the R′ group is named like any hydrocarbon group, with a *-yl* suffix. The RCO— part is derived from the name of the parent acid, with the *-ic acid* suffix replaced by the suffix *-ate*.

Examples

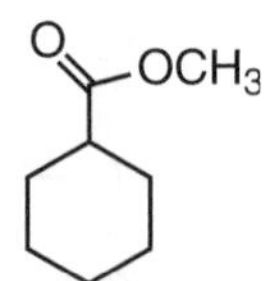

Methyl cyclohexanecarboxylate

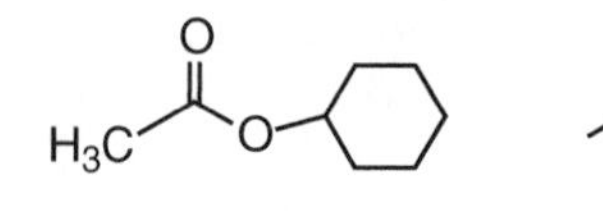

Cyclohexyl acetate

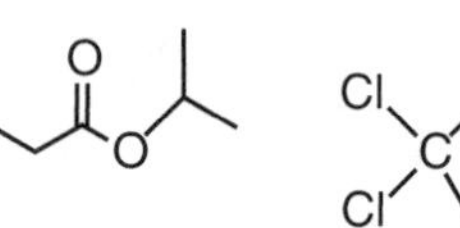

Isopropyl butanoate

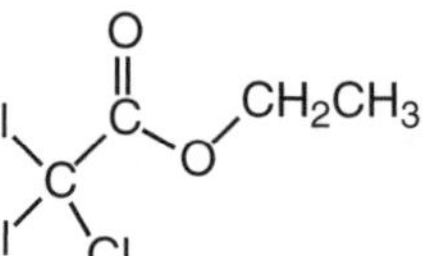

Ethyl trichloroacetate

Amides of the type RCO—NH_2 (termed **primary amides**) are named by taking the suffix *-ic acid* or *-carboxylic acid* of the corresponding carboxylic acid and replacing it with the suffix *-amide* or *-carboxamide*, respectively. When there are other carbon groups on the nitrogen (RCO—NHR, termed **secondary amides**, or RCO—NRR′, termed **tertiary amides**), they are named by citing the groups R and R′ as prefixes using the *N*- prefix as described for amines. When the amide group is a substituent in a molecule with a higher-priority functional group, its presence is noted by the prefix *carbamoyl-*.

Examples

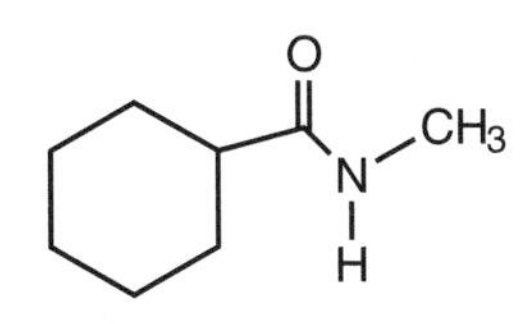

N-Methylcyclohexanecarboxamide

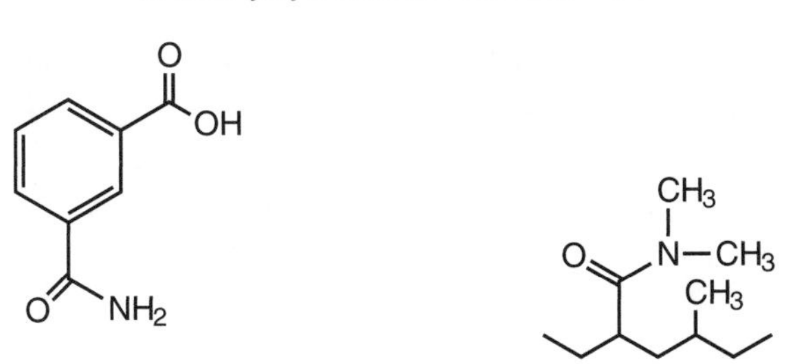

3-Carbamoylbenzoic acid *N,N*-Dimethyl-2-ethyl-4-methylhexanamide

Although they do not contain an acyl group, nitriles are chemically related to acyl derivatives. Their names are derived by considering the —CN group to have replaced the —COOH group of a carboxylic acid and the *-oic acid* suffix replaced by the suffix *-nitrile* or the *-carboxylic acid* suffix replaced by *-carbonitrile*. For acids with common names, such as acetic acid, CH_3COOH, the *-ic acid* suffix is changed to *-onitrile* for the nitrile (acetonitrile, CH_3CN). When a —CN group is not the principal functional group, the prefix *cyano-* is used.

EXERCISES

Provide IUPAC names for the following molecules.

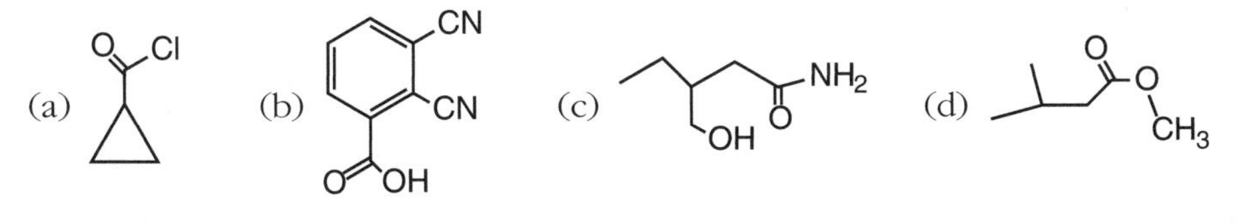

(e) 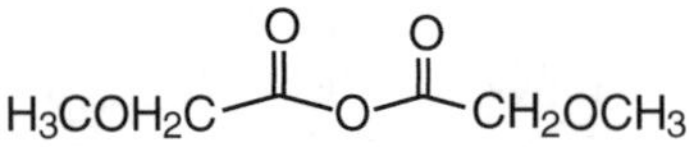 H_3COH_2C–C(=O)–O–C(=O)–CH_2OCH_3 (f)

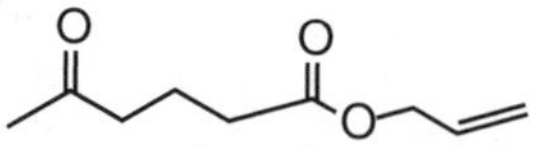

Solutions

(a) Cyclopropanecarbonyl chloride; (b) 2,3-dicyanobenzoic acid; (c) 3-(hydroxymethyl)pentanamide; (d) methyl 3-methylbutanoate; (e) methoxyacetic anhydide or methoxyethanoic anhydride; (f) allyl 5-oxohexanoate or 3-propenyl 5-oxohexanoate

4
ACIDITY AND INDUCTIVE EFFECTS

BRØNSTED AND LEWIS ACIDS

Acid-base chemical concepts are as important in organic chemistry as they were in introductory general chemistry. Normally, the fastest of all chemical reactions are reactions of acids with bases. Thus, when two reactants are mixed, the first thing that happens, when possible, is an acid-base reaction. The ability to recognize what species is the strongest acid and what species is the strongest base is central to understanding organic chemistry.

Two definitions are used for acids and bases in organic chemistry, one more general than the other. The simpler definition, due to Brønsted and Lowry, states that an **acid** is a proton donor and a **base** is a proton acceptor. Proton is another name for a hydrogen ion, H^+. Using the Brønsted-Lowry definition, an acid-base reaction is a **proton transfer** reaction. The American chemist G. N. Lewis formulated a more inclusive definition of acids and bases. What we now call a **Lewis acid** is an electron pair acceptor, and a **Lewis base** is an electron pair donor. A proton is one type of Lewis acid, but many other kinds of Lewis acids exist. Some examples of acid-base reactions are shown here.

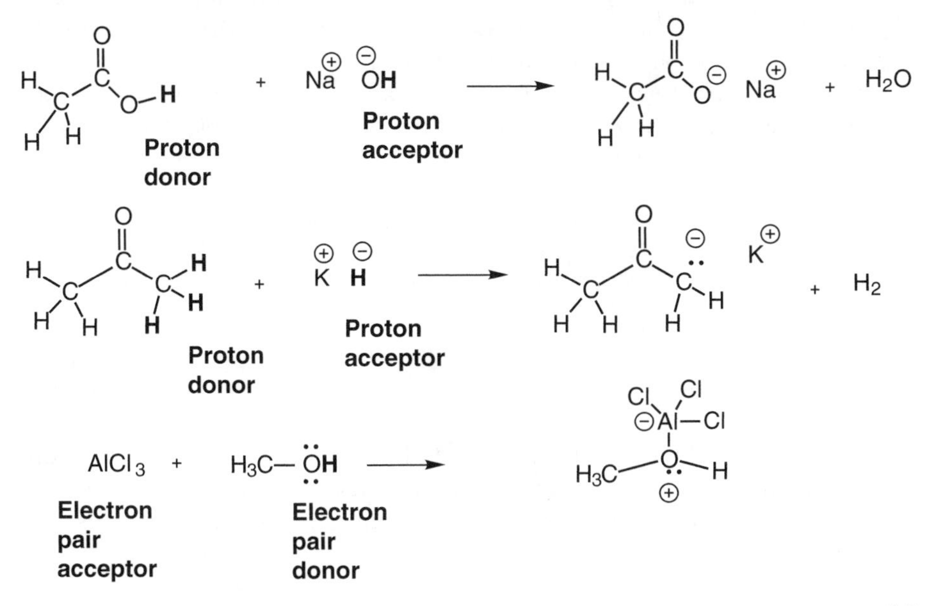

Brønsted Conjugate Acids and Bases

Every chemical reaction is, to some degree, reversible. So when a Brønsted acid, generally referred to as HA, transfers a proton to a base, the new protonated species can act like an acid in the reverse direction:

$$H{-}A \; + \; :B \; \rightleftharpoons \; :A^{\ominus} \; + \; H{-}B^{\oplus}$$

Acid	Base	Conjugate base	Conjugate acid

The new protonated species is called the **conjugate acid** of the base, and the deprotonated form of the original acid is called the **conjugate base** of the acid.

A strong Brønsted acid is a compound that transfers a proton to almost any base. A strong base is any species that can accept a proton from most acids, HA. There is another, very important, way to describe acid and base strength in terms of the strengths of the conjugate acids and conjugate bases: *A strong acid has a weak conjugate base, and a strong base has a weak conjugate acid.* Conversely, a weak acid has a strong conjugate base, and a weak base has a strong conjugate acid.

These two statements make sense once one thinks about what is important in a proton transfer reaction. For any acid, HA, a proton is a proton. *The stabilities of the conjugate bases, A^-, distinguish one acid from another.* HCl is a strong acid because the conjugate base, Cl^-, does not accept a proton from most conjugate acids. Methane, CH_4, is a very weak acid because its conjugate base, CH_3^-, is an extremely strong base. For virtually all bases, the equilibrium $CH_4 + B^- \leftrightarrow CH_3^- + HB$ lies to the left, favoring CH_4, because CH_3^- deprotonates almost any conjugate acid HB.

Since a weak base has a strong conjugate acid, we can infer base strengths from acid-base reactions as well as acid strengths. For example, the following reaction has its equilibrium as shown.

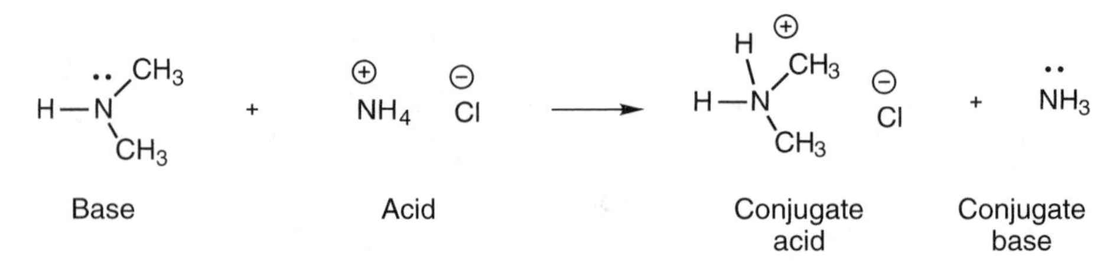

Since a stronger acid and a stronger base always react to give a weaker conjugate acid and a weaker conjugate base (because otherwise the equilibrium would point in the other direction), the equilibrium position tells us that $HN(CH_3)_2$ is a stronger base than NH_3.

EXERCISES

1. Draw the conjugate bases of the following acids. Include all lone electron pairs and charges.
   (a) HF (b) H_3CCH_3 (c) NH_3 (d) $CH_3OH_2^+$

2. Draw the conjugate acids of the following bases. Include all lone electron pairs and charges. Normally lone pairs or negatively charged atoms accept a proton.
   (a) H_2S (b) HO^- (c) $H_2C{=}O$

Solutions

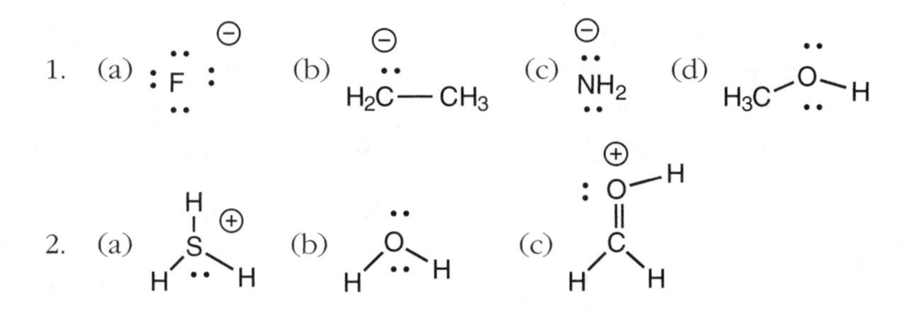

Important Brønsted Acids and Bases in Organic Chemistry

In addition to common inorganic acids, such as HCl, H_2SO_4, and HNO_3, a number of organic compounds have important acidic or basic characteristics.

ORGANIC ACIDS

Compound Type	General Structure	Conjugate Base
Sulfonic acid	R—S(=O)(=O)—OH	R—S(=O)(=O)—O⊖
Carboxylic acid	R—C(=O)—O—H	R—C(=O)—O⊖
Phenol	C6H5—O—H	C6H5—O⊖
Protonated amine	R—N⊕(R′)(R″)—H	R—N(R′)(R″):
Alcohol	ROH	RO⊖

The ordering of the acids in the accompanying table is from strong to weak, although, as we shall discuss, the exact structure of a molecule can influence its acidity.

Except for the conjugate bases of sulfonic acids, which are very weak bases, the other conjugate bases are common organic bases. In addition, if one needs a very strong base, the conjugate bases of amines (such as $Li^+R_2N^-$, called **lithium amides**), or the lithium salts of hydrocarbons (called alkyl lithium reagents, RLi), are used. Potassium hydride, KH, which acts as if it were a source of the strong base H^-, is another strong base useful in organic chemistry.

Lewis Acids and Bases in Organic Chemistry

Any species with a lone pair of electrons or a negative charge can be a Lewis base. Therefore, any oxygen-containing group, including an alcohol, ether, aldehyde, ketone, or ester, can act as a Lewis base. Likewise, nitrogen-containing molecules, particularly amines but also nitriles, can be Lewis bases. The lone electron pairs of organic halides are additional examples of weak Lewis bases.

Most of the important Lewis acids in organic chemistry are inorganic compounds, except for carbocations, which are very reactive Lewis acids. Compounds of group 3 elements, such as BF_3, $AlCl_3$, and R_3B, are Lewis acids because these elements have a vacant *p* orbital that can accept an electron pair. Most salts of transition elements, especially those in higher oxidation states, act as Lewis acids. $TiCl_4$, $FeCl_3$, $ZnCl_2$, and $CuCl_2$ have numerous applications as Lewis acids in organic chemistry. Since Brønsted acids are Lewis acids, H^+ and the conjugate acid of water, H_3O^+, are important Lewis acids.

Examples The following reactions are examples of Lewis acid–Lewis base complex formation.

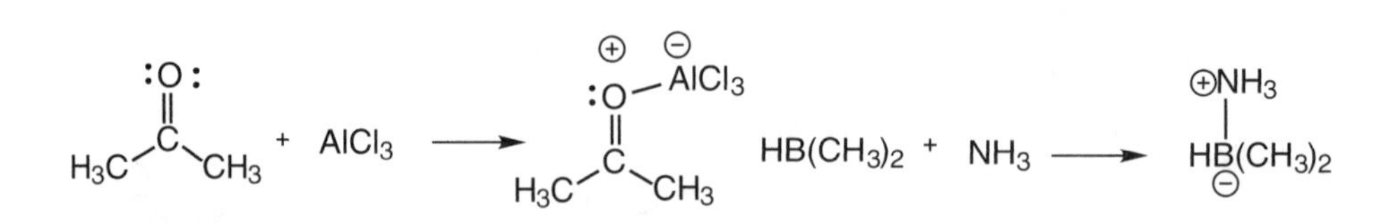

It is important to understand the acid-base properties of organic molecules because they can be used to control reactivity. For example, in the first example shown above, the ketone (acetone) has a certain level of reactivity. When acetone forms a complex with the Lewis acid $AlCl_3$, its reactivity toward many reagents greatly increases.

A QUANTITATIVE DESCRIPTION OF ACIDITY USING pK_a

The Definition of pK_a

Unlike the general chemistry course that normally precedes it, a course in introductory organic chemistry is largely qualitative. However, when discussing Brønsted acids, it is possible to be quantitative about acid strength using the concept of an acid's pK_a. The pK_a of an acid is a number that indicates how strong or weak a particular acid is (it is used only for Brønsted acids because it involves proton transfer). For example, the strong acid HCl has a pK_a of −7. Methane, CH_4, which is not a proton donor at all, has a pK_a of approximately 50. It is useful to memorize these two numbers in order to remember how the pK_a scale runs. Strong acids have negative or slightly positive pK_a values, and the larger the pK_a value, the weaker the acid.

The **acid dissociation constant K_a**, is based on the ionization of a particular acid in water: $HA + H_2O \leftrightarrow H_3O^+ + A^-$. One can write the equilibrium constant for this proton transfer reaction, K_{eq}, as

$$K_{eq} = \frac{[H_3O^+][A^-]}{[H_2O][HA]} \quad \text{and} \quad K_a = K_{eq}[H_2O] = \frac{[H_3O^+][A^-]}{[HA]}$$

Under certain conditions, the acid HA is one-half dissociated, that is, one-half of the molecules are in the form HA and one-half are dissociated into A^- and H_3O^+. When this happens, $[A^-] = [HA]$ and $K_a = [H_3O^+]$. Taking the negative logarithm of both sides of this equation gives pK_a = pH, under these conditions, since pK_a is the negative logarithm of K_a. Thus, we can think of the **pK_a** of an acid as the pH at which it is one-half dissociated. A very weak acid needs very basic conditions in order to generate A^-. Conversely, a strong acid can generate A^- even when the medium is acidic; A^- will not be protonated to give HA.

Establishing the Equilibrium Position of an Acid-Base Reaction

Knowing the pK_a values of an acid and the conjugate acid of a base allows one to determine the equilibrium position of any acid-base reaction. For example, consider the reaction

$$H_2O + NH_3 \rightleftharpoons HO^- + NH_4^+$$
$$pK_a = 15.7 \qquad\qquad pK_a = 9.2$$

Since a strong acid always reacts to give a weaker acid, the equilibrium lies to the left for this reaction. In general, the equilibrium always lies in the direction of the larger pK_a value. Recall that knowing the equilibrium position of a reaction does not establish how fast the reaction will occur. However, proton transfer reactions are some of the fastest reactions; so if an acid-base reaction can take place, it usually will.

EXERCISES

Indicate where the equilibrium lies for the following acid-base reactions. For reaction (c), which compound is a stronger base, NH_3 or $HN(CH_3)_2$?

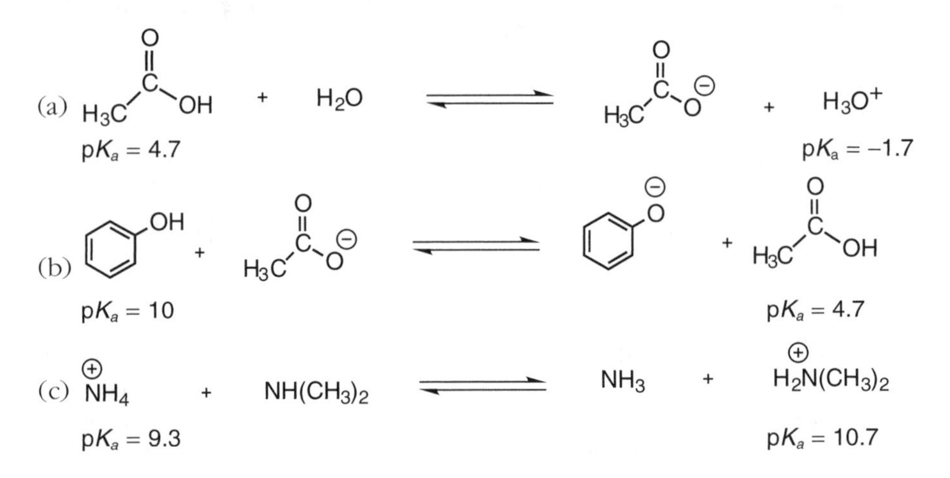

Solutions

In reactions (a) and (b) the equilibrium lies to the left. In reaction (c) it lies to the right. Since a strong base has a weak conjugate acid and because $H_2N(CH_3)_2^+$ is a weaker acid than NH_4^+, $HN(CH_3)_2$ is the stronger base.

The Henderson-Hasselbach Equation

An equation frequently used in biochemistry is the **Henderson-Hasselbach equation**:

$$pK_a = pH + \log([HA]/[A^-])$$

Biological fluids are buffered to a specific pH. Blood, for example, is buffered by carbonic acid, H_2CO_3, to a pH of 7.2. Using the Henderson-Hasselbach equation one can calculate what form an acid takes at a given pH. For example, lactic acid, $CH_3CHOHCO_2H$, is produced by muscles during exercise. Lactic acid has a pK_a of approximately 4. At biological pH 7.2, essentially all the acid is ionized as $CH_3CHOHCO_2^-$.

FACTORS THAT INFLUENCE ORGANIC ACIDITY

From Table 4.1, it is clear that a tremendous range of acidities is found in organic molecules. Since pK_a is a logarithmic scale, there is more than a 10^{60}-fold difference in acid strengths. Because it is the stability of the acid's conjugate base, A^-, that determines acid strength, a study of pK_a values can illustrate what factors stabilize negative charge.

TABLE 4.1. pK_a VALUES OF COMMON MOLECULES

Acid	pK_a	Acid	pK_a
$\mathbf{H}I$	−10	C_6H_5–OH	10
$\mathbf{H}Br$	−9	$(CH_3)_3\overset{+}{N}H$	10.6
$\mathbf{H}Cl$	−7	$Et\overset{+}{N}H_3$	11
$EtO\mathbf{H}_2^+$	−2.4	$CH_3O\mathbf{H}$	15.7
$\mathbf{H}_3O^+$	−1.7	$\mathbf{H}_2O$	16
CF_3–C(=O)OH	0.2	CH_3–C(=O)–NH_2	16
CH_3–C(=O)OH	4.7	CH_3–C(=O)–H	17
$C_5H_5\overset{+}{N}$–H	5.2	H_3C–C(=O)–CH_3	20
$\mathbf{H}_2S$	7.0	CH_3–C(=O)–OR	24.5
$CH_3C\mathbf{H}_2NO_2$	8.6	H–C≡C–H	25
H_3C–C(=O)–CH_2–C(=O)–CH_3	8.9	$N\mathbf{H}_3$	36
$\mathbf{H}_4N^+$	9.4	C_6H_5–CH_3	41
		H_2C=CH_2	46
		$C\mathbf{H}_4$	50

The factors that affect acidity—**atomic size**, **electronegativity**, **resonance**, **inductive effects**, and **hybridization**—are important to understand because they show up in many different areas of organic chemistry. When a chemical reaction takes place, bonds are broken and new bonds are made. Because bonds are just electron pairs, the factors that stabilize the unshared electron pair in A^- also help to stabilize bonding electron pairs. Thus, understanding trends in acidity will be useful later in understanding trends in reactivity.

Atomic Size and Acidity

The pK_a values of the hydrohalic acids are HI, −10; HBr, −9; HCl, −7; and HF, 3.5. Also, for group 16 hydrides, H_2O has a pK_a of 15.7 and H_2S a pK_a of 7. In general, the heavier atom in the same group of the periodic table is the stronger acid. The orbitals are larger for heavier atoms, and the charge density is lower, resulting in a more stable anion.

Electronegativity

Electronegativity is the tendency of an atom or group to attract electrons. Electronegative atoms act like electron sponges. Electronegativities of common atoms were given in Chapter 1.

The most electronegative atoms are in the upper right corner of the periodic table. The effect of electronegativity is seen in the acidities of the second-period hydrides, all of which are approximately the same size:

Hydride:	CH_4	NH_3	H_2O	HF
pK_a:	50	36	15.7	3.5

Stronger acids have more electronegative central atoms because they stabilize the negative charge on the conjugate base of the acid.

Some common functional groups also have electron-withdrawing properties. The nitro group, $—NO_2$, the cyano group, —CN, and functional groups with a carbon-oxygen double bond (**carbonyl groups**) such as aldehyde groups, —CHO, ketone groups, —COR, and ester groups, —COOR, are electronegative groups.

Resonance

Because of its ability to delocalize charge over several nuclei, resonance has a significant effect on acidity:

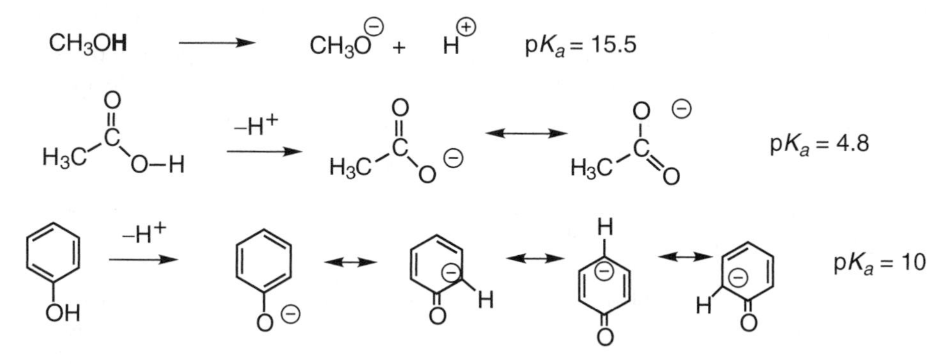

In the above examples, in each case deprotonation results in an oxygen anion. For alcohols, the charge is localized on a single oxygen atom, resulting in an acidity about the same as that of water. For carboxylic acids, there are two equivalent resonance structures, each of which places the negative charge on the electronegative atom, oxygen. The result is a significant increase in acidity compared to that of alcohols. Finally, phenols, which are OH-substituted benzene rings, have an acidity between that of carboxylic acids and alcohols. There are four contributing structures for the conjugate base, but three of them have the negative charge located on the less electronegative atom, carbon. The nature, as well as the number of resonance structures, influences acidity.

Inductive Effects

Resonance is a phenomenon of π bonds and lone electron pairs. **Inductive effects** result from the polarization of electrons in σ bonds. An example of the inductive effect is seen in the acidities of chlorine-substituted carboxylic acids:

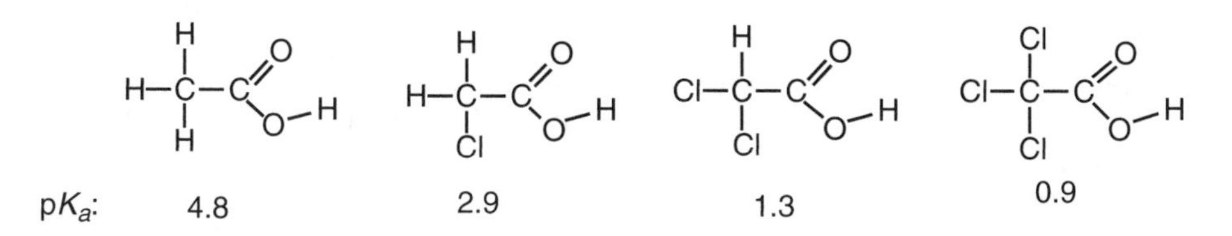

Since chlorine is an electronegative atom, it polarizes the electrons in a C—Cl bond, causing the carbon atom to be at the positive end of a dipole and the chlorine atom to be at the negative end. The partial positive charge on carbon stabilizes the adjacent negative charge of the conjugate base of the acid. The inductive effect decreases with the number of intervening σ bonds. For example, when a chlorine atom is four bonds removed from a —COOH group, as in the case of $ClCH_2CH_2CH_2COOH$, the pK_a is 4.5.

Closely related to the inductive effect is the **field effect**, where a charge influences reactivity through space and not through bonds. Field effects are clearly seen in the acidities of molecules with two —COOH groups. For example, the biologically important succinic acid, $HOOC—CH_2CH_2—COOH$, exhibits two different pK_a values. Ionization of one of the equivalent —COOH groups occurs with a pK_a of 4.21 because of a weak inductive effect. The second pK_a value is 5.64 since the negative charge of the first —COO$^-$ opposes the formation of the same charge on the second —COOH group.

EXERCISES

For each of the following pairs of acids, underline the more acidic compound.

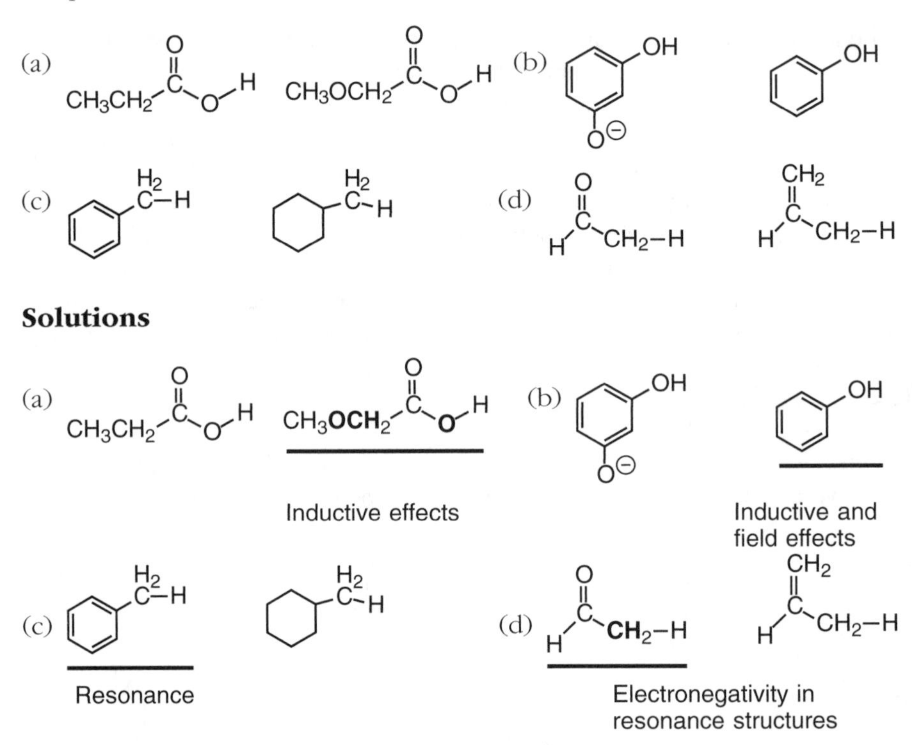

Hybridization

For carbon acids, hybridization can influence acidity:

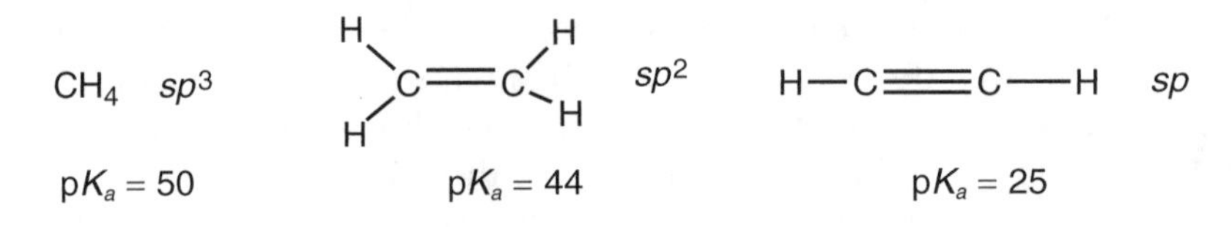

The percentage of an *s* orbital in a hybrid orbital changes from 25 percent in an sp^3 orbital to 33 percent in an sp^2 orbital and to 50 percent in an *sp* orbital. Since an electron in an *s* orbital is closer to the nucleus than an electron in a p orbital, an anionic electron pair is most stabilized in an sp orbital. As a result, acetylene, C_2H_2, a carbon acid, is 10^{11} times more acidic than NH_3 although N is a more electronegative atom.

EXERCISES

Circle the most acidic proton or group of protons in each of the following molecules. Consult Table 4.1 as needed.

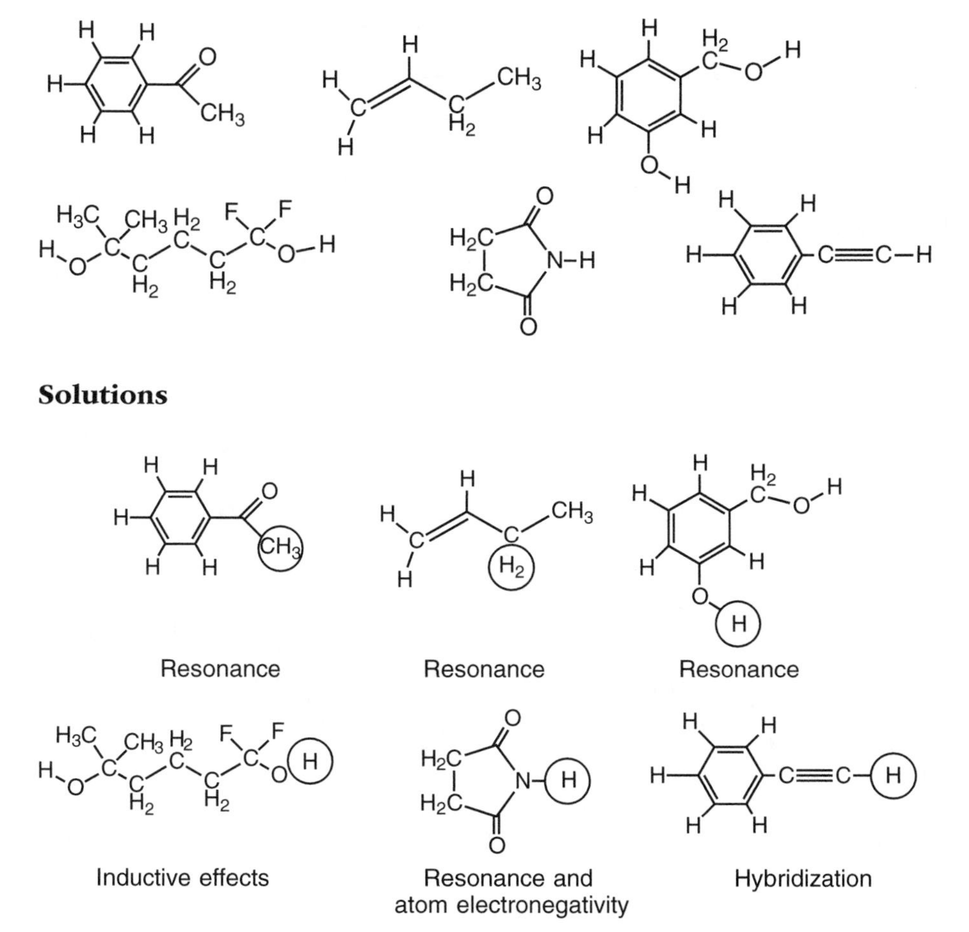

A Note on Organic Chemistry Problems

A major difference between introductory chemistry and organic chemistry is that almost all organic problems are based on qualitative concepts or pattern recognition. For example, while it is possible to give numerical pK_a values for any organic acid, students do not have the tools to calculate, with any accuracy, the acidity of a novel organic molecule. Instead, they must review the general factors that influence acidity and catalog the factors at work in the specific molecule. Normally, one finds that all or almost all the effects favor one answer over another.

Because so many new concepts are introduced so quickly, students may be unsure about which idea to apply to a given problem. Usually it is a good idea to ask what the problem is really about. It may be a resonance problem, a hybridization problem, or a conjugate acid problem. A common mistake students make is to answer hybridization problems with resonance ideas, and so on.

5
STEREOCHEMISTRY

Each kind of hybridized carbon has a characteristic three-dimensional shape. A molecule with *sp* carbons is linear. A molecule with only sp^2 carbons is normally flat. The sp^3 carbon is tetrahedral, and its three-dimensional structure leads to a new kind of isomer, a stereoisomer. Isomers with the same connectivity but different three-dimensional shapes are called **stereoisomers**.

Ordinary language does not have common words to describe intricate spatial relationships. However, the shapes of organic compounds and their symmetry properties are central to understanding organic chemistry. Thus, organic chemists have developed a specialized vocabulary to describe stereochemical relationships. In order to master stereochemistry, the student must first be able to visualize molecules in three dimensions. Second, he or she must understand and be able to apply the specialized vocabulary of stereochemistry, particularly the important terms defined in Table 5.1 at the end of this chapter.

Most students find that the use of molecular models helps to visualize tetrahedral carbon compounds. Numerous brands of molecular model kits are sold, and many college bookstores stock one or more kinds. In addition, many Internet sites provide interactive graphical representations of organic molecules. Using a search engine with terms such as *organic stereochemistry* or *tetrahedral carbon* will produce worthwhile sites.

ALKENE STEREOCHEMISTRY

Cis, Trans, and E,Z Stereoisomers

Before discussing stereoisomers due to tetrahedral carbon, we will introduce isomerism in alkenes. There are three isomers for the compound with the formula $C_2H_2Cl_2$:

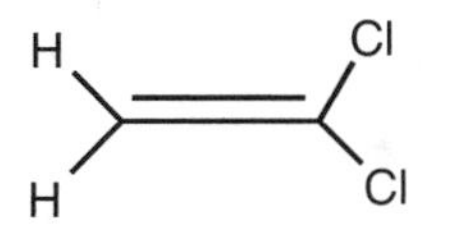

gem-Dichloroethylene

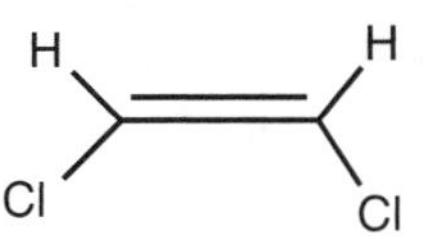

cis-Dichloroethylene

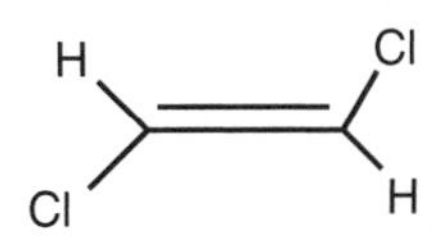

trans-Dichloroethylene

The first isomer has both chlorine atoms on the *same* carbon and is a **geminal** (abbreviated ***gem***) isomer. Since it has a different connectivity from the other two isomers, it is a constitutional isomer of the other two. The second isomer is the **cis** isomer, cis meaning on the same side (in this case the same side of the double bond). The third isomer is a stereoisomer of the cis isomer, and it is a **trans** isomer, trans meaning on opposite sides. The cis and trans isomers are clearly different compounds, and they have different physical properties and somewhat different reactivities toward numerous reagents. Cis and trans isomers are sometimes called geometric isomers. The only way to interconvert the cis and trans isomers is to break the C—C π bond, rotate one carbon 180°, and re-form the π bond. Normally, stereoisomers can be interconverted only by breaking and re-forming bonds.

Examples Note that when naming compounds using cis and trans descriptors, the prefixes are italicized.

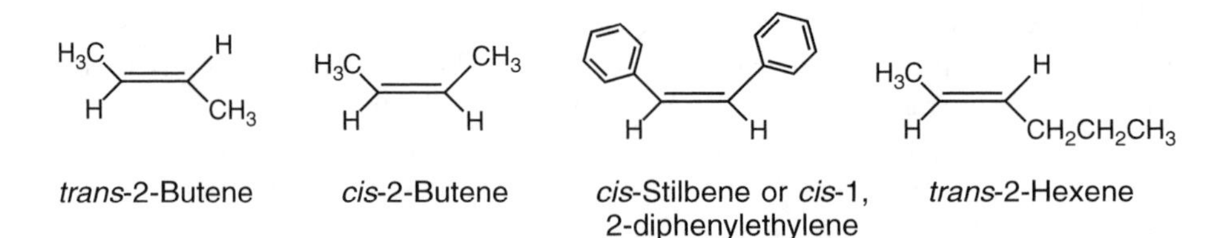

For structurally more complicated alkenes, the cis and trans nomenclature becomes ambiguous.

These two compounds are obviously stereoisomers, but so far we have no clear way to designate which isomer is which. Cis and trans do not work very well for alkenes with four different substituents.

In order to establish the stereochemistry around any double bond, a series of sequence rules were developed that assign priorities to the two groups on each carbon of the double bond. One group is assigned high priority, and one group is assigned low priority:

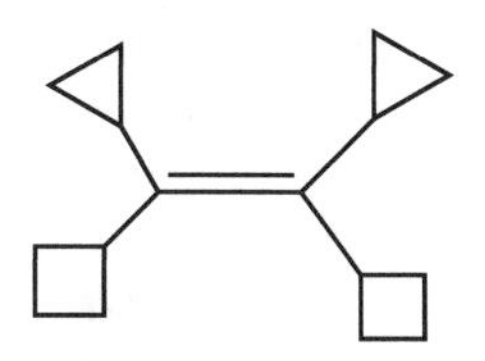

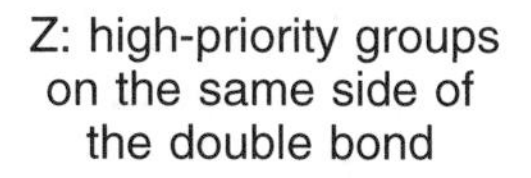

Z: high-priority groups on the same side of the double bond

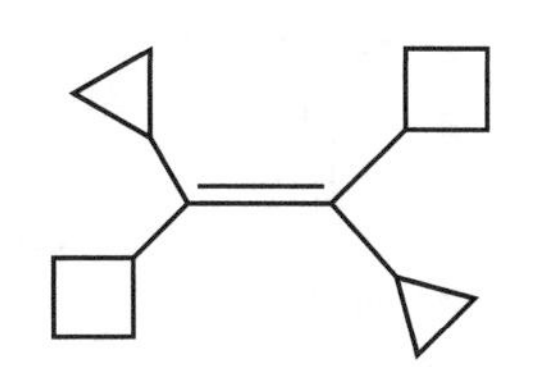

E: high-priority groups on opposite sides of the double bond

□ = **high priority**

△ = **low priority**

When the two high-priority groups are on the same side of the double bond, the compound is the **Z** isomer (from the German *zusammen*, meaning together). When the two high-priority groups are on opposite sides of the double bond, the compound is the **E** isomer (from the German *entgegen*, meaning opposite). Essentially, Z is a more general version of cis and E is a more general version of trans. One may notice that the shapes of the letters Z and E are opposite the ideas they are meant to convey. In the letter Z the horizontal bars are on opposite sides of the central line (but Z means same side), and in the letter E the horizontal bars are on the same side of the central line (but E means opposite sides).

The rules used to assign priorities to organic groups are the **Cahn-Ingold-Prelog sequence rules**, which are named after the three chemists who proposed them. Ultimately, these rules are based on an atom's atomic number. The first rule states that the heavier the atom, the higher the priority. Thus, I > Br > Cl > O > N > C > Li > D > H. Sometimes this ordering alone is sufficient to assign E or Z:

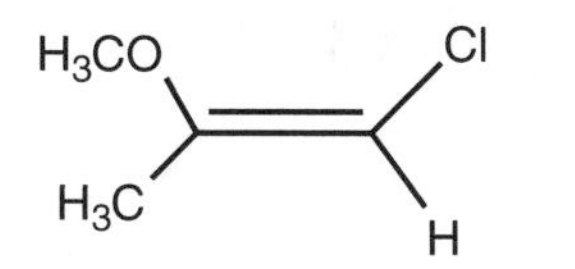

(*Z*)-1-Chloro-2-methoxypropene

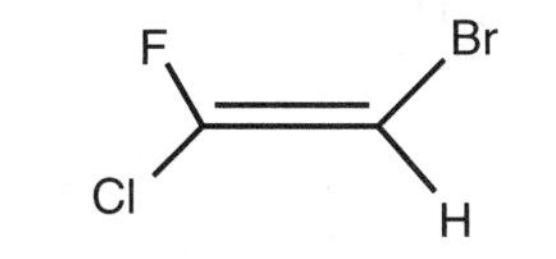

(*E*)-1-Bromo-2-chloro-2-fluoroethene

For each of the above molecules, there are two different atoms attached to each alkene carbon. In the first molecule, O > C and Cl > H, so the isomer shown is Z.

Often, the same type of atom is directly bonded to the sp^2 carbon. With such molecules, one applies the second rule: Compare atoms equally removed from the first atom, assigning priority based on atomic number until the first point of difference is reached. For example, if two groups on an alkene carbon are $—CH_2CH_2CH_2CH_3$ and $—CH_2CH_2CH_3$, the first three groups along the chain are the same. However, in the butyl group one encounters a C and in the propyl group one encounters a H, and C > H, giving the butyl group a higher priority. A higher-priority atom trumps any number of lower-priority atoms. For example, $—CH_2I$ has a higher priority than $—CF_3$ since I > F.

Applying the second rule is somewhat like following routes on a subway map. One can think of the alkene carbon as the main station and the bonds as the tracks connecting it to the branch stations, or groups, along the chain. For example, if $—CH_2CH_2CH_3$ and $CH_2CH(CH_3)_2$ were attached to the same alkene carbon, to assign priority one compares the first groups and finds they are identical. The second groups are not the same. One is $—CH_2CH_3$, and the other $—CH(CH_3)_2$. Looking at the first carbon atom of each group, in the first group the carbon is attached to two hydrogens and one carbon, while in the second group the carbon is attached to one hydrogen and two carbons. In this case, 2C,1H > 2H,1C, and the $—CH_2CH(CH_3)_2$ group has the higher priority.

For groups with double and triple bonds, a third rule is needed: Double and triple bonds are treated as if they are made up of two or three equivalent single bonds, respectively.

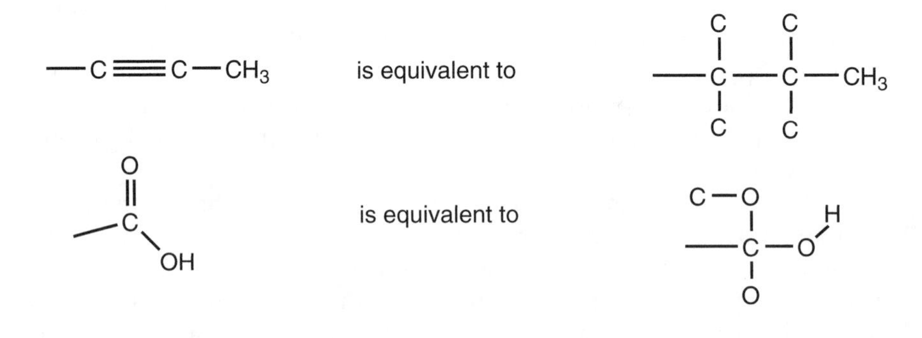

Notice that the added atoms do not have their valences completed. In the case of the carboxylic acid group, one oxygen is added, while the other oxygen is the original C=O oxygen and has its double bond replaced by two single bonds.

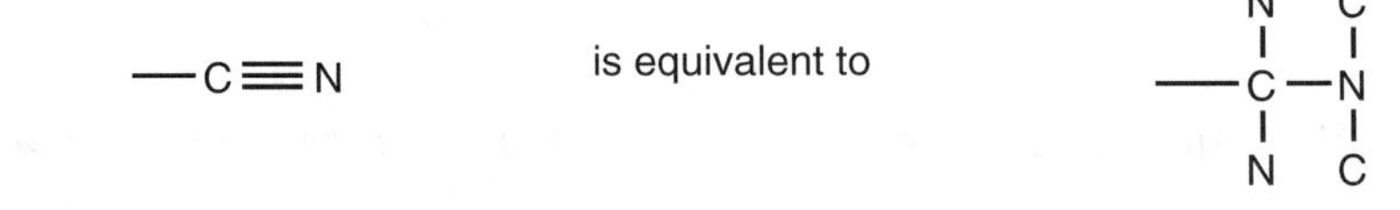

EXERCISES

1. Which group has the higher priority?

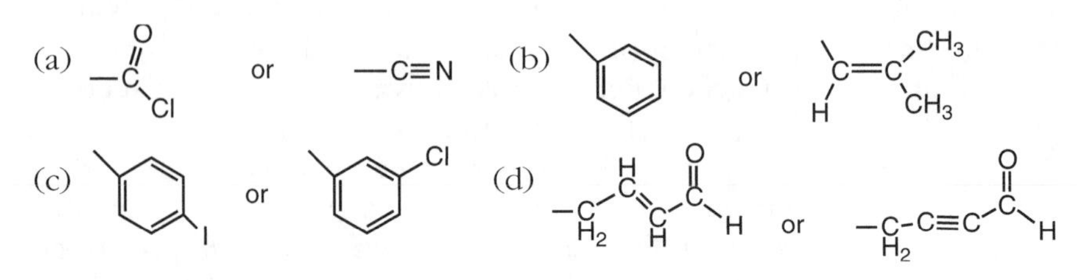

2. Assign each alkene as E or Z.

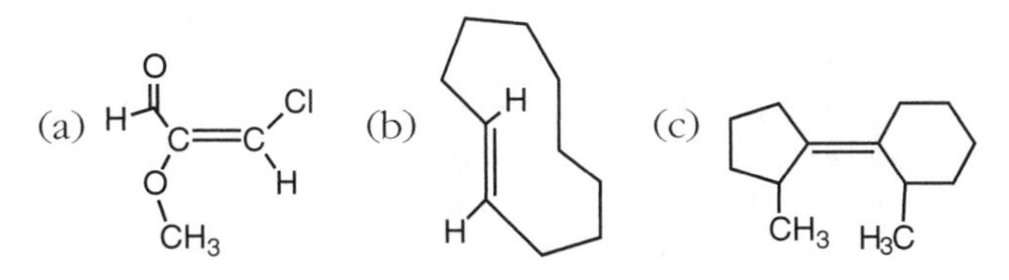

Solutions

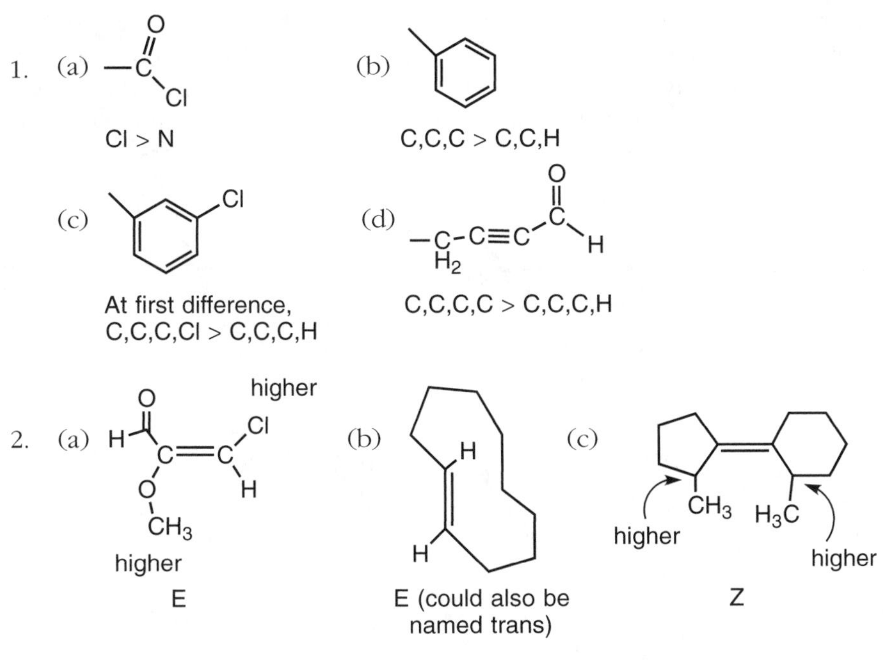

There are additional priority rules for assigning E and Z for more complicated molecules, but for the kinds of molecules encountered in introductory organic chemistry, these three priority rules should be sufficient.

TETRAHEDRAL CARBON STEREOCHEMISTRY

It is obvious that a right hand cannot hand-shake with a left hand in the normal way. Likewise, one learns early in life that left shoes fit only on the left foot, and right shoes only on the right foot. In organic chemistry, many compounds with tetrahedral carbons exist in what can be called right-handed and left-handed forms. This handedness is particularly common among biological molecules. At the molecular level, biology works by molecules fitting together in well-defined ways. A hormone that we could designate as having a right-handed shape may fit tightly with its receptor and initiate physiological changes. If a chemist made a left-handed form of the same hormone, the compound would have no biological activity because it would not fit its receptor. Thus, in order to understand biochemistry and pharmacology at the molecular level, one must master stereochemistry.

Chirality and Enantiomers

Chirality is the property of an object that makes it nonsuperimposible on its mirror image. Right and left hands are examples of chiral objects, and, in fact, chirality is derived from the Greek word for hand. Any everyday object that comes in right and left versions, such as shoes, gloves, and screws, is chiral. A tetrahedral carbon atom bonded to four different groups is also chiral.

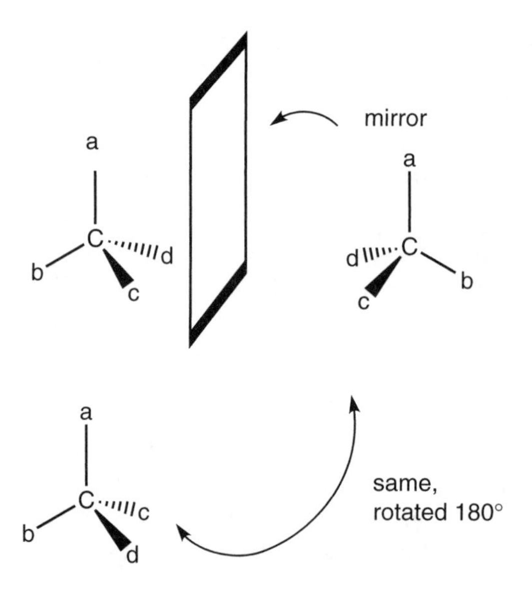

FIGURE 5.1. *Tetrahedral carbon with four different groups, which cannot be superimposed on its mirror image.*

Recall from the section on hybrid orbitals in Chapter 2 that we can draw tetrahedral carbons using wedge and dashed bonds. In Figure 5.1, the central carbon C and groups a and b are in the plane of the paper. For the top structures, group c is pointing out of the plane of the paper and group d is pointing into the plane of the paper. The carbon compound and its mirror image are clearly different. A chiral compound and its mirror image are a pair of

enantiomers. The carbon atom bonded to four different groups is called a **stereogenic center**, and some common synonyms are asymmetric carbon, chiral carbon, and chiral center. See Table 5.1 at the end of this chapter for important definitions.

A tetrahedral carbon atom with two identical groups, such as H_2CBrF, is not a stereogenic center. The molecule is the same as its mirror image (after rotating the molecule a bit to superimpose the atoms) and is thus **achiral**.

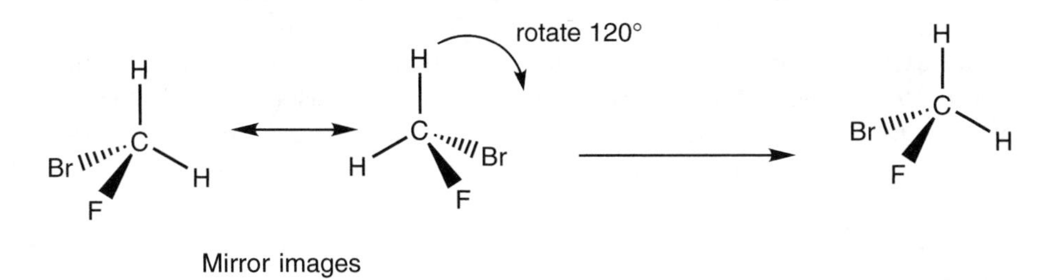

Given the structure of a chiral molecule with one stereogenic center, there is a quick way to draw its enantiomer. One merely interchanges any two groups. The result is always the mirror image:

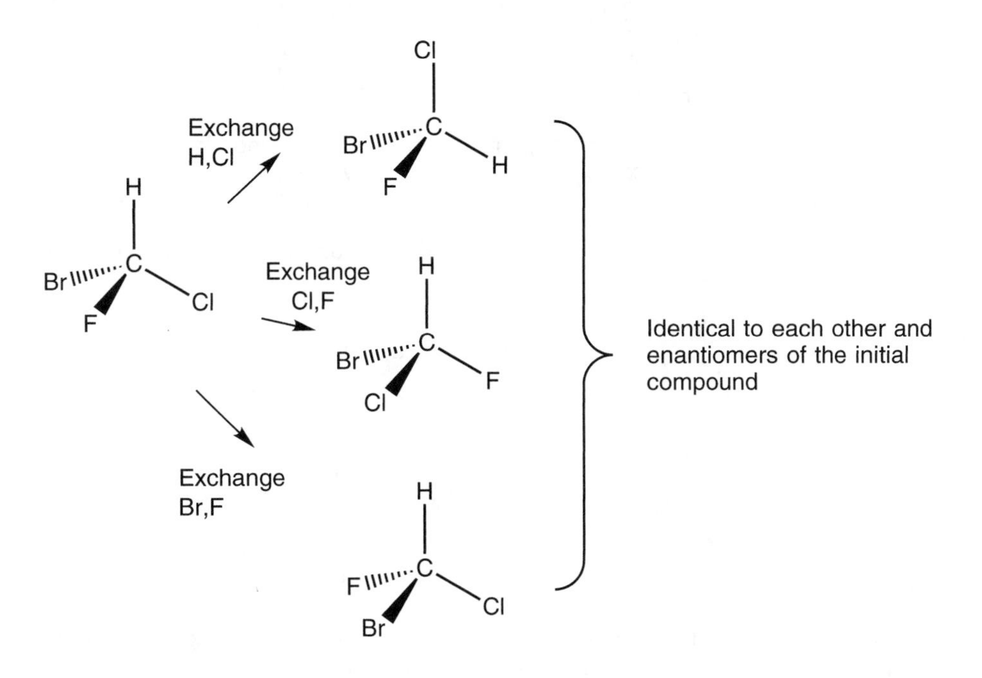

Identifying Chiral Molecules

Several strategies can be used to identify chiral molecules. Most basic is comparing a molecule with its mirror image. If the mirror image is nonsuperimposable on the original molecule, the original molecule is chiral. Often, it is simpler to look for a single stereogenic center. While complications can arise if more than one stereogenic center is present in a molecule (see the discussion of meso compounds later), if a molecule contains a single tetrahedral carbon attached to four different groups, it is chiral. Finally, certain kinds of symmetry properties are incompatible with chirality. Whenever a molecule contains a **plane of symmetry**, which is a mirror plane that passes through a molecule, that molecule is achiral. By this rule, any flat molecule must be achiral since the plane containing the molecule is a plane of symmetry. Looking for a mirror plane can be particularly useful when examining cyclic molecules.

Examples

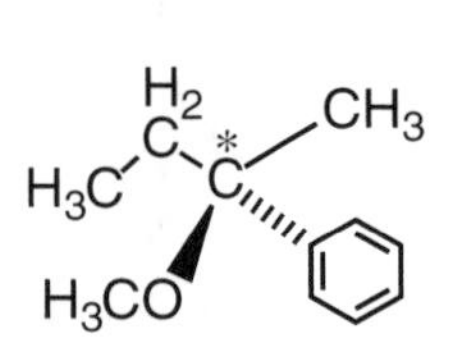

Chiral (* = stereogenic center)

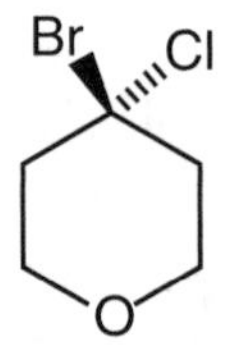

Achiral with plane of symmetry defined by the three atoms Br, Cl, and O

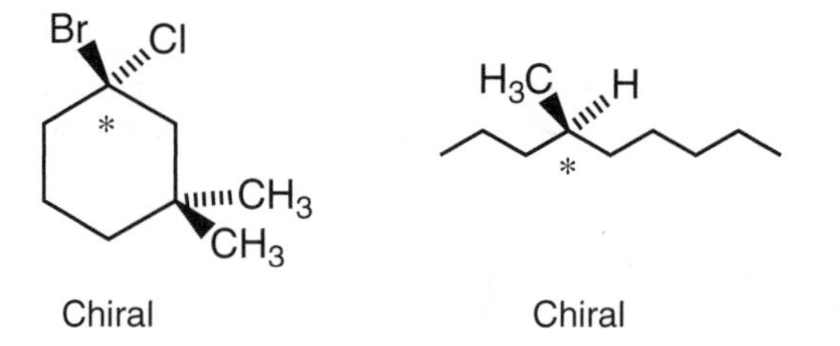

Chiral

Chiral

Notice that when looking for a stereogenic center, one must look at the entire group attached to a particular carbon, not just its local environment. For example, the fourth molecule above has two CH_2 groups bonded to the stereogenic center, but one must look at the complete groups, which are C_3H_7 and C_5H_{11}.

EXERCISES

1. Identify the stereogenic center with an asterisk in each of the following molecules. Write "none" if there is no stereogenic center.

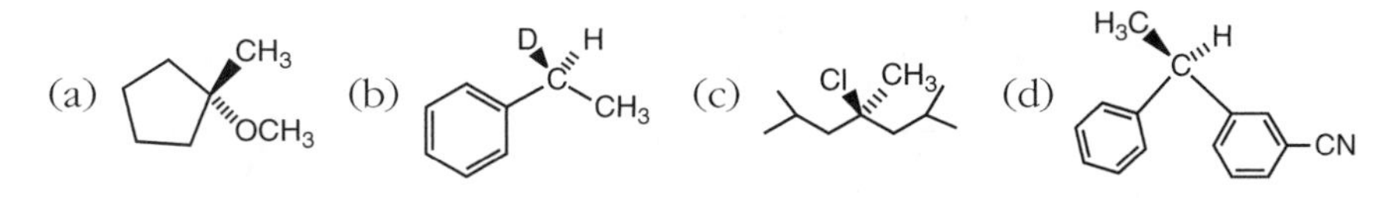

2. Are both members of each pair of molecules the same molecule or enantiomers?

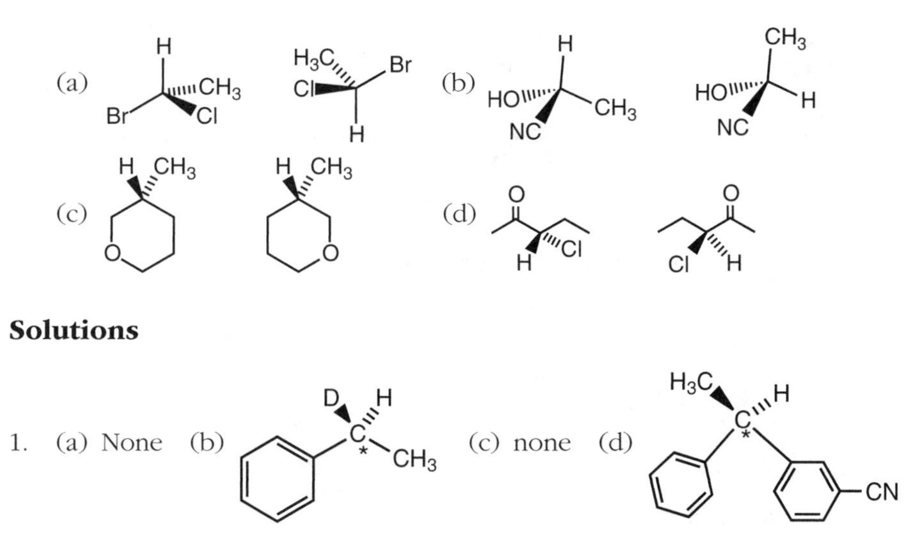

Solutions

1. (a) None (b) (c) none (d)

2. (a) Same
   (b) enantiomers
   (c) enantiomers
   (d) same

The R,S Description of Molecular Configuration

The **configuration** of a molecule is the arrangement of the atoms around its stereogenic center or centers. We have already discussed how the configuration of alkenes can be described using E,Z nomenclature without drawing the molecule. A very similar descriptive method is available to describe the configuration of stereogenic centers.

The assignment of the configuration of stereogenic centers uses the same Cahn-Ingold-Prelog priority rules used for E,Z assignments. First, one identifies the stereogenic center (the carbon with four different groups bonded to it). Second, one assigns a priority to each of the four different groups, 1 through 4, where 1 is the heaviest, or highest priority and 4 is the lightest, or lowest priority. Finally, one orients the molecule in a special way so that the lowest-priority group, group 4, points away from the observer, as shown here.

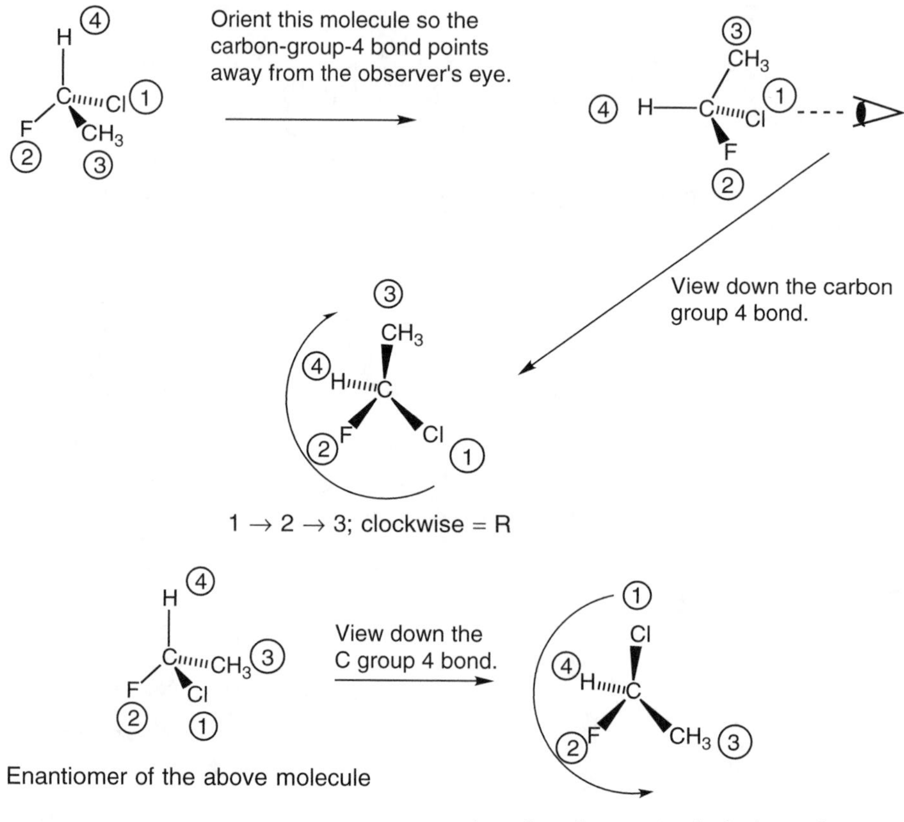

When the observer looks down the carbon group 4 bond, so that the group with the lowest priority points away from the observer, groups 1 through 3 are pointed at the observer. If moving from priority 1 to priority 2 to priority 3 groups is in the *clockwise* direction, the configuration of the stereogenic center is **R** (from the Latin *rectus*, meaning right). If moving from priority 1 to priority 2 to priority 3 groups is in the *counterclockwise* direction, the configuration of the stereogenic center is **S** (from the Latin *sinister*, meaning left).

In the special orientation used to determine R or S, the carbon takes on the approximate shape of a car's steering wheel. Group 4 (often, but not always, a C—H bond) is the steering column. Groups 1, 2, and 3 are on the steering wheel. If moving from 1 to 2 to 3 is clockwise, turning the steering wheel in that direction causes a right turn (R). If moving from 1 to 2 to 3 is counterclockwise, turning the steering wheel in that direction causes a left turn (S).

Examples

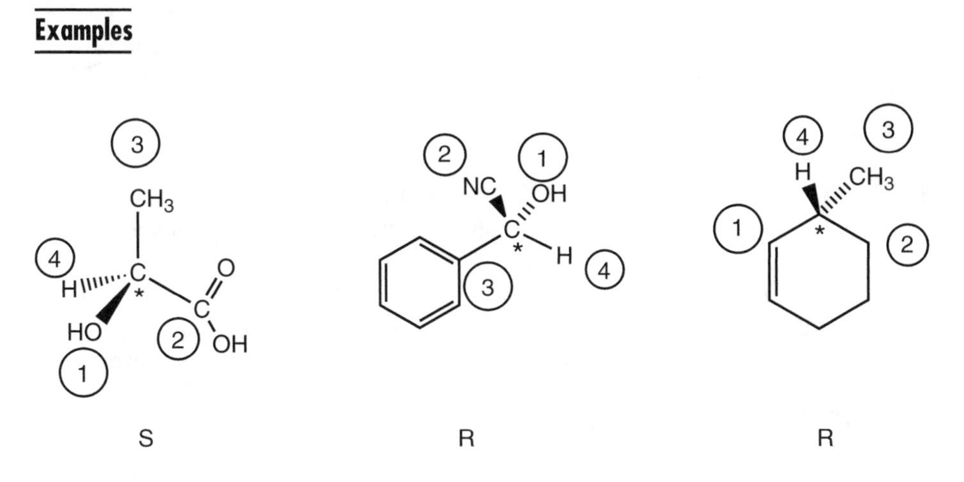

It is common for students initially to have problems assigning R and S configurations to stereogenic centers. It can be difficult to mentally rotate tetrahedral carbons and keep the correct configuration. Realize that, for this task, the tetrahedral carbon resembles an umbrella that has been turned inside out by the wind and the shaft is pointing away from you. For assigning R and S, molecular models are quite helpful. But if models are unavailable, sticking four differently colored pins into a small ball to model an sp^3 carbon can help in visualizing a tetrahedron.

EXERCISES

Identify each stereogenic center and assign the configuration as R or S. There may be more than one per molecule.

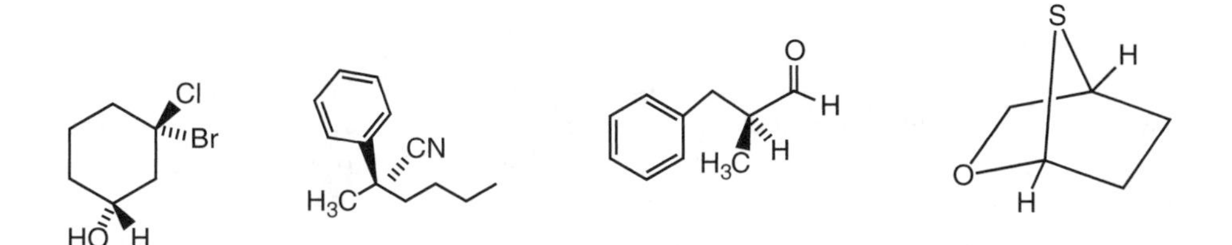

Solutions

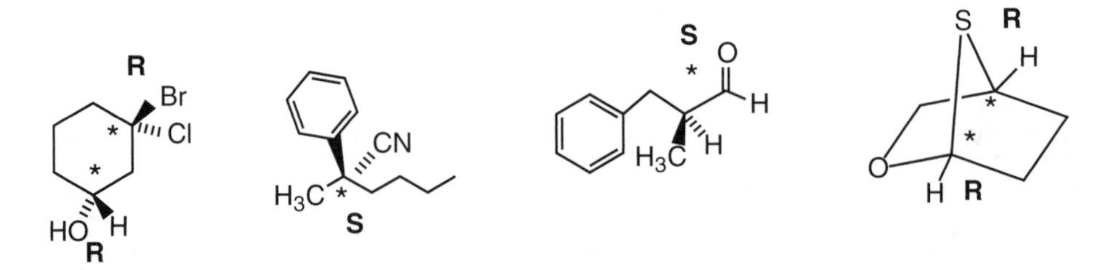

Experimental Detection of Chirality

Because light has wave properties, it can interact with various substances, including certain minerals, to become plane polarized. This means that the light waves are all oscillating in the same plane. Chiral molecules are said to be **optically active** since they can rotate the plane of polarized light. The degree of rotation can be measured using a device called a **polarimeter**.

The measured angle α that the chiral molecule rotates the plane of polarized light depends on the following.

1. The wavelength of light used. Usually the yellow light given off by hot sodium, the sodium D line, is used; so the **specific rotation** is written as $[\alpha]_D$.
2. The temperature, which is usually 25°C.
3. The concentration *c* of a solution of the chiral molecule (in g/ml). (Note that by using these concentration units, the structure of the chiral molecule need not be known, so that unknown compounds isolated from natural sources can be studied.)
4. The pathlength *l* through which the polarized light travels (in decimeters).
5. The **observed rotation** α of the chiral molecule, which is a measure of how strongly or weakly the molecule interacts with polarized light. It is a physical property of a chiral molecule like a melting point or boiling point.

The formula relating these quantities is

$$[\alpha]_D = \alpha/cl$$

While quantum physics is needed to completely explain why chiral molecules rotate the plane of polarized light, a qualitative explanation is possible. Light is electromagnetic radiation, which means it is an electric field moving through space. Molecules are collections of electric charges with their own electric fields of a given shape. When light goes past a molecule, the two electric fields interact and, if the light is polarized, the plane of polarization is rotated a little. In a solution, there are billions and billions of molecules, each in a random orientation. If the molecules are achiral, the sum of all the light-molecule interactions cancels out to zero. However, for chiral molecules, there are no mirror image molecules to cancel out every light-molecule interaction, and the result is a net rotation of the plane of polarized light.

Some chiral molecules rotate polarized light in a clockwise direction, and they are designated (+) or (*d*), for **dextrorotatory**. Other chiral molecules rotate polarized light in a counterclockwise direction, and they are designated (–) or (*l*), for **levorotatory**. If a chiral molecule is (+), its enantiomer is (–), and vice versa. Do not confuse (+)/(–) and R/S because there is absolutely no connection. Some R molecules are (+), and others are (–).

While a chiral molecule can rotate the plane of polarized light, its actual specific rotation may be so small that the observed rotation is 0°. For example, it is possible to have stereogenic centers containing one H and one D group. However, the difference between hydrogen isotopes is so small that the measured optical rotation in such molecules is often zero.

Diastereomers

Many compounds have two or more stereogenic centers. Suppose a molecule has three stereogenic centers and that all three are of the R configuration, RRR. This molecule's enantiomer is SSS. But, as yet, we have no term for isomers that have some of their stereogenic centers the same and others different:

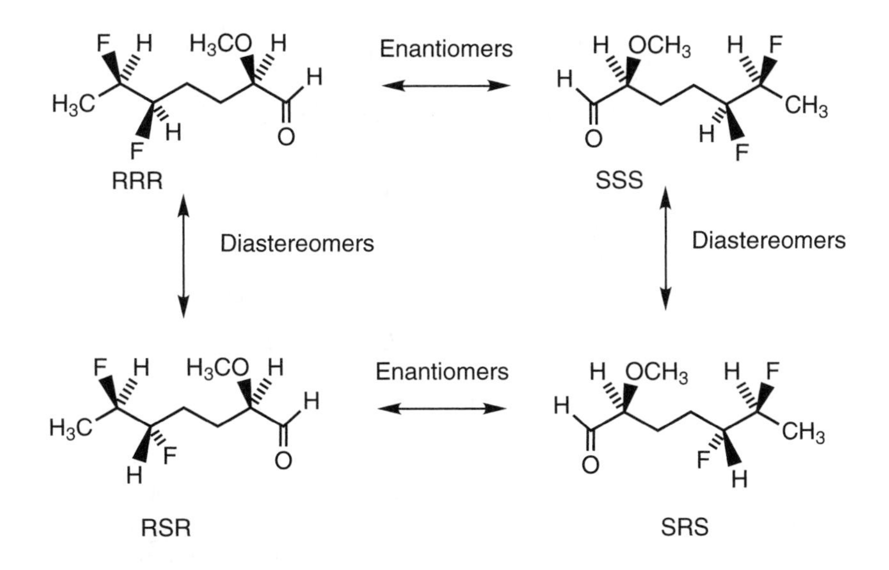

Diastereomers are stereoisomers that are not enantiomers. In the above example, the enantiomer of the RSR compound is the SRS isomer. The other possible isomers, the RRR, SSS, RRS, SSR, RSS, and SSR, are all related to the RSR isomer as diastereomers. One needs at least two stereogenic centers to generate diastereomers. **Optical isomer** is a synonym for molecules that are either enantiomers or diastereomers.

The number of stereoisomers increases rapidly with the number of stereogenic centers. Since each center can be R or S, a molecule with n stereogenic centers can have as many as 2^n stereoisomers. Thus, a molecule with 4 stereogenic centers has 2^4, or 16, possible stereoisomers, with 8 pairs of enantiomers, and each molecule is related to 14 other diastereomers.

EXERCISES

Draw all the stereoisomers for each molecule. Indicate which molecules are pairs of enantiomers. Assign R or S to each stereogenic center.

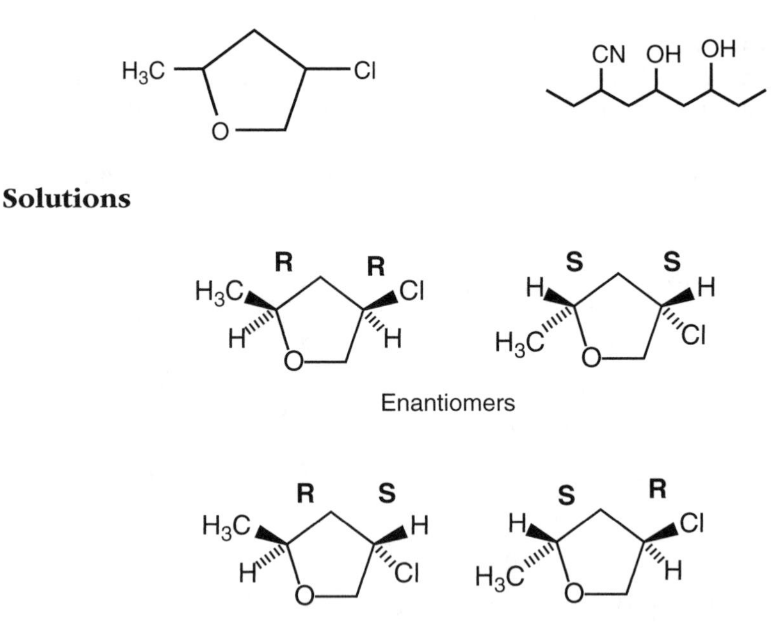

There is a quick way to solve this type of problem. First identify the stereogenic centers. Then take one molecule and carefully assign R and S to each center. Switching the configurations of each stereogenic center produces the enantiomer, and each R center will be S and each S will be R. Then systematically change the centers one at a time to generate the required number of diastereomers. It is a good idea to do quality control at the end of the problem, if time is available, to make sure the R and S configurations are correctly assigned.

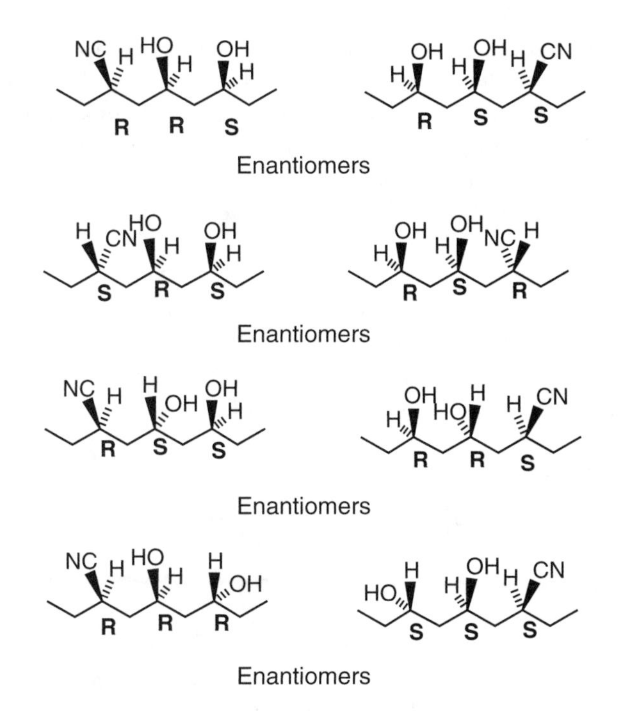

It is extremely important to understand the effect stereoisomerism has on a molecule's physical properties. A molecule and its enantiomer have identical physical properties as long as no additional source of chirality is introduced. Thus, a molecule and its enantiomer have the same melting point, boiling point, solubility, and so on. However, if a source of chirality is present, such as polarized light, a molecule and its enantiomer can differ. We have already pointed out that the specific rotations of a molecule and its enantiomer are equal and opposite because polarized light has chiral properties. Since the odor receptors of the nose are made of chiral molecules, enantiomers can have different smells. A molecule and its enantiomer react identically with an achiral reagent. However, a chiral reagent can distinguish between a molecule and its enantiomer.

Unlike pairs of enantiomers, pairs of diastereomers can, and normally do, have different physical properties. The reason is easiest to understand when examining cyclic compounds:

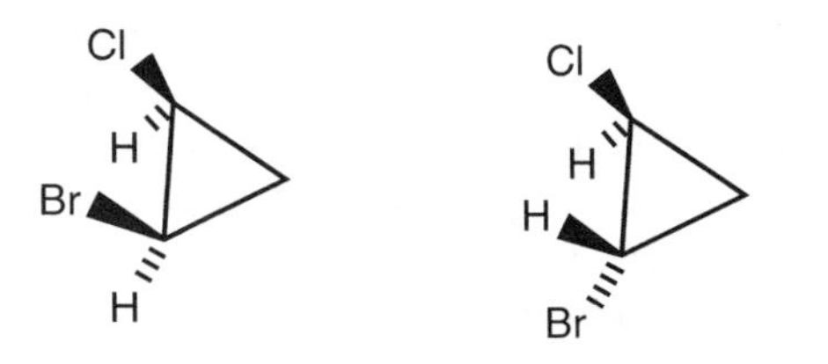

These two diastereomers, (*R*)-1-bromo-(*S*)-2-chlorocyclopropane and (*S*)-1-bromo-(*S*)-2-chlorocyclopropane, have the halogens on different sides of the ring, leading to different dipole moments, different shapes, and different functional group interactions. Because acyclic diastereomers have free rotation around single bonds, they do not maintain a single shape, but such diastereomers have different average shapes and thus different physical properties also.

Racemic Mixtures

A fundamental observation in chemical reactivity is that chirality generates chirality. This can be seen for the following reaction, where a hydride, H^-, reacts with a ketone to give a molecule that, after the addition of dilute acid to protonate the basic O^-, produces an alcohol:

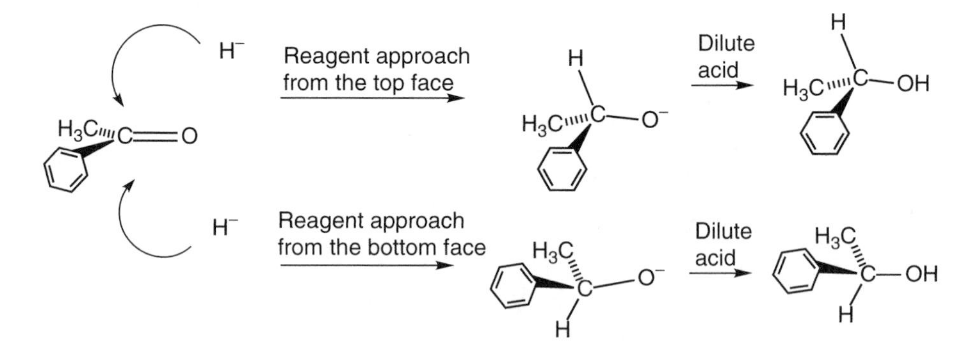

As we will describe in detail later, ketone carbons react with many negatively charged reagents to form new bonds to carbon, in this case, a new C—H bond. The sp^2 carbon of the ketone molecule is planar. The two sides, or faces, of the carbonyl group are equivalent in the absence of chiralty. A reagent can approach from the top or bottom face. In this case, approach from the top generates the S alcohol and approach from the bottom generates the R alcohol. Since there is nothing to bias the direction of the incoming reagent, one obtains a 50:50 mixture of R and S isomers. Such an equal mixture of a molecule and its enantiomer is called a **racemic mixture**. Although each individual molecule in a racemic mixture is chiral, if a racemic mixture is put into a polarimeter, the observed rotation is exactly 0°. The optical properties of each molecule are canceled out by the presence of its enantiomer.

One can think of chirality as information. The R and S or (+) and (–) of individual chiral molecules are like the 1s and 0s used to store information in a computer. You cannot generate information from nothing, and you cannot generate chirality from achiral starting materials. Thus, whenever a new stereogenic center is created from achiral reagents and achiral reactants, the result is always a racemic mixture.

Because humans are chiral, we can look at an achiral molecule, such as the ketone shown on facing page, and imagine the reagent approaching only from the top face or only from the bottom face of the ketone group. One can imagine holding the phenyl, C_6H_5, group in one's right hand and the methyl group in one's left hand, with the oxygen atom pointing away from one's body. In such a binding orientation, H^- approaching from the top face produces the S alcohol. Enzymes, the chiral catalysts used by living organisms, act like our imaginary human and bind many organic molecules in chiral environments. They direct the approach of reagents in such a way that only a single enantiomer is produced from an achiral molecule. The chiral information contained in the enzyme is transferred to the stereogenic center in the product molecule. In other words, chirality generates chirality.

Organic chemists have invented chiral catalysts to mimic enzyme properties. Sometimes the synthetic catalysts are not perfect, and instead of generating 100 percent of an S or R isomer, one obtains a mixture of R and S, with one major enantiomer. **Enantiomeric excess** is the amount by which the major enantiomer predominates over the minor enantiomer. For example, if a reaction gives a 90 percent S, 10 percent R mixture, it has a $90 - 10 = 80$ percent enantiomeric excess (abbreviated 80% ee). A racemic mixture has 0% ee.

Meso Isomers

A molecule with n stereogenic centers has, as noted above, at most 2^n stereoisomers. Sometimes this number is less because of the symmetry properties of the molecule. For example, at first sight the two molecules shown here appear to have two stereogenic centers:

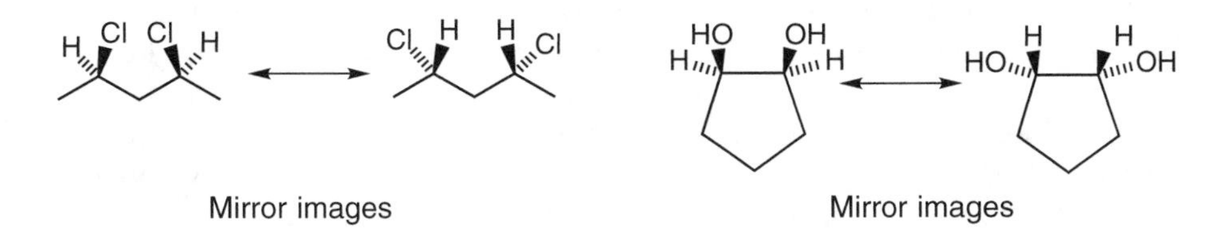

However, merely by turning the mirror images 180°, they are seen to be identical to the original molecule. Thus, these two molecules are achiral although they have tetrahedral carbons with four different groups attached. Each molecule has a plane of symmetry, in the first case through the central CH_2 group and in the second case through a bond between the two alcohol groups. Molecules that from a local view appear to have stereogenic centers but are achiral because of an internal plane of symmetry are called **meso** compounds or meso isomers.

EXERCISE

Draw all the stereoisomers of the following compound. Indicate which isomer is a meso compound. Assign R and S to each stereogenic carbon. *Note*: It is possible to assign R and S to the carbons of a meso compound. If one of the centers is R, the other will be S because of the internal mirror plane.

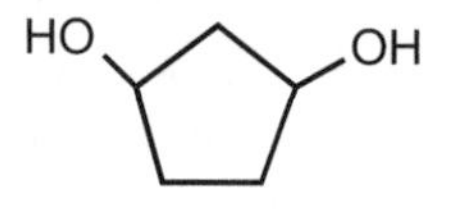

Solution

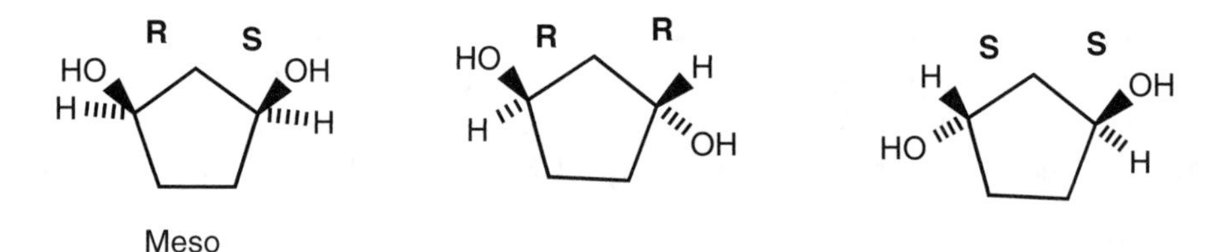

Resolution of Racemic Mixtures

A chemical reaction that generates a new stereogenic center produces, in the absence of any source of chirality, a racemic mixture. Therefore, one needs a way to separate a molecule from its enantiomer. The process of separating a mixture of enantiomers into pure optical isomers is called **resolution**.

The most common chemical method of resolution is the conversion of a mixture of enantiomers to a mixture of diastereomers. Recall that pairs of enantiomers have identical physical properties, such as solubility, but that pairs of diastereomers usually have different physical properties. These differences provide a basis for separation. The precise details of this method vary depending on the functional groups in a particular molecule. It is particularly simple to explain with a racemic mixture of carboxylic acids.

A racemic mixture of carboxylic acids is 50 percent (+) and 50 percent (–) stereoisomers. It may have one or more than one stereogenic center, but one enantiomer is dextrorotatory and the other is levorotatory. The carboxylic acid can undergo an acid-base proton transfer reaction with a chiral base. Many basic chiral organic amines are produced by plants, called alkaloids. Quinine and nicotine are two examples of alkaloids.

When a racemic mixture of carboxylic acids reacts with a chiral base, designated here as $(+)R_3N$, the result is a diastereomeric mixture of salts:

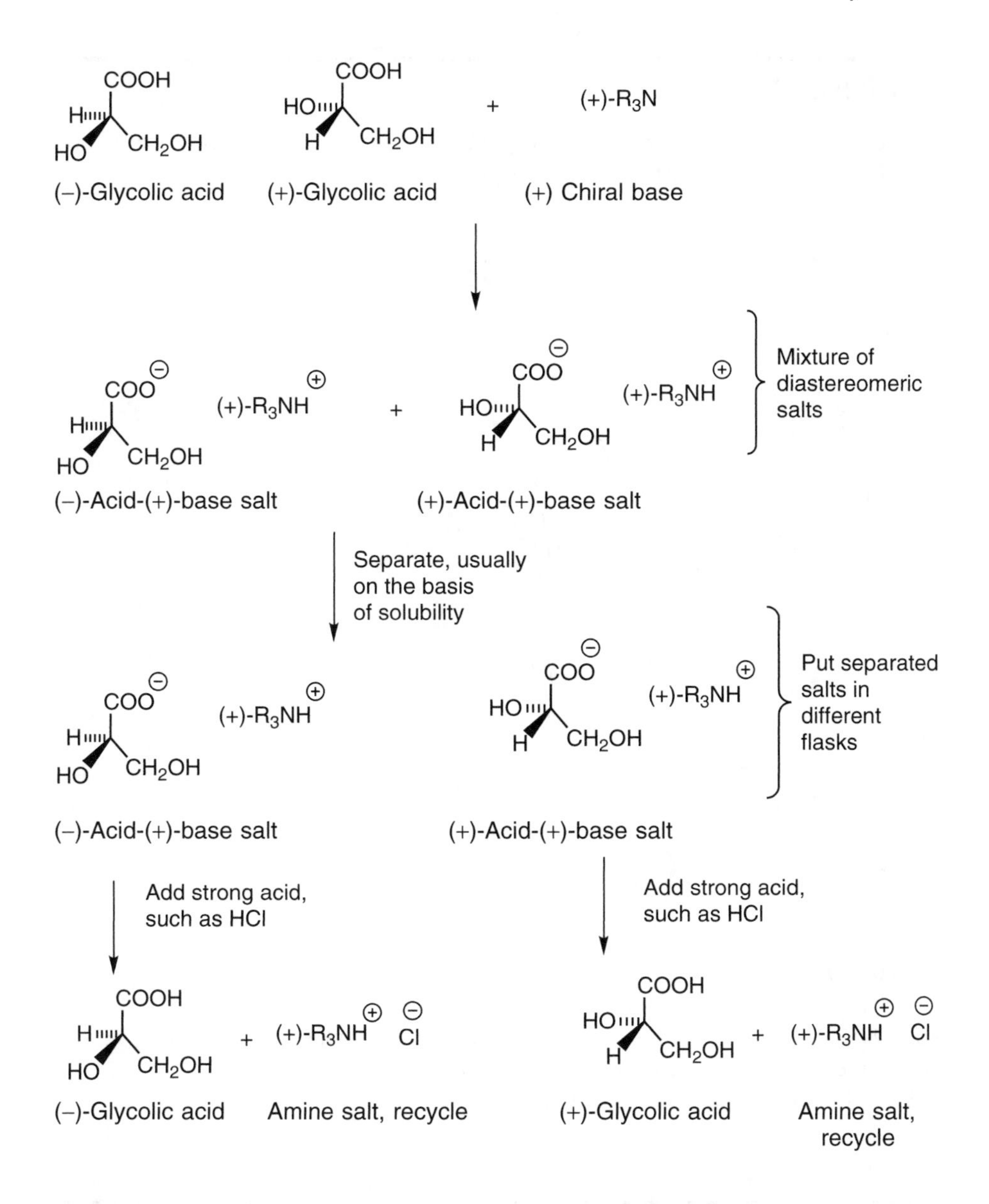

By trial and error, a solvent is found that dissolves one stereoisomer of a diastereomeric mixture of salts much better than the other. A series of recrystallizations is carried out until one has a pure sample of the (+)acid-(+)base salt and the (–)acid-(+)base salt. In order to recover the optically pure samples of the acid, a strong acid, such as HCl, is added. It protonates $RCOO^-$, giving RCOOH and the base salt $(+)R_3NH^+Cl^-$. The resolved organic acid is finally separated from the amine salt by solubility differences, and the amine is recycled.

EXERCISES

Indicate which pairs of compounds have identical physical properties and which can have different physical properties.

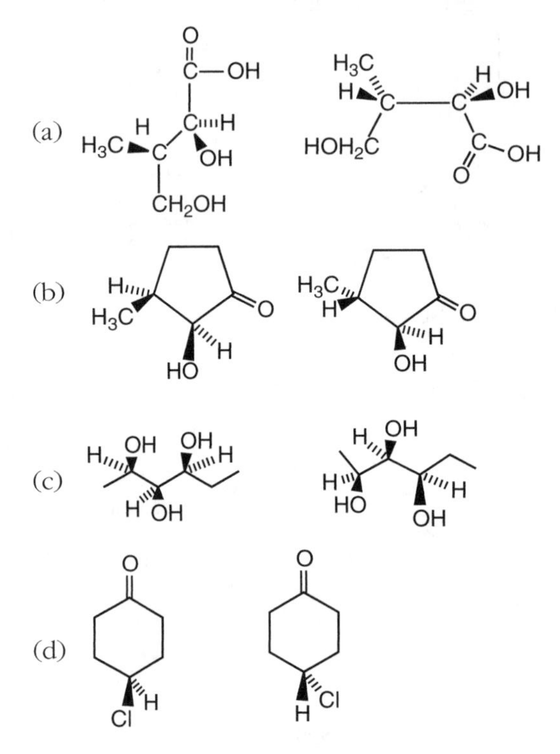

Solutions

(a) Same properties; molecules are enantiomers.
(b) Different properties; molecules are diastereomers.
(c) Same properties; molecules are enantiomers.
(d) Same properties; molecule is achiral.

INHERENTLY CHIRAL MOLECULES

Some molecules are chiral even without containing a tetrahedral carbon bonded to four different groups. A simple example is a compound with a tetrahedral atom other than carbon bonded to four different groups. Compounds with four groups bonded to nitrogen, termed **quaternary ammonium compounds**, $R_1R_2R_3R_4N^+$, can be chiral if the four R groups are different.

Other molecules are chiral because their overall shape is chiral. The molecule hexahelicene, shown below, has a helical shape because atoms must move out of each other's way. It can exist in the form of a right-handed screw or a left-handed screw and is thus chiral.

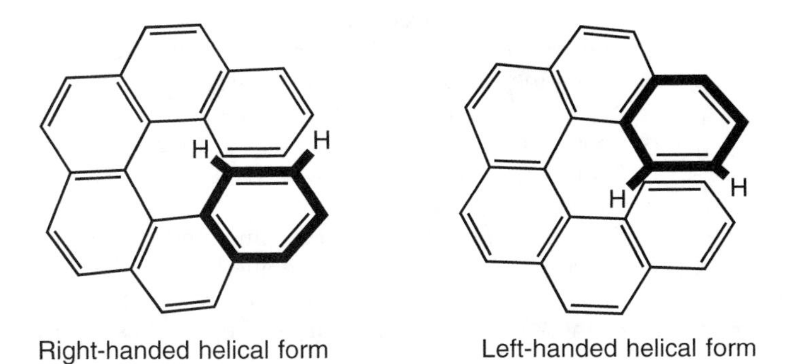

Right-handed helical form Left-handed helical form

Perhaps the most subtle form of chirality is seen in **allenes**, compounds with two double bonds in a row, RR′C=C=CRR′. Because the central carbon atom is *sp* hybridized, the groups at either end of the allene functional group are in planes that are perpendicular to each other. As long as the R and R′ groups are different at each end, the allene is chiral.

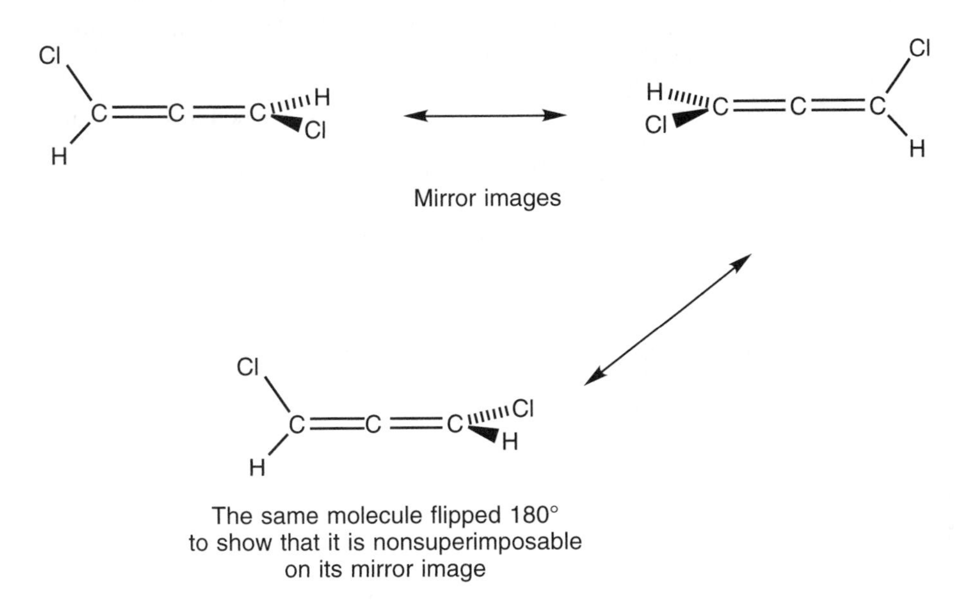

Mirror images

The same molecule flipped 180° to show that it is nonsuperimposable on its mirror image

TABLE 5.1. DEFINITIONS OF STEREOCHEMICAL TERMS

Absolute configuration	The exact three-dimensional arrangement of atoms in space for a stereogenic center.
Cahn-Ingold-Prelog rules	Rules designed to assign a description of absolute configuration, designated R or S, to every stereogenic center in a molecule.
Chiral	Not superimposable on its mirror image. This word comes from the Greek word for hand. Each of one's hands is chiral. Chiral is pronounced with a hard "k" sound, like the *c* in *cat*. The opposite of chiral is **achiral**, which refers to an object that is the same as its mirror image.
Diastereomer	A type of stereoisomer that has the same connectivity as another isomer but is not its mirror image. For two compounds to be diastereomers, they must have at least two stereogenic centers.
Enantiomer	The mirror image of another compound. Together, the two isomers make a pair of enantiomers.
Enantiomeric excess	The amount of one enantiomer in excess of the other enantiomer in a mixture (always reported as a positive number). If a mixture is 80 percent S and 20 percent R, the enantiomeric excess (ee) is $80 - 20 = 60$ percent.
Meso compound	A molecule with stereogenic centers but with an internal mirror plane (one that passes through one of the atoms or one of the bonds of the molecule) that causes the halves of the molecule to be mirror images of each other. Meso compounds are achiral.
Optical activity	The ability of some substances to rotate the plane of polarized light. Optically active molecules are chiral.
Optical isomers	Enantiomers and diastereomers.
Racemic mixture	A 50:50 mixture of enantiomers.
Stereogenic center	An atom that, because of the groups attached to it, generates chirality. In organic chemistry, almost every stereogenic center is a tetrahedral carbon with four different groups bonded to it. The terms **chiral center**, **asymmetric center**, and **chiral carbon** are used synonymously with stereogenic center.
Stereoisomers	Isomers with the same connectivity but with different three-dimensional shapes.

6

ALKANES AND CYCLOALKANES: CONFORMATIONAL ANALYSIS

CONFORMATIONS OF ALKANES

Because of a σ bond's cylindrical symmetry, there is free rotation around a single bond. In effect, a single bond is a molecular axle. Since very little energy is required to rotate around a single bond, compared to the energy available at room temperature, molecules are constantly spinning around their single bonds. For a simple molecule such as ethane, CH_3—CH_3, the two methyl groups have an unlimited number of relative orientations. These different orientations have slightly different shapes and are thus isomers. Molecules that differ by rotation around single bonds are **conformational isomers** or **conformers**. Each shape that results from rotation around a single bond is a specific **conformation**. The study of conformers and their energies is called **conformational analysis**. Unlike other forms of isomerism, where isomers can be interconverted only by the breaking and remaking of bonds, conformational isomers are interconverted simply by rotation around single bonds.

Earlier, when we discussed drawing molecules, we explained that there are many different ways to draw the same molecule. Because of the free rotation around single bonds, a sample of a molecule with many single bonds (e.g., hexane) exists in many different conformations. A few of hexane's possible conformations are

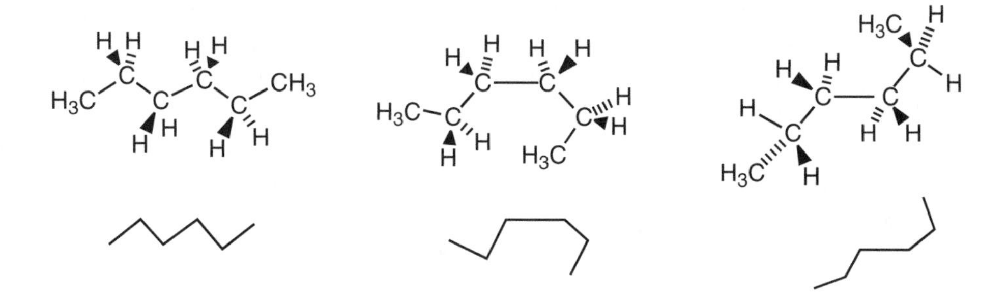

Newman and Sawhorse Projections

The wedge and dashed bond notation we have been using to indicate the three-dimensional shapes of molecules is not always the most convenient for depicting conformers. Another common drawing convention is the **sawhorse projection**. All bonds are drawn with the same solid lines, and the molecules are represented as if the observer is looking down and from the side at the molecule, as shown here for ethane.

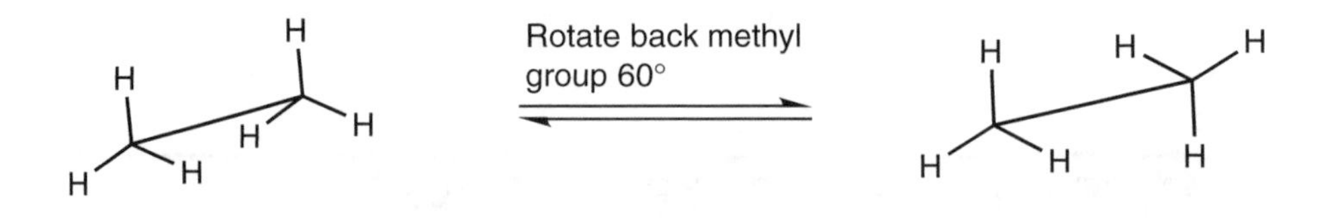

The central carbon atoms are not explicitly written in.

A second convention for drawing conformers was invented by the American chemist Melvin Newman and is called the **Newman projection**. In a Newman projection one looks directly in front of a carbon-carbon bond. The front carbon is a point, and the back carbon is drawn as a large circle. Each carbon has three bonds radiating from it and separated by 120°, as depicted here for ethane.

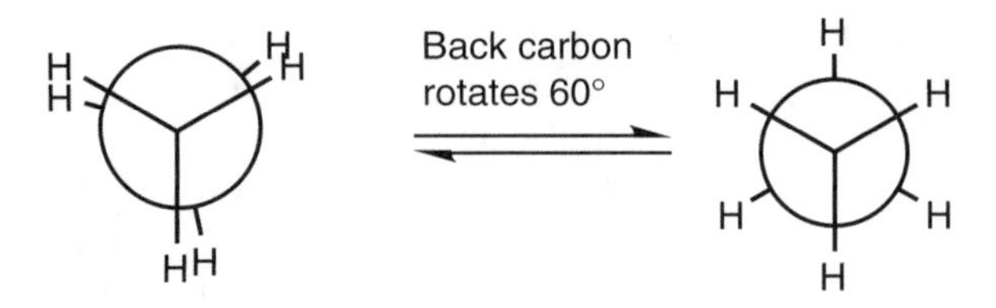

Dihedral Angles

Each conformation of a molecule can be described in terms of dihedral angles. A **dihedral angle** (also called a torsion angle) is the angle a bond must be rotated in order to line it up with a bond in front or in back of it. The two bonds must be attached to adjacently bonded atoms, as shown here. A description of all the dihedral angles in a molecule fixes the particular conformation of the molecule.

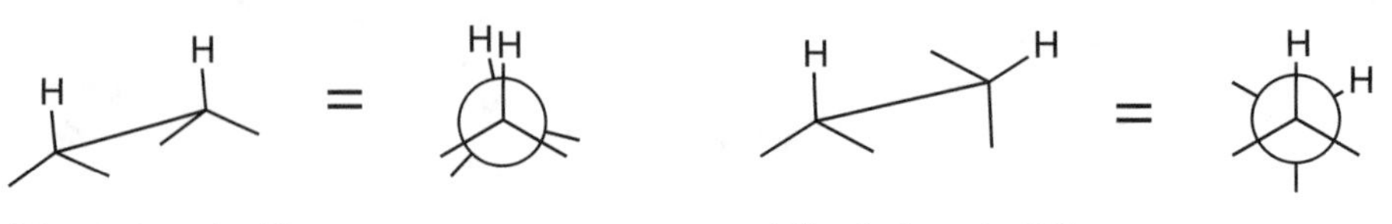

Dihedral angle 0° Dihedral angle 60°

When drawing a Newman projection of a molecule with a 0° dihedral angle, the bonds have to be rotated slightly so that one can see the groups in both the front and the back. If one wanted to draw a molecule with a small dihedral angle, such as 10°, one would have to note this fact in the drawing.

When two bonds are in front of each other, so that the dihedral angle is 0°, this conformation is the **eclipsed conformation**. When a bond is between two other bonds, for example, ethane with a dihedral angle of 60°, this conformation is a **staggered conformation**. In general, staggered conformations are more stable (lower in energy) than eclipsed conformations.

EXERCISES

1. Sketch Newman projections of 1-bromo-2-iodoethane where the halogen atoms have dihedral angles of 0°, 60°, 120°, and 180°.
2. Draw the most stable sawhorse projection of propane.
3. Draw the Newman projection of cyclopropane.

Solutions

1.

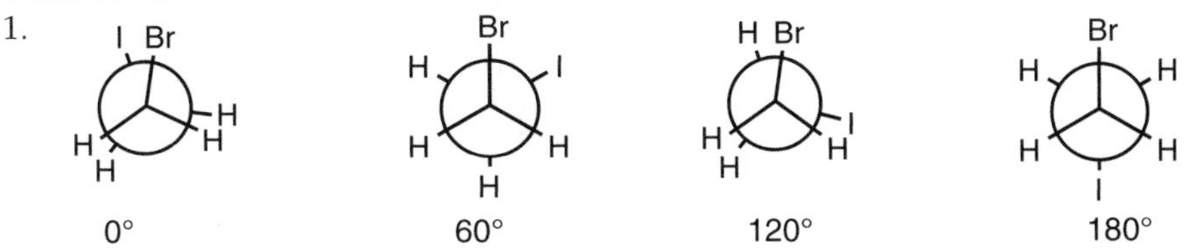

Note: It is just the relative orientation that is important. For example, the 60° structure could be drawn as

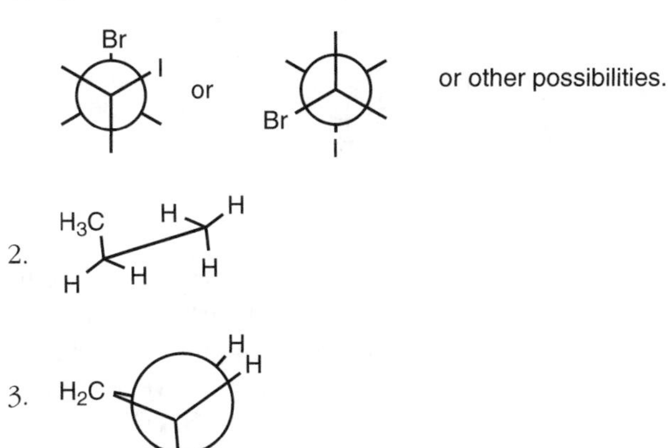

Conformations of Butane

The conformations of butane introduce many concepts that are generally useful in understanding molecular geometry. One can draw an energy diagram, or **conformational energy diagram**, of any acyclic hydrocarbon as it is rotated from a dihedral angle of 0° to 360° around any bond. For ethane, the diagram is very simple.

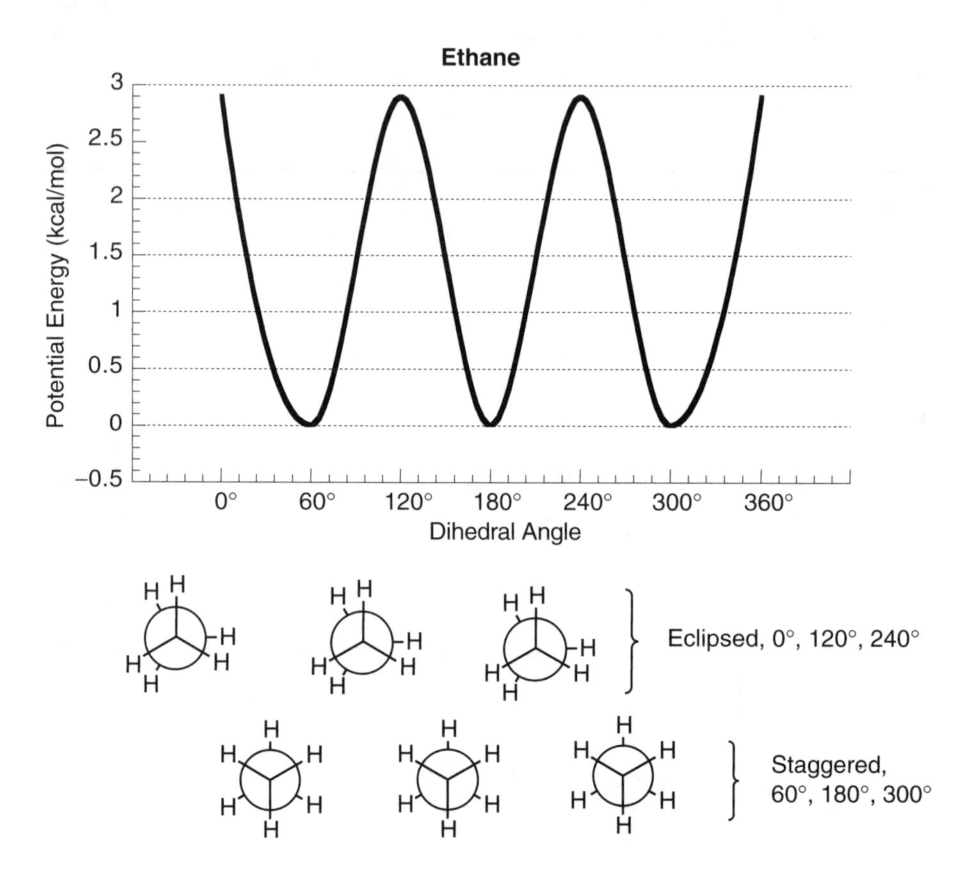

Although all the C—H bonds in ethane are equivalent, one can imagine rotating one methyl group 360° around the other. The staggered conformation represents an energy minimum, and relative to the minimum, the eclipsed conformation represents an energy maximum 2.9 kcal/mol above the minimum (this is an experimentally obtained number). We say the **barrier height** between adjacent minima is 2.9 kcal/mol. The extra energy of the eclipsed conformation is called **torsional strain** or **bond eclipsing strain**. This energy arises because the electrons in the C—H bonds repel one another when they are eclipsed. Since there are three pairs of C—H eclipsing interactions, eclipsing one pair of C—H bonds requires approximately 0.9 kcal/mol.

The conformational energy diagram for butane is more complex than that for ethane. The Newman projections look down the central C—C bond between C-2 and C-3. The 0° starting point is arbitrarily taken as the conformation with the C—CH_3 bonds eclipsed.

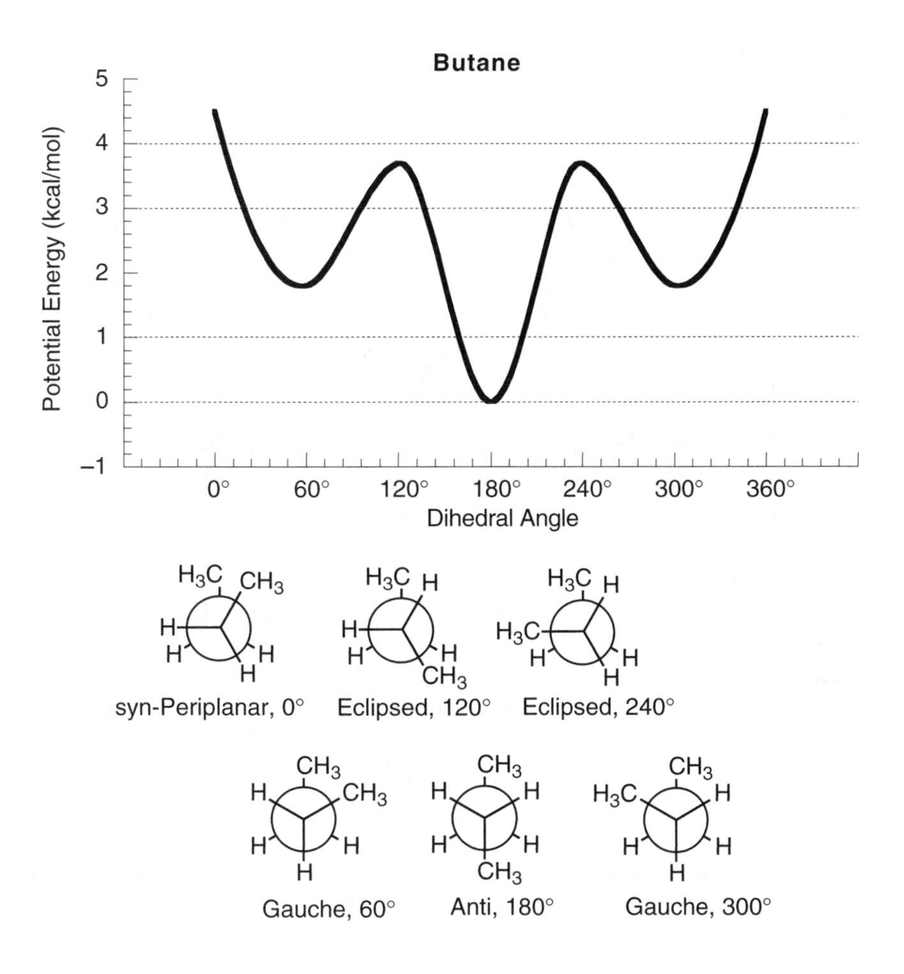

The highest energy conformation has the methyl groups eclipsed (0°). The methyl groups are **syn** to one another, and this conformation is sometimes called **syn-periplanar**. Two kinds of destabilizing interactions are present for this conformation. First is torsional strain, corresponding to two pairs of C—H torsional interactions and one pair of C—C torsional interactions. In addition, this conformation exhibits **steric strain**. Whenever two atoms or groups try to occupy the same region of space (i.e., they are closer than the sum of their van der Waals radii), they create steric strain:

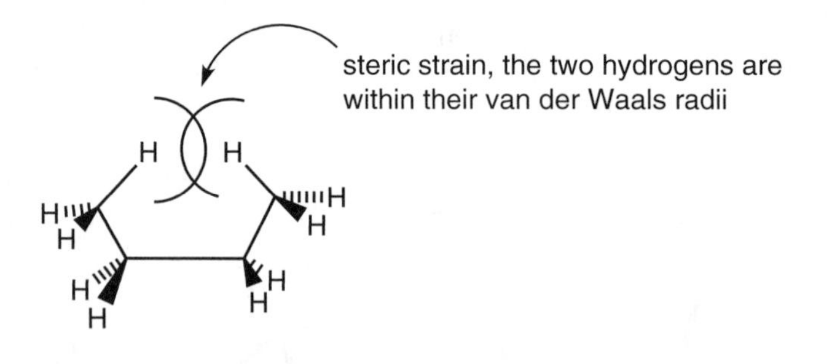

The sum of both these types of strain equals 4.5 kcal/mol relative to the most stable conformer of butane.

The conformers with dihedral angles of 60° and 300° are staggered; so there is no torsional strain. However, the hydrogens on the two methyl groups still suffer from steric strain equal to 0.9 kcal/mol. The Me-Me interaction at 60° and 300° is called a **gauche interaction**. The larger the groups, the more destabilizing the gauche interaction.

In addition to the eclipsed interaction at 0°, there are two more eclipsed interactions at 120° and 240°. They have one pair of C—H torsional strains and two pairs of C—C/C—H torsional strains, which add up to 3.8 kcal/mol. Since the C—H torsional strain is about 0.9 kcal/mol, the C—C/C—H torsional strain is destabilizing by about 1.4 kcal/mol.

The most stable conformation of butane occurs at 180°. The methyl groups are **anti**, or 180°, to one another, and the conformation is sometimes called **anti-periplanar**. At room temperature, there is much more energy than the barrier height between conformations. Thus, butane molecules rapidly change their conformation, and all possible dihedral angle values (not just 60° and 180° but also 1°, 47°, and so on) are populated.

EXERCISES

1. Sketch a conformational energy diagram for propane. Using the energy values for butane, deduce the barrier heights for propane.
2. Sketch a conformational energy diagram for 1-iodopropane using the C-1—C-2 bond. Assume iodine is slightly larger than a methyl group but do not try to give exact energies for each conformation.

Solutions

1. The propane diagram has the same shape as the ethane diagram. The barrier height is 3.3 kcal/mol.

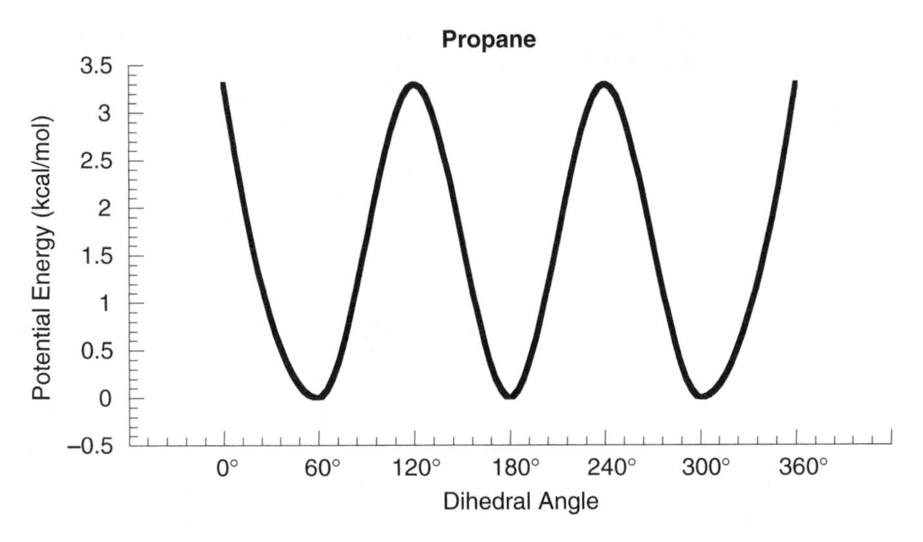

2. The shape of the 1-iodopropane diagram is the same as the butane plot. The barrier heights will be slightly higher for 1-iodopropane, assuming iodine to be larger than a methyl group.

Calculating Conformer Percentages From Energy Differences

Elementary thermodynamics relates the amounts of reactants at equilibrium and their energy difference (the **Gibbs free energy difference, ΔG^0**, referenced to standard conditions) by the equation

$$\Delta G^0 = -RT \ln K_{eq} \qquad \text{or} \qquad \Delta G^0 = -2.303RT \log K_{eq}$$

where K_{eq} is the equilibrium constant for the reaction. For A ↔ B, $K_{eq} = [B]/[A]$, R is the gas constant, $1.986\,\text{cal}\,\text{mol}^{-1}\,\text{K}^{-1}$ (note that the units are calories, not kilocalories), and T is the temperature in kelvins, 298 for 25°C.

If two conformers differ by a known ΔG^0, it is easy to calculate their proportions from the above equation. For example, in ethane, the eclipsed and staggered conformations differ by 2.9 kcal/mol. At 298 K, this corresponds to $K_{eq} = 134$. Since there are only two components, the major, lower-energy staggered component, $B = 100(K_{eq}/(1 + K_{eq})$ or 99.26 percent of the mixture, and the eclipsed conformer is 0.74 percent of the mixture.

Small differences in energy can lead to large differences in equilibrium composition, as summarized in Table 6.1.

TABLE 6.1. RELATIONSHIP AMONG THE FREE ENERGY DIFFERENCE ΔG^0, K_{eq}, AND THE PERCENTAGE OF MAJOR ISOMER %B AT EQUILIBRIUM AT 298 K

ΔG^0 (kcal/mol)	K_{eq}	%B
0	1	50
−0.5	2.3	70
−1.0	5.4	84
−2.0	29	97
−3.0	159	99.4
−5.0	4680	99.98

Example Suppose that a molecule has two conformations with the same amount of torsional strain and steric strain. One conformation forms an internal hydrogen bond (assume the hydrogen bond strength is 1.5 kcal/mol). What are the percentages of hydrogen-bonded and non-hydrogen-bonded conformers at −78°C?

$$-1.5\,(\text{kcal mol}^{-1}) = -2.303\,(1.986\ \text{cal K}^{-1}\text{mol}^{-1})(195\ \text{K})\log K_{eq}$$

which gives $K_{eq} = 48$. The hydrogen-bonded conformer is 98 percent of the mixture, and the other conformer is 2 percent of the mixture.

CONFORMATIONS OF CYCLOALKANES

Conformation of Cyclopropane

A new variety of strain is found in cyclopropane, **angle strain**, **ring strain**, or **Baeyer strain** (named for the chemist Adolph von Baeyer who first proposed this kind of strain). Cyclopropane is an equilateral triangle with internal angles of 60°. However, the tetrahedral angle is 109.5°. Thus, each ring angle is compressed. To compensate for this strain, the electrons in the C—C bond lie outside the C—C bond axis:

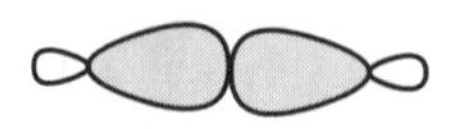

Bonding in a normal σ bond

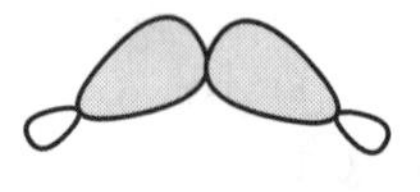

Bonding in a cyclopropane σ bond

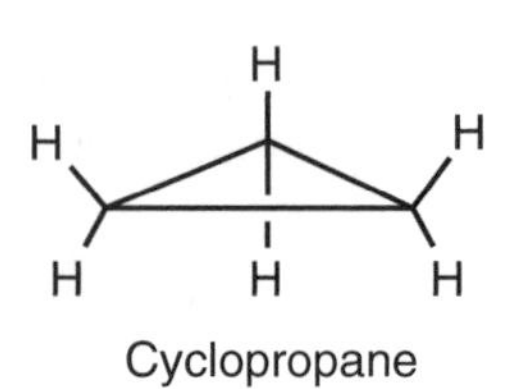

Cyclopropane

The ring bonds in cyclopropane are sometimes called banana bonds because the electron density curves outside the C—C bond axis in the shape of a banana. The poor orbital overlap of the σ bonds in cyclopropane makes them unusually reactive.

In addition to angle strain, cyclopropane also possesses torsional strain. All the C—H bonds are eclipsed. The total amount of strain in cyclopropane can be found using **combustion analysis**, which analyzes the heat released when a compound is burned completely to CO_2 and H_2O.

Cyclohexane is a strain-free cycloalkane, as will be shown below. As $(CH_2)_6$, it can be burned and the heat released divided by 6 to give the **heat of combustion** of a CH_2 group. This number is 157.4 kcal/mol. Cyclopropane, with three CH_2 groups, should release 472 kcal/mol, but 499 kcal/mol is in fact released. Cyclopropane is therefore a high-energy compound. Energy has been stored in angle strain and torsional strain, and it is released when the molecule is burned. The total **strain energy** in cyclopropane is 27 kcal/mol.

Conformation of Cyclobutane

If cyclobutane were planar, it would have significant angle strain (internal angles of 90° instead of 109.5°) and torsional strain due to all the C—H bonds being eclipsed. However, cyclobutane is not planar but puckered.

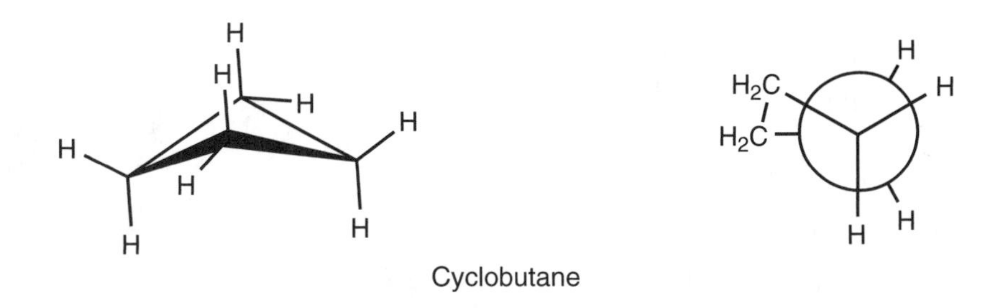

Cyclobutane

This puckering, which a Newman projection shows is just a slight rotation around a C—C bond, causes a small increase in the angle strain but relieves part of the torsional strain. As a result, cyclobutane is strained with a total strain energy of 26 kcal/mol. However, the strain is distributed over four CH_2 groups in cyclobutane.

Conformation of Cyclopentane

If cyclopentane were planar, it would have essentially no angle strain, given the 108° internal angles vs the tetrahedral value of 109.5. However, all the C—H bonds would be eclipsed. Consequently, like cyclobutane, cyclopentane is nonplanar. It adopts a conformation called an **envelope conformation**, where four of the carbon atoms are approximately planar and one carbon lies outside the plane. This shape resembles an envelope:

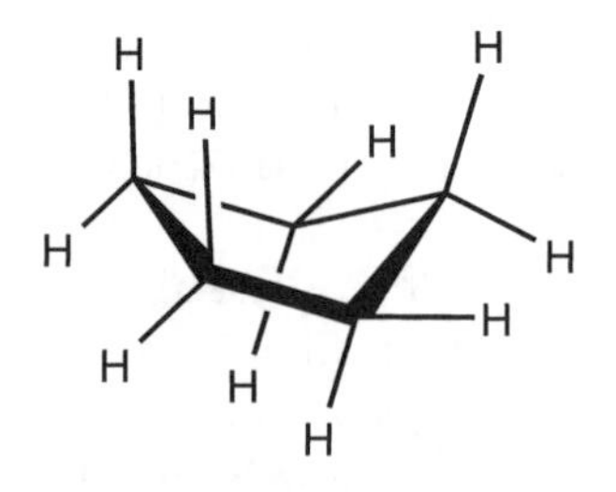

If one makes a model of cyclopentane, it is quite flexible. Each CH_2 group can in turn move out of the plane. This oscillation of the cyclopentane ring is called **pseudorotation**. Cyclopentane has little strain energy, only 6.5 kcal/mol in all.

Conformations of Cyclohexane

Cyclohexane can adopt a strain-free conformation where there is no angle strain, steric strain, or torsional strain. This conformation is called the **chair conformation**. In the chair conformation, four of the carbons are in one plane, one carbon is above this plane, and the other is below it:

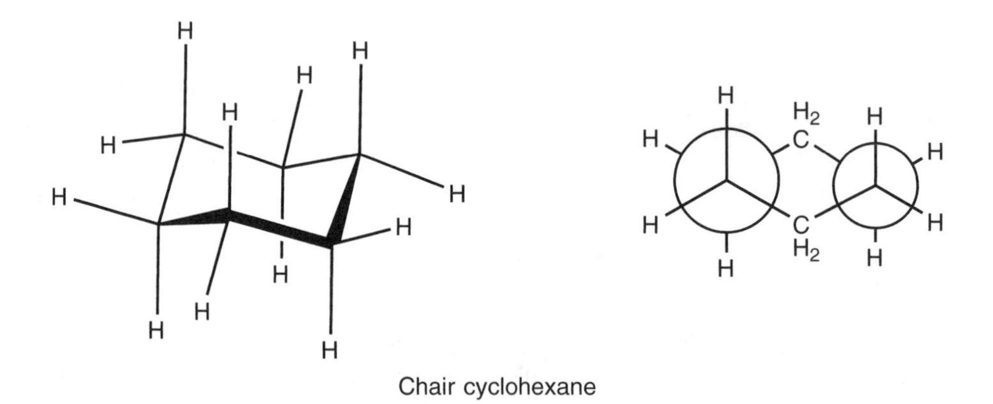

Chair cyclohexane

The depiction of chair cyclohexane shown on the right can be called a double Newman projection since it looks down what could be numbered bonds C-2—C-3 and C-5—C-6. Its advantage is that it shows that all the C—H bonds in cyclohexane are perfectly staggered.

In the chair conformation of cyclohexane, there are two classes of C—H bonds: six that point straight up or down and are called **axial**, and six that point out from the middle and are called **equatorial**.

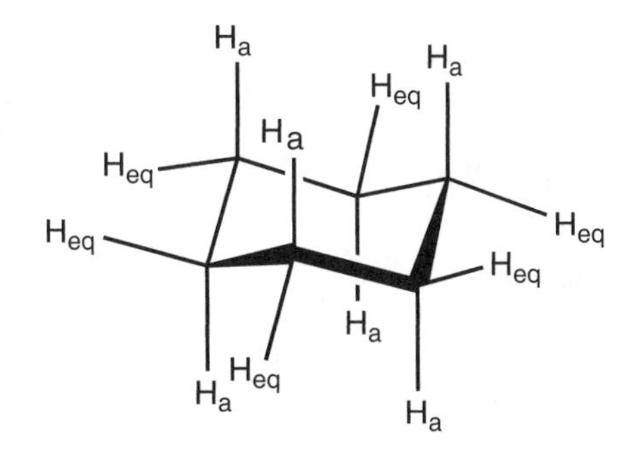

Notice that three of the equatorial bonds are tilted up and three are tilted down.

Drawing the chair conformation of cyclohexane can be difficult, but it is an essential skill in organic chemistry. When drawing cyclohexane, always keep in mind that all the carbons are tetrahedral. It is important to look at each carbon in your drawing and check that its tetrahedral nature is represented. Also, when drawing cyclohexane, keep in mind that each pair of ring bonds opposite one another is parallel. One way to draw cyclohexane is described here.

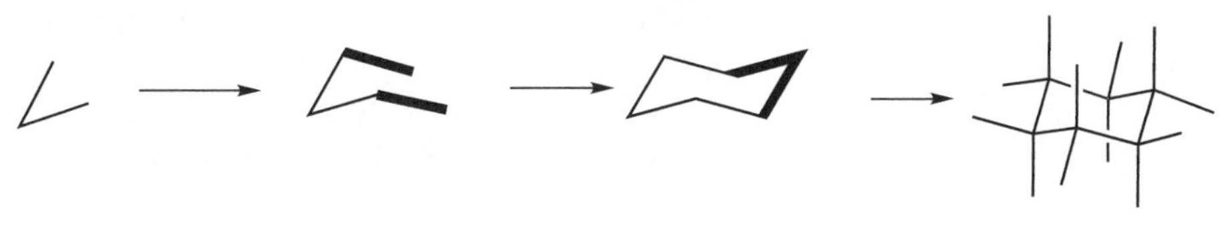

First draw one end of the ring.	Then add two parallel lines at each end.	Next complete the ring by adding two lines parallel to the first two lines.	Finally add the axial and equatorial bonds.

The chair conformation of cyclohexane is not rigid; it can undergo a **ring flip**, converting one chair conformation to another chair conformation. Like all conformational changes, this ring flip involves rotation around single bonds:

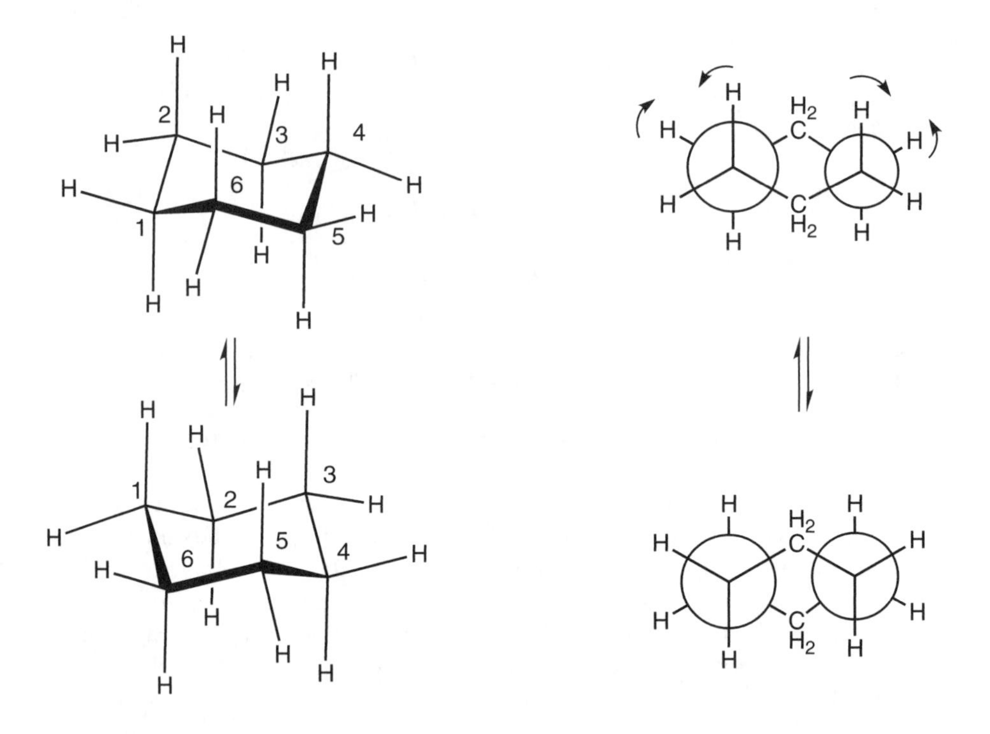

In the ring flip, as shown above (where the carbons are numbered 1 through 6 so that they can be followed), C-1 converts from pointing down to pointing up and C-4 does the opposite. Any pair of opposite carbons in the cyclohexane ring can engage in this flipping, C-1 and C-4, C-2 and C-5, or C-3 and C-6.

On examination of the above figures, it should be clear that a cyclohexane ring flip interconverts all the axial and equatorial bonds. The axial bonds pointing straight up become the equatorial bonds tilted up (and vice versa). The axial bonds pointing straight down become the equatorial bonds tilted down (and vice versa). This interconversion of axial and equatorial is easier to see if some groups replace a few of the hydrogen atoms:

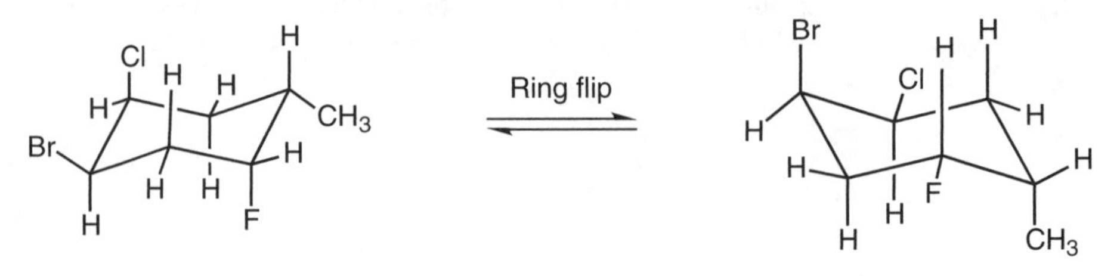

EXERCISES

1. Draw two chair forms of methycyclohexane.
2. Draw the chair conformations of 1,3-dibromocyclohexane. Which conformer is chiral?
3. Draw 1,4-dimethycyclohexane in the chair conformation where both groups are equatorial.

Solutions

1.

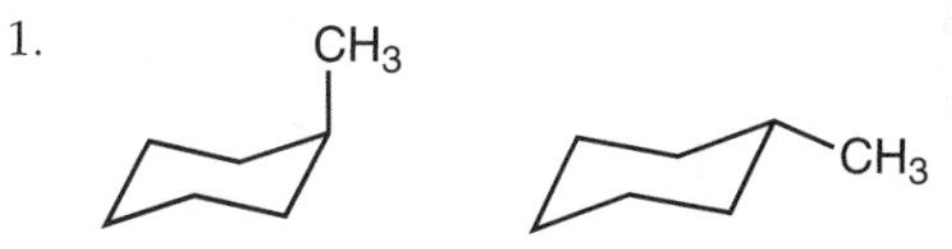

All the ring carbons are equivalent in cyclohexane, so the methyl group can be on any carbon atom. Also, the ring can be drawn as [chair] or [chair].

2.

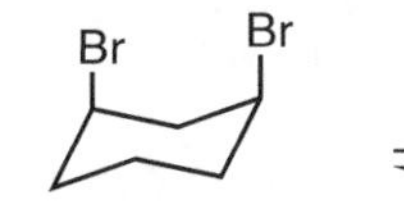

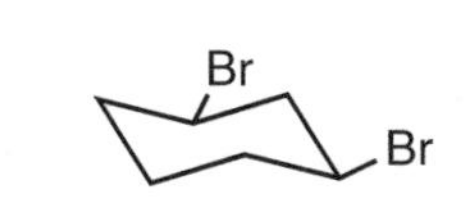

Meso isomer

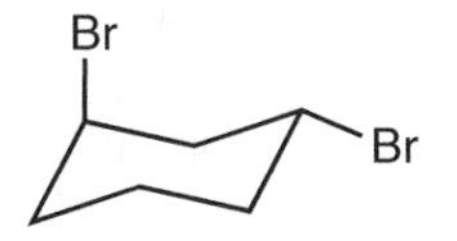

and

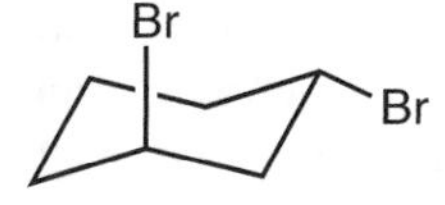

Enantiomers

3.

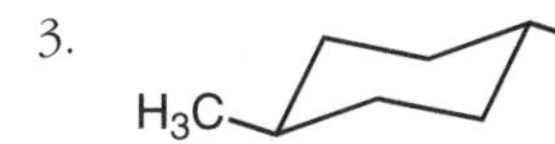

Nonchair Conformations of Cyclohexane

If one uses molecular models of cyclohexane, one finds that other conformations are easy to form. These other forms are floppier than the chair conformation, which is rather rigid. Some of these conformations are on the ring flip pathway from one chair conformation to the other.

Two limiting conformations exist between one chair conformation and the other:

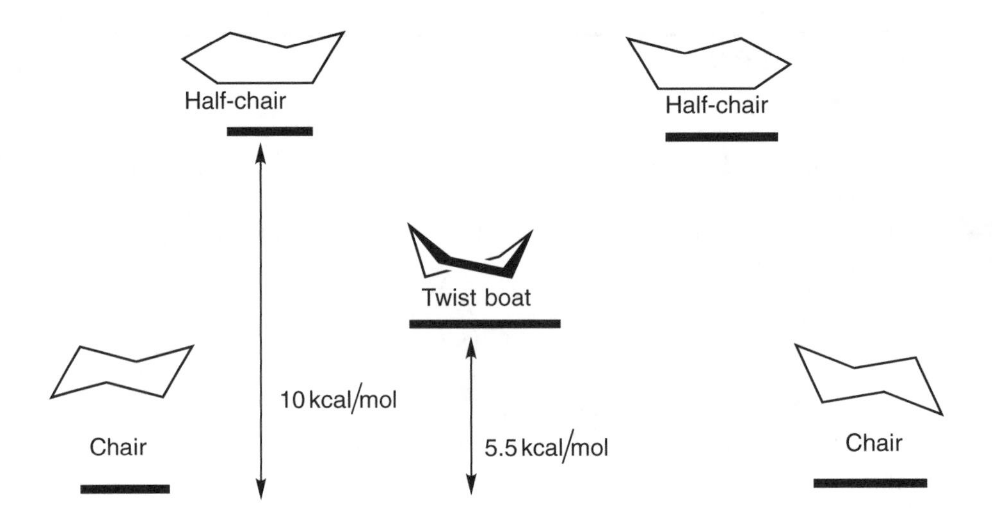

The **half-chair** conformation is the first conformation on the chair-chair ring flip pathway. It is relatively high in energy (10 kcal/mol above the energy of the chair conformation) because flattening one end of the chair introduces angle strain and torsional strain. The **twist boat** conformation experiences some torsional strain. In addition to the twist boat conformation, there is a third important nonchair conformation, the **boat** conformation. It has significant torsional strain, and there is the possibility of steric interactions if there are larger groups at the flagpole positions. The boat conformation of cyclohexane lies 6.9 kcal/mol higher in energy than the chair conformation. To convert the boat to the twist boat conformation, imagine taking the flagpole positions and twisting one toward you and the other away from you.

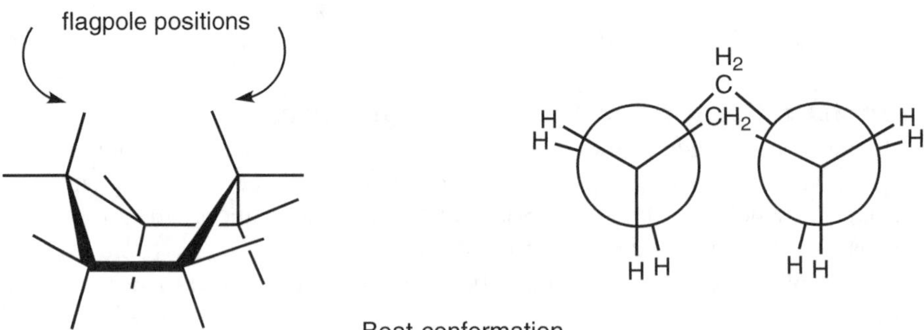

Boat conformation

Monosubstituted Cyclohexanes

Substituents in an axial position on a cyclohexane ring experience steric hindrance compared to groups in an equatorial position. Thus, with any group, a monosubstituted cyclohexane exists predominately as the equatorial conformer. The type of steric hindrance an axial group experiences is called a **1,3-diaxial interaction**. The axial group interacts sterically with the axial groups at C-3 and C-5:

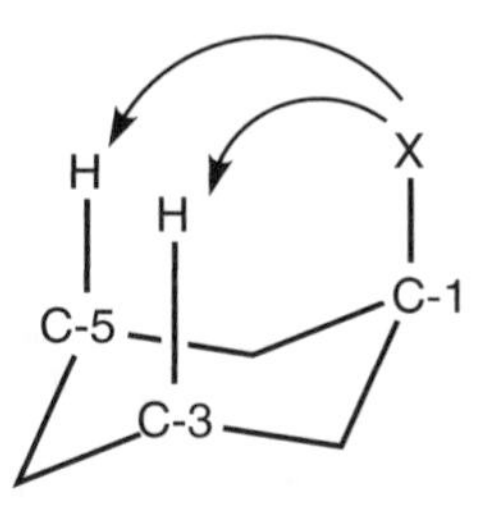

1,3-Diaxial interactions

The 1,3-diaxial interaction is exactly the same kind of interaction as the gauche interaction in butane. The geometry is the same:

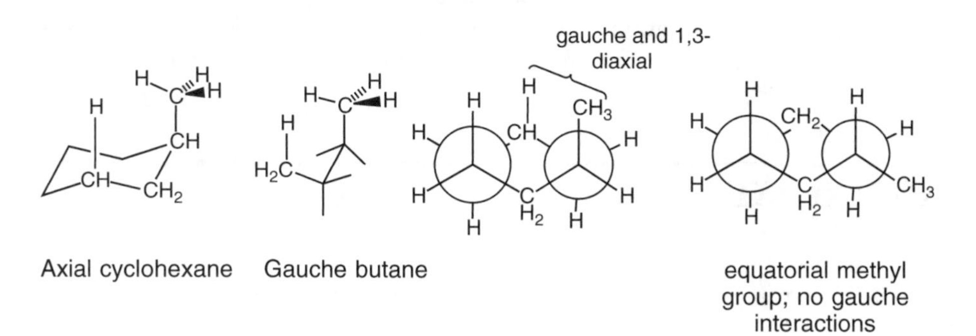

Notice that, unlike an axial group, an equatorial group experiences no gauche interactions.

In butane the gauche conformer is 0.9 kcal/mol higher in energy than the most stable, anti, conformation. There are two gauche interactions in the axial conformation of methyl cyclohexane, one with the axial hydrogen on C-3 and one with the axial hydrogen on C-5. Thus, one would expect axial methyl cyclohexane to be about 1.8 kcal/mol higher in energy than equatorial methyl cyclohexane. Experimentally, the value is 1.7 kcal/mol.

The larger the group, the greater the energy difference between the axial and equatorial conformations.

As seen in Table 6.2, the methyl, ethyl, and isopropyl groups appear to be about the same size since they have similar axial-equatorial energy differences. This is because each group can rotate so that at least one C—H bond points over the cyclohexane ring, interacting with the axial groups at C-3 and C-5. However, the *t*-butyl group must have one methyl group pointing over the

TABLE 6.2. ENERGY DIFFERENCE BETWEEN THE AXIAL AND EQUATORIAL CONFORMERS OF SELECTED MONOSUBSTITUTED CYCLOHEXANES

Group	Ax-Eq Energy Difference (kcal/mol)
$—C{\equiv}N$	0.2
—F	0.2
—Cl	0.5
—OH	1.0
$—CH_3$	1.7
$—CH_2CH_3$	1.8
$—CH(CH_3)_2$	2.1
$—C(CH_3)_3$	5.4

ring. Consequently, it has a strong preference for the equatorial conformer. This preference is so great that it is sometimes said that the *t*-butyl group locks the ring in one conformation. However, although more than 99% of all *t*-butylcyclohexane molecules have an equatorial *t*-butyl group, the ring is not locked in one conformation. Since more than 20 kcal/mol of energy is available at room temperature, the molecule still undergoes rapid ring flips.

EXERCISES

1. Sketch the most stable conformation of isopropylcyclohexane.
2. What is the percentage of axial cyclohexanol at 298 K?

Solutions

1. $CH(CH_3)_2$

   There are many other ways to draw this molecule, depending on one's preference: $(H_3C)_2HC$, $CH(CH_3)_2$, and many others.
2. 16 percent. See Table 6.1.

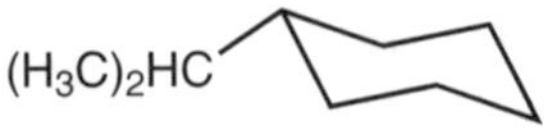

Disubstituted Cyclohexanes

When cyclohexane possesses two or more substituents, it can show conformational isomerism and stereoisomerism. When two groups are bonded to different carbons in cyclohexane, they can be on the same side of the ring (either both above it or both below it) or they can be on opposite sides of the ring. When two groups are on the same side of the ring, they are

cis stereoisomers, and when they are on opposite sides of the ring, they are **trans** stereoisomers. This cis and trans nomenclature was also used for alkene stereoisomers.

It is easy to see that the diaxial conformer of 1,2-dimethylcyclohexane is a trans isomer. However, because a ring flip interconverts axial and equatorial bonds, the diequatorial conformer of 1,2-dimethycyclohexane is also a trans isomer. Recall that conformational isomers interconvert by rotating around single bonds, while to interconvert stereoisomers, covalent bonds have to be broken and remade.

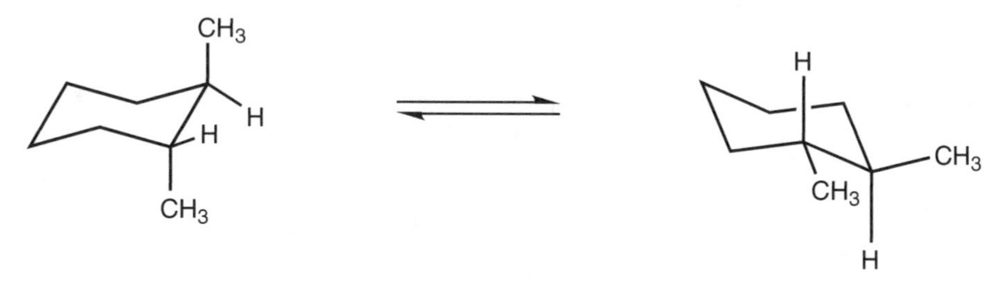

trans-1,2-Dimethylcyclohexane

In the diequatorial conformer, one methyl group is tilted down and the other methyl group is tilted up. Thus, the two groups are still directed toward opposite sides of the ring, just as in the case of the diaxial conformer. Because groups are lower in energy in the equatorial position, the diequatorial conformer is favored over the diaxial conformer in *trans*-1,2-dimethylcyclohexane.

In *cis*-1,2-dimethylcyclohexane, one methyl group is axial and the other is equatorial. The two ring flip conformers are, in this case, chemically equivalent. (Actually, and this is a subtle point, the two conformers are enantiomers. If one could somehow make one enantiomer of this molecule, at room temperature rapid ring flips would quickly generate a racemic mixture).

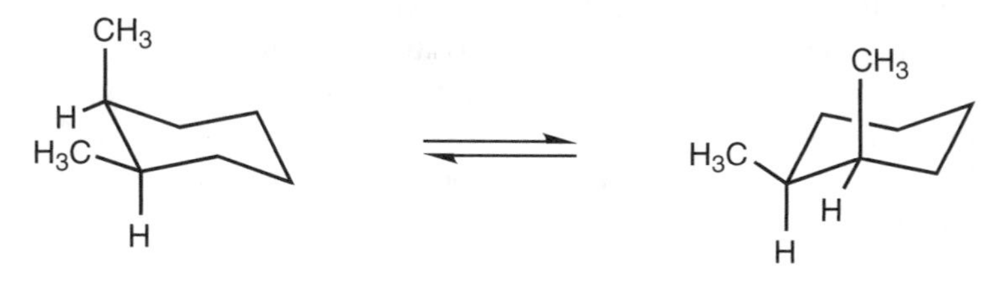

cis-1,2-Dimethylcyclohexane

Cis and trans isomers are also seen in 1,3- and 1,4-disubstituted cyclohexanes. All the relationships are summarized in Table 6.3.

TABLE 6.3. THE RELATIONSHIP BETWEEN CONFORMATIONAL ISOMERS AND STEREOISOMERS IN DISUBSTITUTED CYCLOHEXANES

Stereoisomer	Conformational Isomer
Cis-1,2	Ax, Eq or Eq, Ax
Trans-1,2	Ax, Ax or Eq, Eq
Cis-1,3	Ax, Ax or Eq, Eq
Trans-1,3	Ax, Eq or Eq, Ax
Cis-1,4	Ax, Eq or Eq, Ax
Trans-1,4	Ax, Ax or Eq, Eq

Example Draw all the conformational isomers for *cis*- and *trans*-1,3- and *cis*- and *trans*-1,4-difluorocyclohexane.

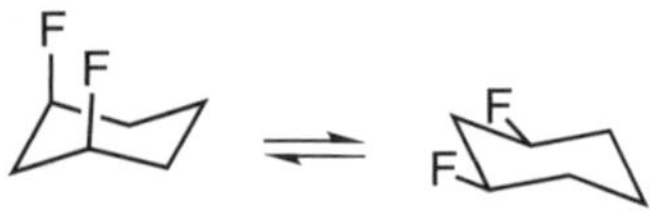

cis-1,3-Difluorocyclohexane

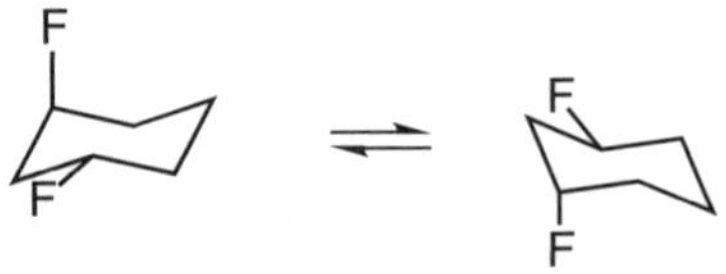

trans-1,3-Difluorocyclohexane

cis-1,4-Difluorocyclohexane

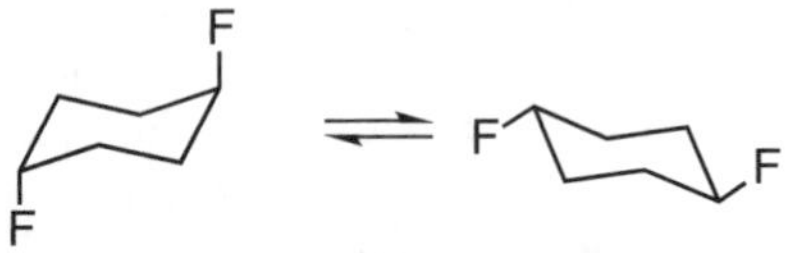

trans-1,4-Difluorocyclohexane

When two different substituents are on a cyclohexane ring, the larger group predominately occupies the equatorial position. Because a methyl group is bigger than a fluorine group (from the data in Table 6.2), in *trans*-1-fluoro-3-methylcyclohexane, for example, the methyl group is equatorial and the fluorine group is axial. Since the *t*-butyl group is larger than any other common group, it tends to be equatorial and forces a second group to be predominately axial.

Wedge and dashed bonds are often used to denote cis and trans in cyclohexane molecules, as illustrated here.

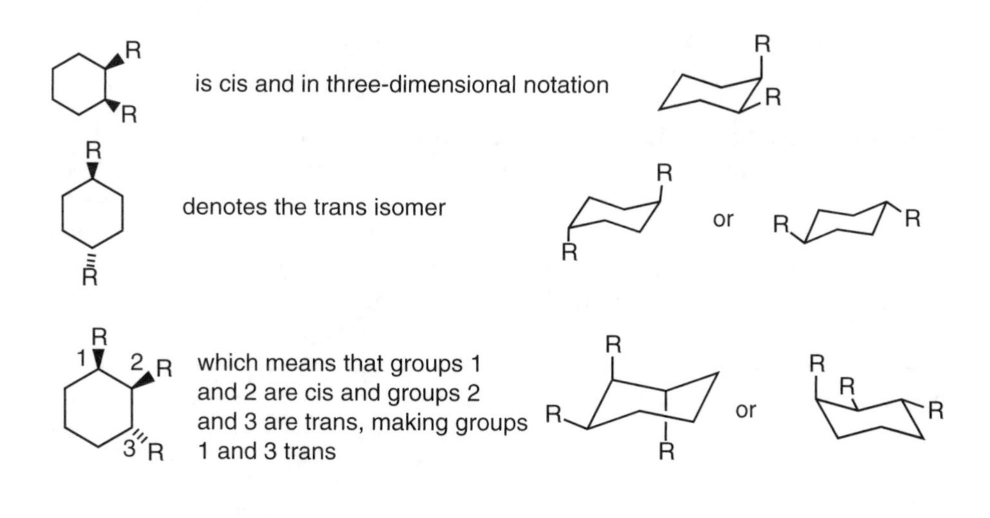

EXERCISES

1. Draw the most stable chair conformations of the following molecules.
   (a) *trans*-1,2-Diethylcyclohexane (b) *trans*-3-Methylcyclohexanol
   (c) *cis*-1-*t*-Butyl-4-phenylcyclohexane (d) *cis*-2-Propylcyanocyclohexane
2. Draw *cis*-1,2-*trans*-1,4-trimethylcyclohexane using both wedge and dashed bonds and in three-dimensional (chair) form.
3. Which molecules in Exercise 1 are chiral?

Solutions

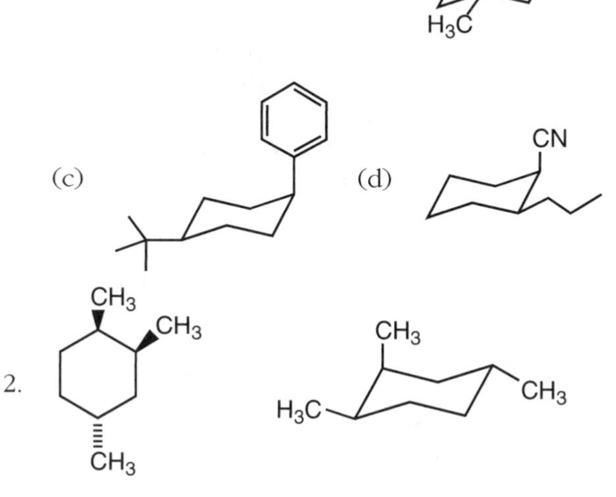

3. Compounds (a), (b), and (d) are chiral. Compound (c) has a plane of symmetry and is achiral.

Conformations of Larger Rings and Fused Rings

The most common ring sizes in organic chemistry are five and six, and both are relatively strain-free. There are many compounds with three-membered rings, too. Although this ring size is strained, bringing together two carbons attached to the same carbon to form a three-membered ring is favored by entropy. Four-membered rings are less common, being strained and not having a significant entropic advantage for formation. Rings larger than six-membered have problems with torsional strain and often steric strain from hydrogens on opposite sides of the ring bumping into each other in the interior of the ring. There are also entropic penalties in closing a large ring since the ends are unlikely to be close. Finally, many larger rings have numerous different conformations of similar energy.

Unlike larger rings, the properties of **fused rings** are an appropriate subject for introductory organic chemistry. Rings can be fused at a single carbon, along a bond, or along a face. Two rings that share a common atom are called **spiro compounds**. Such compounds are named as spiro[X.Y]*alkane*, where X and Y are the numbers of atoms in each ring (not counting the common atom) and *alkane* is the name of the hydrocarbon corresponding to the total number of carbon atoms in the two rings:

Spiro[2.2]pentane

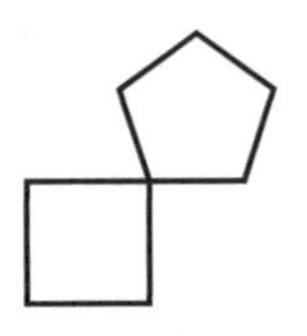

Spiro[3.4]nonane

Two rings that share an edge are known as fused ring compounds. These molecules are extremely important in biochemistry and pharmacology. Steroid hormones, for example, belong to this class. The fused ring system **Decalin** has two cyclohexane rings sharing an edge. Its IUPAC name is bicyclo [4.4.0]decane, and this naming system will be explained below. There are two stereoisomers of Decalin, *cis*-Decalin and *trans*-Decalin:

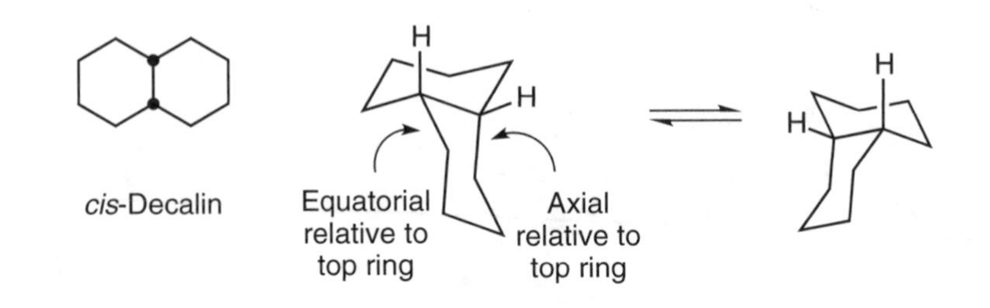

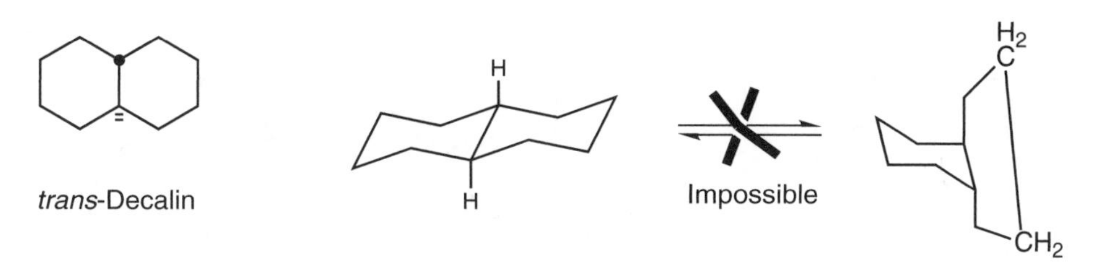

In *cis*-Decalin, one ring is attached to the other ring through one equatorial bond and one axial bond, hence the cis designation. The two hydrogens where the rings fuse, called the **ring junction**, are cis to one another. *cis*-Decalin can undergo ring flips; it is no more rigid than cyclohexane itself. *trans*-Decalin has two six-membered rings fused so that the bonds coming off one ring are both equatorial and the two hydrogens at the ring junction are both axial. *trans*-Decalin is a rigid molecule; it cannot undergo ring flips. If it did, one of the six-membered rings would be attached to the other ring via two axial bonds. Carbon-carbon bonds are too short to close a ring with this geometry.

The class of steroid molecules, which include such molecules as estrogen, cortisone, and cholesterol, has a common nucleus of three six-membered rings and one five-membered ring fused together. Normally, they are all trans-fused, which produces a very rigid framework that fits tightly into many biological receptors.

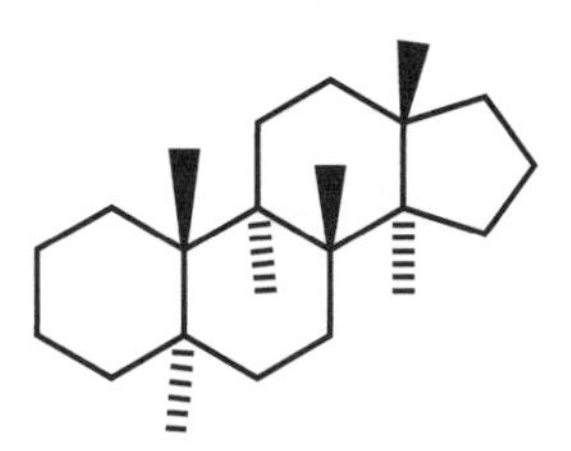

Steroid nucleus

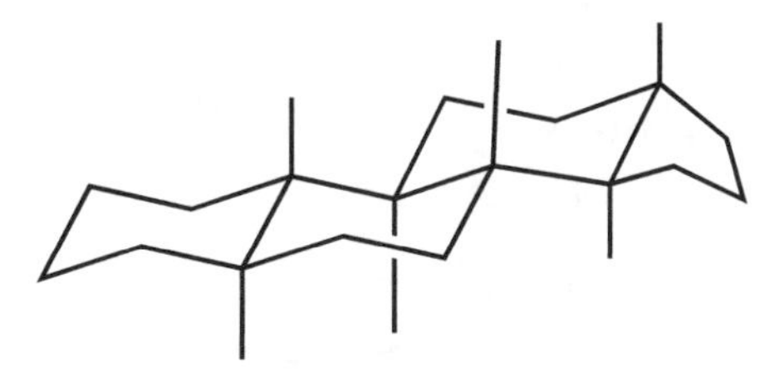

Three-dimensional representation

Rings that share a face are called **bridged bicyclic** rings. They are characterized by **bridgehead carbons**, carbons common to all rings. Examples of these bicyclic molecules, are the following (where bridge head carbons are indicated by CH).

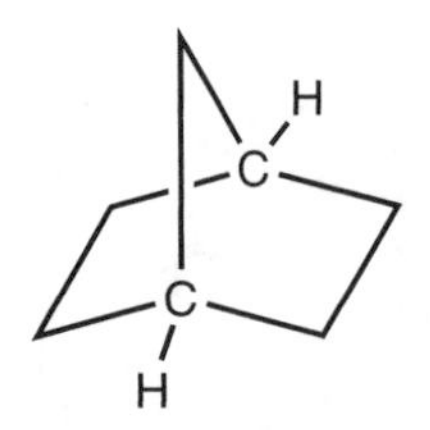

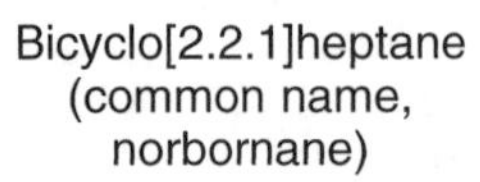
Bicyclo[2.2.1]heptane
(common name,
norbornane)

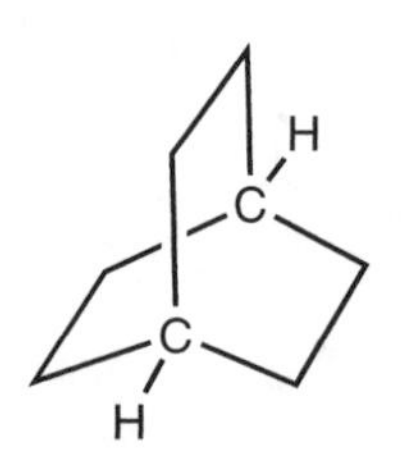

Bicyclo[2.2.2]octane

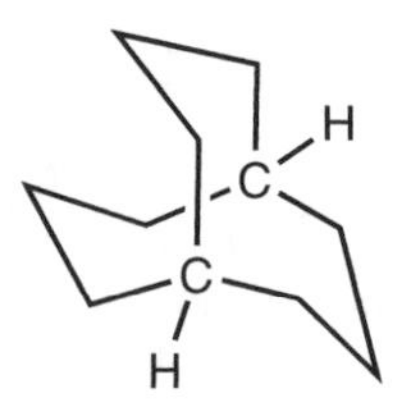

Bicyclo[3.3.3]undecane

These compounds are named using the system bicyclo[X.Y.Z]*alkane*, where X, Y, and Z are the numbers of carbons in each bridge and *alkane* corresponds to the name of the hydrocarbon corresponding to the total number of carbons in the molecule. Many natural plant oils, such as those from the pine tree, are bicyclic hydrocarbons.

EXERCISES

Draw the three-dimensional representation of the following fused ring systems.

(a) 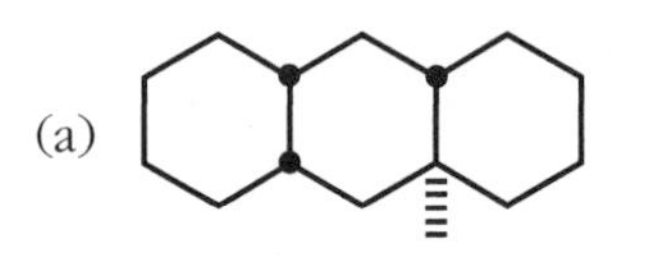

(b) Bicyclo[2.2.1]hexane

Solutions

(a) 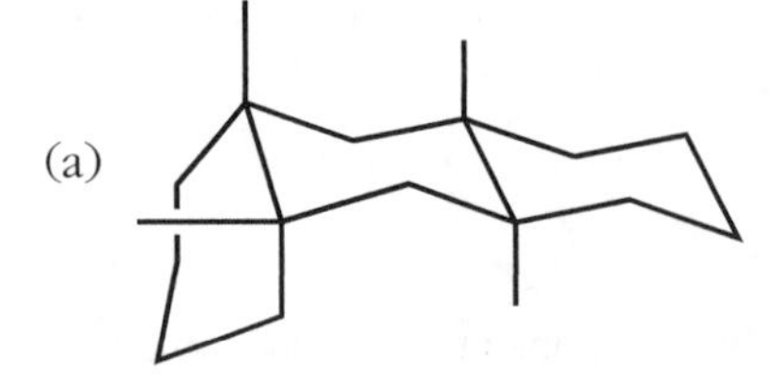

(b)

7 INTRODUCTION TO ORGANIC REACTIONS

There are tens of thousands of different organic reactions. Hundreds of them are so well known that they are named after the chemist or chemists that invented them and are called **name reactions**. In order to remember even the most important examples of such a vast amount of data, one needs an efficient classification scheme. One approach classifies common organic reactions into four types: **addition reactions**, **elimination reactions**, **substitution reactions**, and **redox reactions**. As in any classification scheme, some reactions do not easily fit into one of these categories. However, this scheme can provide a beginning for understanding organic reactions.

A given organic reaction can occur on any one of a million different organic substrates. In order to predict the reaction of a molecule one has never encountered before, one needs to understand the mechanism of important organic reactions. A **reaction mechanism** is a molecular level understanding of how a starting compound (the **starting material**) is converted to a product. The mechanism describes the order in which bonds are broken and formed in going from reactants to products, and what **reactive intermediates** (such as carbocations, carbanions, or carbon radicals), if any, are generated during the reaction. As subsequent chapters will illustrate, there can be numerous different mechanisms within each of the four reaction types.

ADDITION REACTIONS AND REACTION SYSTEMATICS

Single bonds are more stable than the π component of double or triple bonds. Thus, it is often energetically favorable for unsaturated molecules to add reagents and become saturated. For example, bromine, Br_2, reacts with ethylene, $CH_2{=}CH_2$, to form 1,2-dibromoethane, $BrCH_2{-}CH_2Br$. Or acetylene, HC_2H, can react with hydrogen in the presence of certain catalysts to give the single addition product, ethylene, or the product from the addition of two molecules of hydrogen, ethane. Any sort of unsaturated functional group can undergo addition reactions, including a carbon-carbon double or triple bond, a carbon-oxygen double bond, and a carbon-nitrogen double or triple bond.

Whenever one encounters a reaction of any type, the student should ask the following five questions about it.

1. What is the mechanism of the reaction?
2. What is the stereochemistry of the reaction?
3. What is the regiochemistry of the reaction?
4. What is the overall transformation?
5. Is there another way to carry out the overall transformation?

The meaning of a reaction's stereochemistry and regiochemistry is given below as we discuss the mechanism of the Br_2-ethylene addition reaction.

Reaction Mechanisms and the Curved Arrow Notation

During a reaction, electron pairs move as bonds break and re-form. Chemists have developed a way to keep track of electron pair movement using **curved arrow notation**. A curved arrow shows the movement of an electron pair. The tail of the arrow starts at the place of electron density, and the arrowhead ends up at a place of electron deficiency. For the addition of bromine to ethylene, the electron flow for this addition mechanism is as depicted here.

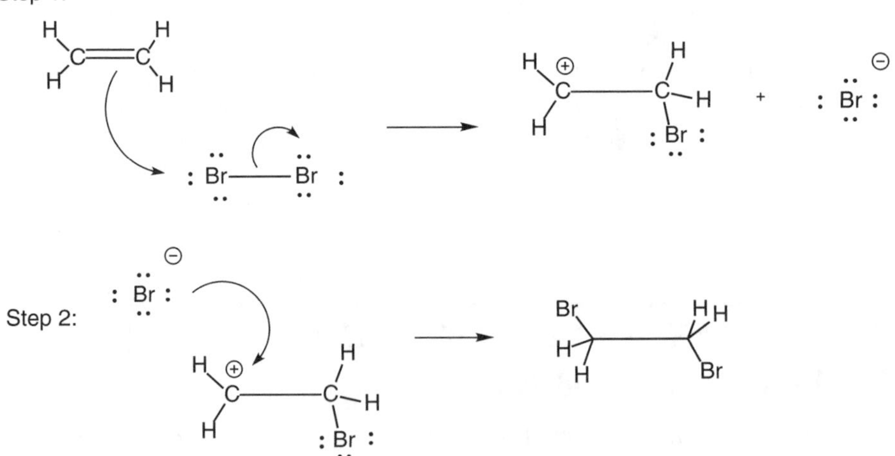

In the first step, electrons in the π bond move to form a C—Br σ bond. The electron pair in the Br—Br bond becomes a lone pair on a Br^-. The curved arrows provide a method of electron bookkeeping for the reaction mechanism. When drawing curved arrows, the direction of the arrow is very important. In step 2 of the mechanism, for example, the arrow must point to the positive charge. Reversing the direction of the arrow so that it starts at the positive charge is wrong.

In step 1, since electrons flow out of the double bond and are not replaced, the carbon molecule must become positively charged. Second, the octet rule must be obeyed. Whenever electrons flow into a molecule with an octet of electrons, electrons must also flow out. Otherwise, a molecule would go from having eight electrons to having ten electrons. In this mechanism, as the carbon-bromine bond forms, the bromine-bromine bond must break to obey the octet rule.

The mechanism for this addition reaction informs us that a reactive intermediate is formed during the course of this reaction, a carbocation. (The exact structure of this carbocation is somewhat more complicated than our drawing suggests, but we will cover it in detail in Chapter 10.) Since a charged intermediate is created from uncharged reagents, this mechanism is a **polar** mechanism. Whenever there is more charge separation in the intermediates in a reaction than in its reactants or products, the reaction is polar. Whenever there is no significant charge separation during the course of a reaction, the reaction is **nonpolar**.

EXERCISES

Add curved arrows to show electron flow in the following reactions.

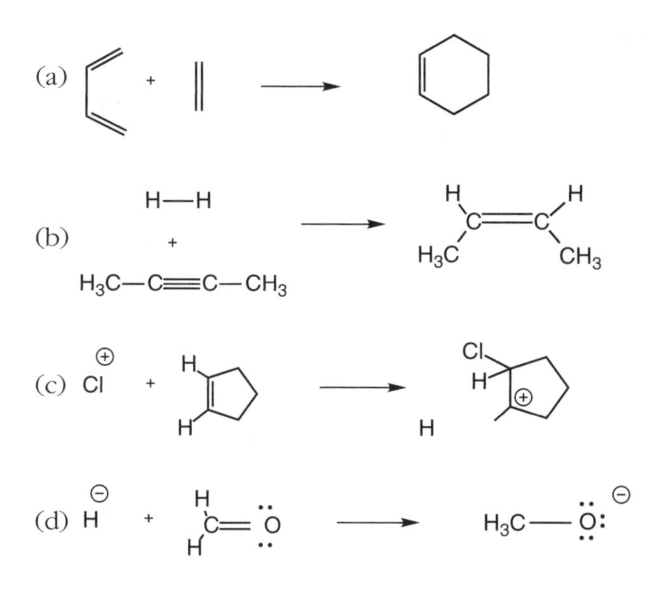

Solutions

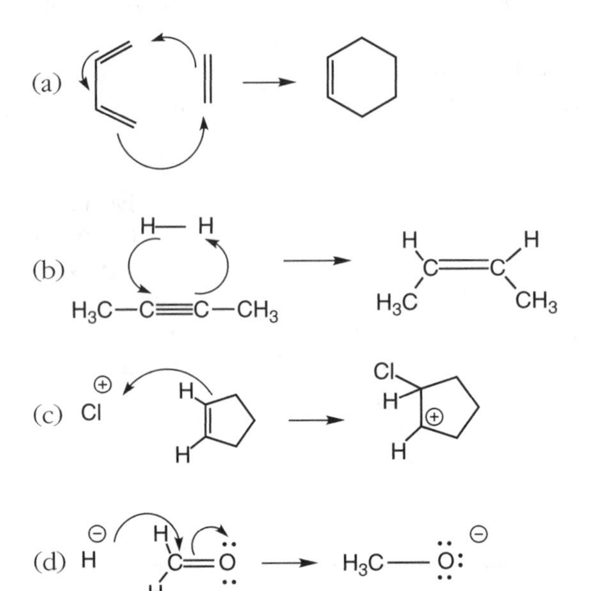

The Stereochemistry of a Reaction

In many reactions, new stereocenters are formed. For example, in Exercise 1(b) above, the addition reaction as shown generates the Z alkene and not the E alkene. Another example is the addition of Br_2 to a cyclic alkene, like cyclopentene, which produces the trans dibromide and not the cis dibromide:

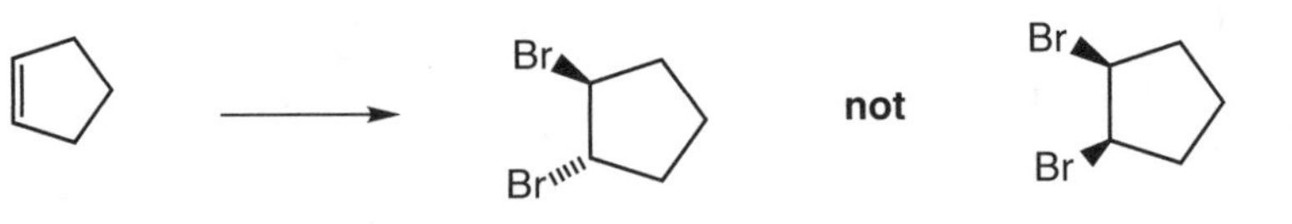

The **stereochemistry** of a reaction refers to what stereoisomers are produced in a reaction from a given reactant. If a reaction gives only one of several possible stereochemical products, the reaction is **stereospecific**. Remember, however, that without a source of chirality one gets a stereoisomer and its mirror image (a racemic mixture). Another way to add bromine to a double bond might give only the cis dibromide; this, too would be stereospecific. If a certain reaction gives mostly one stereoisomer but a small amount of other stereoisomers, the reaction is called **stereoseselective**. If a reaction gives a random mixture of stereocenters, the reaction is considered **stereorandom**.

Knowing the mechanism of a reaction usually enables one to predict its stereochemistry. For example, the molecule BH_3 reacts with a double bond to add a B and an H to the same side of the double bond:

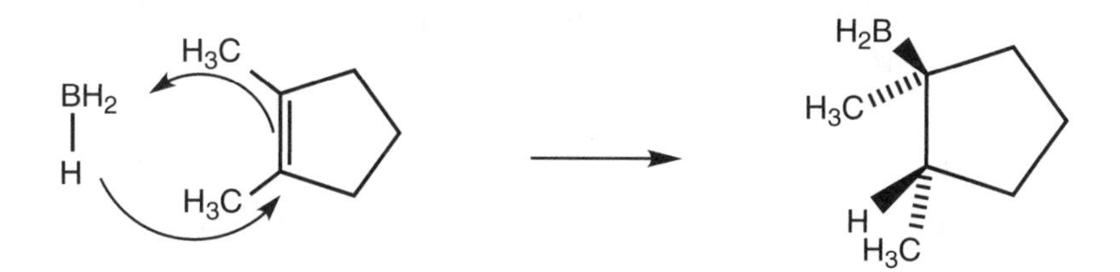

This type of reaction, where there is a simultaneous making and breaking of bonds, is called a **concerted reaction**. Since the boron and the hydrogen are never far apart, they must end up on the same side of the ring, giving the stereochemistry shown.

Some substrates do not result in the generation of any new stereocenters. For example, when BH_3 adds to $CH_2{=}CH_2$, the product is $H_2BCH_2{-}CH_3$, which has no stereocenters. Nevertheless, the mechanism is the same, a concerted addition reaction. Whenever you are learning a new reaction, make sure to practice it with some substrates that will generate new stereocenters. Think about what features of the mechanism are responsible for the particular stereochemistry of the reaction.

Example The first two reactions are stereospecific; the third is stereoselective.

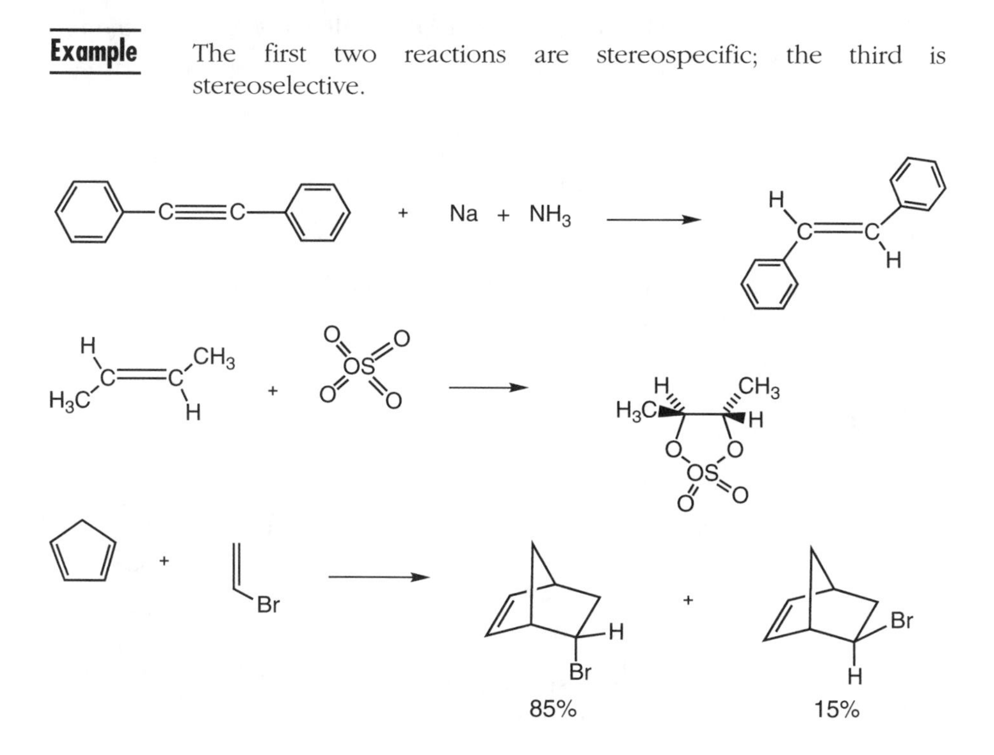

Regiochemistry of Organic Reactions

The **regiochemistry** of a reaction refers to the orientation of an unsymmetric reagent when it adds to a substrate. For example, H—Br can add to many double bonds, creating a new C—H and a new C—Br bond. If the alkene is also unsymmetric, several products are possible:

$$H{-}Br + H_2C{=}CMe_2 \longrightarrow H_3C{-}C(Me)_2Br \text{ and/or } H_2CBr{-}CH(Me)_2$$

Regioisomers

These are constitutional isomers, not stereoisomers. Since they result from different regions of the reagent and substrate coming together to form bonds, they are called **regioisomers**. Normally, knowing the mechanism of a reaction and the factors that stabilize reactive intermediates allows one to understand and predict the regiochemistry of a given reaction. For every reaction one studies, one should ask what the possible regiochemical outcomes are and why the reaction proceeds the way it does.

EXERCISES

Show which regioisomers are possible for the addition of H—Br to the following unsaturated compounds:

$H_3C{-}C{\equiv}C{-}H$

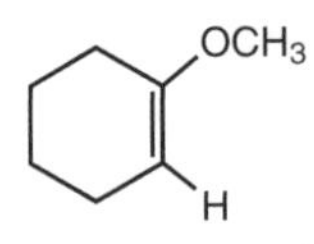

(*Note*: the double bonds in a phenyl ring are unusually unreactive and do not undergo most addition reactions.)

Solutions

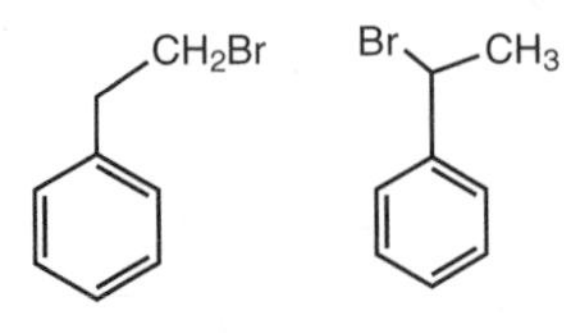

$H_3CHC{=}CHBr$

$H_3CBrC{=}CH_2$

E and Z isomers are possible in the top case, which are stereoisomers.

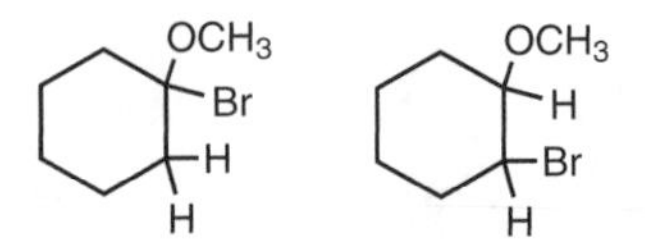

cis and trans isomers are possible in the second case, which are stereoisomers.

Classifying Reactions Through Functional Group Changes

Most organic reactions, including addition reactions, are examples of **functional group interchange** reactions. That is, one functional group in the starting material is transformed into another functional group in the product. The following are a few of the reactions we will cover in later chapters. For now, notice which functional group is in the starting material and which functional group is in the product. Which functional groups are being interchanged is the first thing you should learn about any new reaction.

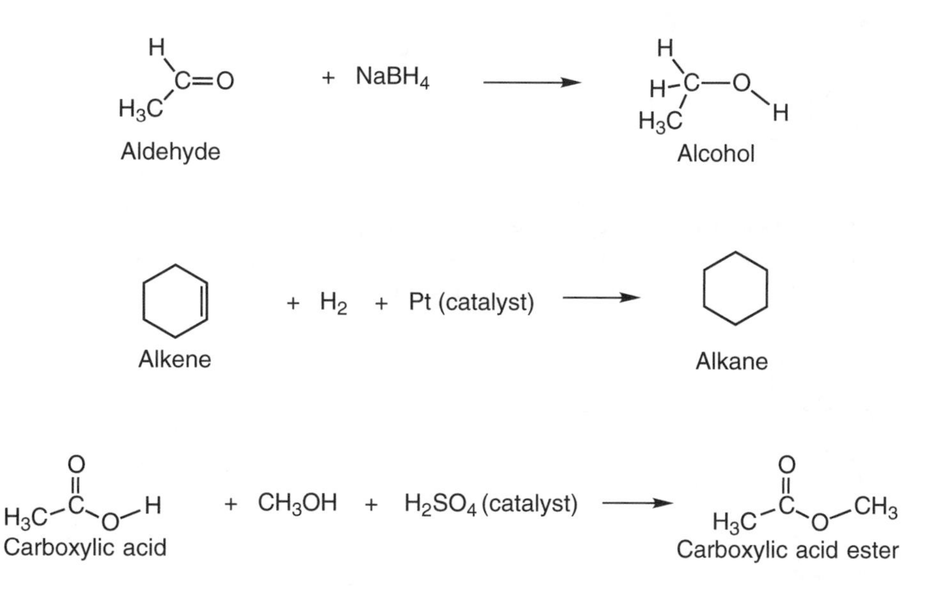

In addition to reactions that interchange functional groups, some reactions also change the carbon framework of a substrate by forming or (rarely) breaking carbon-carbon bonds. Reactions that form new C—C bonds turn smaller, simpler molecules into larger, more complex ones. Carbon-carbon bond-forming reactions are probably the most important organic reactions. For these kinds of reactions, you need to note which functional groups one starts with, which functional groups are present in the product, and how the carbon framework changes. Shown below are examples of a few C—C bond-forming reactions. Note how the carbon framework changes in each case.

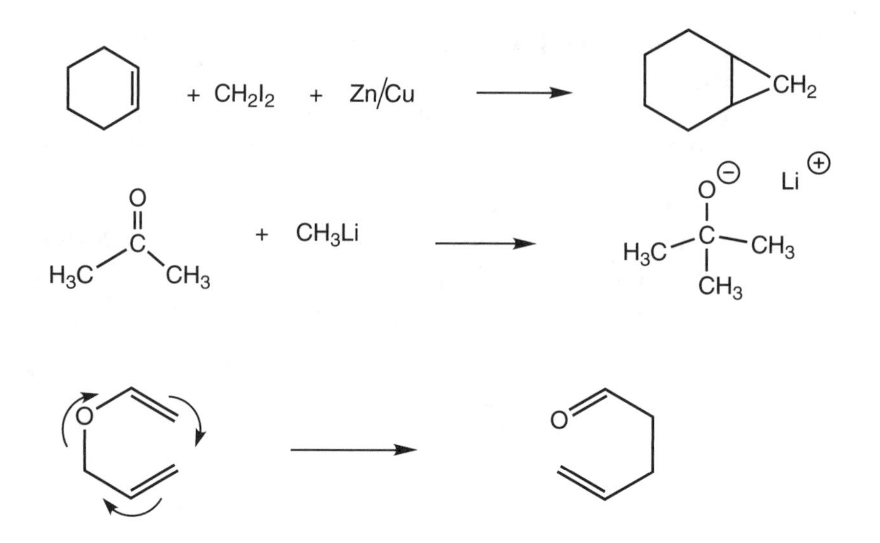

Finally, whenever one encounters a new organic reaction, it is useful to ask if the same transformation can be carried out in a different way. For example, here are three ways to convert certain kinds of ketones to hydrocarbons.

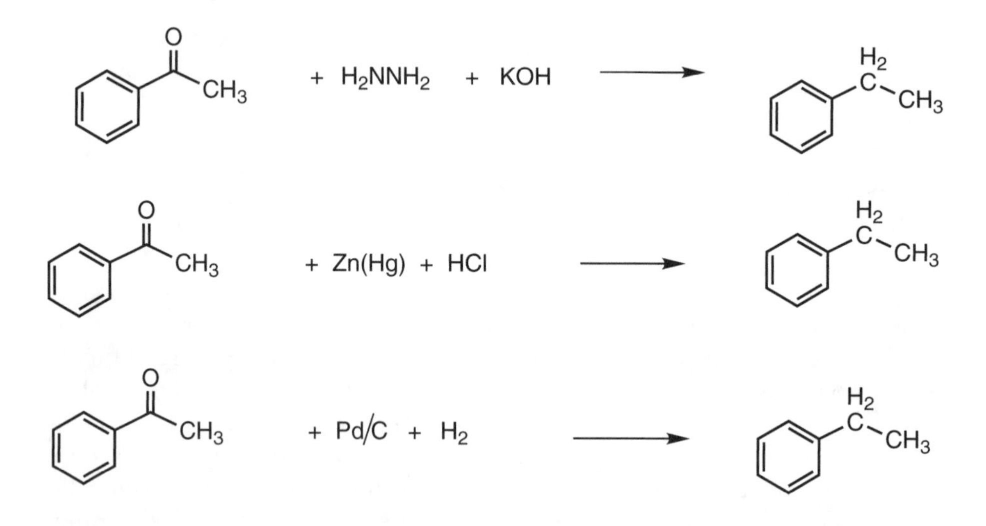

All of them work quite well, so it seems redundant to have three ways to carry out the same kind of functional group interchange. However, the first reaction takes place under basic conditions, the second under acidic conditions, and the third under neutral conditions. Whichever one chooses depends on what other functional groups are present in the molecule.

Since, starting with Chapter 8, many organic reactions will be introduced in virtually every chapter, the student should start immediately to organize this material. Many students find that it is useful to make cards or files for each reaction and record the reaction's mechanism, stereochemistry, regiochemistry, and functional group interchange properties, as well as other ways available for carrying out this transformation, if any. A second type of study aid that is very useful is a functional group interchange chart. One lists the common functional groups along the top and left side of the chart, the top functional groups being the reactant groups and the side groups the product functional groups. Where each row and column intersect, one writes in the reagent or reagents that carry out this particular functional group interchange. For example, given the information on the conversion of ketones to hydrocarbons shown opposite, a portion of such a chart would look like the following.

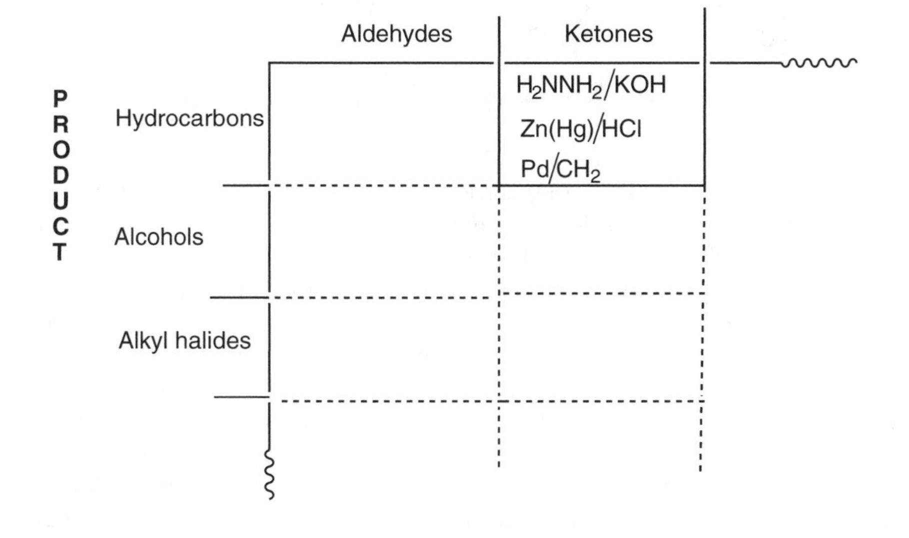

ELIMINATION REACTIONS

The elimination reaction is the second general class of organic reactions. In an addition reaction, two molecules come together to form a new molecule. In an elimination reaction, a molecule breaks apart into two or more fragments, producing a new double or triple bond. For any reaction, the free energy change depends on the change in the enthalpy and entropy of the reaction, $\Delta G^0 = \Delta H - T\Delta S$. The entropy term is generally unfavorable for addition reactions, so they occur at low to moderate temperatures. Elimination reactions have favorable entropy changes, so they tend to be carried out at higher temperatures. Examples of a few elimination reactions, with the electron flow indicated using curved arrow notation, are

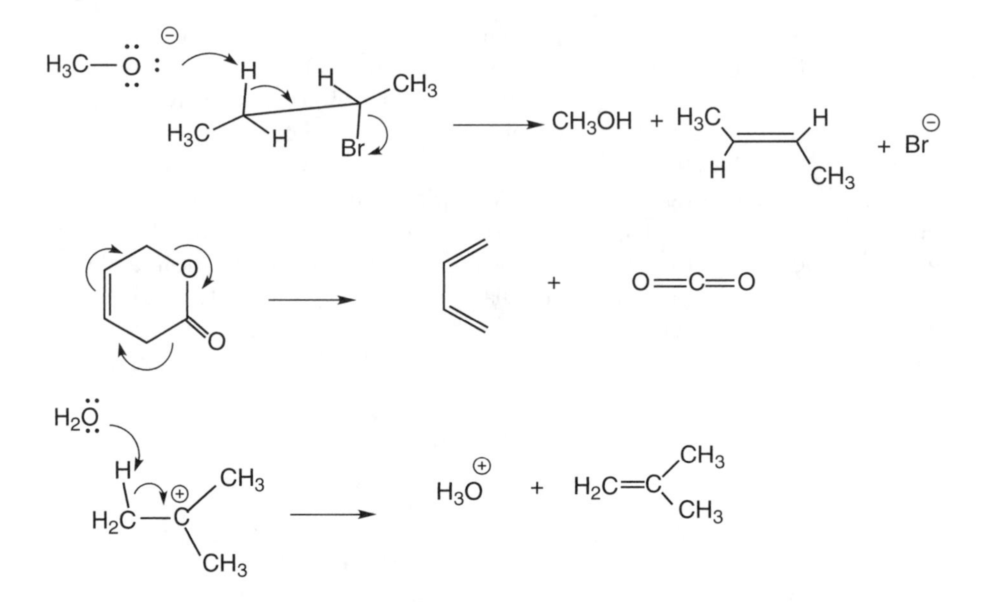

The temperature at which one carries out a reaction is one example of what organic chemists call **reaction conditions**. The temperature, the solvent, the concentration of reagent and substrate, the pH, the ionic strength, and the pressure are some of the factors an organic chemist can change in order to increase the yield of a reaction. Knowing the mechanism of a reaction permits the chemist to rationally change these variables. For example, the first elimination reaction shown above looks a lot like an acid-base reaction where the organic compound is the acid and CH_3O^- is the base. A solvent that enhances the basic properties of CH_3O^- might be expected to facilitate this reaction, and indeed this is the case. Conversely, the second elimination reaction is a nonpolar, concerted reaction. It would not respond to changes in reaction conditions the same way that the first reaction would, although both are elimination reactions.

SUBSTITUTION REACTIONS, NUCLEOPHILES, AND ELECTROPHILES

Addition and elimination reactions involve the reactions and formation of multiple bonds. Substitution reactions largely involve the interconversion of single bonds:

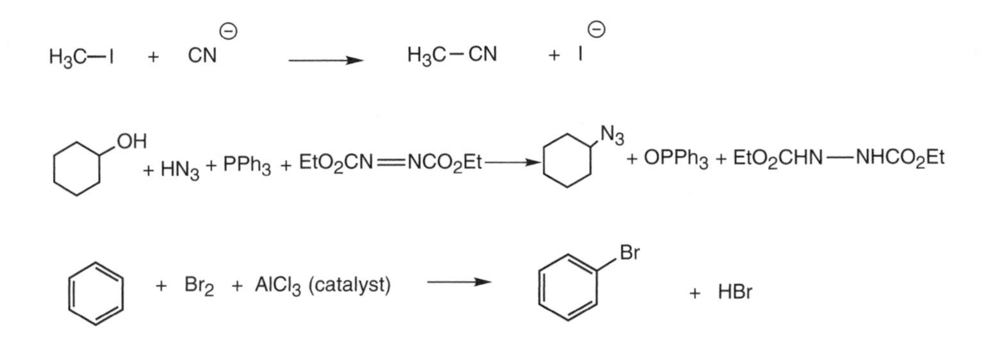

Most substitution reactions are polar reactions in which a partly or fully negatively charged region of one molecule reacts with a partly or fully positively charged region of another molecule. Molecules with a high density of negative charge are called **nucleophiles**. This term comes from a Greek word meaning nucleus lover since the nucleus is the source of positive charge. Molecules with a high density of positive charge are called **electrophiles**, which means electron lover. Nucleophiles react with electrophiles, which is another way of saying that opposite charges attract. Whenever one is faced with the problem of how two molecules might react, one should look for the most electrophilic site and the most nucleophilic site in the system. Often these two sites are the regions that come together.

Common nucleophiles in organic chemistry include anions such as halide ions (especially Br^- and I^-), RO^-, CN^-, double bonds, amine nitrogens, R_3N, and carbon bonded to electropositive atoms such as Li, RLi. Electrophiles are species such as H^+, the carbon of carbonyl groups, $R_2C{=}O$, carbons attached to halogens, such as CH_3Br, and Lewis acids, including BF_3 and $AlCl_3$.

Example Some examples of nucleophiles reacting with electrophiles are

H^+ +

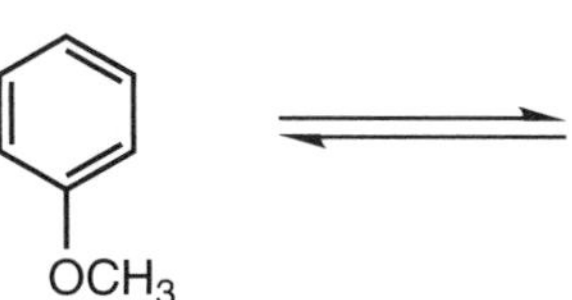

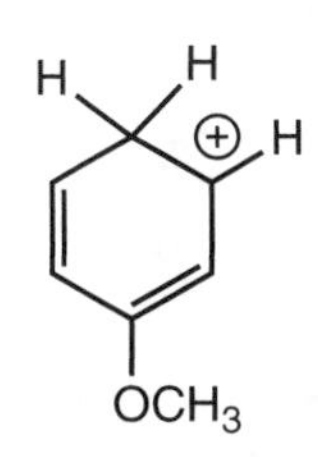

Electrophile Nucleophile

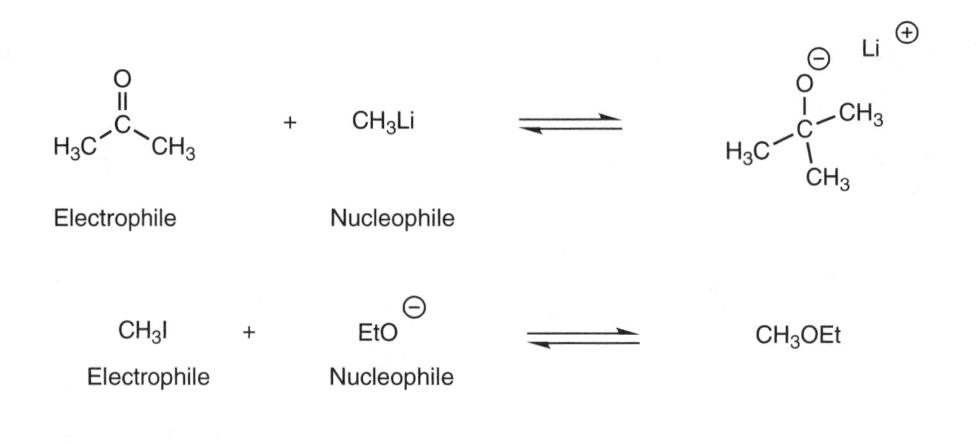

EXERCISES

For the above examples, show electron flow using curved arrow notation.

Solutions

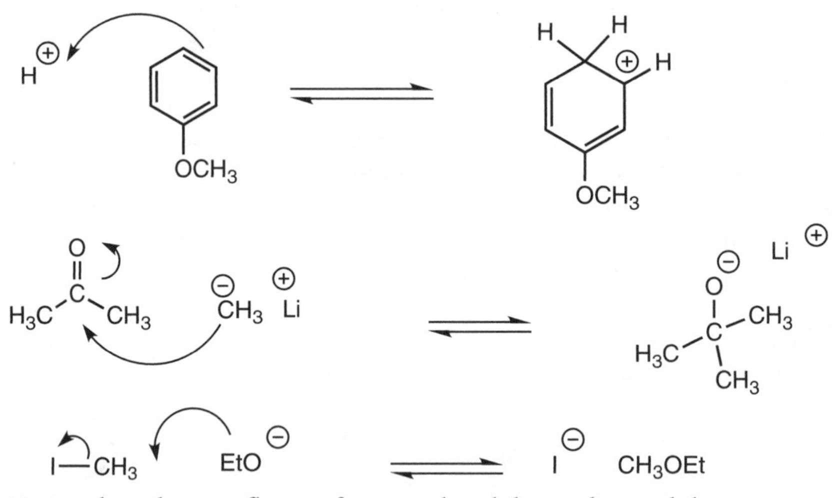

Notice that electron flow is from nucleophile to electrophile.

REDOX REACTIONS AND CARBON OXIDATION STATES

In introductory chemistry, one learns that many reactions can be classified as oxidation-reduction, or redox, reactions. For example, $2CrCl_2$ + Cl_2 gives $2CrCl_3$. In this reaction, the Cr compound is oxidized and loses electrons, and the Cl_2 is reduced and gains electrons. The oxidation state of Cr changes from +2 to +3, and that of Cl changes from 0 to −1. Just as for inorganic compounds, it is possible to assign oxidation states to various carbon functional groups. Several different schemes can be used to describe oxidation states in carbon compounds. Here, we will describe carbon as having the oxidation states 0 to +4, where the 0 state has no electronegative atoms bonded to a carbon and the +4 state has four electronegative atoms bonded to a carbon.

OXIDATION LEVELS OF CARBON FUNCTIONAL GROUPS

0	+1	+2	+3	+4
Saturated hydrocarbons:	CH_3OH	$R_2O{=}O$	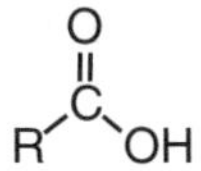 $RC(=O)OH$	CO_2
CH_4, 	CH_3Br CH_3SH	$RC{\equiv}CR$ H_2CCl_2	$HCCl_3$ 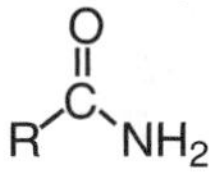 $RC(=O)NH_2$	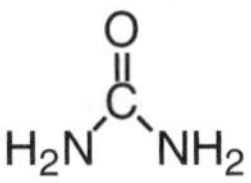 CCl_4 $H_2NC(=O)NH_2$
	$R_2C{=}CR_2$	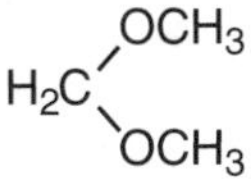 $H_2C(OCH_3)_2$	$R{-}CN$	
and so on				

← Reduction Oxidation →

For example, an alcohol has one electronegative atom bonded to it, so we arbitrarily give it an oxidation state of +1. The carbonyl carbon of a carboxylic acid has three bonds to electronegative atoms (one double bond and one single bond to oxygen), giving it a formal oxidation state of +3. Interchanging a functional group across the diagram left to right, say from an alcohol to a ketone, requires an oxidizing agent. Going from right to left, say from a nitrile to an amine, $R{-}CH_2NH_2$, requires a reducing reagent. Going up and down a column, however, does not require any redox chemistry. Some of the transformations up and down a column require the addition of water (a **hydrolysis reaction**) or the elimination of water (a **dehydration reaction**). Since we do not consider water an oxidizing agent or a reducing agent, such reactions are not redox reactions. In our arbitrary oxidation scheme, changing oxidation levels by one unit involves a gain or loss of two electrons. The reduction of a ketone to an alcohol requires the addition of two electrons, and the oxidation of a carboxylic acid to CO_2 involves the removal of two electrons.

Example The following are examples of oxidation, reduction, and hydrolysis reactions.

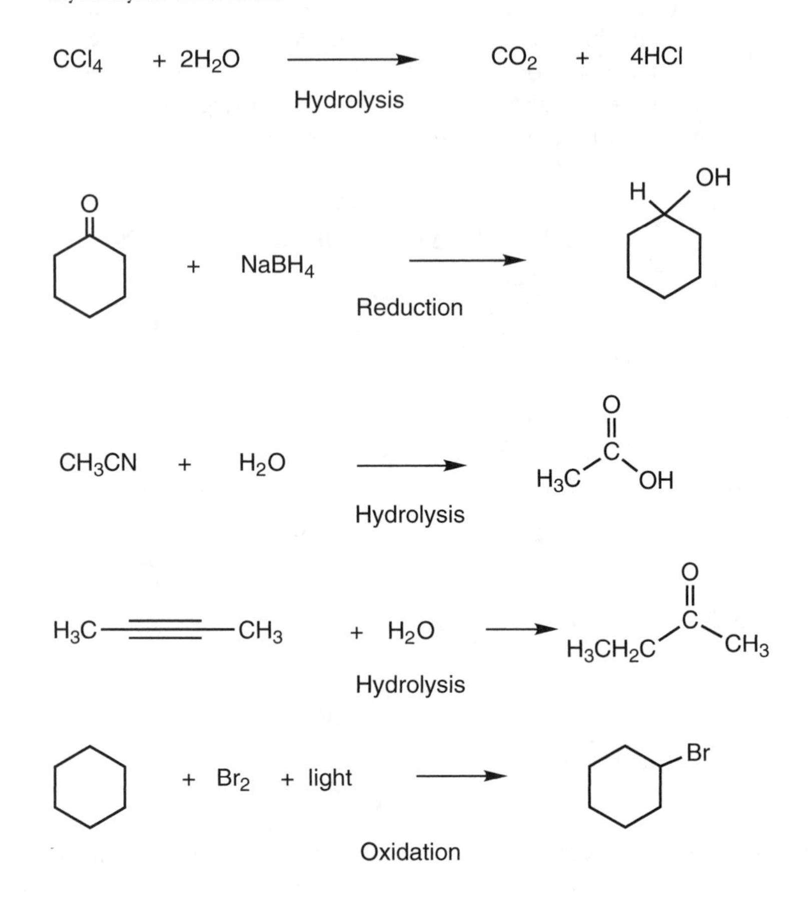

8 NUCLEOPHILIC SUBSTITUTION

NUCLEOPHILIC SUBSTITUTION REACTIONS

The first reaction that we will examine is the nucleophilic substitution reaction. An example is the reaction of sodium iodide, a source of I^-, with ethyl bromide in the solvent acetone to give ethyl iodide and sodium bromide:

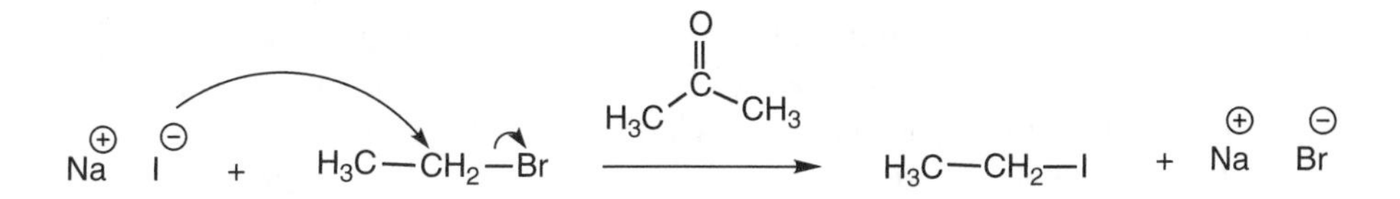

Here the iodide anion is the nucleophile, the carbon bonded to the halide is the electrophile, and the group being substituted or displaced, the **leaving group**, is Br^-. Normally, in writing organic reactions the solvent and any other reaction conditions (such as the temperature, a catalyst, and so on) are drawn above or below the reaction arrow. For every nucleophilic substitution reaction there are a nucleophile, an electrophile, and a leaving group.

To understand this reaction, we must deduce its mechanism. Clues to the mechanism will come from a study of the reaction's kinetics, its stereochemistry, and how changes in the nucleophile, electrophile, leaving group, and reaction conditions affect the reaction. Organic chemists can make literally millions of molecules capable of undergoing a nucleophilic substitution reaction. Our goal is to understand all the general factors that influence this reaction so that we can decide if a specific nucleophilic substitution reaction will proceed as written.

Kinetics of Nucleophilic Substitution

Ultimately, science is based on observables. A feature of a nucleophilic substitution reaction often easy to observe is the rate at which the starting materials are used up or the products are formed. For the ethyl bromide reaction shown above, the solvent acetone was chosen because NaI is soluble in acetone but NaBr is not. Thus, one could follow the rate of the reaction by collecting and weighing the amount of NaBr precipitate formed at specific times. Because the concentration of one of the products of this reaction is made to equal zero by this solvent trick, the equilibrium of the reaction lies completely to the right. By **Le Châtelier's principle**, this reaction is made *irreversible.*

Chemists now have sophisticated instruments to measure concentration changes of organic molecules in solution, so we do not have to weigh salts as they precipitate out of solution. When the rate at which the concentration of ethyl bromide decreases is measured for the nucleophilic substitution reaction of EtBr with NaI, the **rate law** found is

$$\frac{-d[\mathrm{H_3C{-}CH_2Br}]}{dt} = k[\mathrm{H_3C{-}CH_2Br}]\,[\mathrm{NaI}]$$

The rate at which the concentration of the electrophile decreases (i.e., the rate at which it is turned into products) is a function of the concentration of the electrophile and the concentration of the nucleophile. Doubling the concentration of EtBr forms product at twice the previous rate. Doubling the concentrations of both EtBr and NaI increases the rate by a factor of 4. The reaction is kinetically **second order**, which is the sum of the exponents of the concentration terms on the right hand side of the equation (1 + 1). The factor k is the **second-order rate constant** and has the units $M^{-1}s^{-1}$. It is a constant only for a specific set of reaction conditions. It increases with temperature and can increase or decrease with a change in solvent.

The reaction of EtBr with NaI is an example of an **S_N2 reaction**, which stands for substitution, nucleophilic, kinetically second order. Any time we use the term S_N2 to describe a reaction, its kinetics are second order. Some other examples of S_N2 reactions are shown below. The nucleophile appears first in each case.

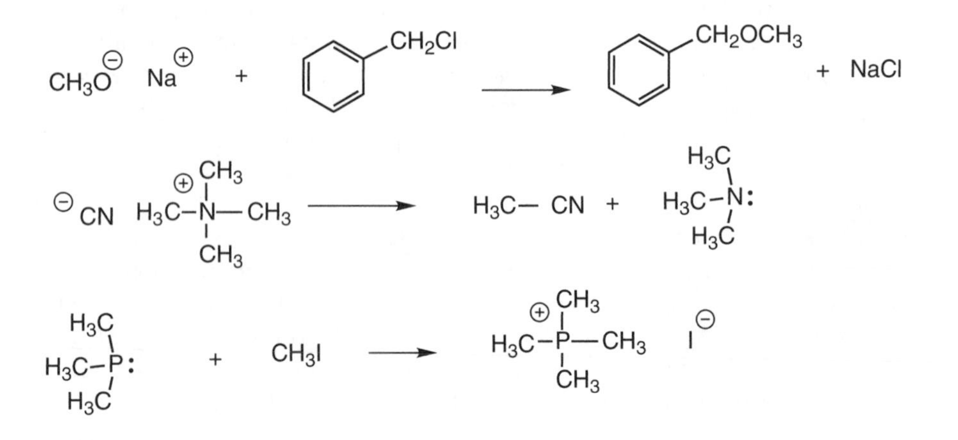

Notice that a neutral nucleophile and a neutral electrophile react to give a salt, as in the third reaction, and that the anion and cation of an organic salt can react to give two neutral molecules, as seen for the second reaction.

Not all nucleophilic substitution reactions are kinetically second order. For example, when the kinetics of the following acid-catalyzed reaction were studied, the rate law was found to be first order:

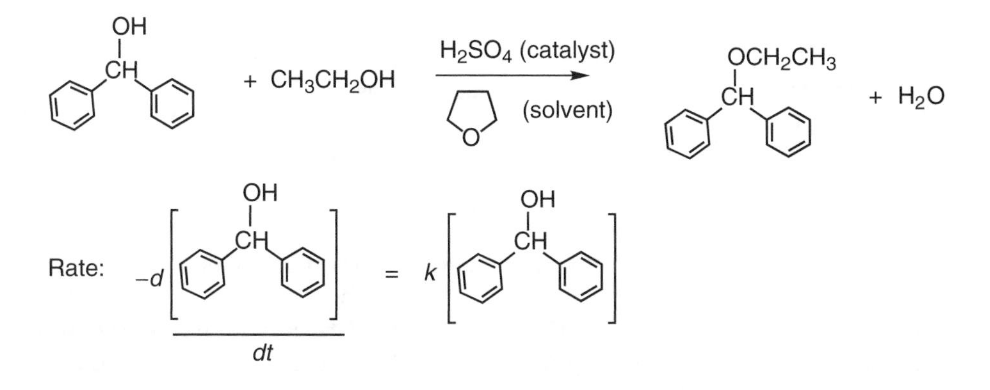

It so happens that many nucleophilic substitution reactions are kinetically **first order**; that is, the rate depends only on the concentration of the electrophile. The concentration of the nucleophile is not part of the rate law, and changing its concentration has no effect on the rate. For a first-order reaction, the rate constant has the units s^{-1}. The term **S_N1** is used to describe substitution, nucleophilic, kinetically first-order reactions.

The species present as concentration terms in a rate law are the species present in the **rate-determining step** of the reaction mechanism. The rate-determining step of a reaction is like the bottleneck of a reaction. An analogy is often made with the flow of water through a system of pipes. If pipes of different diameters are joined together, the rate at which water flows through the system depends on how fast it can flow through the smallest-diameter pipe. The smallest-diameter pipe determines the rate of flow just like the rate-determining step of a reaction determines the rate that molecules move through a multistep reaction.

Since S_N2 and S_N1 reactions have the nucleophile and electrophile, or only the electrophile, respectively, present at their rate-determining steps, they must have different mechanisms. In fact, the differences are so numerous that they are best considered separately, although they are both nucleophilic substitution reactions. The S_N2 reaction is somewhat simpler mechanistically and is treated first.

Stereochemistry of the S_N2 Reaction

Knowing the stereochemistry of a reaction limits the possible mechanisms one can put forward for a given reaction. For an S_N2 reaction, there is a very precise relationship between the stereochemistry of the starting material and the product. An S_N2 reaction proceeds via **inversion** at the electrophilic carbon:

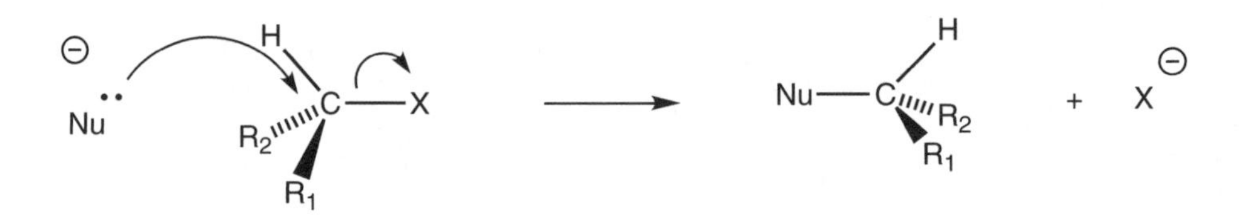

Inversion is a shape change for the electrophilic carbon undergoing an S_N2 reaction. It is exactly like what happens to an umbrella in a high wind. The nucleophile comes in on a trajectory 180° opposite the leaving group, staging a **backside attack**. The other groups move toward the opposite side of the carbon, while the bond to the leaving group breaks. Inversion was experimentally demonstrated by the chemist Paul Walden, so it is sometimes called **Walden inversion**. As the orbital diagram below indicates, in the transition state for an S_N2 reaction, the nucleophile, electrophilic carbon, and leaving group are *colinear*.

Orbital changes:

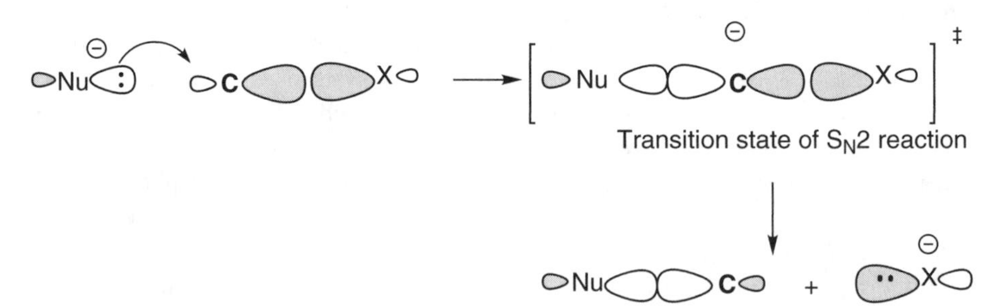

During the discussion of hybrid orbitals, the presence of the tail of the sp^3 orbital was mentioned. It is this part of the orbital that initially interacts with the incoming nucleophile. When the transition state structure for an S_N2 reaction is reached, the orbital has come to resemble an unhybridized *p* orbital. The geometry of the transition state is called **trigonal bipyramidal**:

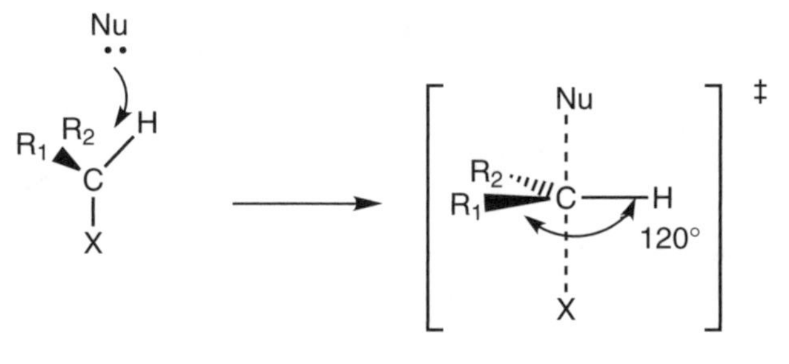

Trigonal bipyramidal transition state for an S_N2 reaction

The octet rule is not violated in the transition state because the bond to the nucleophile is only partly formed and the bond to the leaving group is partly broken.

For chiral electrophiles, one can assign the absolute configuration of the carbon undergoing reaction as R or S. Often the R isomer gives the S product, and vice versa. However, the R and S relationship is independent of the inversion geometry and depends only on the atomic weight of the nucleophile and the leaving group. In addition, carbons with no stereogenic centers, even molecules like CH_3I, still undergo the geometric shape change of inversion.

EXERCISES

Sketch the mechanism of the following S_N2 reactions using curved arrows and showing inversion at a carbon. For chiral molecules, assign R and S to all stereogenic centers.

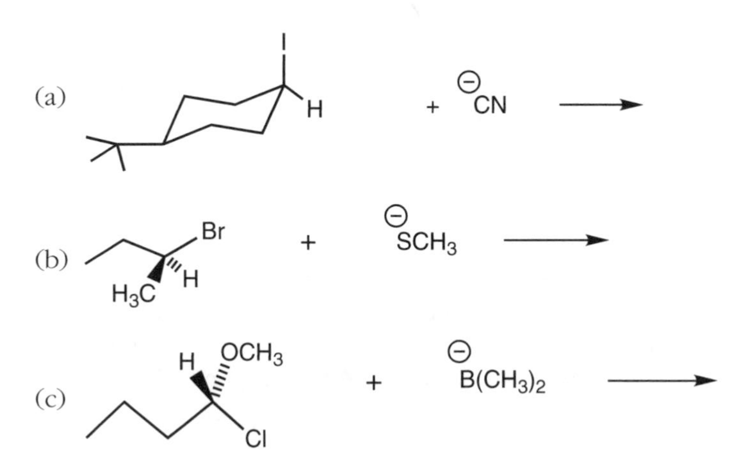

Solutions

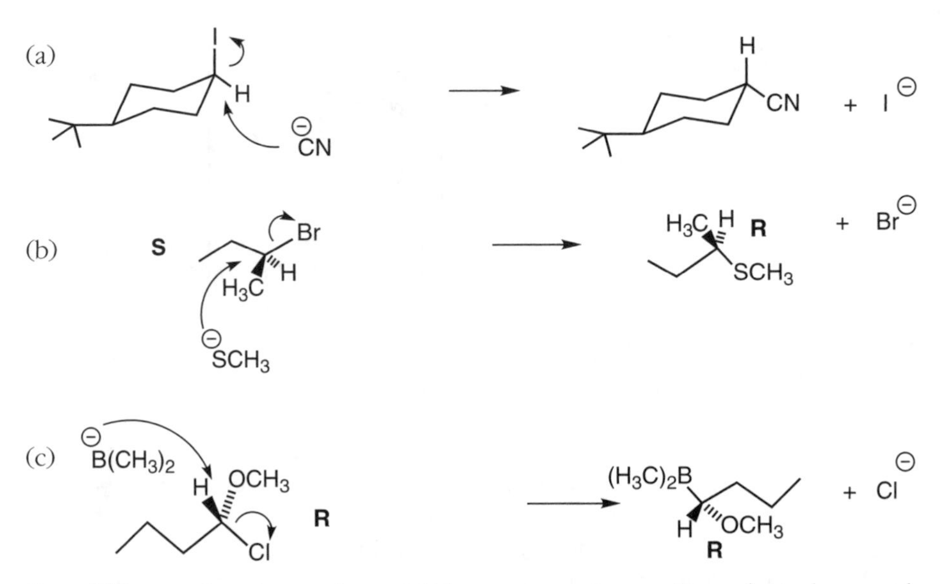

Note: This reaction shows that an R isomer can give an R product along with inversion. These particular reagents might not react this way in real life.

Example

Explain why a halide at a bridgehead position cannot undergo an S_N2 reaction.

Solution: The answer is in the geometry of the S_N2 reaction. A bridgehead halide, such as the bicyclo[2.2.2]octane molecule

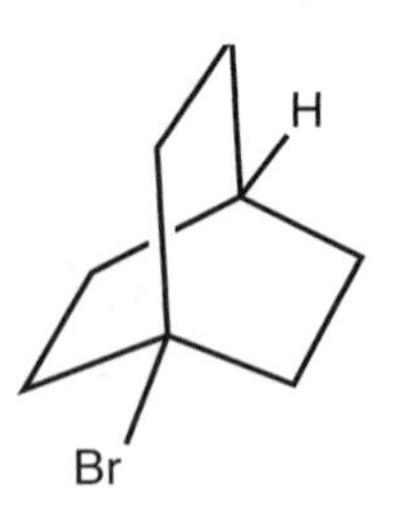

has a another bridgehead carbon directly behind it. No nucleophile is small enough to occupy the space behind the C—Br bond. Thus, the S_N2 transition state geometry cannot be attained. This is an example **steric effects** preventing an S_N2 reaction.

Reaction Coordinate Diagram for the S_N2 Reaction

During the course of a reaction, the energy of the system changes as bonds break and re-form. A **reaction coordinate diagram** plots the total free energy of the reaction vs the reaction coordinate, which measure the progress of the reaction. The reactants are placed on the left side of the diagram, and the products on the right. The reaction coordinate diagram for an S_N2 reaction is given here.

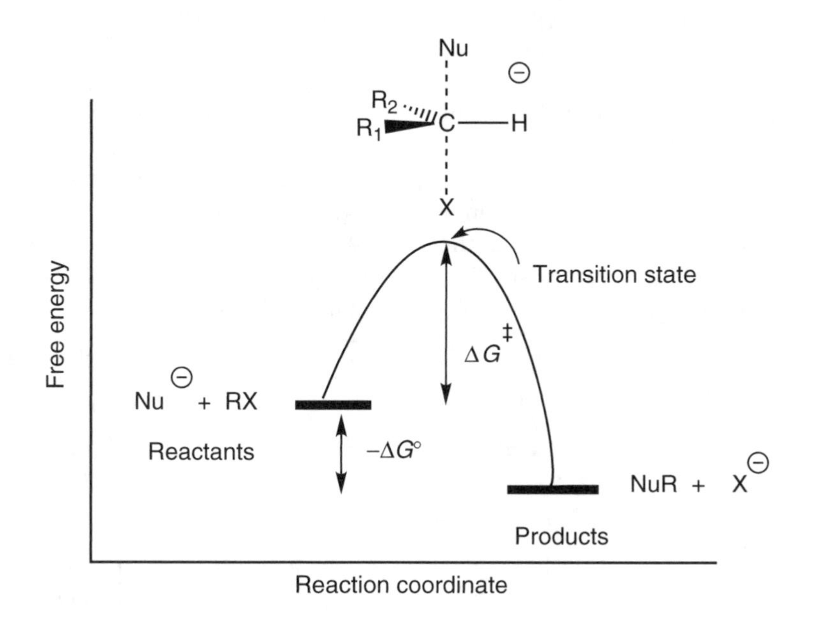

In an S_N2 reaction, there is a single **transition state** between the reactants and products. The transition state represents the maximum energy point between two energy minima. It exists for only one molecular vibration, and any change, either one that leads back to reactants or one that leads to products, is stabilizing. The energy difference, $\Delta G^{\ddagger}$, between the reactants and the transition state is the **energy of activation**. This energy determines the rate of the reaction. For this particular diagram, the products are lower in energy than the reactants by $-\Delta G^0$, so the reaction is **exergonic**. The equilibrium lies to the right. If the products were higher in energy than the reactants, the equilibrium would lie to the right, the free energy change would be positive, and the reaction would be **endergonic**.

A reaction coordinate diagram is a graphical depiction of the information in the rate law. The rate-determining step for an S_N2 reaction is formation of the transition state. In general, the step in which the transition state is highest in overall energy is the rate-determining step. Since the rate law is bimolecular, both the nucleophile and the electrophile must be present in the transition state of the rate-determining step.

A Classic Experiment Proving That Every S_N2 Reaction Proceeds with Inversion

There is compelling evidence that every S_N2 event proceeds with inversion at a carbon. A classic experiment that proved this fact used (*S*)-2-iodobutane as the electrophile and radioactive I^- from NaI as the nucleophile. This reaction is **degenerate**; that is, the products are the same as the starting materials (except for the radioactivity and the inversion of configuration). For this S_N2 reaction, one can measure two things: the rate at which radioactive iodine, I*, appears in the product and the rate of change in the observed optical activity of the (*S*)-2-iodobutane. The first rate is the rate of the S_N2 reaction since the nucleophile is radioactive. It was found that the measured optical activity of the (*S*)-2-iodobutane decreased to zero, and the rate of this decrease was *twice* the rate of the S_N2 reaction as measured by I* incorporation. These observations prove that every S_N2 event proceeds with inversion. While a mathematical derivation is possible, it is easiest to see what these rates mean with a specific example.

Suppose one did a very small scale reaction with only four (*S*)-2-iodobutane molecules. Imagine what happens if half of them react with radioactive iodide and the reaction proceeds with inversion:

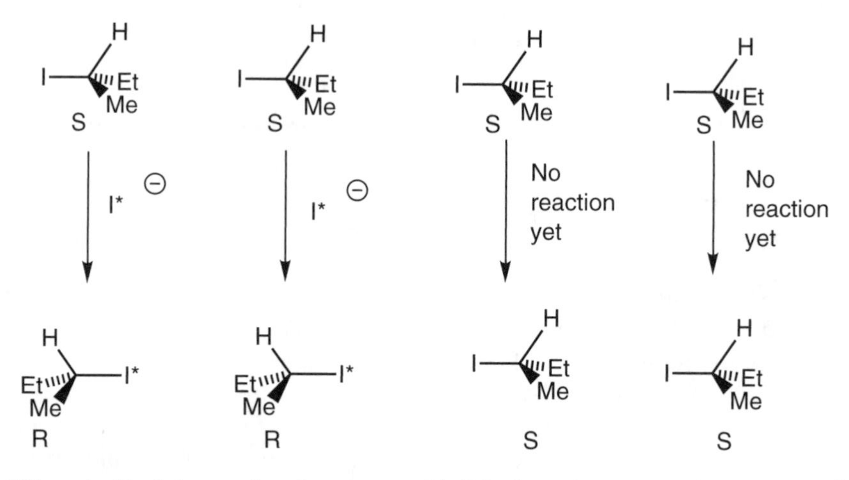

When half of the molecules react with I^{-*}, there is a racemic mixture of 2-iodobutane if and only if each S_N2 event proceeds with inversion. At this point, the observed optical rotation is 0° since one has a racemic mixture. The observed rotation continues to be 0° as the rest of the molecules react because there is an equal probability of R and S isomers reacting. Thus, for an S_N2 reaction with inversion, the rate of reaction is one half the rate at which the optical rotation goes to zero.

It is important for the student to realize that almost every S_N2 reaction with a sample of optically active molecules gives optically active products. Only in the very unusual case where the nucleophile and the leaving group are the same atom does an S_N2 reaction give a racemic mixture.

Measuring Nucleophilicity

There is a strong similarity between a nucleophile and a Lewis base and between an electrophile and a Lewis acid. A nucleophile donates an electron pair, and an electrophile accepts an electron pair. The major difference between the two concepts is that acidity is normally measured on an equilibrium basis. One assumes that acid-base reactions are fast and that the position of the equilibrium determines acid and base strengths. Conversely, nucleophiles are normally measured on a kinetic scale. The faster a given nucleophile reacts with a standard electrophile, the stronger that nucleophile is said to be.

One can rank nucleophiles by their rate of reaction with CH_3I as shown below. Species that are more nucleophilic react faster in an S_N2 reaction with this substrate.

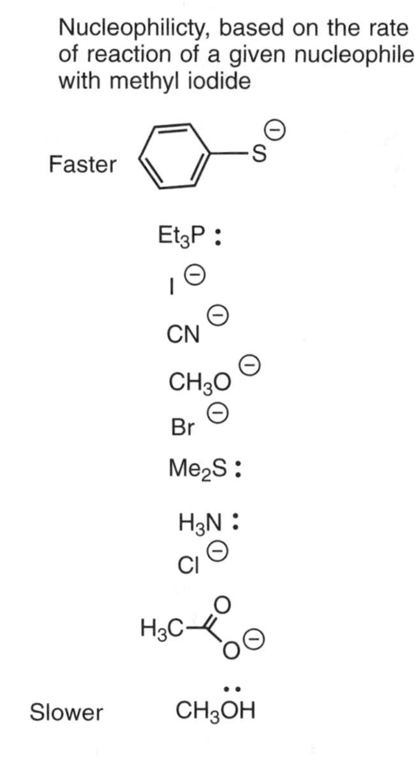

Several factors have been found to enhance nucleophilicity. First, charged nucleophiles are more nucleophilic than uncharged species with the same central atom. Thus, CH_3O^- is more nucleophilic than CH_3OH. Second, nucleophiles with heavier atoms are more nucleophilic than those with lighter atoms. This effect is seen with halogens, where $I^- > Br^- > Cl^-$, and is thought to be due to the **polarizability** of larger atoms. Their electron clouds can deform so that they can better interact with the partly positively charged region of the electrophile. Third, for the same central atom, strong bases are more nucleophilic than weak bases. Thus, the conjugate base of an alcohol, CH_3O^-, is more nucleophilic than the conjugate base of a carboxylic acid, CH_3COO^-. Finally, steric effects can modify nucleophilicity. CH_3O^- is more nucleophilic than $(CH_3)_3CO^-$ since the oxygen of methoxide is sterically accessible. The oxygen atom of *t*-butoxide is buried in a crowd of methyl groups and is sterically blocked from getting close to an electrophilic carbon.

EXERCISES

For each group of nucleophiles, circle the strongest nucleophile and underline the weakest nucleophile. Give a brief explanation for your choices.

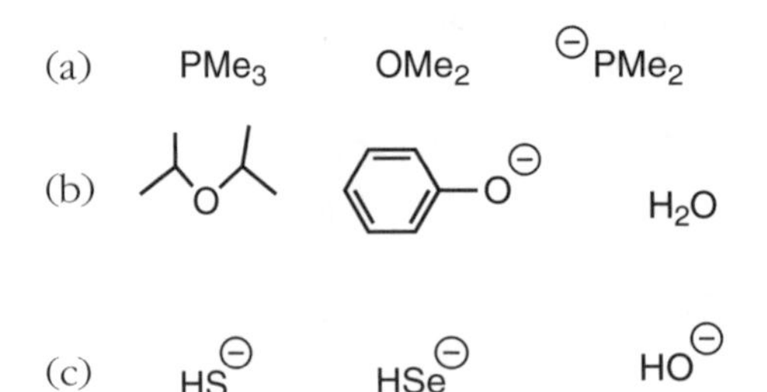

Solutions

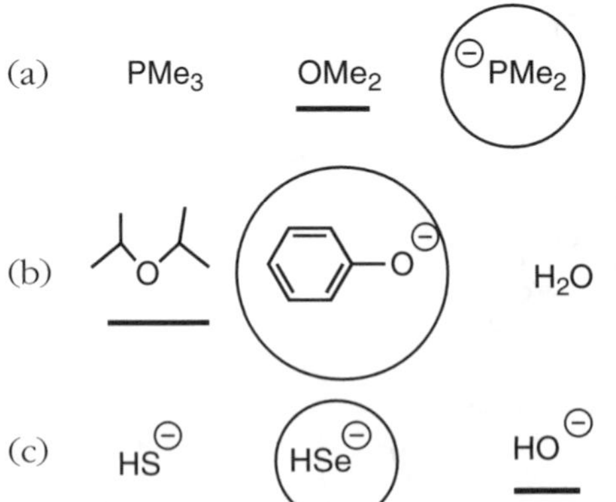

(a) P is heavier than O, and the charge favors $\dot{P}Me_2$.

(b) The central atoms are all the same, so the charge leads to a stronger nucleophile. Steric effects lower the nucleophilicity for ether.

(c) Atom size determines nucleophilicity.

Leaving Groups in Nucleophilic Substitution Reactions

Nucleophiles are more important in S_N2 reactions than in S_N1 reactions because the nucleophile shows up only in the rate law of an S_N2 reaction. However, leaving groups are important in both types of nucleophilic substitution reactions. Unless a group leaves, no reaction takes place. Although several factors influence nucleophilicity, it is easy to recognize a good leaving group. A good leaving group is a weak base.

Table 4.1 lists the pK_a values for a number of acids. In general, the conjugate base of any acid with a pK_a of 3 or less can act as a leaving group under mild reaction conditions. Other factors, such as bond strengths, make minor perturbations in the order of leaving-group ability. However, one expects that, in comparing two bases, the weaker base will be a better leaving group. The relative leaving-group abilities in nucleophilic substitution reactions are as follows.

$$\text{Sulfonates, } {}^{\ominus}O-\overset{O}{\overset{\|}{\underset{\underset{O}{\|}}{S}}}-R \; > \; H_2O \; > \; I^{\ominus} \; > \; Br^{\ominus} \; > \; Cl^{\ominus}$$

Examples of sulfonates are

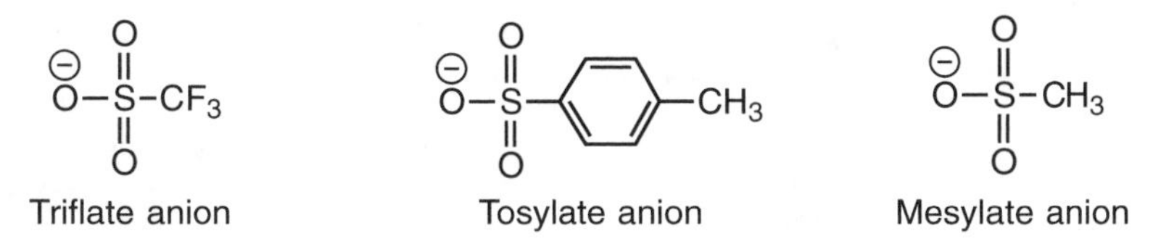

Sulfonates are organic derivatives of sulfuric acid, H_2SO_4, in which one —OH bond is replaced by a carbon group. They are easily prepared from alcohols, as will be discussed later. Water is the leaving group when the electrophile is of the form ROH_2^+, that is, a protonated alcohol group. Notice that strong bases are not leaving groups in nucleophilic substitution reactions. In particular, CH_3^-, NH_2^-, and OH^- (and their derivatives) are not leaving groups because they are strong bases.

Examples In the following S_N2 reactions of sulfonates and protonated alcohols, pay close attention to which bond is broken in the electrophile.

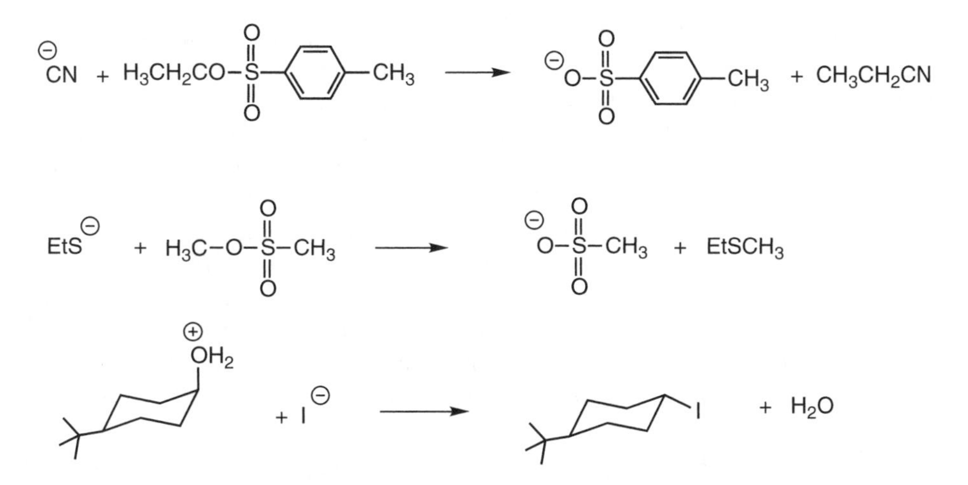

Formation of a Carbocation Intermediate in the S_N1 Reaction

There are no intermediates in an S_N2 reaction, only continuous bond making and bond breaking in going from starting materials to products. For an S_N1 reaction, the rate-determining step is the formation of a carbocation. This reactive intermediate then reacts quickly with the nucleophile to form the substitution product. The accompanying reaction coordinate diagram summarizes this process.

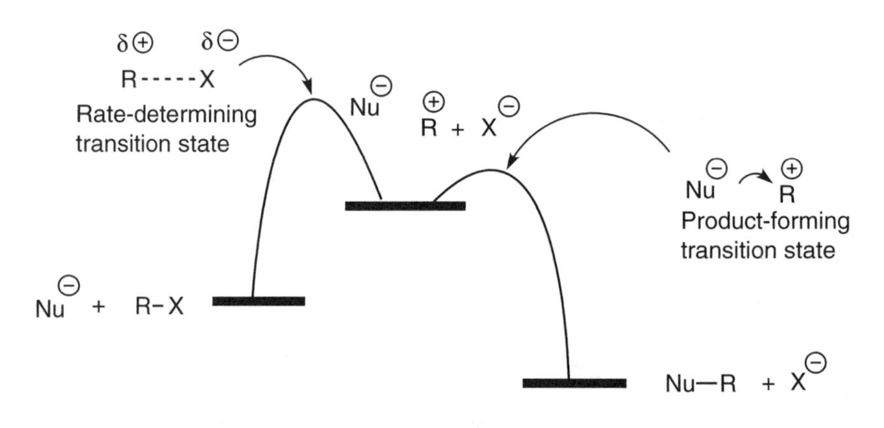

An S_N1 reaction differs from an S_N2 reaction in that the former is a two-step reaction. The first step is the formation of a carbocation, and the second step is the reaction of the electrophilic cation and the nucleophile. The formation of the carbocation is endergonic, while the second step is exergonic. One should be clear about the difference between a transition state and a reactive intermediate such as a carbocation. A reactive intermediate represents a **local minimum** in a reaction coordinate diagram. Although it is relatively high in energy, there is still a barrier for it to react further. A transition state exists at the highest point of a reaction step, and there is no energy barrier for it to change.

Because only the electrophile is involved in the rate-determining step of an S_N1 reaction, the rate law is first order. The nucleophile is involved only in the fast product-forming step, which is kinetically irrelevant. It is easy to understand why the first step in an S_N1 reaction has a relatively high-energy transition state, since a bond is broken. However, there is also a lower-energy transition state for the anion and cation to come together to form the product. The source of this energy barrier is the **reorganization energy** for the reaction. The geometry of the ions is different from their geometry in the product. In addition, the solvent molecules interact differently with charged and neutral species, and it takes energy for them to move and accommodate the bonding changes.

Ionization, the formation of ions from molecules, is common in inorganic chemistry and takes place in organic chemistry as well. The main distinction between molecules that react by an S_N2 and by an S_N1 mechanism is that S_N1 substrates can form stable carbocations. Understanding which molecules can form carbocations and what features of molecular structure stabilize carbocations are key to understanding organic chemistry.

While students can accept that ionic solids like NaCl dissociate into ions, some are bothered by the ionization of covalent organic molecules. We will see that most S_N1 substrates are sterically hindered, so ionization relieves strain. Also, the solvent is important in an S_N1 reaction, so that only solvents that stabilize ions promote S_N1 reactions. Finally, entropy favors ionization since two particles are being created from one.

Stabilization of Carbocations by Electron Donation and Resonance

The more alkyl groups bonded to a carbocation, the more stable the carbocation. There are specific class names for alkyl substituted functional groups and carbocations. The structures of methyl, primary, secondary, and tertiary compounds and carbocations are shown in the accompanying table.

Methyl	Primary, 1°	Secondary, 2°	Tertiary, 3°
CH_3X	RCH_2X	R_2CHX	R_3CX
$\overset{\oplus}{CH_3}$	$\overset{\oplus}{RCH_2}$	$\overset{\oplus}{R_2CH}$	$\overset{\oplus}{R_3C}$

For example, a specific example of a tertiary (3°) chloride is $(CH_3)_3CCl$, *t*-butylchloride. When this tertiary chloride ionizes, it gives Cl^- and the tertiary *t*-butyl carbocation $(CH_3)_3C^+$. A secondary alcohol would be $(CH_3)_2CHOH$. If the OH group is protonated and leaves as H_2O, it leaves behind a 2° carbocation, $(CH_3)_2CH^+$.

Alkyl groups stabilize the electron-deficient carbon of a carbocation through a special kind of resonance interaction called **hyperconjugation**. In hyperconjugation, there is an overlap of the vacant *p* orbital of the carbocation carbon with the electrons of adjacent σ bonds:

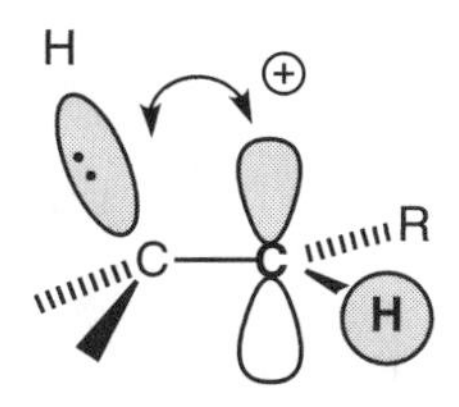

Hyperconjugation between the electrons in a σ bond and a vacant *p* orbital of a carbocation

Valence bond depiction of hyperconjugation

Hyperconjugation is sometimes called "no bond resonance" as suggested by the valence bond depiction of hyperconjugation. The σ bond on the alkyl group can be either a C—H or a C—C bond. It must be aligned approximately parallel with the axis of the vacant *p* orbital for overlap to take place. A C—H bond attached to the carbocation center cannot provide hyperconjugative stabilization because the hydrogen 1*s* orbital used in bonding is in the nodal plane of the vacant *p* orbital. Thus, alkyl groups, but not hydrogens, stabilize positively charged carbon atoms. *The stability of carbocations is 3° > 2° >> 1° > Me.* This order should be memorized. Methyl or primary carbocations are too unstable to be formed in ordinary organic reactions.

Because of their ability to stabilize electron-deficient carbon, alkyl groups are said to be electron-donating. Another way to stabilize carbocation centers is through resonance. **Allylic** and **benzylic** are types of substrates that show resonance stabilization:

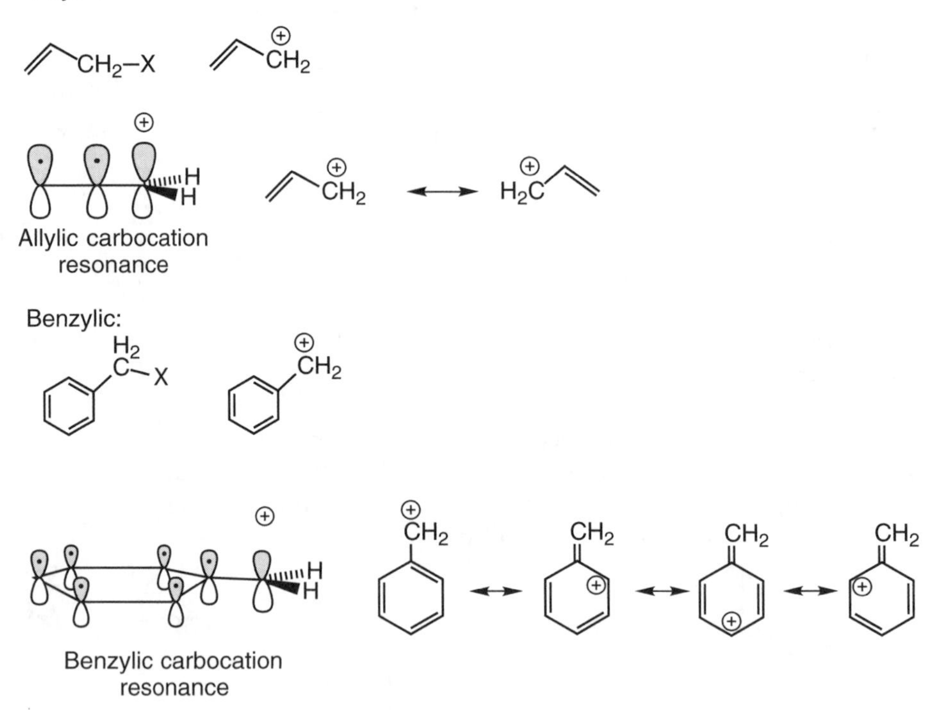

Allylic and benzylic carbocations are somewhat more stable than 3° carbocations. When alkyl groups are added to the allylic or benzylic center, for example, $CH_2{=}CH_2C(Me)_2^+$, the effects of hyperconjugation and resonance work together to further stabilize the carbocation.

EXERCISES

In each group of compounds which compound is most likely to undergo an S_N1 reaction?

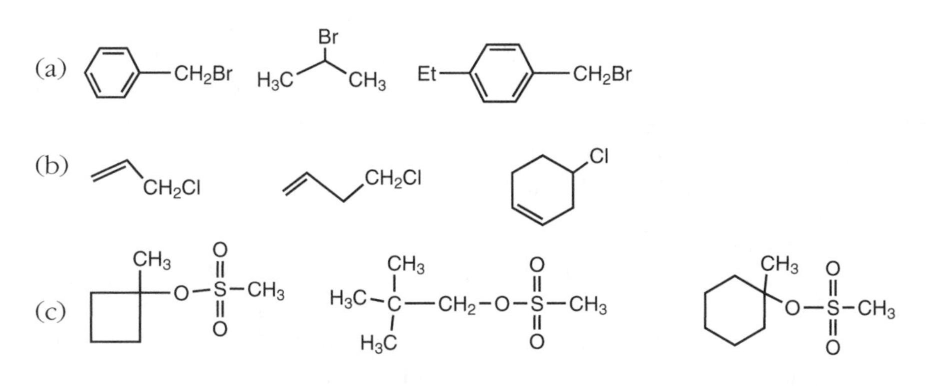

Solutions

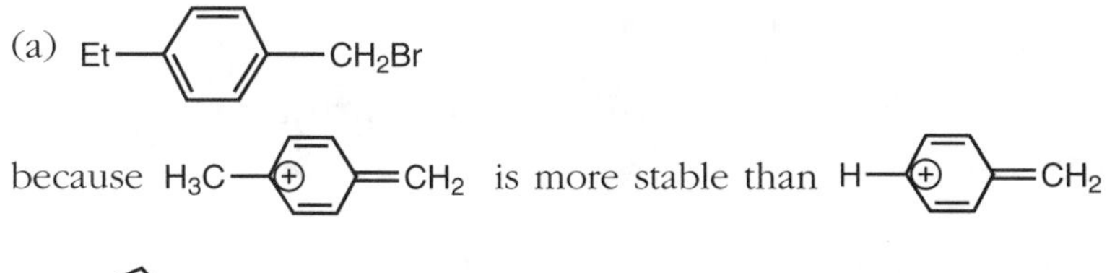

(b) CH_2Cl because there is only one allylic substrate

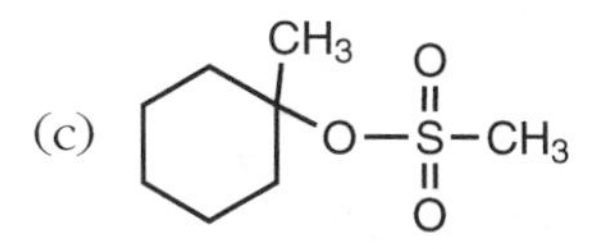

A carbocation has, ideally, 120° angles around the positively charged carbon. Cyclobutane has 90° ring angles, while cyclohexane has tetrahedral, 109.5° angles. There is more angle strain in a cyclobutyl carbocation.

Information Provided by the Hammond Postulate on the Transition State for an S_N1 Reaction and on Reactions in General

Since a transition state has the shortest possible lifetime on the molecular scale, it is extremely difficult to obtain experimental information on its structure and properties. One approach to understanding the transition state was put forward by the American chemist George Hammond, who proposed that a transition state resembles the species to which it is closest in energy. This suggestion has become known as the **Hammond postulate**. For an S_N1

reaction, the rate-determining transition state is closest in energy to the carbocation intermediate.

Anything that lowers the energy of a transition state causes a reaction to go faster. Since we know that 3° carbocations are more stable (lower in energy) than 2° carbocations, the Hammond postulate says that 3° halides react faster in S_N1 reactions than 2° halides. As another example, carbocations have trigonal geometry; the angles around carbon are 120°, while the angles around a tetrahedral carbon are 109.5°. Anything that tends to increase the bond angles around a tetrahedral carbon makes it a better substrate for an S_N1 reaction. **Steric crowding** has this effect, so that the Hammond postulate correctly predicts that a molecule like *t*-butyl$(Me)_2CBr$ will form a carbocation faster than $(Me)_3CBr$.

Stereochemistry of the S_N1 Reaction

The S_N1 reaction proceeds through a carbocation intermediate. Since carbocations are planar, they cannot be chiral. Thus, starting with a chiral electrophile in an S_N1 reaction leads to a racemic product. Often, however, the product is not completely racemic. The leaving group tends to be associated with one side of the carbocation, partly blocking the approach of the nucleophile from that side. However, the product ratio is so close to a racemic mixture that S_N1 reactions effectively give racemic products.

The Structure of the Electrophile is Predominantly Responsible for a Molecule Reacting by the S_N1 or S_N2 Mechanism

There are limiting structures for S_N1 and S_N2 reactions as indicated in the accompanying table.

SUBSTRATES FOR NUCLEOPHILIC SUBSTITUTION

	S_N2		S_N1
Methyl	CH_3I		
1°,	CH_3CH_2Br	3°,	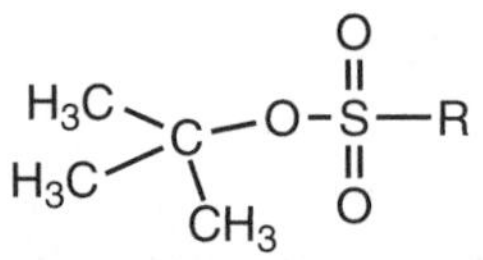 $(H_3C)_2C(CH_3){-}O{-}S(=O)_2{-}R$
2°,	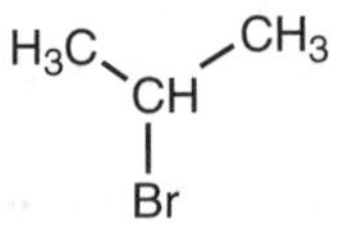 $(H_3C)(CH_3)CH{-}Br$	Allylic,	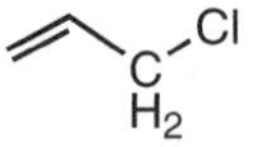 $CH_2{=}CH{-}CH_2{-}Cl$
		Benzylic,	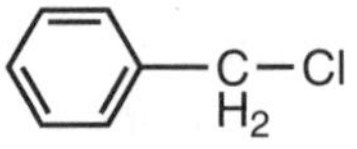 $C_6H_5{-}CH_2{-}Cl$

Methyl and primary substrates *always* undergo nucleophilic substitution through an S_N2 mechanism. Tertiary substrates *always* undergo nucleophilic substitution through an S_N1 mechanism. Normally, 2° substrates react via an S_N2 mechanism, but reaction conditions can be found where they react as S_N1 substrates. Also, allylic and benzylic substrates can undergo S_N1 or S_N2

reactions, depending on the reaction conditions, but it is more common for them to react through an S_N1 mechanism.

Some substrates cannot react by an S_N1 or S_N2 mechanism. Treating them with a nucleophile often results in the recovery of starting material. These substrates include leaving groups bonded to sp^2 carbons, such as **aryl halides** (benzene and substituted benzene rings), **vinyl halides** (halides bonded to alkenes), bridgehead halides, and **neopentyl substrates**:

Aryl substrates Vinyl substrates Bridgehead substrates Neopentyl substrates

Steric factors hinder the S_N2 reaction for all these substrates. In neopentyl substrates (which are 1°), the methyl hydrogens of the *t*-butyl group block a backside attack. Carbocations do not form at sp^2 carbons because the geometry does not permit hyperconjugative stabilization and because sp^2 carbons have a good deal of *s*-orbital character (which is electron-withdrawing). Although the bridgehead position is tertiary, the carbon framework prevents a carbocation from adopting a favored planar geometry. In addition, the geometry does not allow for hyperconjugation. Thus, the S_N1 reaction is disfavored.

EXERCISES

1. Give the expected product, including its stereochemistry, and indicate if the reaction is S_N1 or S_N2. If no reaction takes place, write NR.

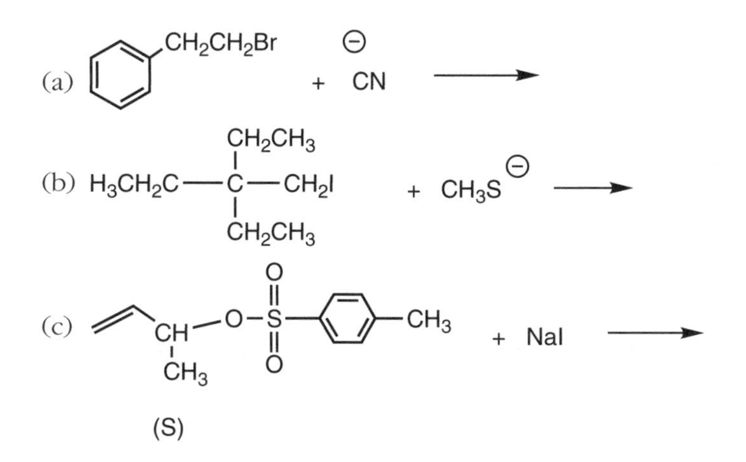

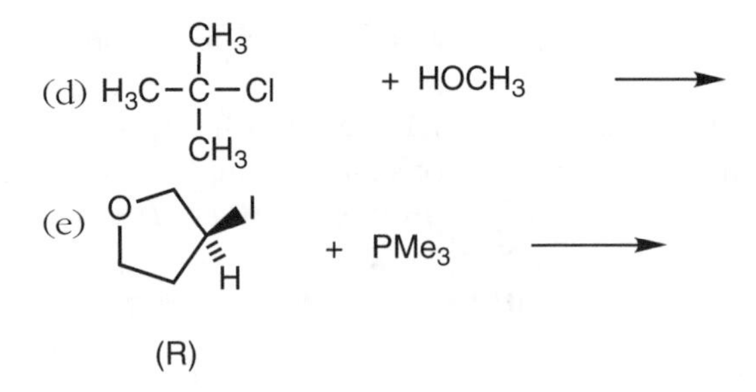

2. Explain how the following reaction takes place.

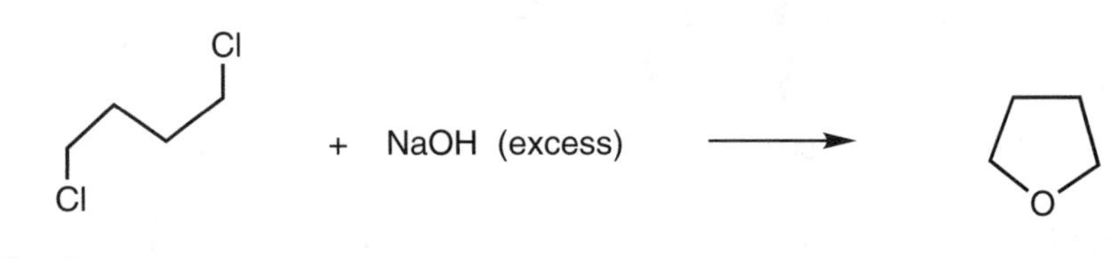

Solutions

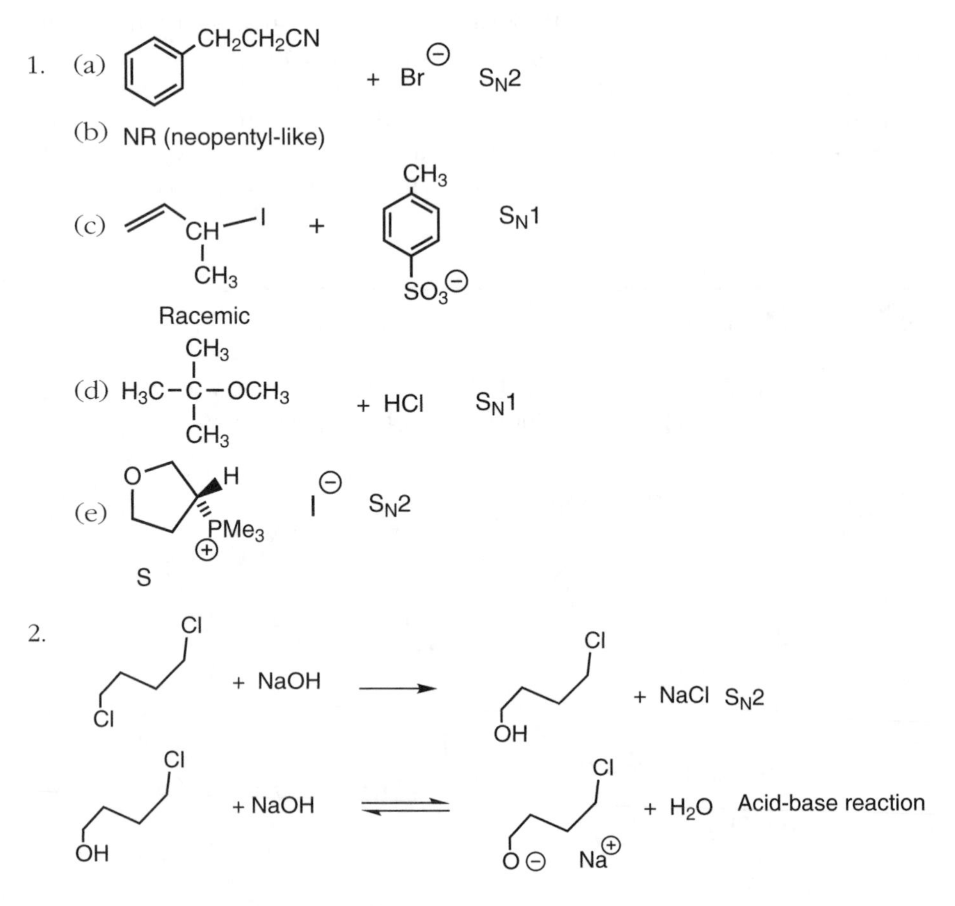

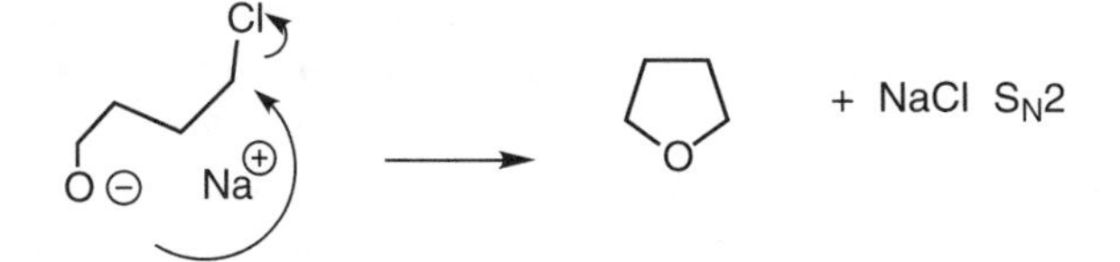

(*Note*: Whenever the nucleophile and the electrophile are in the same molecule, reaction produces a ring.)

Strong Influence of Solvent Conditions on Nucleophilic Substitution Reactions

The solvent for a nucleophilic substitution reaction can strongly influence the outcome of the reaction. This is particularly true of protic solvents and polar aprotic solvents. A **protic solvent** is a hydrogen-bonding solvent and has an OH (or less often an NH) group. Alcohols, water, and carboxylic acids are the most common examples of protic solvents, and ammonia, NH_3, is occasionally used. **Polar aprotic solvents** have significant dipole moments but no hydrogen bond-donating groups:

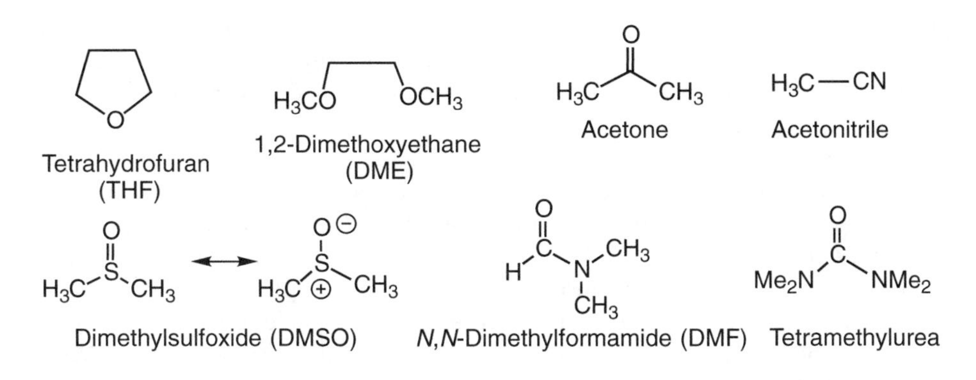

Each type of solvent has its own effects on S_N1 and S_N2 reactions.

Protic solvents retard an S_N2 reaction with an anionic nucleophile. In a protic solvent, an anion forms hydrogen bonds to the protic solvent. For example, for the nucleophile HO^- in water, the negative charge interacts with the many HO bonds surrounding it. This partly neutralizes the charge and creates a solvent shell around the nucleophile that stabilizes it and retards any S_N2 reaction. The only time one uses a protic solvent in an S_N2 reaction is when the nucleophile and electrophile are uncharged. The product of this kind of reaction is a salt, and being more polar than the starting materials, a polar solvent enhances this reaction.

In contrast, protic solvents favor S_N1 reactions because the large dipole moment of the solvent helps to stabilize the transition state for the reaction. Very often, in an S_N1 reaction the solvent acts as a nucleophile with the carbocation formed during ionization. This kind of substitution reaction, where the solvent is the nucleophile, is called a **solvolysis reaction**. In a solvolysis reaction, the first intermediate formed is usually a protonated form of the final product, and in the final step a proton is transferred to another weak base in the medium. The solovolysis of benzyl bromide in acetic acid is shown here.

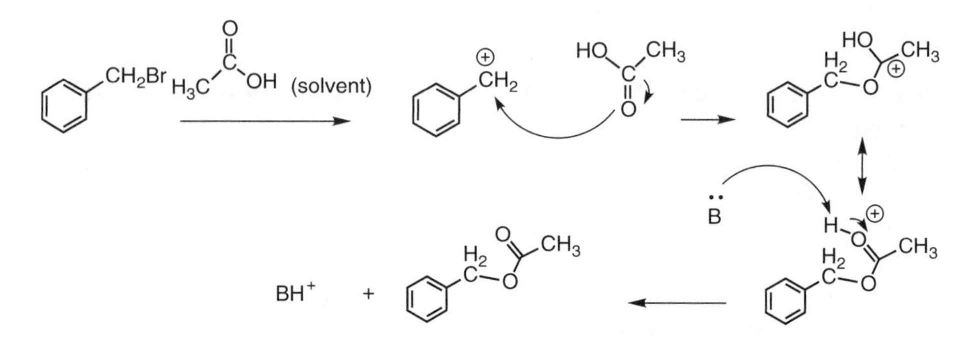

In this case, the base, B, can be a solvent molecule or Br^-, and the carbonyl oxygen of the acid is used as the nucleophile rather than the oxygen of the OH group because the resulting carbocation is stabilized by resonance.

Polar aprotic solvents are excellent solvents for S_N2 reactions. They are sufficiently polar to dissolve many salts, especially lithium salts. For most of these solvents (which are sometimes called dipolar aprotic solvents because of their large dipole moments), their structure puts the negative end of a dipole on the exterior of the solvent molecule. The positive end of the dipole is buried in the interior of the solvent molecule. When a nucleophile is an anion, it is destabilized. Electron-electron repulsion between the nucleophile and the negative end of the solvent dipole speeds up the rate of an S_N2 reaction. DMSO is particularly effective in speeding up this kind of an S_N2 reaction.

Polar aprotic solvents are not often used in S_N1 reactions. Some of them can react with the carbocation intermediate of an S_N1 reaction. Sometimes solvent mixtures of protic and aprotic solvents are encountered, for example, THF/water or acetone/water. Normally, these mixtures are used because most organic molecules do not dissolve in pure water or a salt does not dissolve in an organic solvent and a **cosolvent** is added to produce a homogeneous mixture. When an aqueous solvent mixture is used, it usually implies a solvolysis reaction with water as the nucleophile.

In addition to the solvent, other reaction conditions can affect nucleophilic substitution. Since an S_N2 reaction is second order, increasing the nucleophile concentration increases its reaction rate. Because of entropic factors, increased temperatures favor a dissociative S_N1 reaction and disfavor an associative S_N2 reaction.

EXERCISES

For the following reactions, indicate the expected product and list the factors that favor that reaction by an S_N1 or an S_N2 mechanism. If no reaction takes place, explain why.

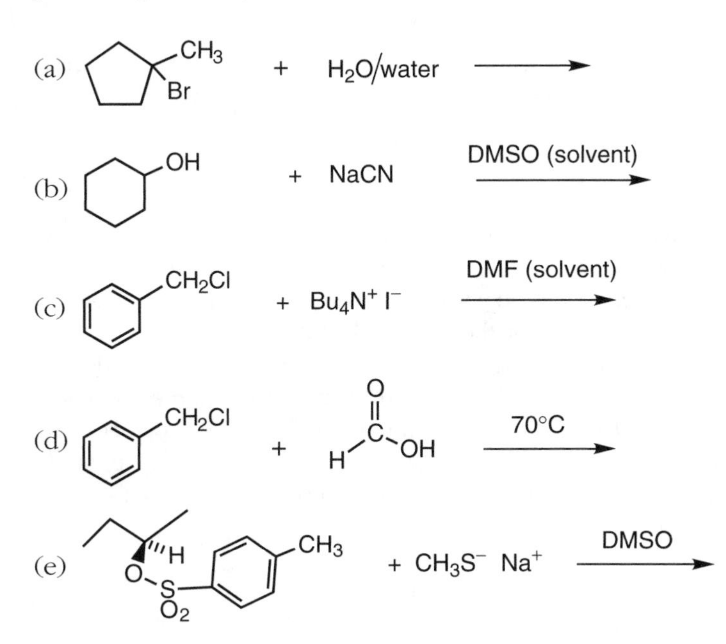

Solutions

(a) S_N1: substrate 3°, protic solvent.

(b) No reaction. HO^- is not a leaving group.

(c) S_N2: substrate can react either way, excellent nucleophile and polar aprotic solvent.

(d) 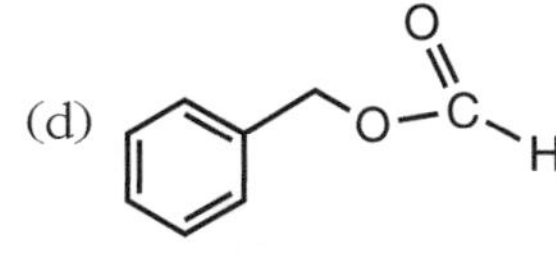

S_N1: substrate can react either way, protic solvent, elevated temperature.

(e) S_N2: substrate can react either way, excellent nucleophile, polar aprotic solvent.

Different Reactions of Carbocation Intermediates in S_N1 Reactions

The carbocation intermediate in an S_N1 reaction can undergo several different reactions in addition to substitution. The most important is **elimination**, which is loss of H^+ to form a double bond:

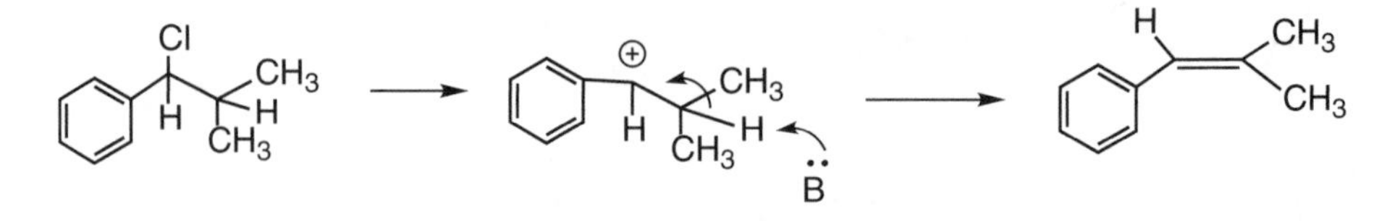

A carbocation is an extremely strong acid (since the conjugate base, an alkene, is a very weak base), and either the solvent or the leaving group can act as a base to generate an alkene. This reaction is covered in depth in Chapter 9.

Carbocations also can rearrange if the new carbocation is more stable than the original structure. Rearrangements can involve C—H or C—C bonds and result in new structural frameworks. Normally, introductory organic chemistry students are not expected to predict a rearrangement but rather to recognize that one has taken place. Some examples of carbocation rearrangements are shown here. Rearrangements always produce carbocations more stable than the starting carbocation:

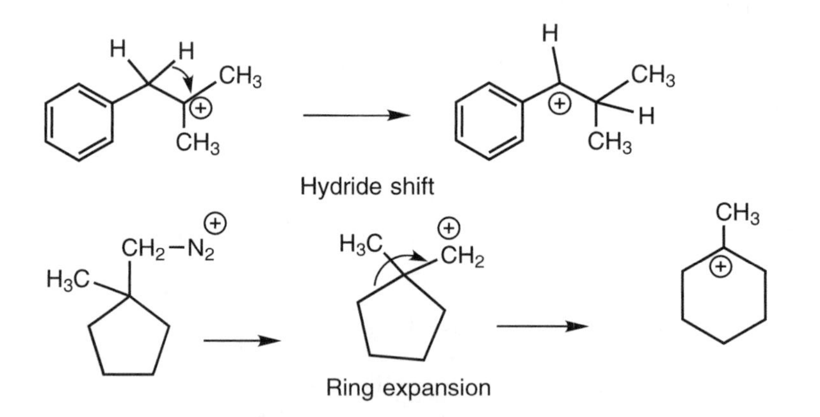

Hydride shift

Ring expansion

Note: N_2 is one of the rare leaving groups that can produce a 1° carbocation

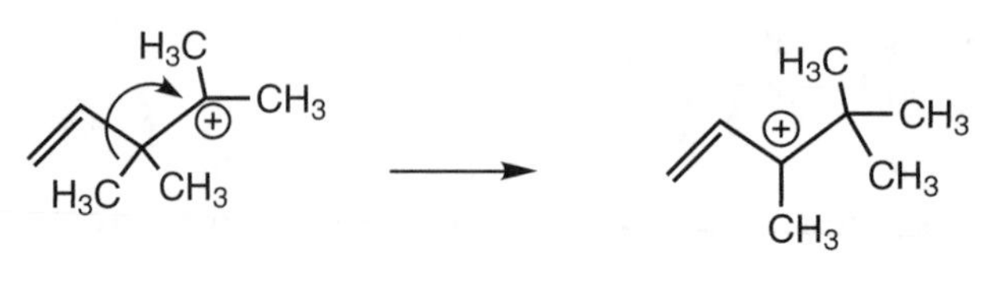

Alkyl group migration

H^- is a hydride group, so when H moves with its associated electrons, it is called a **hydride shift**. Recall that molecules are always in motion, undergoing vibration and bending. If a motion moves an atom or group, and its associated bonding electrons, close to the empty *p* orbital of a carbocation, interactions can begin between the empty *p* orbital and the filled σ orbital. A very small shift in nuclear position can result in a rearranged structure if it results in a lower-energy structure.

Because carbocations are reactive electrophiles, they can react in an **intermolecular** reaction, with nucleophiles in solution, or in an **intramolecular** reaction, with nucleophiles present in the same molecule. For intramolecular reactions, an important nucleophile is a double bond. Many plant-derived compounds are biosynthesized by these **cation-olefin cyclizations**, such as the reaction shown here.

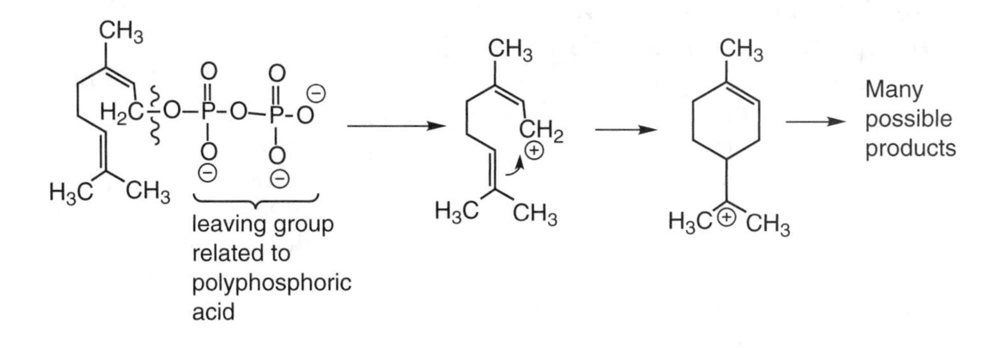

Alcohols Do Not Undergo Nucleophilic Substitution Directly But Can Be Converted To Good Leaving Groups

The —OH group is not a leaving group because HO^- is a strong base. However, several reactions can turn an —OH group into a good leaving group. Protonation of an alcohol —OH, conversion of an alcohol to a sulfonate, or conversion of an —OH group to a halide are three common reactions for activating alcohols to substitution. The addition of a strong acid to an alcohol, such as HCl, HBr, or HI, forms $R—OH_2^+$ that can react with the halide anion to form RX. Normally such a reaction is carried out with anhydrous acid so that water is not a competing nucleophile.

Alcohols react with alkylsulfonyl chlorides to give sulfonates. For example, tosylates are formed from the reaction of alcohols with *para*-toluenesulfonyl chloride (tosyl chloride):

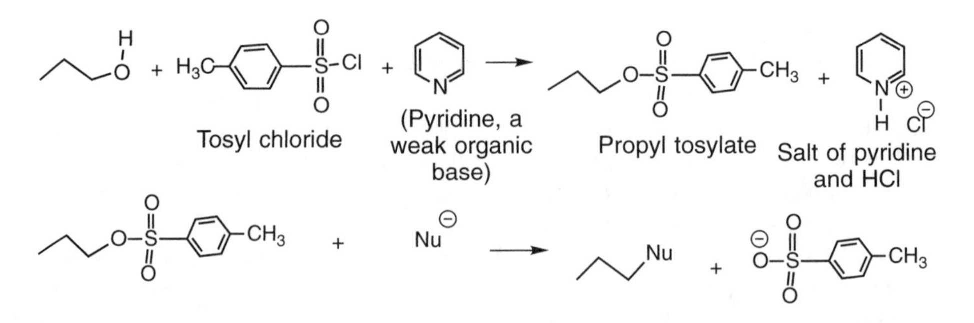

Notice that the O—H bond is broken in making a tosylate and that the C—O bond of a tosylate is broken in a substitution reaction. Depending on the substrate, tosylates can undergo S_N1 or S_N2 reactions. Reactions that interchange —OH and halide functional groups are discussed in Chapter 16.

The Conversion of Alcohols to Many Other Functional Groups by the Mitsunobu Reaction

While few introductory organic chemistry textbooks discuss the **Mitsunobu reaction**, it is a dominant method for performing S_N2 reactions on alcohols. This name reaction, invented in the 1980s, requires four components: an alcohol able to undergo an S_N2 reaction, a nucleophile, triphenyl phosphine (PPh_3), and diethyazidodicarboxylate (DEAD), $EtO_2CN{=}NCO_2Et$. The nucleophile must be an acid with a pK_a value less than 14, that is, a weak acid. In a process that need not be detailed here, the four components are mixed and react to generate an alkylphosphonium salt from the alcohol, $RO{-}PPh_3^+$, and the conjugate base of the nucleophile:

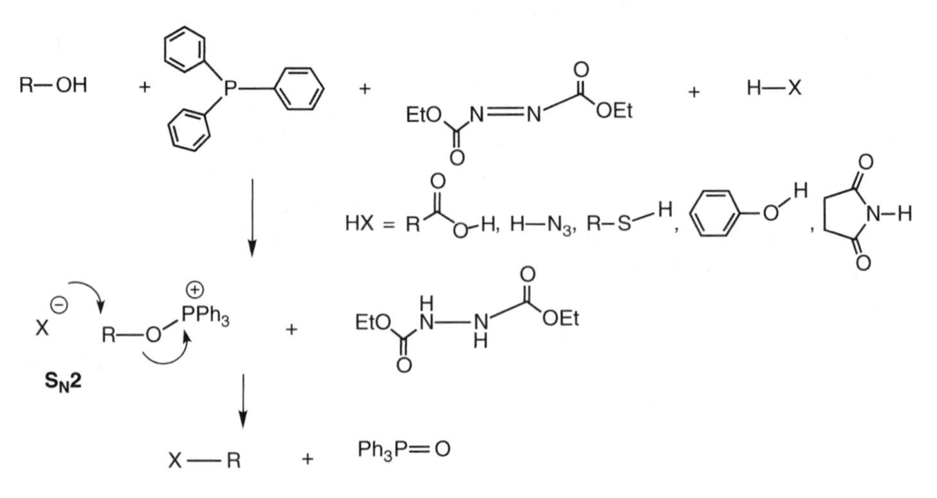

The leaving group in the Mitsunobu reaction is triphenylphosphine oxide, a very stable molecule.

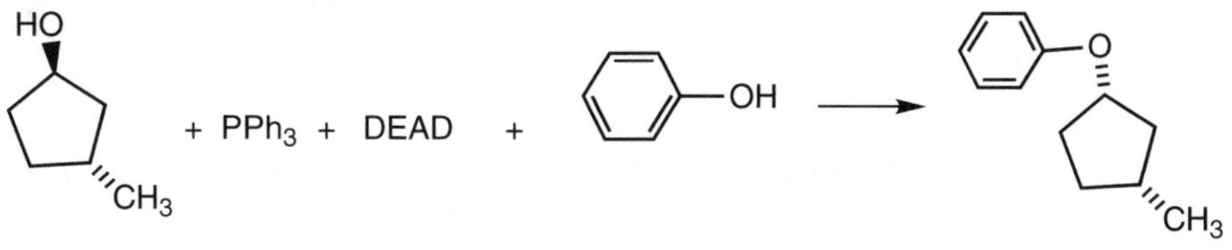

HO ... CH_3 + PPh_3 + DEAD + (phenol) OH → (phenoxy) O ... CH_3

9
ELIMINATION REACTIONS

The conditions used for nucleophilic substitution reactions frequently lead to the formation of alkenes. Some alkenes are formed in reactions that are kinetically second order, and they are called **E2 eliminations**. Other reactions follow first-order kinetics and are called **E1 eliminations**. In this chapter, we cover the mechanism, stereoselectivity, and regioselectivity of each reaction and review conditions that favor either substitution or elimination.

E1 ELIMINATIONS

Reaction Conditions That Favor E1 Reactions

E1 and S_N1 reactions are closely related. The first, rate-determining, step is the same for both reactions: the ionization of a substrate to produce a carbocation. The carbocation can either react with a nucleophile to give a substitution product or lose a H^+ to give an alkene:

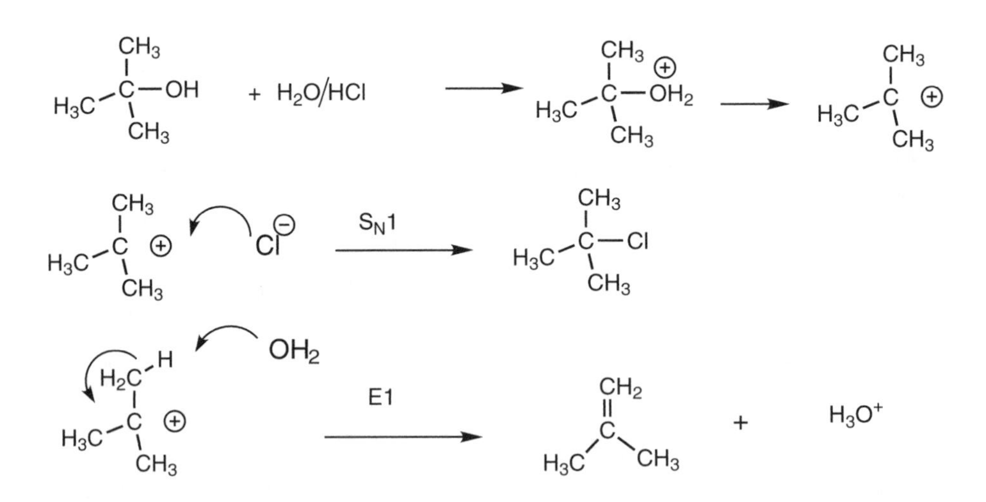

Since carbocations are intermediates, primary substrates or other species that do not form carbocations do not undergo the E1 reaction.

In order for an S_N1 reaction to take place, a nucleophile must become attached to the interior of a molecule. Conversely, an E1 reaction involves removal of a proton from the periphery of the molecule. Thus, larger nucleophiles usually give more elimination than substitution products because of steric factors. A good nucleophile is more likely to give an S_N1 reaction, and a poor nucleophile tends to act as a base and leads to elimination. Since a carbocation is such a strong acid, virtually any species with a lone pair of electrons can act as a base to give E1 products. Nucleophiles that are stronger bases give more E1 products, and weaker bases give more S_N1 product. Finally, higher temperatures favor E1 over S_N1 products. The best strategy for maximizing E1 elimination is to use a polar solvent with weak, hindered nucleophiles at elevated temperatures.

Examples The following substrates and reaction conditions should favor E1 elimination.

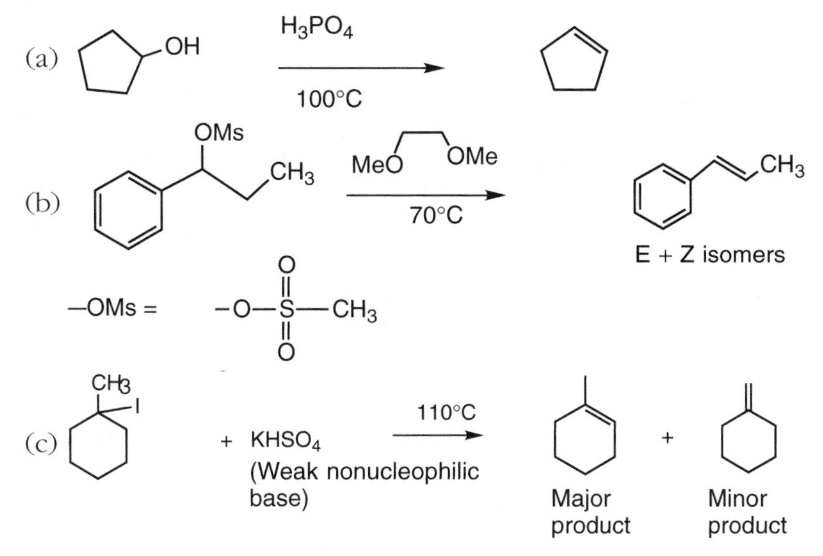

Regiochemistry in E1 Reactions

Example (c) shows that, of the two possible regioisomers that could result from the elimination of HI, methylcyclohexene forms in preference to methylenecyclohexane. This illustrates a general principle of E1 (and E2) reactions: The alkene with the greatest number of alkyl substituents is the major product. This observation was first noted by the Russian chemist Alexander Saytzeff (also spelled Zaitsev), who summarized his studies of eliminations reactions in the following way: Elimination of H—X proceeds by removal of a hydrogen on the carbon with the fewest number of hydrogens. Today, we formulate **Saytzeff's rule** in a different way: Elimination reactions produce the most stable alkene isomer.

EXERCISES

Using the original formulation of Saytzeff's rule, predict the major elimination product from the following substrates.

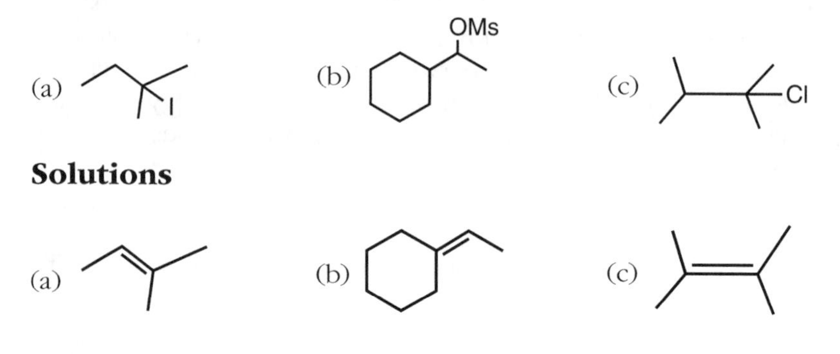

Solutions

Alkene Stability

The stability of a group of alkenes can be compared by measuring their **heats of hydrogenation**. Various metal catalysts, such as finely divided platinum deposited on carbon (Pt/C) catalyze the addition of H_2 to an alkene double bond, producing an alkane. This is an exergonic reaction, and the heat can be measured. Less stable alkenes have a higher energy content and thus release more heat on hydrogenation. 1-Butene, *cis*-2-butene, and *trans*-2-butene all give the same product, butane, on hydrogenation. Thus, their stability can be directly compared, as shown here.

HEATS OF HYDROGENATION OF BUTENE ISOMERS

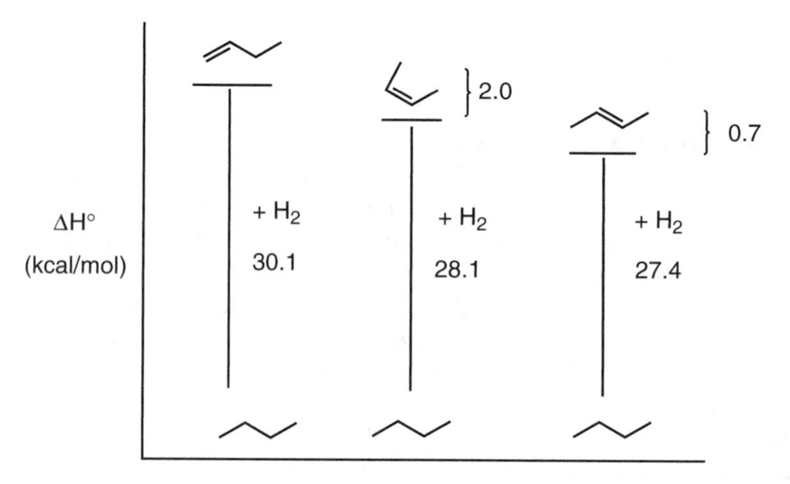

From a series of experiments, the order of alkene stability is found to be

$CH_2{=}CH_2$	<	$RCH{=}CH_2$	<	$RCH{=}CHR$	<	$R_2C{=}CHR$	<	$R_2C{=}CR_2$
Ethylene		Monosubstituted		Disubstituted		Trisubstituted		Tetrasubstituted

Furthermore, cis-disubstituted isomers are less stable than trans-disubstituted isomers by a small amount. The sp^2 carbons of alkenes act as if they are slightly electron-deficient. Thus, alkyl groups, which are electron-releasing relative to hydrogen atoms, stabilize double bonds.

A second important source of alkene stability is **conjugation**. A double bond next to another unsaturated group (such as a double bond, triple bond, aromatic ring, or carbonyl group) is said to be conjugated with that group and is more stable than an isolated double bond. The reason conjugation stabilizes double bonds is resonance. One can draw many charge-separated resonance structures when double bonds are conjugated:

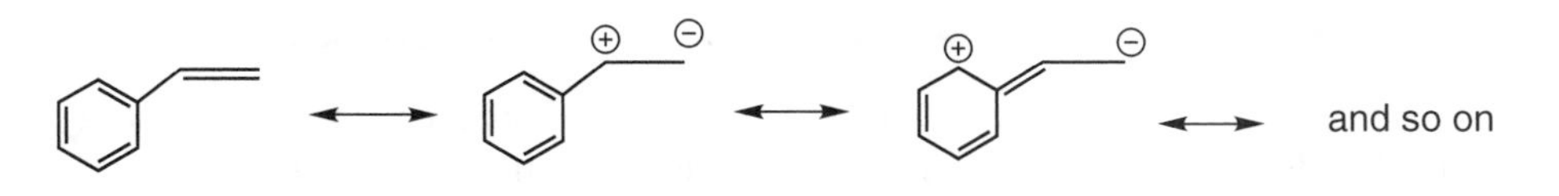

Usually, elimination reactions form conjugated isomers whenever possible.

Examples The following elimination reactions give the more stable regioisomer. One normally gets a mixture of cis and trans isomers on elimination, with the more stable trans predominating. The symbol Δ is used to represent heating.

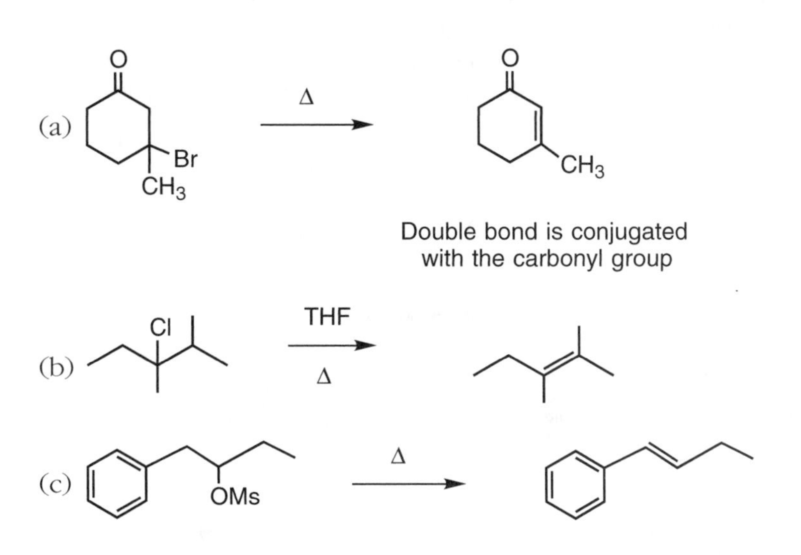

Themodynamic vs Kinetic Reaction Products

The reaction coordinate diagram for an E1 reaction capable of giving different regioisomers has a single carbocation intermediate leading to two different products as shown here.

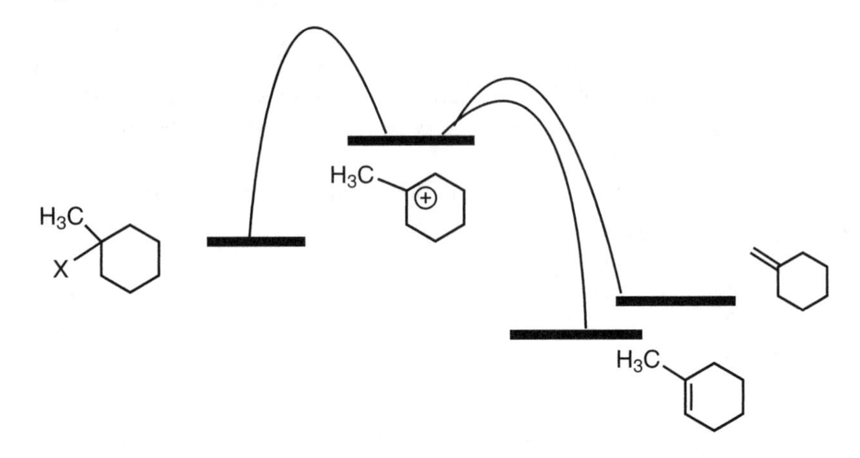

This diagram shows that the trisubstituted alkene product is lower in energy than the disubstituted alkene isomer, consistent with the modern interpretation of Saytzeff's rule. Whenever a reaction can give several isomeric products, the lowest-energy product is called the **thermodynamic product.**

According to this diagram, the product-forming transition state, the one connecting the carbocation intermediate and the alkene product, is lower in energy for the trisubstituted alkene. Since the rate of any single reaction step depends on the energy of the corresponding transition state, the carbocation forms the trisubstituted alkene faster than it forms the disubstituted alkene. The product that forms fastest from a common intermediate is called the **kinetic product.** If the above diagram is correct, methylcyclohexene is, for this reaction, both the kinetic and the thermodynamic product.

Other reactions can have the thermodynamic and kinetic products as different molecules:

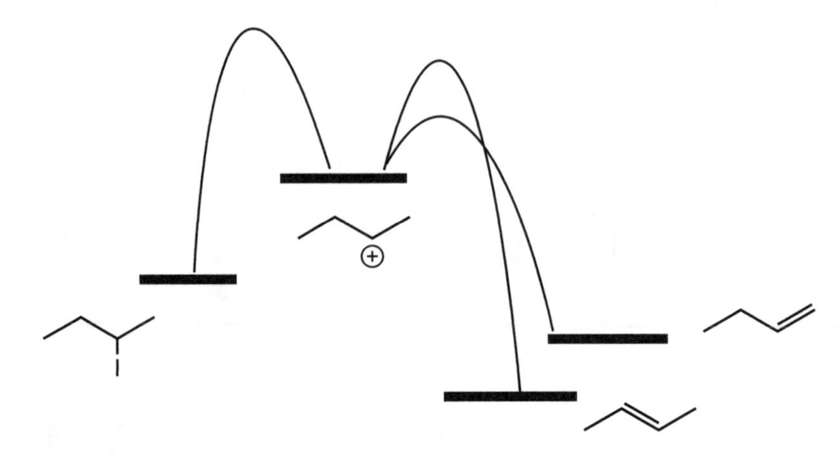

Assuming the preceding diagram is correct, 2-butene is the thermodynamic product and 1-butene is the kinetic product in this reaction. Controlling the reaction conditions can favor the thermodynamic product or the kinetic product. Thermodynamic products are formed under **reversible conditions**. For elimination reactions, this means that the loss of H^+ from a carbocation to give an alkene and the addition of H^+ to an alkene to give a carbocation happen simultaneously. The reaction drains into the lowest energy species. Kinetic products are formed under **irreversible conditions**. The reaction passes over the lowest barrier, and there is not enough energy to pass back up the barrier. Often, kinetic products form at low temperatures and thermodynamic products form at higher temperatures.

It is easy to see that Saytzeff's rule should hold for eliminations carried out under thermodynamic conditions. However, the more substituted alkene also can be the kinetic product. One needs to look at the different σ bonds that can stabilize a carbocation by hyperconjugation:

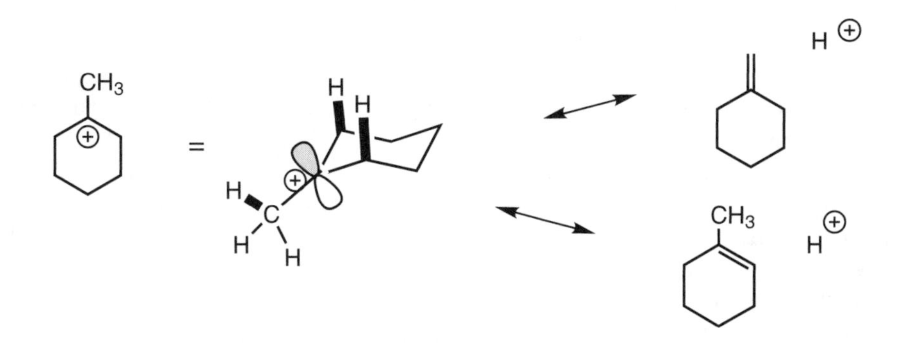

Using the no bond resonance formalism to depict hyperconjugation, we see that a methyl C—H bond or the axial C—H bonds can be involved in hyperconjugative stabilization. Hammond's postulate says that the transition state resembles the species it is closest to in energy. Here the lower-energy transition state resembles the lower-energy resonance form, which is the more substituted alkene. The higher-energy transition state resembles the less substituted alkene. Thus, the more substituted alkene can be the kinetic product as well as the thermodynamic product.

E2 ELIMINATIONS

Reaction Conditions for the E2 Elimination

E2 elimination is kinetically second order. The two chemical species in the rate law are the substrate, RX, and a base that can be negatively charged or neutral. E2 elimination, like an S_N2 reaction, is concerted, with simultaneous making and breaking of bonds:

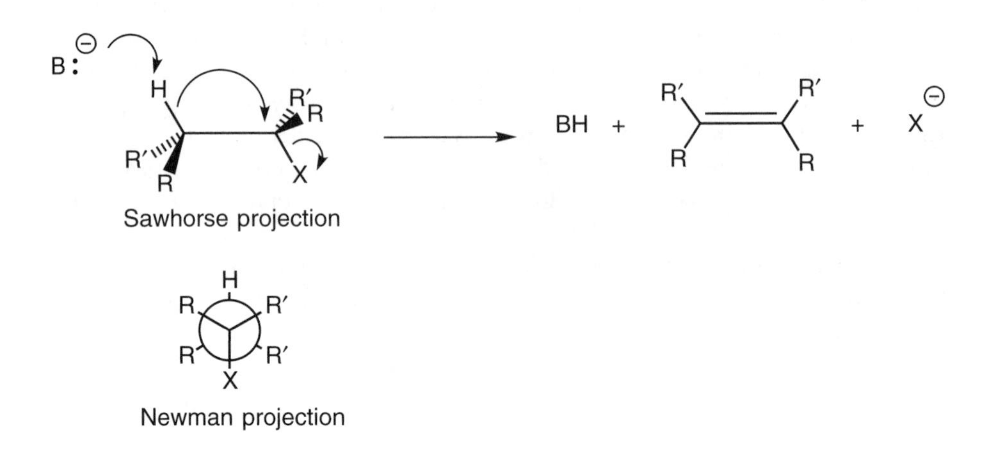

Sawhorse projection

Newman projection

An E2 reaction requires that the leaving group and the H being removed be anti-periplanar to one another. As the reaction progresses, the C—H and C—X σ orbitals transform into *p* orbitals that form a π bond. They must be parallel to one another, as they are in the anti-periplanar conformation, for bonding interactions to develop smoothly. Although the C—H and C—X bonds could be syn to one another and experience the same overlap, this geometry is sterically disfavored.

The substrate for an E2 reaction can be primary, secondary, or tertiary. As long as there is a leaving group and a proton that can become anti periplanar to the leaving group on the adjacent carbon, an E2 reaction is possible. The other species in the rate law, the base, is very important in making E2 elimination possible. As pointed out in Chapter 8, nucleophiles are also Lewis bases. The more basic a nucleophile, the less likely it is to carry out an S_N2 reaction and the more likely it is to promote E2 elimination.

Strong hindered bases, such as **potassium *t*-butoxide**, K^+ $(CH_3)_3O^-$, or lithium diisopropylamide (the suffix *-amide* is used for the salts of amines as well as for the amide functional group) (**LDA**), Li^+ $[(CH_3)_2CH]_2N^-$, are the most reliable bases for E2 eliminations. They are extremely basic and so hindered that even S_N2 reactions on 1° substrates are sterically disfavored. Another excellent base for E2 eliminations is NaOH or KOH in an alcohol solvent. Protic solvents decrease nucleophilicity but not basicity. Finally, certain neutral nitrogen bases are strong enough bases to give E2 elimination. Diisopropylethyamine, $[(CH_3)_2CH]_2EtN$, and 1,8-diazabicyclo[5.4.0]undec-7-ene (**DBU**), are frequently used in E2 reactions, especially when a good leaving group like a sulfonate is involved.

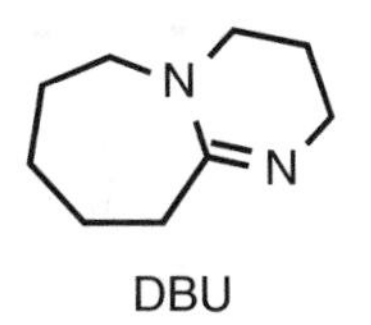

When deciding the product formed from an E2 reaction, the Saytzeff rule that the most stable alkene is formed is usually followed. However, the anti-periplanar relationship between the C—H bond and the C—X bond involved in the elimination is an absolute requirement. This means that different diastereomers can produce different products through an E2 mechanism:

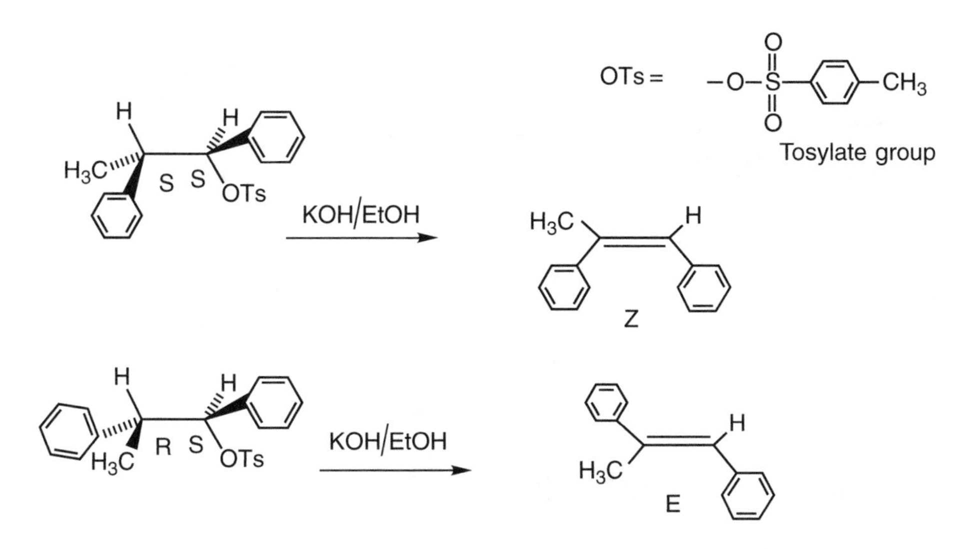

Primary alcohols in strongly acidic solutions, such as H_2SO_4 and H_3PO_4, undergo elimination reactions. Since 1° carbocations are too unstable to form, this cannot be an E1 elimination. Instead, the strong acid protonates the —OH group, turning it into the good leaving group $—OH_2^+$. Then, whatever base is available, either H_2O or the conjugate base of the acid, initiates an E2 elimination.

An E2 mechanism is almost always a better way to carry out an elimination reaction than an E1 reaction. A carbocation can undergo rearrangement and substitution reactions in addition to elimination reactions, so yields from E1 reactions can be poor. Also, the range of substrates able to undergo E2 elimination is much greater than that of those that can undergo E1 elimination.

EXERCISES

Indicate the major product from each of the following E2 elimination reactions.

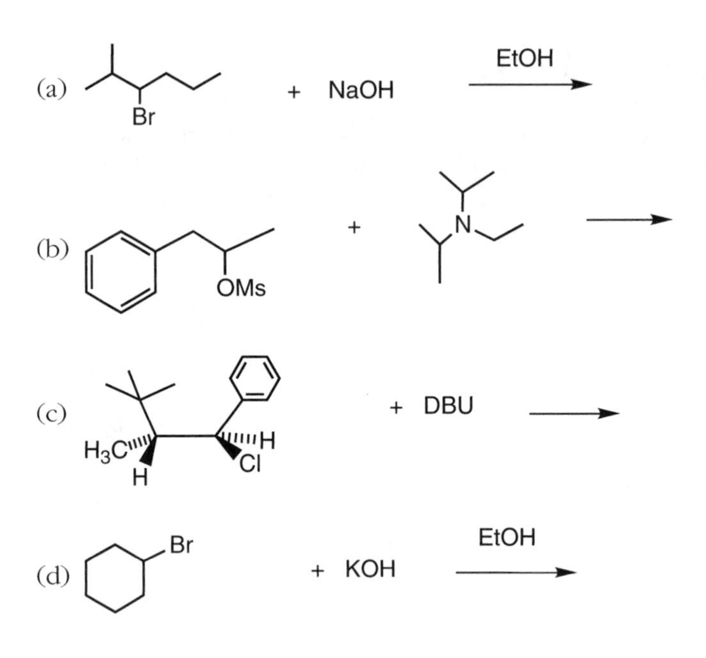

Solutions

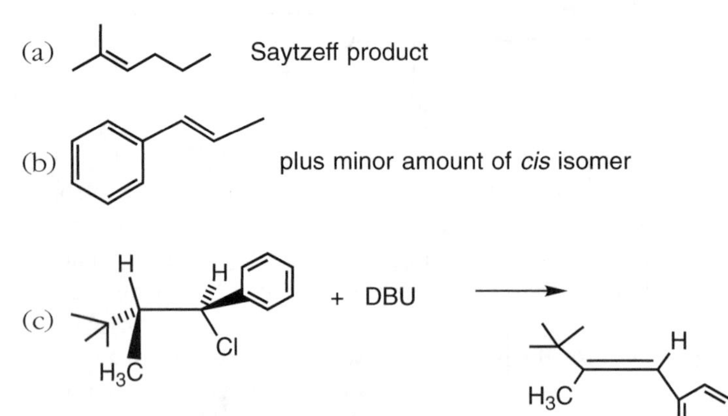

(d) 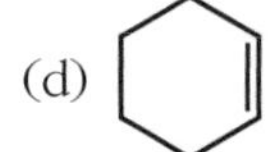 Do not expect much S_N2 product when a strongly basic nucleophile reacts with a 2° substrate.

E2 Elimination Reactions in Cyclohexyl Systems

In a cyclohexyl ring, the only way the proton and leaving group can be anti to one another is if both groups are trans and diaxial. In order to predict the product of E2 eliminations in six-membered rings, it is usually necessary to draw the chair form of the ring and consider both ring flip conformers to see what elimination products are possible:

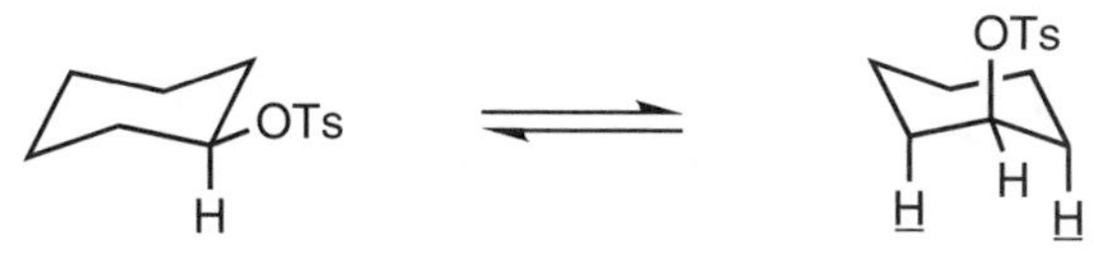

Only the axial tosylate conformer undergoes E2 elimination, with either of the underlined hydrogens

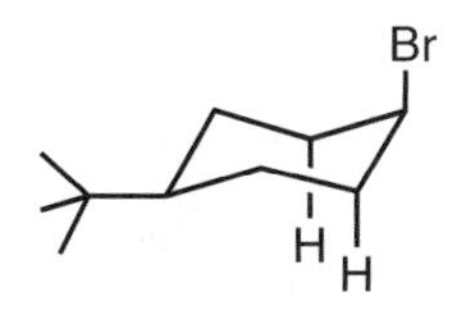

cis-4-Bromo-*t*-butylcyclohexane

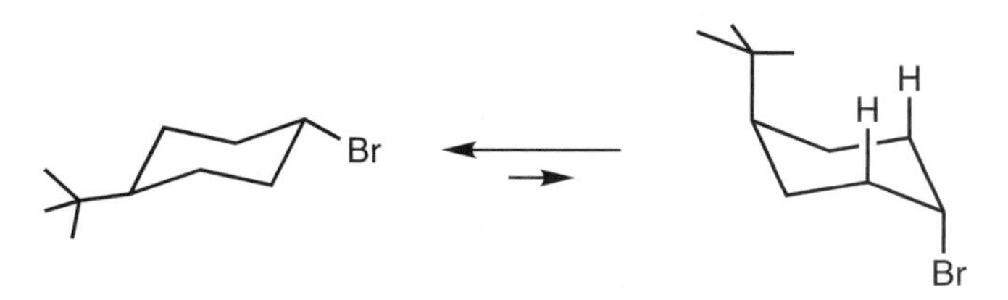

trans-4-Bromo-*t*-butylcyclohexane

For example, *cis*-4-bromo-*t*-butylcyclohexane is set up to undergo E2 elimination since the *t*-butyl group biases the bromide to the axial position. However, E2 eliminations of *trans*-4-bromo-*t*-butylcyclohexane are very slow because they can occur only from a conformer where both large groups are axial, which makes up less than 0.01 percent of the mixture.

Example A popular kind of question in organic chemistry asks for the expected product from the elimination of *trans*-1-bromo-2-methylcyclohexane and *cis*-1-bromo-2-methyl cyclohexane. One has to draw the isomers correctly and then consider what trans-diaxial elimination reactions are possible.

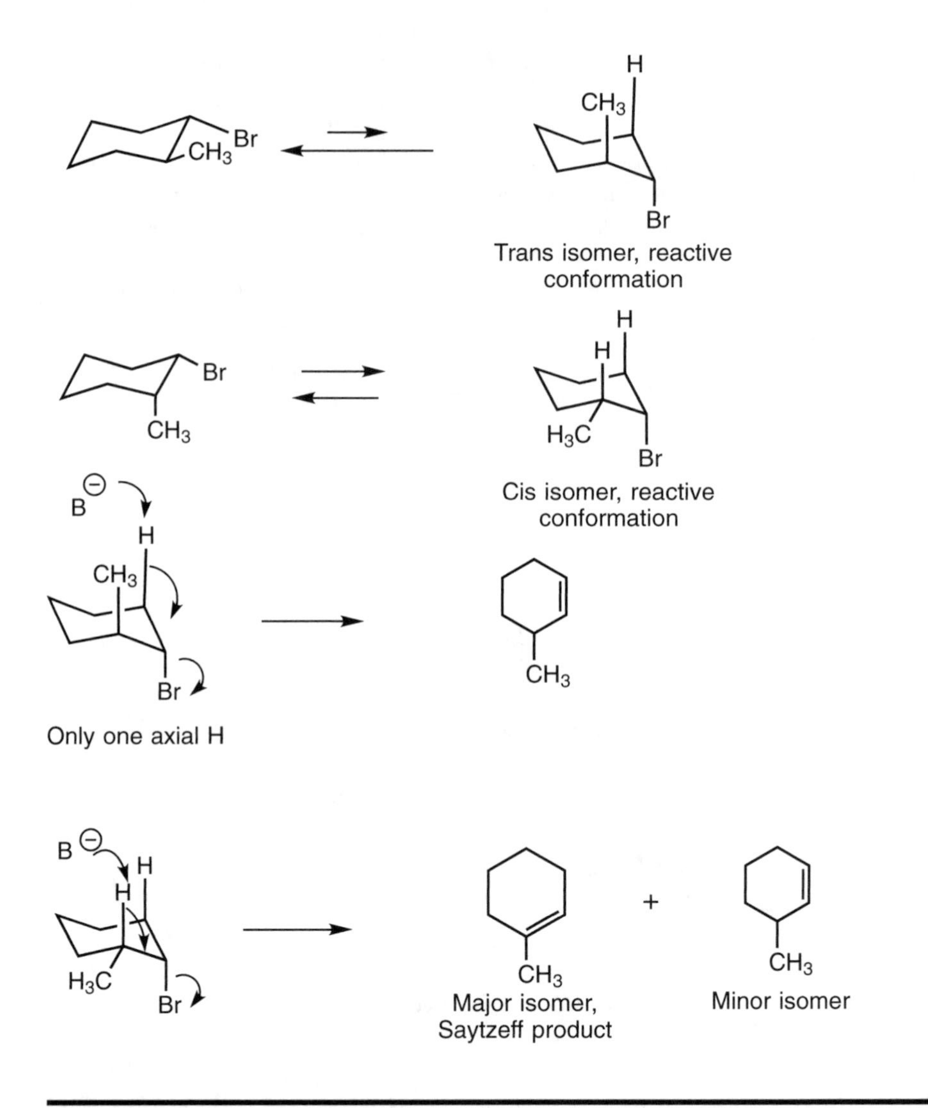

COMPARISON OF S_N1, S_N2, E1, AND E2 REACTIONS

Students normally worry about being able to predict how a given RX substrate will react under specific conditions. Since they have not been given many quantitative rules to predict reactivity, they must rely on general patterns of reactivity. While it is possible to formulate questions that are impossible to answer, well-formulated questions have most or all of the substrates and reaction conditions favoring one mechanism.

S_N2 reactions work best with methyl or 1° substrates, although 2° substrates can react under favorable conditions. The nucleophile should be nonbasic and as unhindered as possible. Polar aprotic solvents strongly favor S_N2 reactions, and protic solvents disfavor S_N2 reactions, often leading to E2 eliminations with basic nucleophiles.

E2 reactions can occur with any RX substrate, although 1° substrates need a strong, hindered base like LDA. Higher temperatures favor elimination reactions over substitution. Tertiary substrates give elimination only with basic nucleophiles. Because of the stereospecificity of E2 elimination, it is often used in organic chemistry.

S_N1 reactions are normally carried out as solvolysis reactions with 3°, allylic, or benzylic substrates. Because a simple benzylic compound, $PhCH_2X$, cannot undergo elimination, it must undergo substitution. Good nucleophiles like HS^- and CN^- give S_N2 reactions, while poor nucleophiles like formic acid give S_N1 reactions.

E1 reactions are not very useful in organic chemistry because carbocations can give several different products. Also, a mixture of E and Z isomers is formed in an E1 reaction, and such isomers are difficult to separate. High temperatures and polar solvents that stabilize carbocations lead to E1 products with 3°, allylic, and benzylic substrates or 2° substrates with good leaving groups.

EXERCISES

Suggest conditions for carrying out the following reactions. Review Chapters 8 and 9.

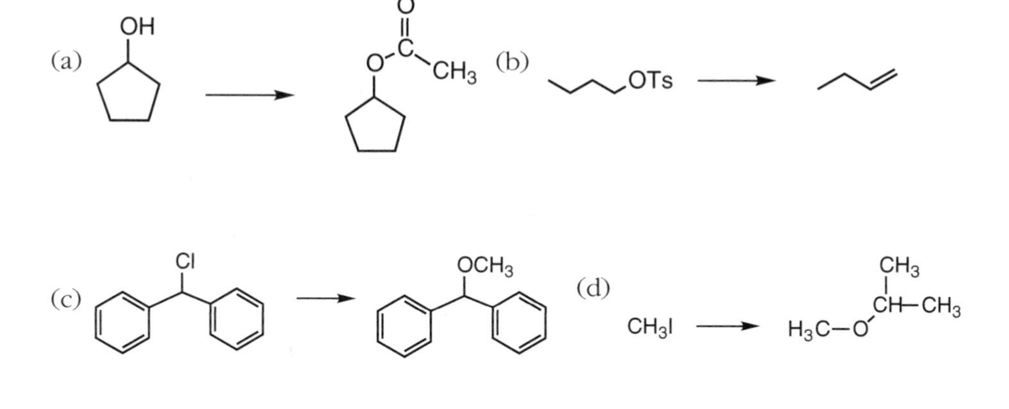

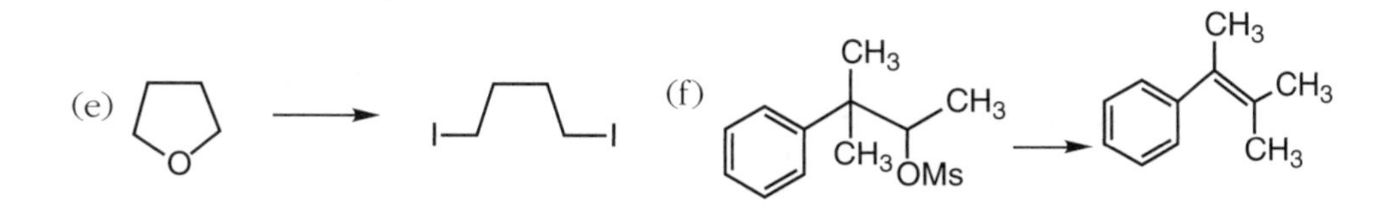

Solutions

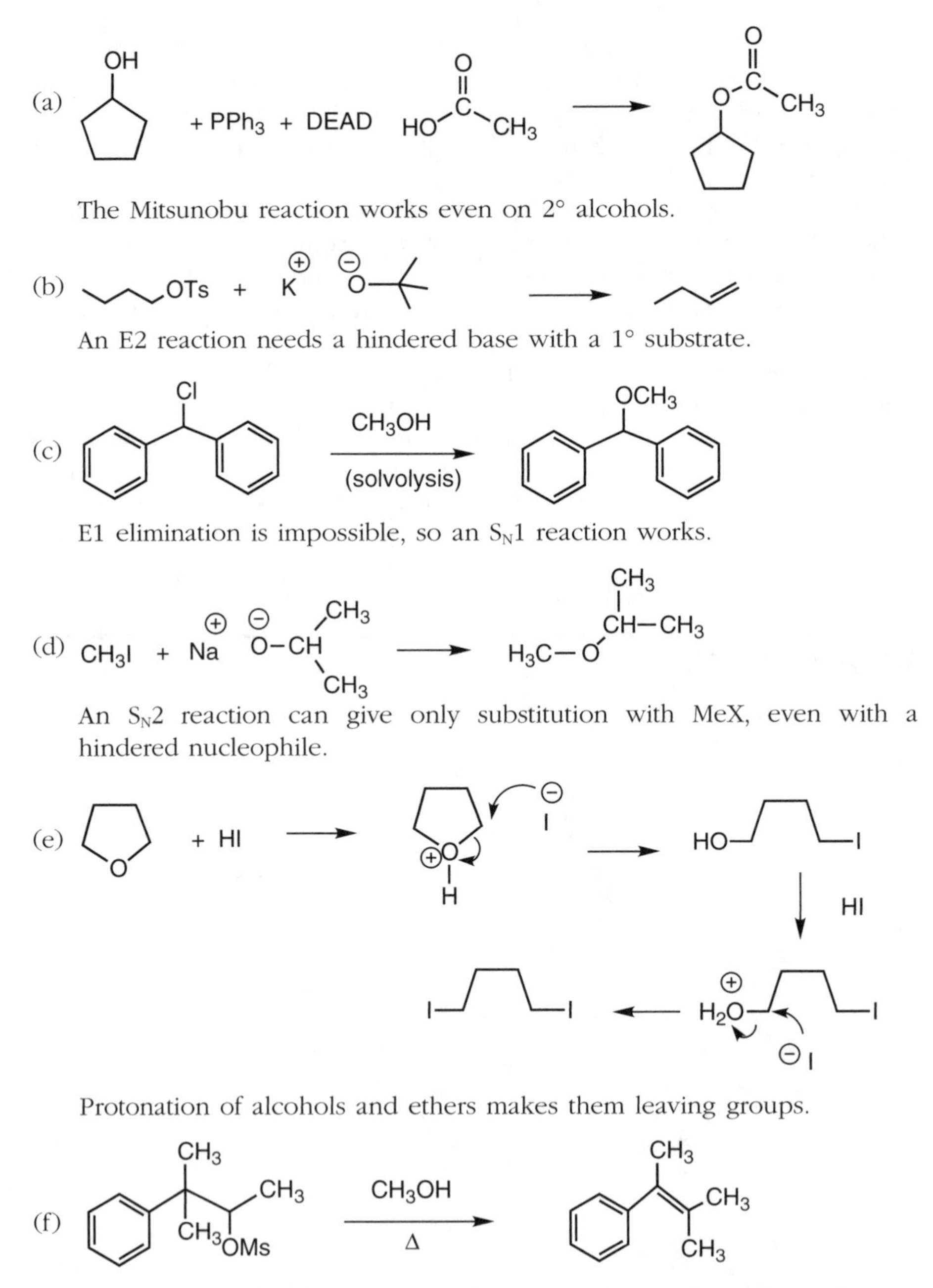
(a)
OH
+ PPh3 + DEAD
HO C CH3
O
O C CH3
The Mitsunobu reaction works even on 2° alcohols.
(b)
OTs + K O
An E2 reaction needs a hindered base with a 1° substrate.
(c)
Cl
CH3OH
(solvolysis)
OCH3
E1 elimination is impossible, so an SN1 reaction works.
(d)
CH3I + Na O–CH
CH3
CH3
H3C–O
CH–CH3
CH3
An SN2 reaction can give only substitution with MeX, even with a hindered nucleophile.
(e)
+ HI
O
H
I
HO
I
HI
H2O
I
I
Protonation of alcohols and ethers makes them leaving groups.
(f)
CH3
CH3
CH3
OMs
CH3OH
Δ
CH3
CH3
CH3
This is an E1 elimination where the intermediate carbocation undergoes a Me migration followed by loss of H+.

10 ADDITION REACTIONS OF ALKENES

MECHANISTIC AND STEREOCHEMICAL ASPECTS OF ALKENE ADDITION REACTIONS

The alkene double bond is a nucleophile. It takes part in addition reactions with electrophiles to convert the relatively weak π component of the double bond to stronger σ bonds. Both the stereochemistry and regiochemistry of the addition reaction must be understood in order to predict the products from a given reaction. Because there are a staggering number of possible nucleophile-electrophile combinations, only by understanding the general mechanisms of these reactions can one predict which specific products will form.

Reactions of Symmetric Electrophiles with Symmetric Alkenes

Bromine, Br_2, is an example of a symmetric electrophile, while HCl is an asymmetric electrophile. For simplicity, the reaction of Br_2 with ethylene will be discussed first. A brief mention of this reaction was made in Chapter 7.

Single bonds between two atoms that have lone electron pairs tend to be weak, especially the single bonds of halogens. Thus, when a Br_2 molecule approaches a double bond, the electron clouds polarize, one end of the Br_2 molecule becoming partly positively charged and the other end partly negatively charged. The net result is formation of a C—Br bond and heterolytic cleavage of the Br—Br bond:

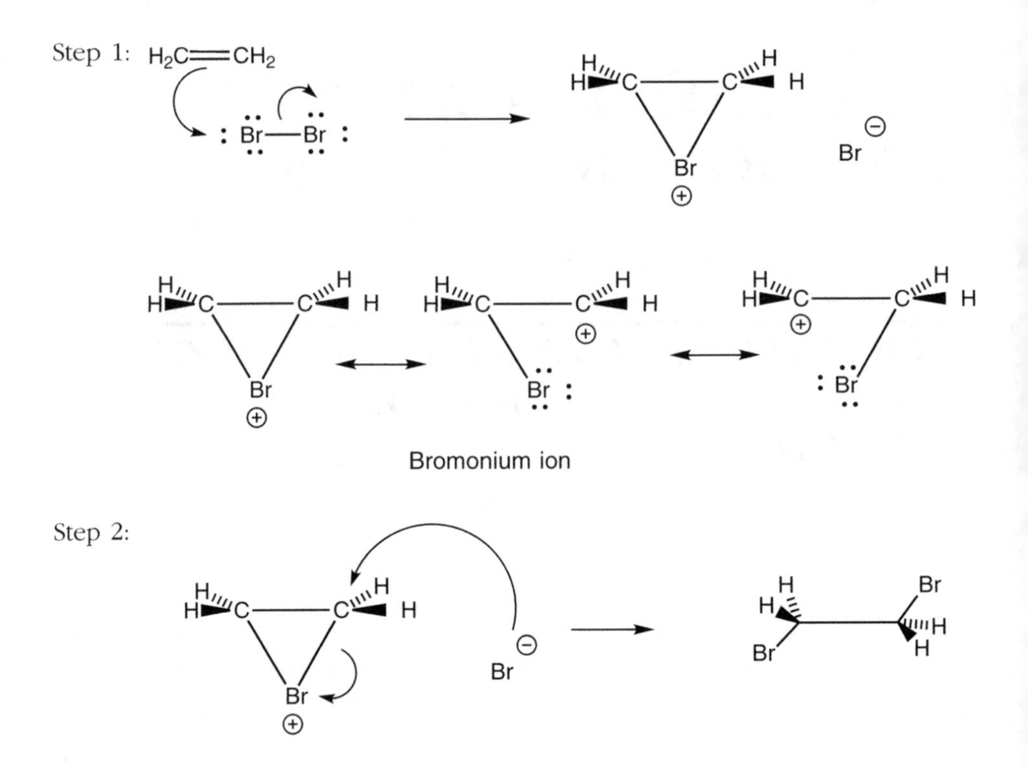

The first, rate-limiting, step in the reaction of an alkene with Br_2 generates a **bromonium ion**. This is an example of a **bridged ion** in which the bromine interacts with both carbons of the alkene. Bromine can do this because it is a very large atom compared to carbon and one of its lone electron pairs is close enough to donate two electrons to the empty *p* orbital on the adjacent carbocation. The second step is essentially an S_N2 reaction. The net result is **anti addition** with the two groups ending up on opposite sides of what was the double bond. The 1,2 dibromides are **vicinal** dibromides, which means they are on adjacent carbons.

The anti stereochemistry is clearly seen with cyclic alkenes:

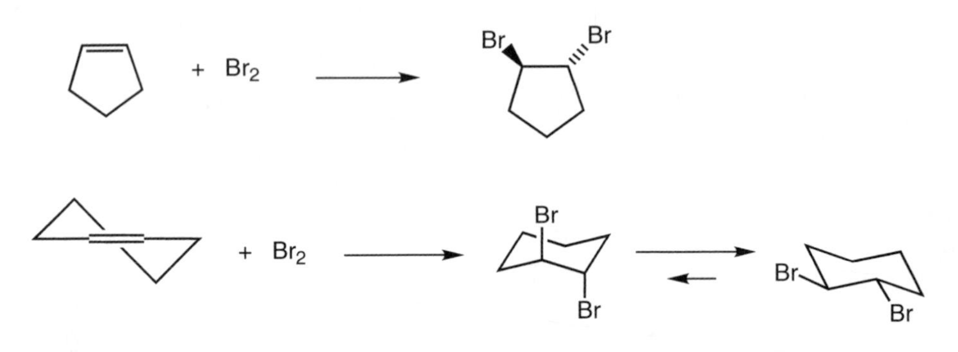

Just as an E2 elimination reaction takes place through trans-diaxial geometry, the addition of bromine proceeds through trans-diaxial geometry. The initially formed cyclohexyl diaxial bromide undergoes a ring flip to give the trans-diequatorial conformer.

Chlorine adds to alkenes through a **chloronium ion** intermediate, similar to the Br_2 reaction. F_2 is so reactive that when it is exposed to alkenes, they burst into flame. I_2 forms such weak C—I bonds that no vicinal diiodides can be isolated.

EXERCISES

1. Explain why 2,3-dimethyl-2-butene reacts with Br_2 approximately 1 million times faster than ethylene.
2. Draw all the products of the addition of Br_2 to cyclohexene and assign any stereogenic centers as R or S.

Solutions

1. The bromonium ion intermediate has some carbocation character. Alkyl groups stabilize carbocations. Therefore, the transition state leading to the 2,3-dimethyl-2-butene bromonium ion is lower in energy than the one leading to the ethylene bromonium ion intermediate, according to Hammond's postulate.
2. 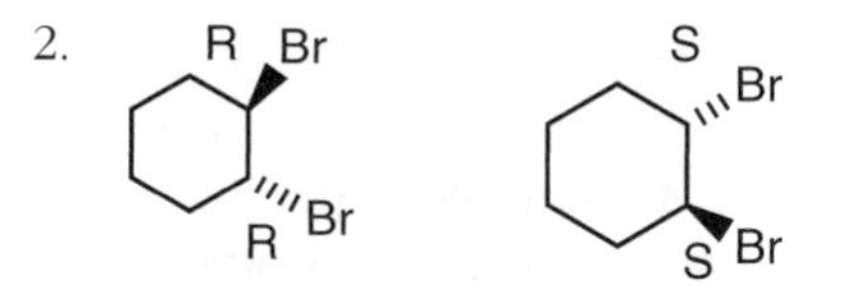

Since there are two stereogenic centers, there is a maximum of four stereoisomers. However, the mechanism of the addition reaction allows for the formation of only two of them. The other isomer, the meso isomer, has syn bromines, and it does not form under these reaction conditions.

Reaction of Unsymmetric Electrophiles With Alkenes

H—Br is an unsymmetric electrophile. As an acid, it is a source of H^+, which is the specific electrophile that reacts with an alkene to generate a carbocation:

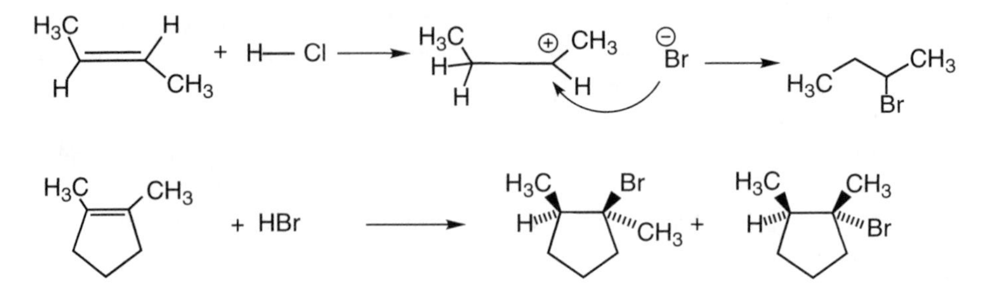

The carbocation then rapidly reacts with the conjugate base of the acid, here Br^-, to give the addition product. The reaction is stereorandom because the carbocation can react with the nucleophile from its top face or bottom face.

When the alkene is unsymmetric, different regioisomers can potentially form on reaction with an acid of the general formula H—X. Experimentally, only one regioisomer is formed:

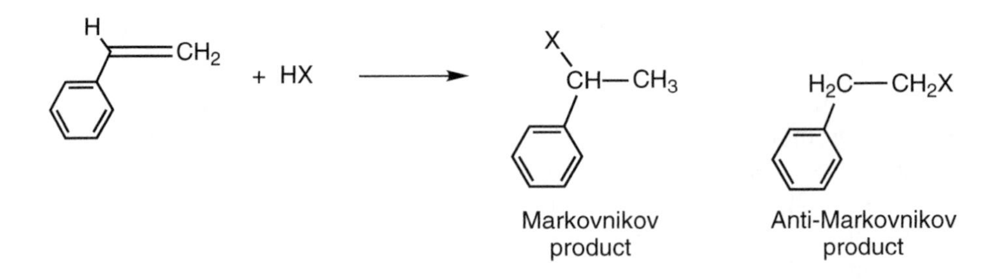

The observed product has the H^+ adding to the CH_2 carbon and the X^- adding to the CH carbon. This type of product is called the **Markovnikov product**, and the other regioisomer is known as the **anti-Markovnikov product**. The Russian chemist Vladimir Markovnikov formulated a rule (now known as **Markovnikov's rule**) stating that in the addition of acids, HX, to an alkene, the H adds to the carbon atom with the greater number of hydrogens, and the anion, X, adds to the carbon with the fewer number of hydrogens.

With the current knowledge of carbocation stability, the physical basis of Markovnikov's rule is now understood. H^+ adds to an alkene double bond to give the more stable carbocation. The resulting carbocation has the lowest number of hydrogens and the most alkyl groups bonded to it. A more modern way to formulate Markovnikov's rule is to say that the proton adds to an alkene to give the more stable carbocation. This way of expressing Markovnikov's rule explains the regiochemistry of HX addition when one group is better at stabilizing a charge than another. For example, in CHPh=CHMe, the phenyl group is much better at stabilizing a charge than a methyl group. Therefore, H—Cl reacts to give Cl—CHPh—CH_2Me and not CH_2Ph—CHMe—Cl, although both carbons in the alkene have the same number of hydrogens bonded to them.

Examples The following reactions illustrate Markovnikov's rule. Since carbocations are formed when HX adds to alkenes, note that rearrangements are possible.

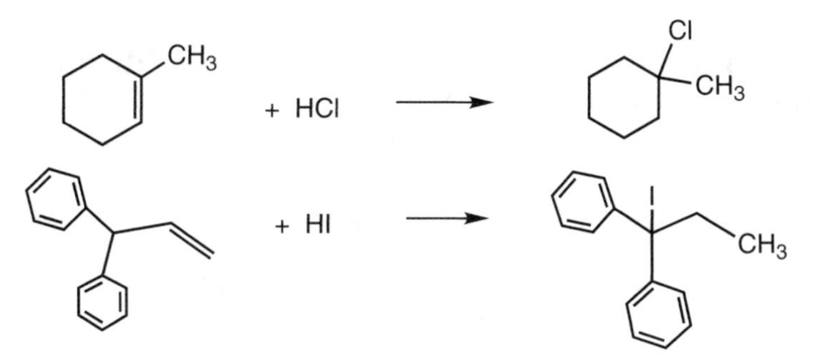

Note the hydride shift of the initially formed carbocation.

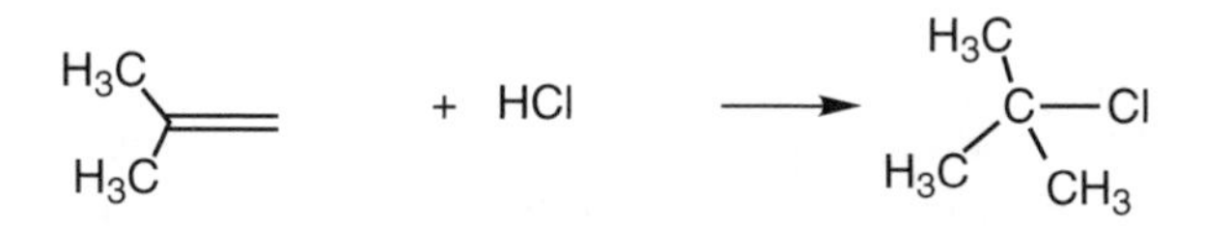

Because carbocations are intermediates in the direct addition of HCl and related acids to alkenes, the yields are frequently poor since competing reactions can be many. In addition, the rates of these addition reactions are often slow. Frequently it is better to add water to an alkene and then convert the C—OH bond to a C—X bond through an S_N2 or S_N1 reaction.

Direct Addition of Water and Alcohols

If a strong acid such as H_2SO_4 is added to a mixture of alkene and water, the net result is Markovnikov addition of H—OH across the double bond:

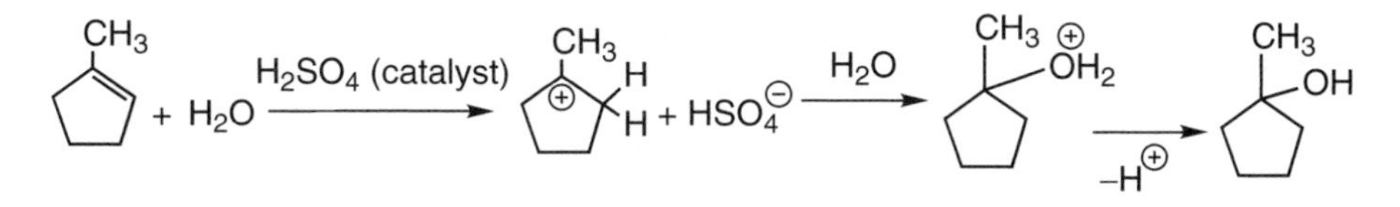

The conjugate base of sulfuric acid, HSO_4^-, is such a poor nucleophile that it does not react with the carbocation. Instead, water forms a C—O bond and in the final step the protonated alcohol loses a proton to some base, either water or HSO_4^-. This **hydration** of an alkene is the reverse of the dehydration of an alcohol mentioned in Chapter 9. Lower temperatures favor addition, and higher temperatures favor elimination. This reaction can be used to add alcohols to a double bond as well as water. For example, the gasoline additive methyl tertiarybutyl ether (MTBE), $(CH_3)_3C—OCH_3$, is made from the acid-catalyzed addition of CH_3OH to $(CH_3)_2C{=}CH_2$. The same problems of side reactions and slow reaction rates make the direct acid-catalyzed addition of water or alcohols to alkenes of limited use in the laboratory, although industrial processes use this route to make many tons of simple alcohols and ethers.

The Production of Halohydrins by the Addition of Halogens to Alkenes in Water

The elements of Cl—OH or Br—OH can be added to an alkene when Cl_2 or Br_2 is added to an alkene in the presence of water. The product of such a reaction is called a **halohydrin** (**chlorohydrin** or **bromohydrin**, respectively):

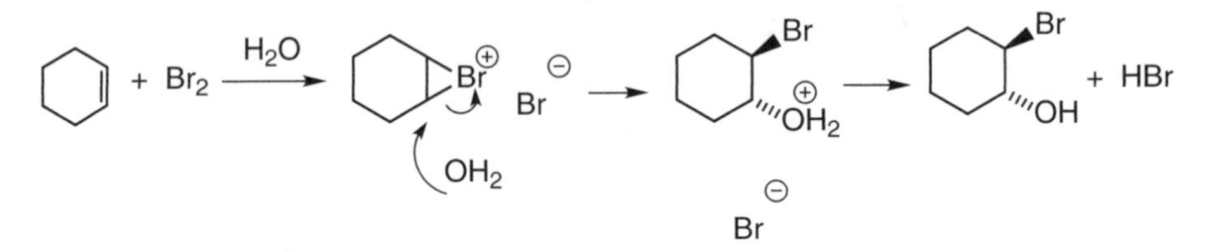

It may seem odd that the uncharged water molecule is the nucleophile instead of the bromide anion. However, the solvent molecules have a tremendous concentration advantage. The bromonium ion is a reactive intermediate, and at any moment its concentration is extremely small, which means the concentration of Br^- is likewise small compared to that of water.

The regiochemistry of halohydrin formation could in principle be dominated by steric or electronic factors (by electronic factors we mean factors that stabilize charge):

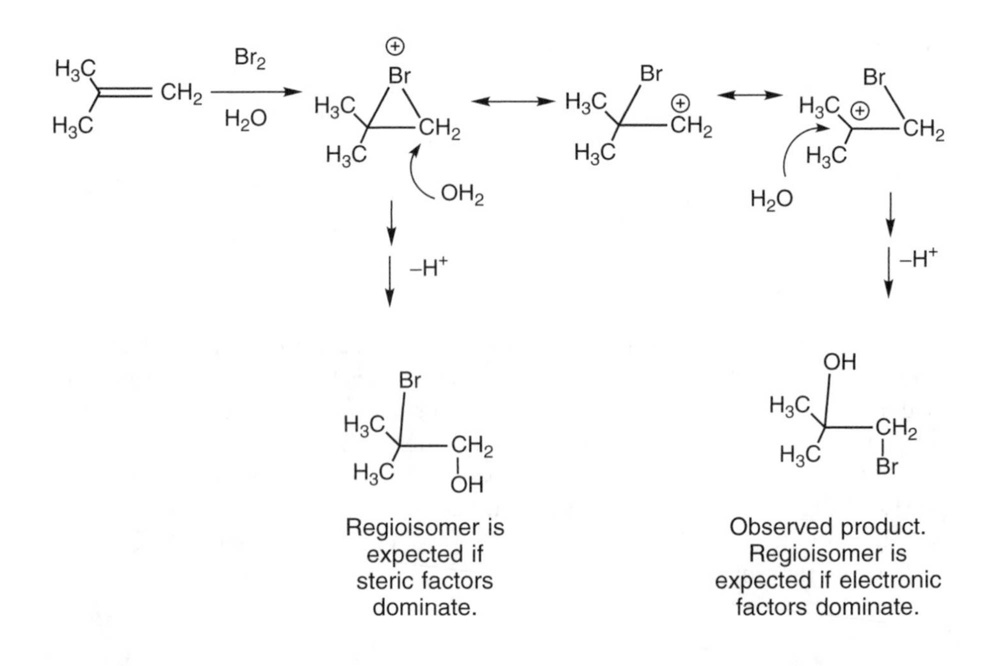

While one could not be expected to predict which regioisomer would predominate, in fact, the product is the one derived from an attack of the nucleophile on the carbon best able to stabilize a positive charge.

Bromohydrin formation is an example of a useful organic synthetic transformation. It has predictable stereochemistry (anti addition), it has predictable regiochemistry (adds as $Br^{+}HO^{-}$), and two new functional groups are created in good yield. Furthermore, bromohydrins can be easily converted to **epoxides** (cyclic three-membered ring ethers) through an intramolecular S_N2 reaction:

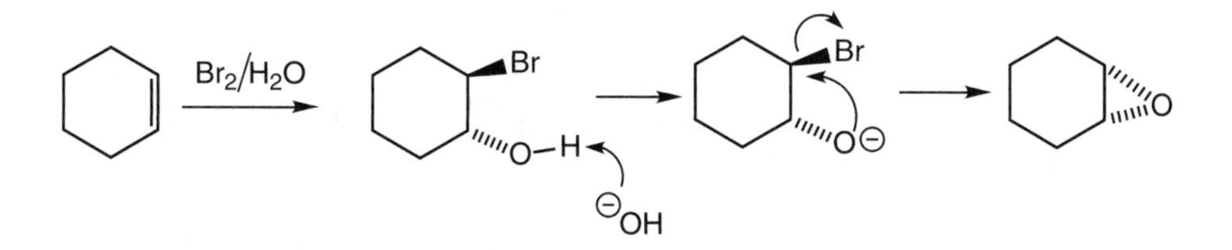

Treating the bromohydrin with a base such as NaOH to reversibly deprotonate the alcohol generates a nucleophile next to a leaving group. Although the product epoxide is strained, entropy favors the reaction of neighboring functional groups. Since acid-base reactions are normally the fastest of reactions, possible competing reactions, such as E2 eliminations, can be avoided.

EXERCISES

Give all the products from the reaction of *cis*-1-phenylbutane and *trans*-1-phenylbutane with Br_2/H_2O. Label any stereogenic centers as R or S.

Solutions

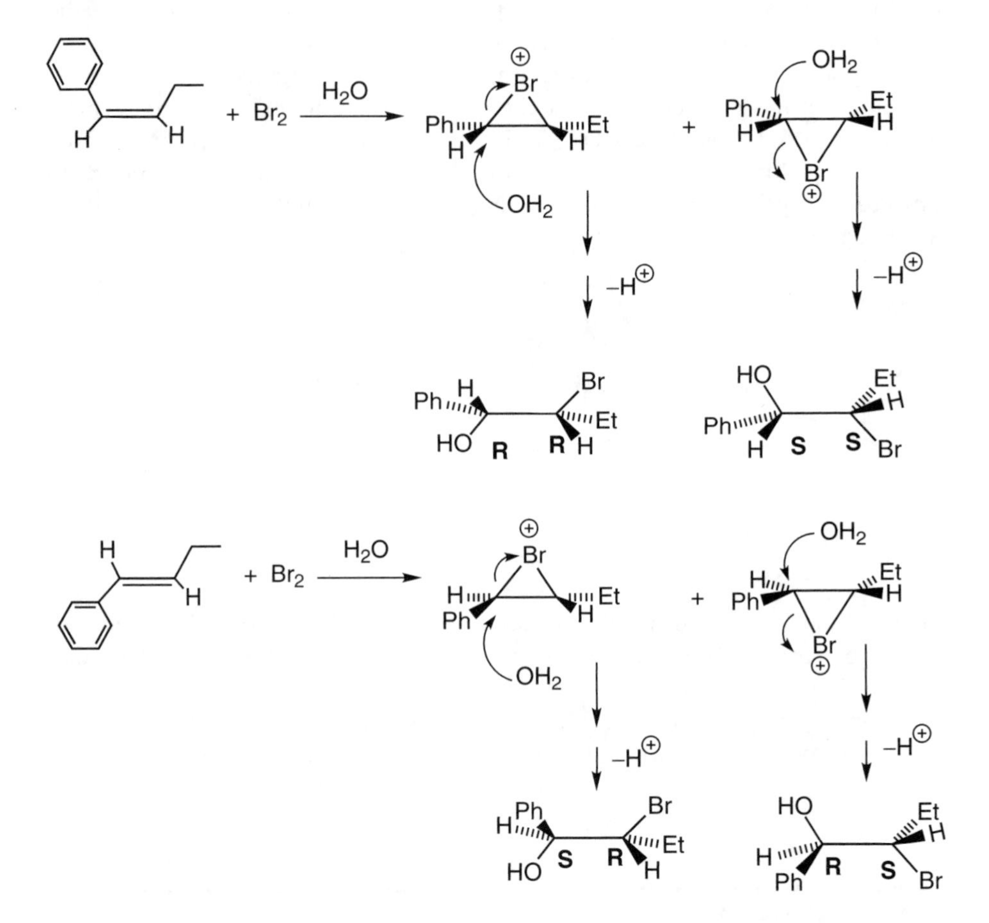

The Solvomercuation Reaction

In order to add water or an alcohol to an alkene, a **multistep** reaction called **solvomercuration** is often used. In one step a mercury salt reacts with an alkene in the presence of a nucleophilic solvent, analogous to the way bromohydrins are formed. In a subsequent step, the C—Hg bond is converted to a C—H bond. The overall transformation is the addition of H—OH (in a Markovnikov sense) to an alkene:

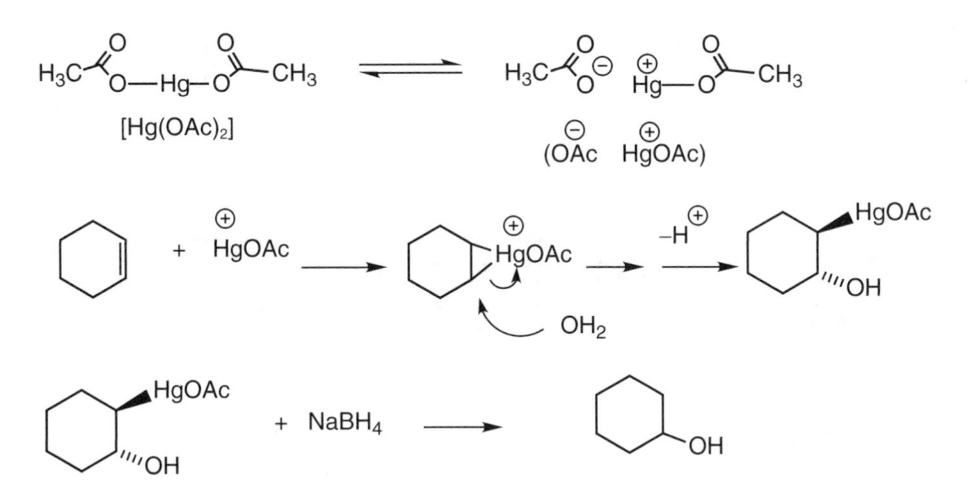

Mercuric acetate is commonly used to initiate the reaction because it can ionize to give a mercurium ion, HgX^+. This ion reacts like Br^+ to form a bridged ion that reacts with a nucleophilic solvent (water, an alcohol, or a carboxylic acid) and produces an organomercury compound. These compounds are stable and can be isolated. In a second step the reducing agent sodium borohydride, ($NaBH_4$), reduces the C—Hg bond to a C—H bond. This step is stereorandom, so the overall reaction is regiospecific (Markovnikov addition) but not stereospecific.

Example A common type of organic chemistry problem asks how one would prepare a given compound in two different ways. For example, one could be asked to prepare 1-methylcyclohexyl methyl ether in two different ways. The answer could involve an S_N2 reaction or solvomercuration:

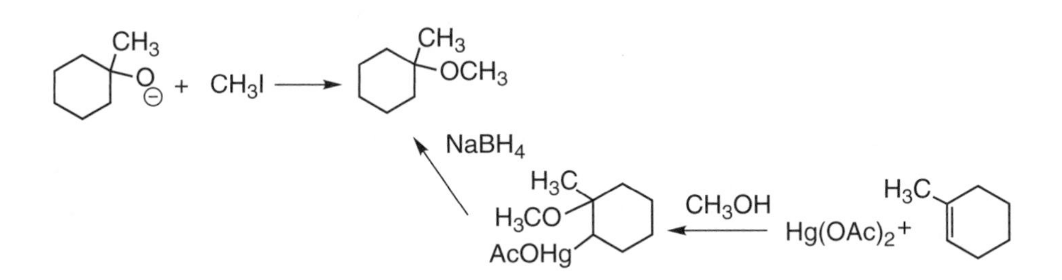

A **synthesis problem** must be distinguished from a **mechanism problem**. In a synthesis problem you are asked what should be added to a flask to get a given product. You have to show only intermediates that you can isolate and put in a bottle. In a mechanism problem you are asked to show all species, including reactive intermediates, on the path from starting materials to products. Normally you must show electron flow with curved arrows. You must not add any other reagents or catalysts except those given explicitly in the problem.

ALKENE ADDITION REACTIONS USEFUL IN ORGANIC SYNTHESIS

Alkenes are reactive, available molecules. They serve as starting materials for many useful organic reactions that reduce, oxidize, or hydrate double bonds. The first reaction discussed in this section, hydroboration, has proven so useful in organic chemistry that its inventor, Herbert C. Brown, was awarded the Nobel Prize in chemistry in 1979.

Hydroboration of Alkenes

Borane, BH_3, a strong Lewis acid, does not exist as such but dimerizes to B_2H_6, whose hydrogen-bridged structure is usually discussed in inorganic chemistry. In the presence of a Lewis base, such as the cyclic ether THF or dimethyl sulfide, a reasonably stable complex is formed that acts as a source of BH_3:

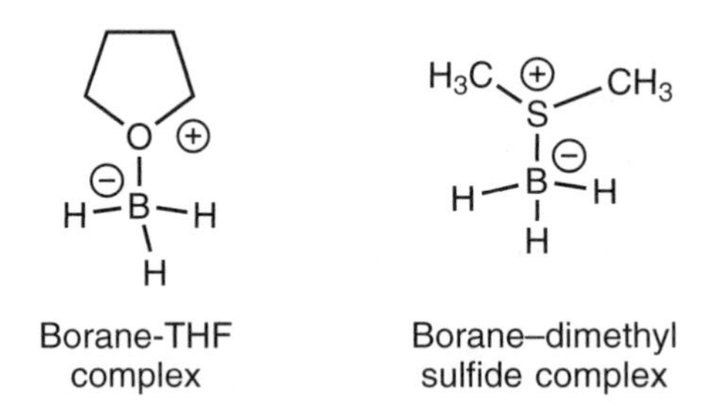

The addition of an alkene to one of these borane complexes results in the transient formation of a Lewis acid–Lewis base complex between the π bond of the alkene and the empty orbital on boron. This complex collapses to an **alkylborane** in which a B—H bond has added across the double bond:

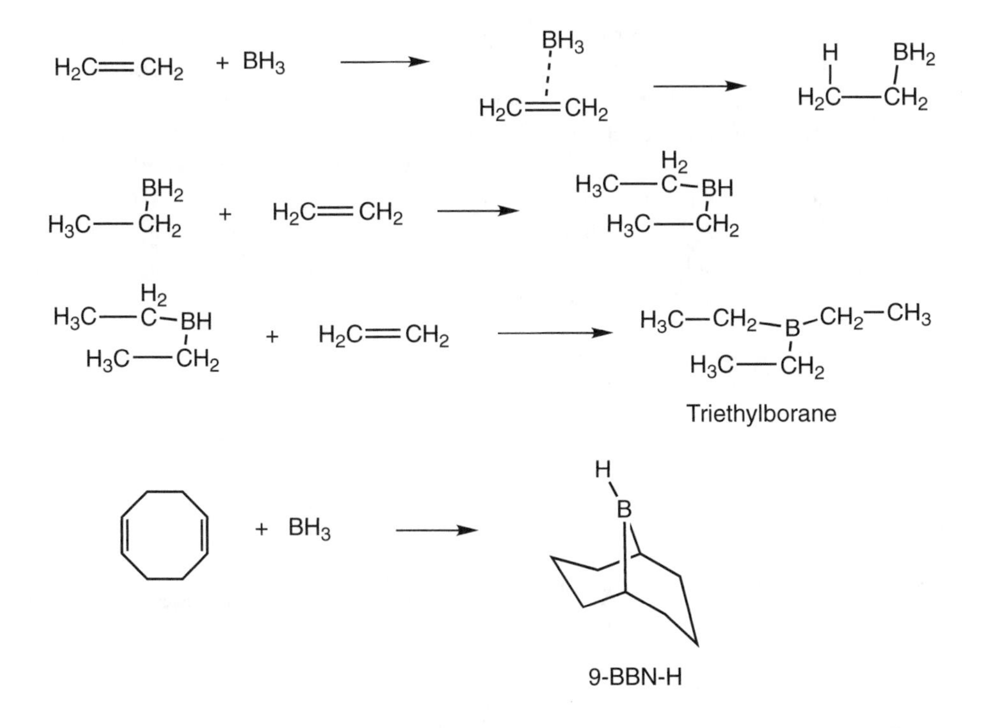

The addition of one B—H bond leaves two B—H bonds left over. For unhindered alkenes, two more alkenes react to give a **trialkylborane**. More hindered alkenes can add just one or two B—H bonds. For example, 1,5-cyclooctadiene reacts with borane to give 9-borabicyclo[3.3.1]nonane (**9-BBN-H**). This **dialkylborane** can add the third B—H bond to an unhindered alkene.

The stereochemistry of addition of the B—H bond is **syn**: Both atoms approach from the same side of the double bond. Because the B—H bond is polarized $B^{\delta+}$—$H^{\delta-}$ and because B is larger than H, the regiochemistry of B—H addition is **anti-Markovnikov**: The H bonds to the carbon with the fewer number of hydrogens:

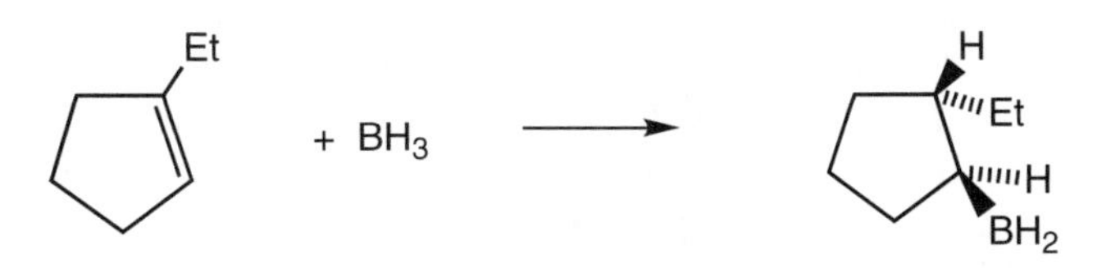

In the above example, only the first addition product is shown. The addition of brorane to a double bond is regioselective rather than regiospecific, but reagents like 9-BBN-H are more selective.

Because alkylboranes tend to be **pyrophoric** (they burst into flame in the presence of air), they are intermediates in the synthesis of other functional groups. The most important transformation is oxidation. When an alkylborane is treated with the reagent combination NaOH/HOOH/H_2O (a source of HOO^-), known as basic hydrogen peroxide, the C—B bond is replaced with a C—OH bond. The stereochemistry is retained; that is, the C—B and C—OH groups have the same stereochemistry. The mechanism of this oxidation is complex and is unlike that of most other reactions covered in introductory organic chemistry, so it will not be given here. The important thing to note about the hydroboration-oxidation process is that it provides a way to add water to a double bond with anti-Markovnikov regiochemistry:

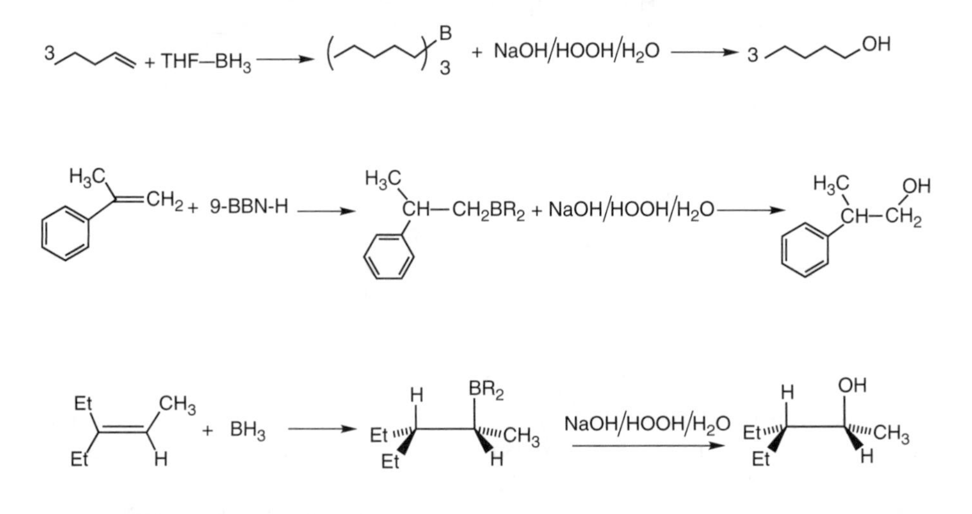

Catalytic Hydrogenation

The addition of H—H across a double bond is an exergonic reaction but does not take place without a catalyst. Normally, platinum group metals are catalysts, especially high-surface-area platinum or palladium metal by itself or deposited on carbon or $BaSO_4$. Treating PtO_2 (**Adams catalyst**) with H_2 gives finely divided Pt metal and water.

The atoms in a metal interact strongly with one another, which accounts for the strength and hardness of the metal. However, the metal atoms at a surface have no bonding partner in one direction, so they interact strongly with any external molecule with accessible electrons. Molecular hydrogen adsorbs on a platinum group metal surface and is dissociated into individual atoms, while an alkene associates through its π bond. The adsorbed species can come together to form an alkane which, since an alkane does not interact strongly with the surface, diffuses away.

Usually, the stereochemistry of H_2 addition is syn to the double bond. Catalytic hydrogenation is selective for alkenes (and alkynes, as discussed in Chapter 11). The C=O double bond of a carbonyl group does not react under typical hydrogenation conditions (1–2 atm H_2, room temperature) nor does a benzene ring:

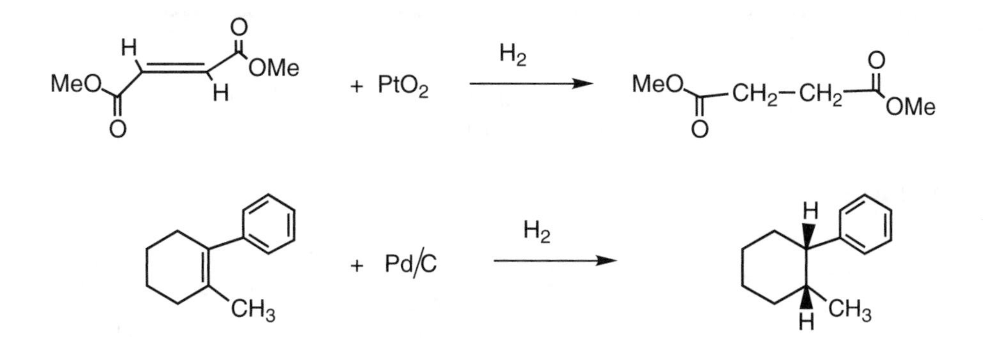

Formation of Vicinal Diols With Osmium Tetroxide

Alkenes are reliably oxidized to vicinal diols with osmium tetroxide, OsO_4. The first step of this reaction is a concerted addition of OsO_4 to an alkene to form an **osmate ester**:

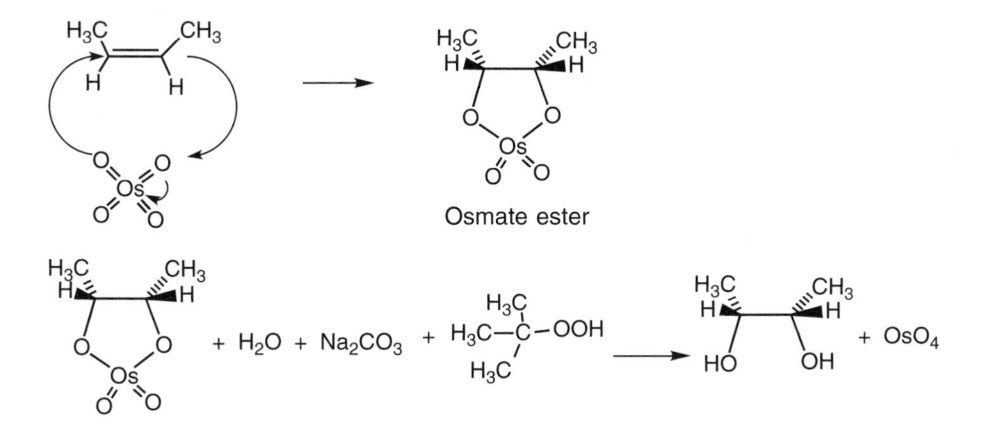

Formation of an osmate ester creates vicinal C—O bonds and reduces the osmium to Os(VI). In the second step of the reaction, the osmate ester is hydrolyzed under basic conditions to break the O—Os bonds and form alcohol groups. Normally this is done in the presence of a oxidizing agent that converts the osmium back to the Os(VIII) oxidation state. Osmium is more expensive than gold or platinum, so it is used in catalytic amounts. One useful oxidant that does not also oxidize alcohols is *t*-butyl hydroperoxide, $(CH_3)_3COOH$. Other, less expensive, reagents can convert alkenes to vicinal diols, including $KMnO_4$. The same concerted mechanism is followed. However, yields with the osmium reagent are usually higher.

EXERCISE

Draw the product of the reaction of cyclohexene with osmium tetroxide in its chair form and the hydrolysis product.

Solution

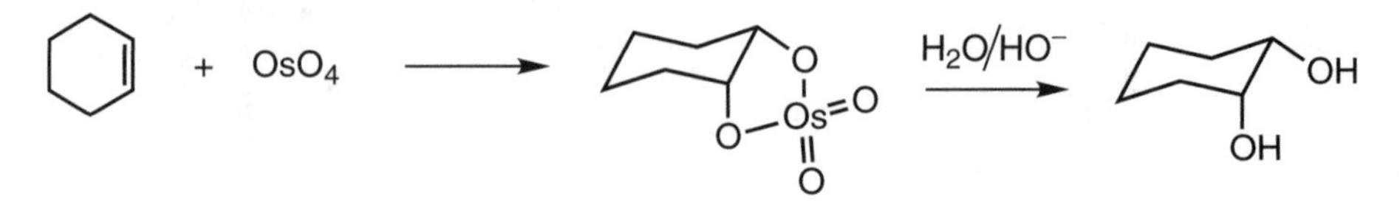

Recall that in cyclohexyl systems, cis-1,2 groups are axial and equatorial.

11
SYNTHESIS AND REACTIONS OF ALKYNES

The alkyne functional group is one of the more reactive functional groups in organic chemistry. The carbon-carbon triple bond is nucleophilic since there are six electrons packed into a small region of space. The linear geometry makes the triple bond sterically accessible. Many of the reactions of alkynes mimic those of alkenes, although there are some unexpected differences.

Acetylene, C_2H_2, was for a time a central starting material for the organic chemical industry. When limestone, $CaCO_3$, and coke (carbon, derived from coal) are heated, calcium carbide, CaC_2, is formed. Adding this to water produces acetylene. Today, acetylene is made by heating ethylene (derived from natural gas) to high temperatures, which forms H_2 and acetylene in an entropy-driven process.

Alkynes with one ≡C—H bond are called **terminal alkynes**. If alkyl groups are attached to both ends of the triple bond, the molecule is an **internal alkyne**. Just like alkenes, alkyl groups stabilize triple bonds.

SYNTHESIS OF ALKYNES

Elimination Reactions That Give Alkynes

If one has a molecule that contains either **vicinal halides** (1,2-substituted) or **geminal halides** (1,1-substituted), a double elimination can take place to give an alkyne. Normally a strong base, such as LDA [LiN(*i*-Pr)$_2$] or $NaNH_2$ in liquid ammonia is needed for such an elimination. The first elimination generates a **vinyl halide** that undergoes a second loss of H—X to give the alkyne. In this second elimination, the H and X groups can be cis or trans to one another. If these **double-dehydrohalogenation** reactions produce a terminal alkyne, the slightly acidic C—H bond is deprotonated by the strong base to give a salt. Such a salt is protonated by water or mild acid to give the desired alkyne.

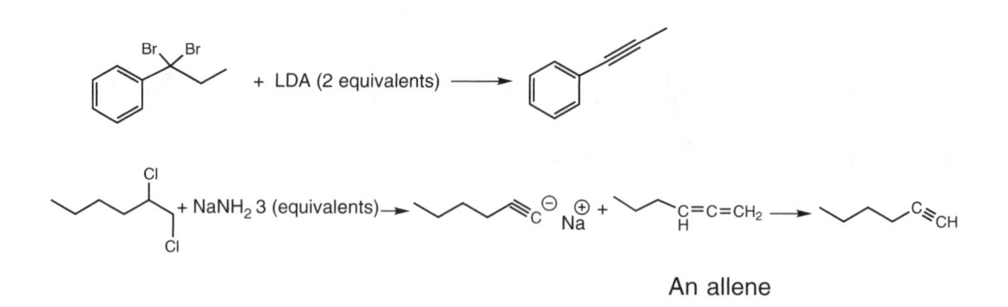

Often, there is more than one way to eliminate pairs of H—X. Then the elimination produces 1,2 dienes, called **allenes**, in addition to alkynes. A mixture of products is formed in such cases, and one must separate the desired compound from the by-products.

Synthesis of Alkynes by Alkylation Reactions

Terminal alkynes have a relatively acidic C—H bond, with a pK_a of approximately 26. Strong bases, including organic lithium reagents, RLi, and $NaNH_2$ or LDA, readily deprotonate terminal alkynes to give a carbon nucleophile that undergoes S_N2 reactions with the usual substrates:

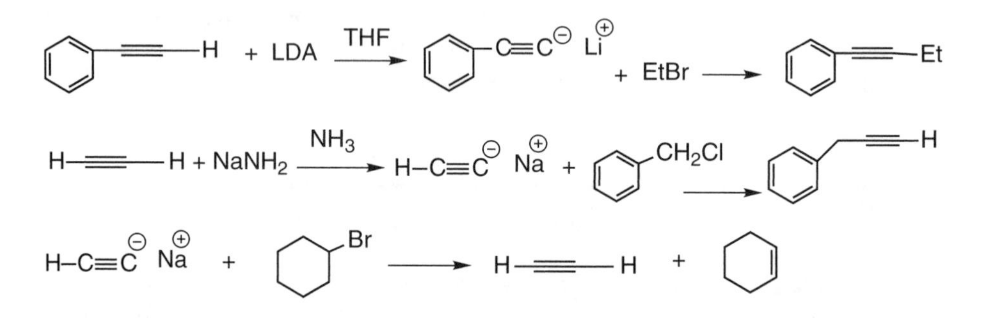

Methyl, allyl, benzyl, and primary substrates react with alkyne anions but, because these anions are much more basic than HO^-, they give mostly elimination products with secondary substrates.

EXERCISES

Propose a reaction that would give the following products.

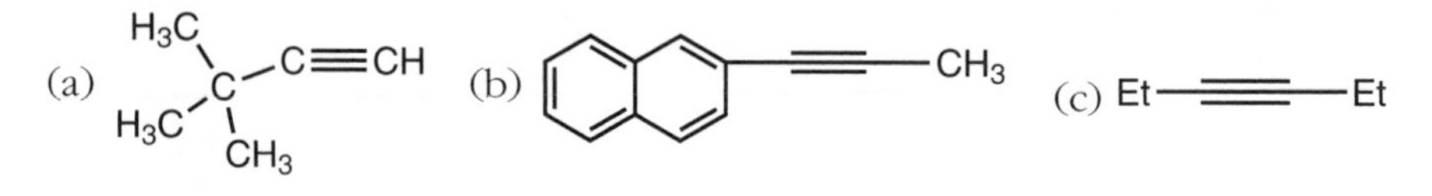

Solutions

(a)

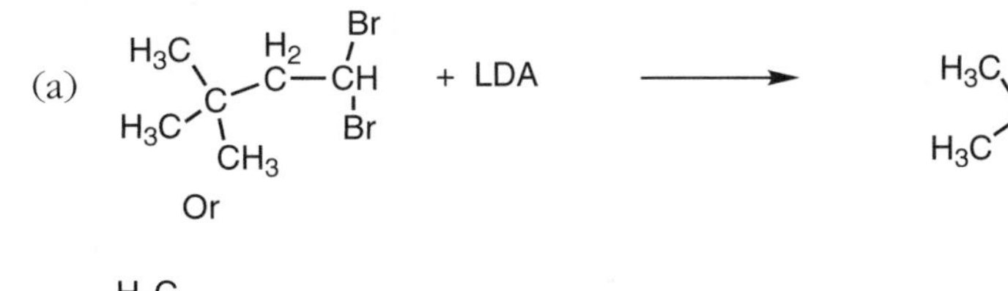

H_3C—C(CH_3)$_2$—C≡CH

$(CH_3)_3C$—CHBr—CH_2Br

An S_N2 reaction will not work.

(b)

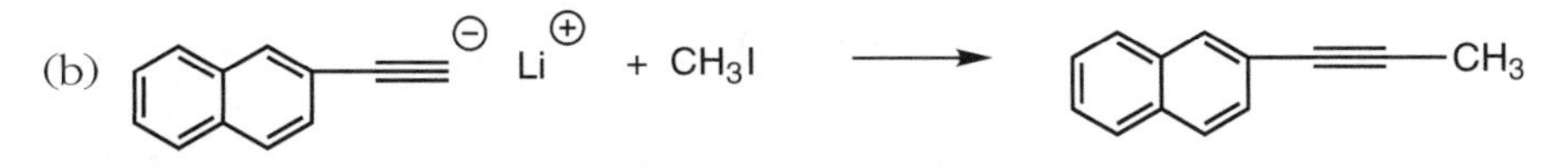

Elimination routes are also possible.

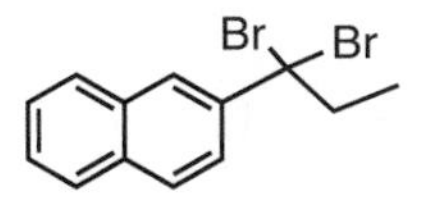

Elimination of this dibromide would avoid allene formation.

(c) Et—C≡C$^{\ominus}$ Li$^{\oplus}$ + EtCl ⟶ Et—C≡C—Et

Any possible elimination reaction would give some allene.

REACTIONS OF ALKYNES

Catalytic Hydrogenation

A triple bond is reduced much faster than a double bond in catalytic hydrogenation. If normal catalysts, such as PtO_2 or Pd/C react with a triple bond in the presence of H_2, two equivalents of hydrogen are absorbed and the alkane is isolated. Normally, one cannot stop the reaction at the alkene stage. However, special catalysts are available that are so unreactive that they can reduce only alkynes and stop after one equivalent of H_2 is taken up. The best known catalyst is the **Lindlar catalyst**, named after its inventor. It is Pd deposited on $BaSO_4$ with added quinoline. Quinoline is an organic amine base that is a **catalyst poison** which means it strongly deactives a catalyst:

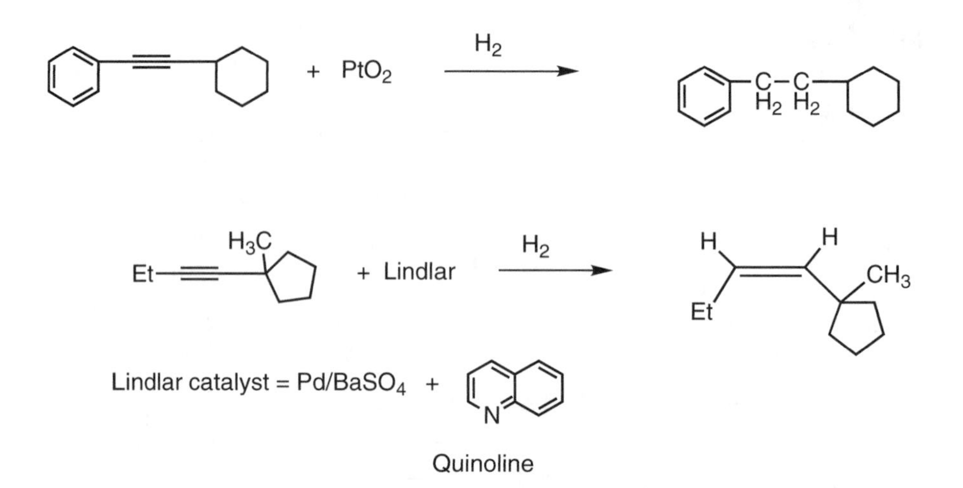

The Lindlar catalyst produces the cis isomer. If one wants to synthesize the trans alkene from an alkyne, a different reaction is available.

Dissolving Metal Reductions

Reduction of alkynes with sodium metal in the solvent ammonia (which liquefies below –33°C) gives the trans alkene. The mechanism of this reaction is rather complicated and involves types of intermediates that are new.

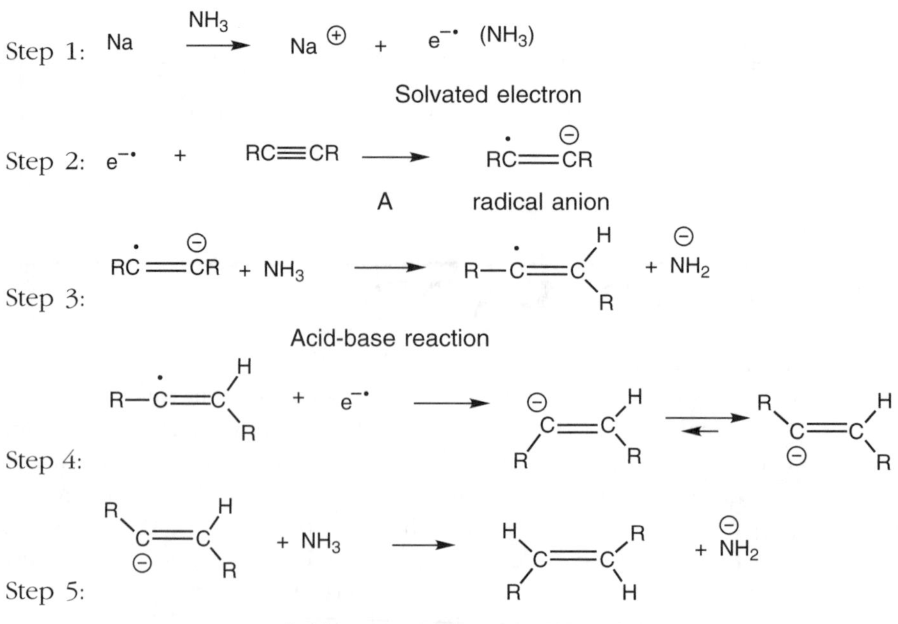

Sodium ionizes in ammonia to give up an electron, which is solvated by the ammonia molecules. An electron is both negatively charged and an unpaired spin, so it is represented as $e^{\bullet-}$. In step 2, the electron adds to one of the π bonds. Since a bonding orbital can hold only two electrons, the bond breaks. Two of the three electrons go to one carbon, leading to an anion, and the third electron is on the other carbon, making it a radical. This intermediate is called a **radical anion**. Step 3 is an acid-base reaction leading to one new C—H bond. In step 4, the radical picks up another electron to form an anion at an sp^2 carbon. Such an anion undergoes cis-trans isomerization, and the trans isomer is more stable because of steric effects. In step 5 the predominate isomer undergoes an acid-base reaction to form the product, the trans alkene. The action of sodium in ammonia is called a **dissolving metal reduction** since the sodium is dissolved in ammonia. Many functional groups are reduced by this reagent, so the formation of trans alkenes is usually limited to simple alkynes.

Hydration of Alkynes

Water adds to alkynes in the presence of a mercury catalyst, usually $HgSO_4$ in the presence of H_2SO_4. The mechanism of this reaction is more complex than one might think, but the overall transformation is Markovnikov addition of water to the triple bond:

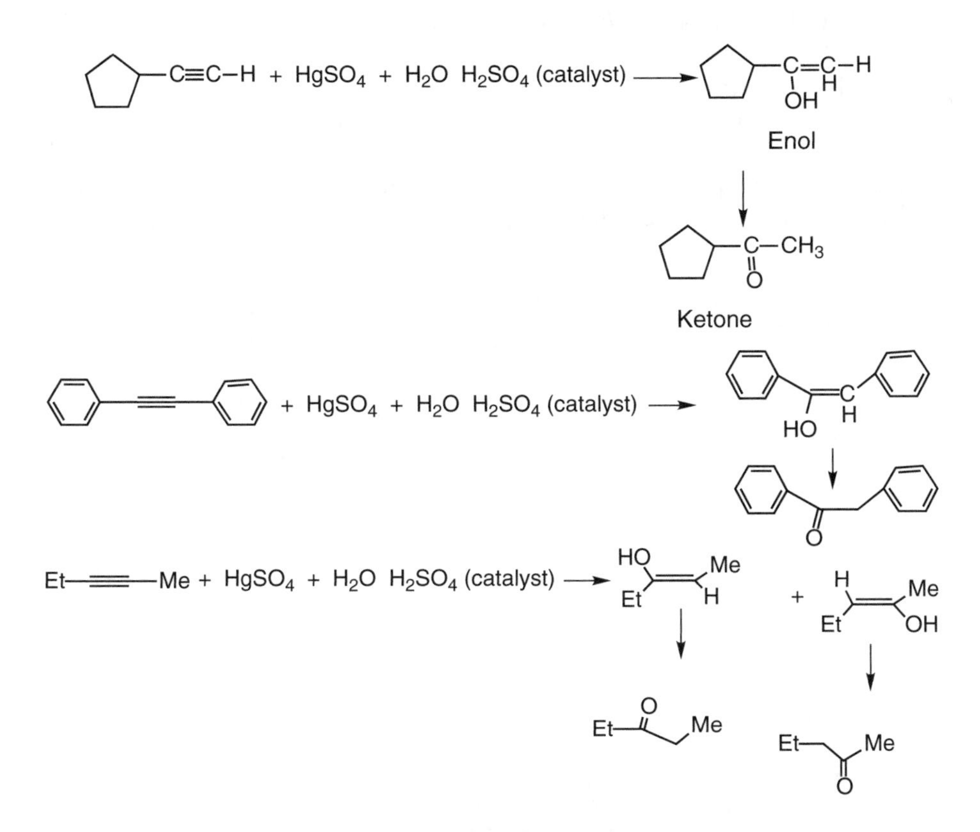

When water adds to a triple bond, the product is an **enol**. An enol is a compound with an OH group on a double bond. Enols are unstable and rapidly undergo **tautomerization**, a specific rearrangement of enols, to a carbonyl compound. In tautomerization, the C—O single bond becomes a C=O double bond and the adjacent sp^2 carbon picks up a H to become an sp^3 carbon. Enols and tautomerization are covered in detail in Chapter 19. As shown in the last example, if both groups on an alkyne are equally able to stabilize a carbocation, the Markovnikov addition gives a mixture of ketone products.

Anti-Markovnikov addition of water to alkynes can be carried out using hydroboration, just as with alkenes:

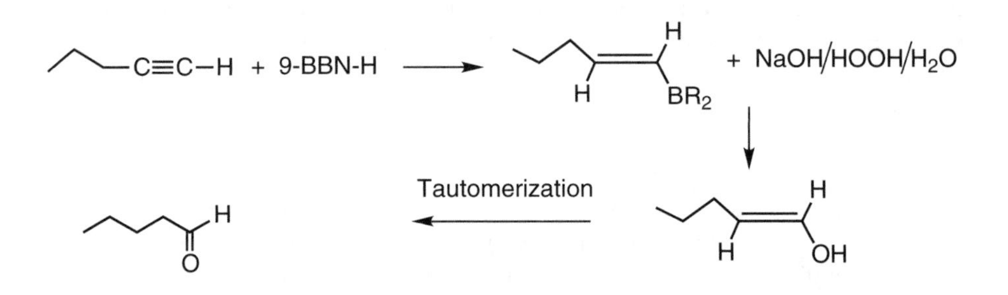

The **vinyl boranes** produced when alkynes are hydroborated can also be converted to cis alkenes in a stereospecific reaction. The C—B bond is cleaved with carboxylic acids, producing a C—H bond. This provides an alternate way to make cis alkenes in addition to Lindlar hydrogenation. If the carboxylic acid has a O—D or O—T acidic hydrogen, one can label the product alkene with deuterium or tritium. Such molecules are extremely useful for studies of drug metabolism, as well as for other purposes.

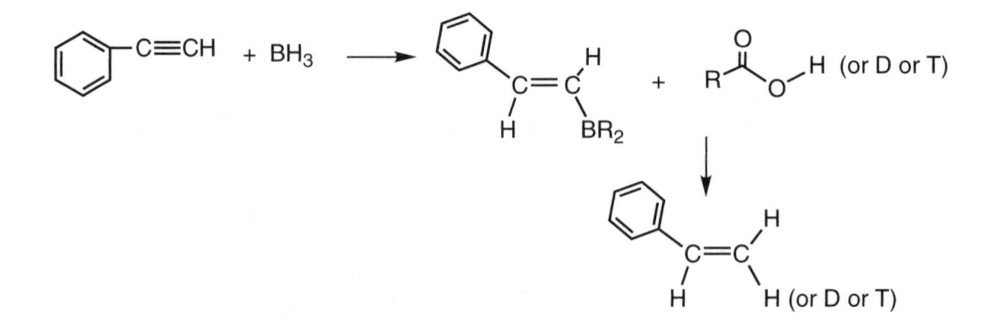

Addition of Halogens to Alkynes

Triple bonds add Br_2 and Cl_2 just as double bonds do. However, when X^+ adds to a triple bond, it generates an sp^2 carbon with a positive charge. This is an intermediate that is high in energy. Thus, although alkynes are more reactive than alkenes in most addition reactions, the addition of a halogen to a triple bond occurs at a rate about the same as that of addition to a double bond. The E isomer is obtained when one equivalent of halogen is added to an alkyne:

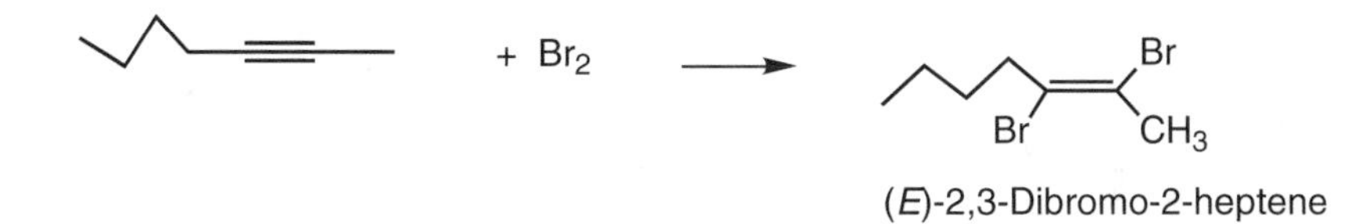

(*E*)-2,3-Dibromo-2-heptene

For the reasons outlined above, the addition of hydrogen halides to alkynes is relatively slow. The reaction is a Markovnikov addition, but the initially formed **vinyl halide**, RCH=CXR, is often more reactive than the starting alkyne. The lone electron pairs on the halogen, X, can stabilize a carbocation, so vinyl halides can react with a second equivalent of HX to give $RCH_2—CX_2R$, a geminal dihalide.

EXERCISES

Fill in each blank with the correct reagent combination, starting material, or product.

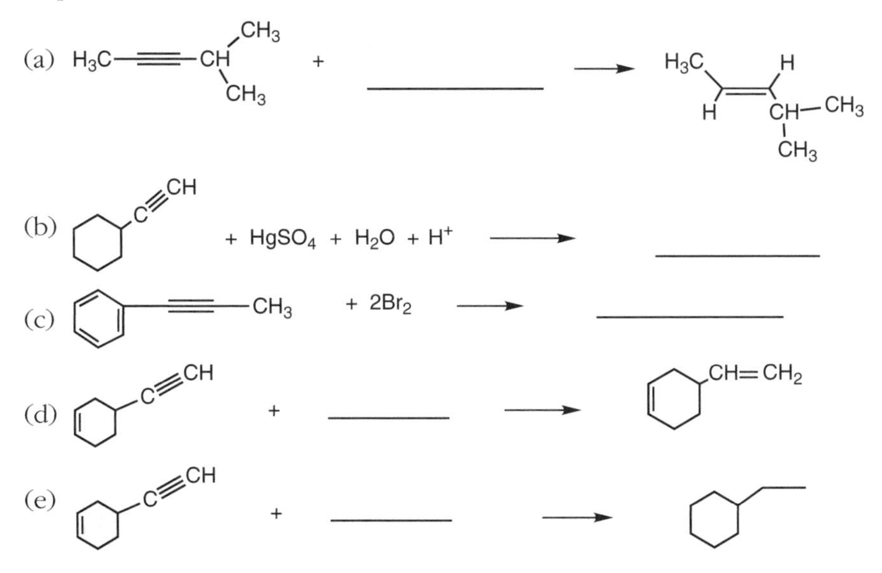

Solutions

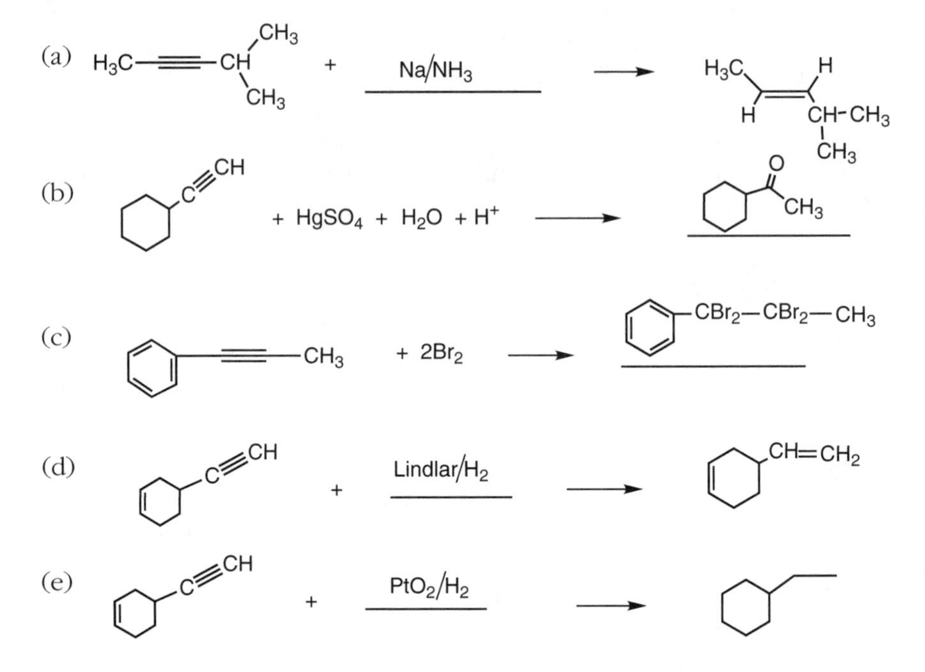
(a)
H3C—≡—CH
CH3
CH3
+
Na/NH3
H3C
H
H
CH-CH3
CH3
(b)
C≡CH
+ HgSO4 + H2O + H+
O
CH3
(c)
CH3
+ 2Br2
CBr2—CBr2—CH3
(d)
C≡CH
+
Lindlar/H2
CH=CH2
(e)
C≡CH
+
PtO2/H2

12
PROPERTIES AND REACTIONS OF DIENES

Often, a system exhibits new properties when components of the system interact. We have seen that isolated double bonds have a characteristic stability and reactivity that make alkenes an important class of molecules in organic chemistry. When two double bonds in a **diene** are next to one another, new kinds of reactivity emerge. It is useful to classify dienes into three types, depending on the degree of interaction: **isolated**, **cumulated**, and **conjugated**. This chapter will focus on conjugated or **1,3 dienes**. These compounds possess unusual stability, and the overlap of *p* orbitals leads to new kinds of reactivity. The basis of the extra stability of conjugated dienes is explained by a simple application of molecular orbital theory.

A triene with three isolated double bonds

2,3-Pentadiene, a cumulated diene

1,3-Cycloheptadiene, a conjugated cyclic diene

1,3-Butadiene, an acyclic conjugated diene

MOLECULAR ORBITAL THEORY OF BONDING IN CONJUGATED DIENES

In a conjugated diene like 1,3-butadiene, a total of four atomic *p* orbitals (one on each carbon) make up the π system. Using the linear combination of atomic orbitals (LCAO) approach to molecular orbital theory, we can combine these four atomic orbitals to create four molecular orbitals:

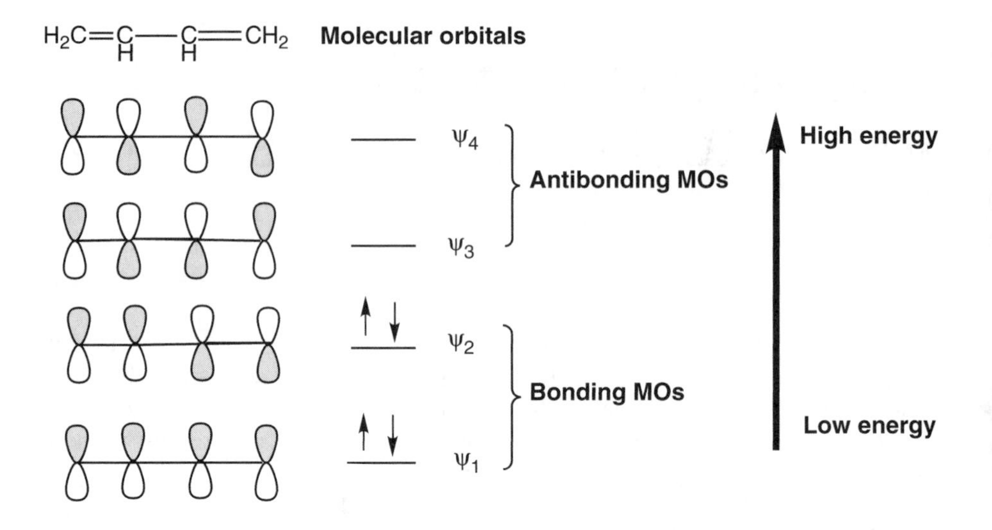

The lowest-energy MO, ψ_1, holds two electrons (represented by the two arrows). These two electrons are delocalized across all four *p* orbitals. The only node is the one enforced by the nature of the *p* orbital. In this orbital this is a bonding interaction between C-2 and C-3, as well as between C-1 and C-2 and between C-3 and C-4. The second-lowest energy MO, ψ_2, has a node between C-2 and C-3. In valence bond terms, it corresponds to a double bond between C-1 and C-2 and between C-3 and C-4. Since there are only four π electrons, the antibonding orbitals are empty.

The way to interpret this MO picture is to realize that the bond between C-2 and C-3 is not a pure σ bond but that it has some double-bond character. The two double bonds are not isolated but interact because of ψ_1. Because of this interaction, or delocalization, a conjugated diene is more stable than two isolated double bonds. Because of ψ_1, two electrons can interact with four positively charged carbon nuclei.

1,2 AND 1,4 ADDITION TO CONJUGATED DIENES

When a conjugated diene, such as 1,3-butadiene, reacts with one molar equivalent of an electrophile, such as Br_2, two products can form, the 1,2 dibromide and the 1,4 dibromide:

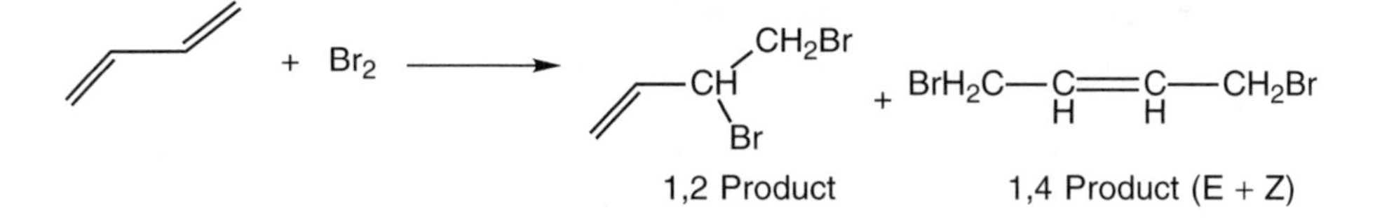

These two products result from the structure of the allylic cation that forms as an intermediate. The electrophile, in this case Br^+, adds to one end of the conjugated diene to give an allylic cation:

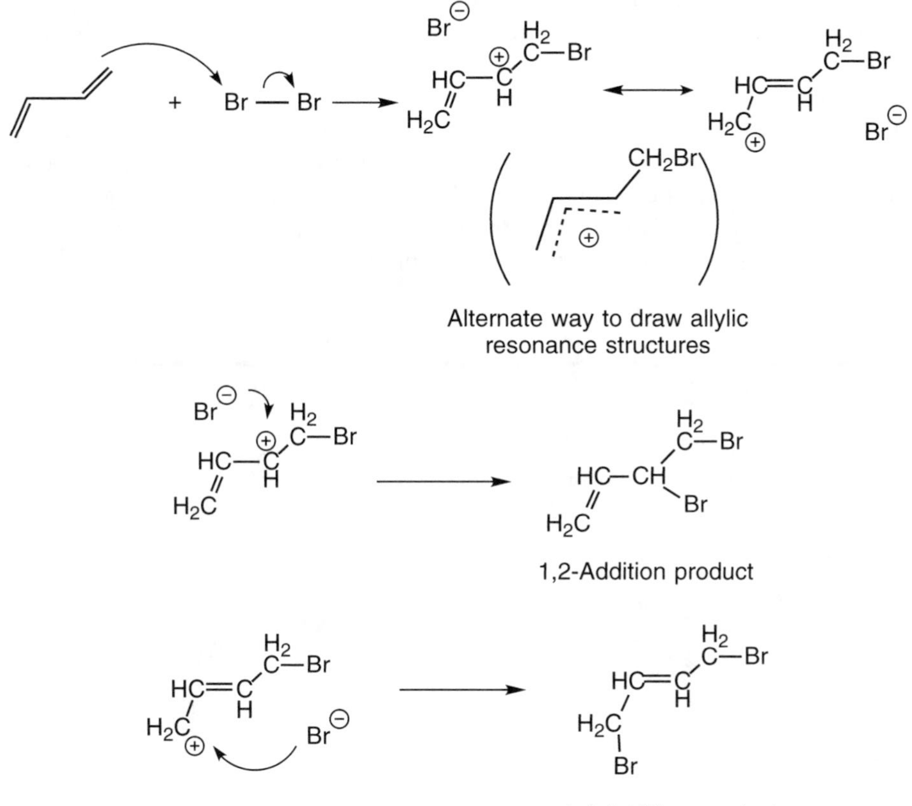

The addition of Br^+ does not take place at C-2 because the resulting carbocation would have no resonance stabilization. Because allylic resonance is so effective at stabilizing a charge, the carbocation in this case is best drawn not as a bridged bromonium ion but as a simple allyl cation.

An allyl carbocation has a positive charge density at C-2 and C-4. The anion can react at either site. Experimentally, it is found that the 1,4-addition product predominates at higher temperatures. At low temperatures (–80°C), the major product is the 1,2-addition product. This change of product ratio is another example of **thermodynamic** vs **kinetic control** as discussed in Chapter 9.

The 1,4-addition product is the more stable, or thermodynamic, product because it contains a double bond substituted with two alkyl groups. The 1,2-addition product is the kinetic product because there is more positive charge density on the secondary carbon of the allylic cation than on the terminal, primary carbon. However, at higher temperatures the allylic bromide of the 1,2 product can reversibly ionize. It can then collapse to the 1,2 or 1,4 product. Since the 1,4 product is more stable, eventually most of the molecules become trapped in the lower-energy, or thermodynamic, 1,4 product.

The structure of the diene determines what product is the thermodynamic product and what product is the kinetic product. For example, 2,4-hexadiene, $CH_3CH{=}CH{-}CH{=}CHCH_3$, adds Br_2 to give a 1,2-addition product, $CH_3CHBr{-}CHBr{-}CH{=}CHCH_3$, and the 1,4-addition product $CH_3CHBr{-}CH{=}CH{-}CHBrCH_3$. Neither of these is appreciably more stable than the other, and neither would be formed faster than the other since the charge density in the allyl intermediate is symmetric. Thus, because of the structure of the diene, there is no kinetic or thermodynamic product.

EXERCISES

What are the kinetic product and the thermodynamic product for the following electrophilic addition reactions?

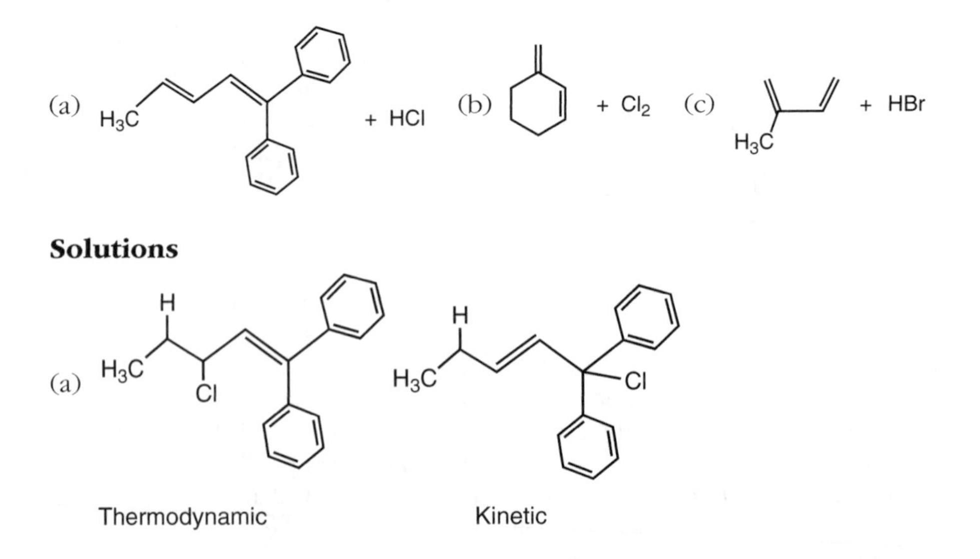

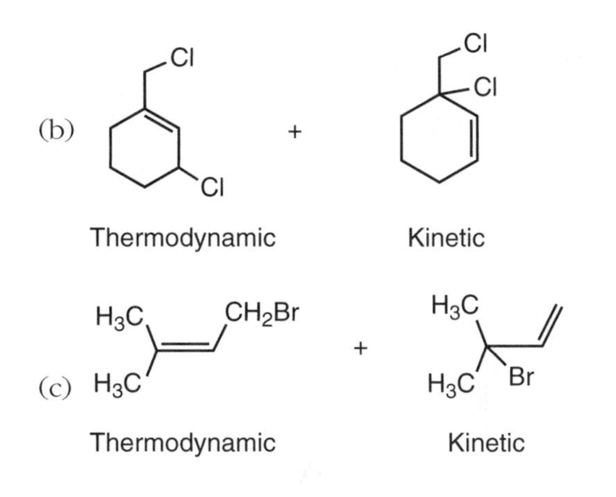

THE DIELS-ALDER REACTION

Dienes and Dienophiles

One of the most powerful reactions for forming new carbon-carbon bonds is the Diels-Alder reaction (discovered by the German chemists Otto Diels and Kurt Alder in 1928). In this reaction, a 1,3 diene reacts with a double-bonded or triple-bonded partner called a **dienophile** (lover of dienes) to produce a new cyclohexene ring:

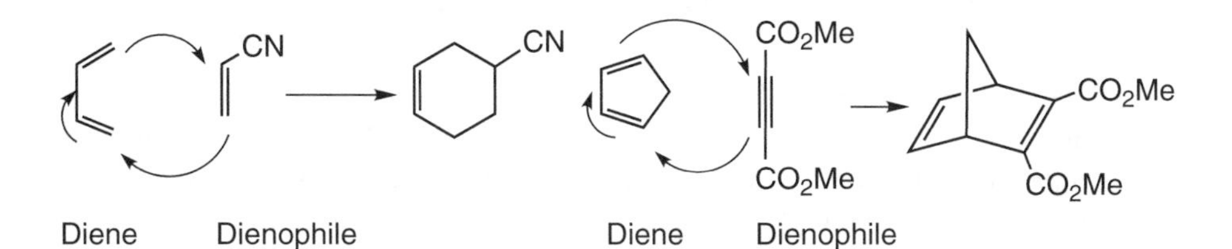

This is another example of a concerted reaction. Three pairs of electrons move in a closed cycle. Two π bonds become σ bonds, and the 2,3 bond of the 1,3 diene becomes a double bond.

The structure of the 1,3 diene (normally just called the diene) and the dienophile determine how efficient the Diels-Alder reaction will be in a specific case. In order for the diene to work, it must adopt a conformation where C-1 and C-4 at the ends of the diene are close to one another, called the **s-cis** conformation (for cis around a single bond) and not adopt the **s-trans** conformation:

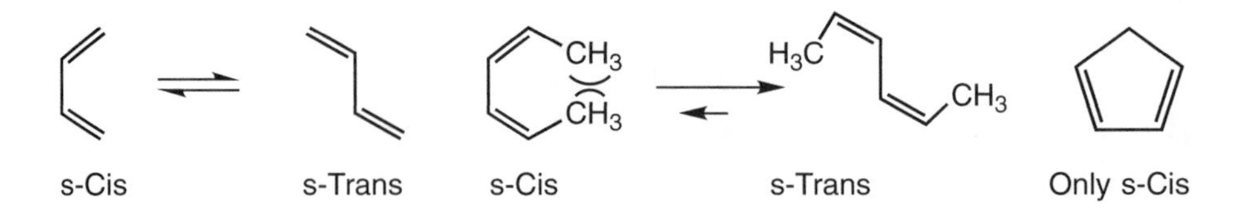

A diene like 2(*Z*),4(*Z*)-hexadiene is sterically unable to adopt an s-cis conformation. The ends of the diene are too far apart to react with a dienophile, so no Diels-Alder reaction takes place. The electronic nature of groups on a diene also influence the rate of a Diels-Alder reaction. Groups that donate electron density to a π system, such as alkyl groups, —OR groups, and $—NR_2$ groups, favor the Diels-Alder reaction. Electron-withdrawing groups such as carbonyl groups, —CN, and halogens retard the Diels-Alder reaction when on the diene.

Reactive dienophiles have electron-withdrawing groups on double or triple bonds. Halogens and carbonyl groups such as esters, aldehydes, ketone groups, and nitriles all enhance the reactivity of dienophiles. Some examples of commonly used dienophiles are shown here.

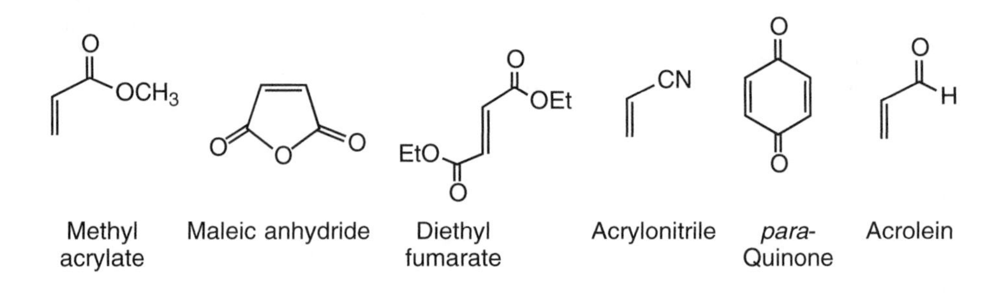

An important feature of the Diels-Alder reaction is that stereochemistry is maintained in going from starting materials to products. If two groups are cis on a dienophile, they will be cis in the product, and the same is true for trans groups:

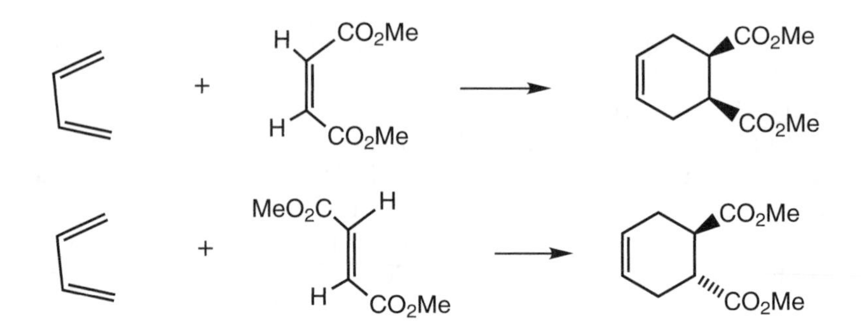

As usual, we show only one enantiomer, although the product from 1,3-butadiene and dimethyl fumarate is chiral, so that a racemic mixture is produced.

Any stereochemistry in the diene is also maintained, although this is harder to see. It is perhaps easiest to understand if one first looks at the reaction of 1,3-cyclopentadiene with a dienophile:

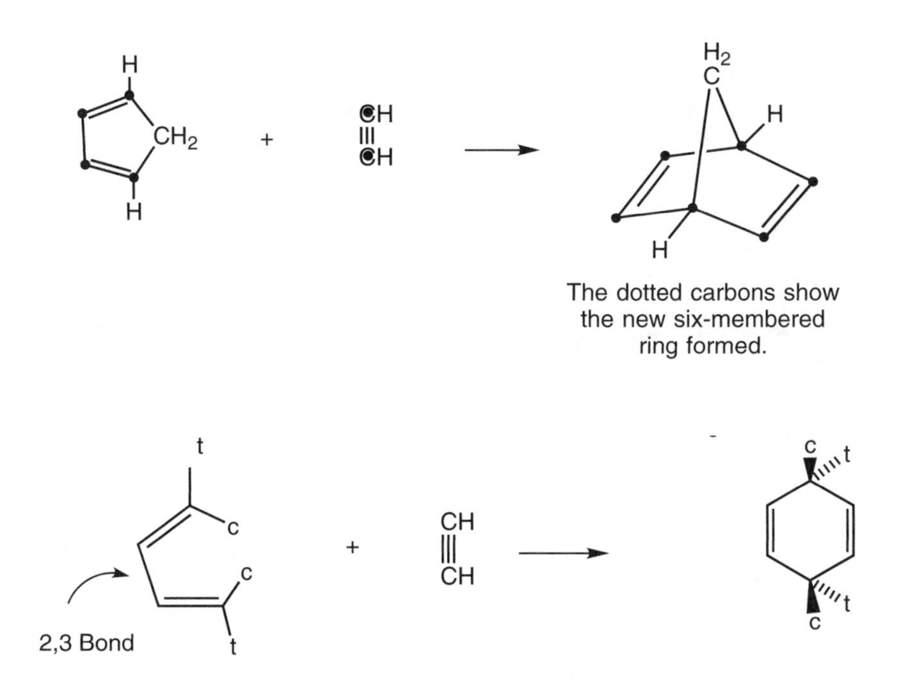

In the above reaction, compare the positions of the CH_2 group in the cyclic diene and the two c groups in the acyclic diene. The c groups are both cis to the 2,3 bond of the diene, and they end up in the same relative orientation in the product. Likewise, the t groups are trans to the 2,3 bond in the diene, and they have the same relative orientation in the product. Therefore, the stereochemical relationships in the diene are maintained in the six-membered ring product.

EXERCISE

For compounds (a) through (d) give the diene and dienophile that reacted to give the observed product. For reactions (e) and (f), give the expected product.

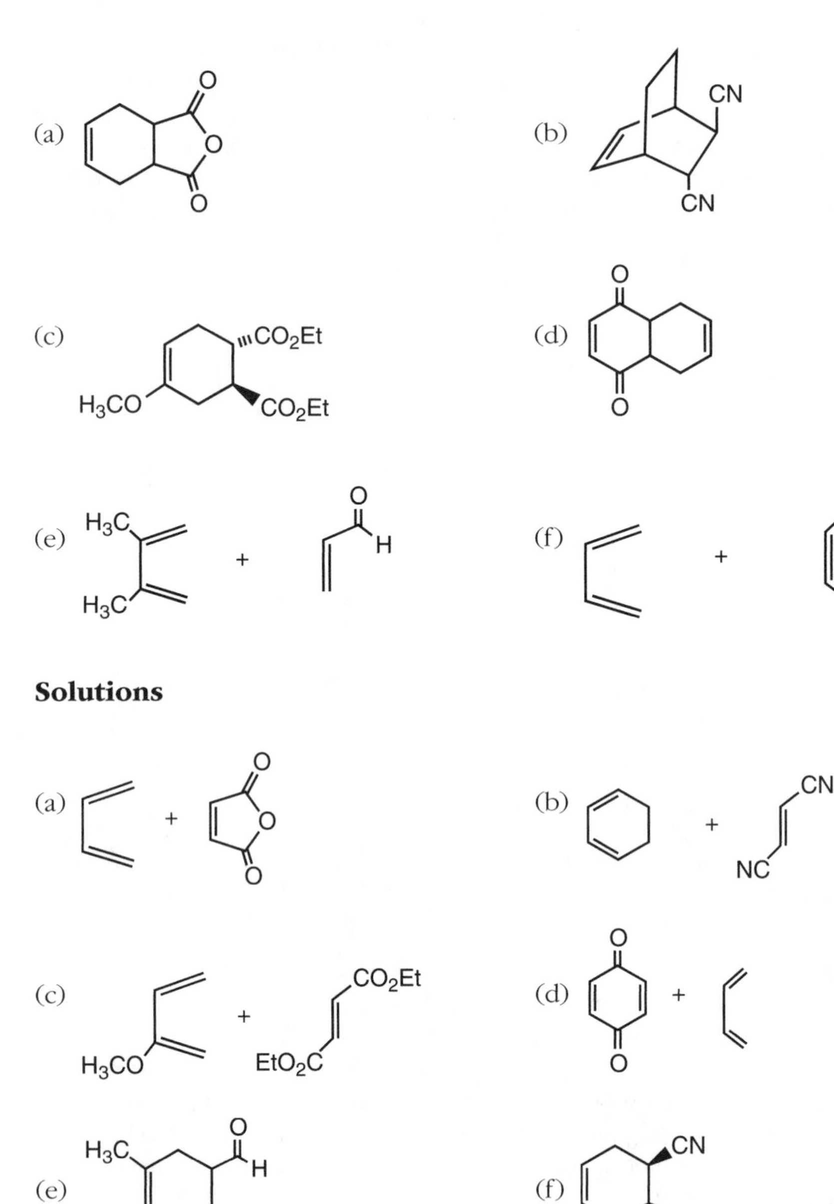

The Endo Rule

There is another stereochemical aspect of the Diels-Alder reaction that is known as the **endo rule**. When a diene and a dienophile react, the electron-withdrawing group on the dienophile can point under the diene toward the diene 2,3 bond. Or the group can point away from the diene:

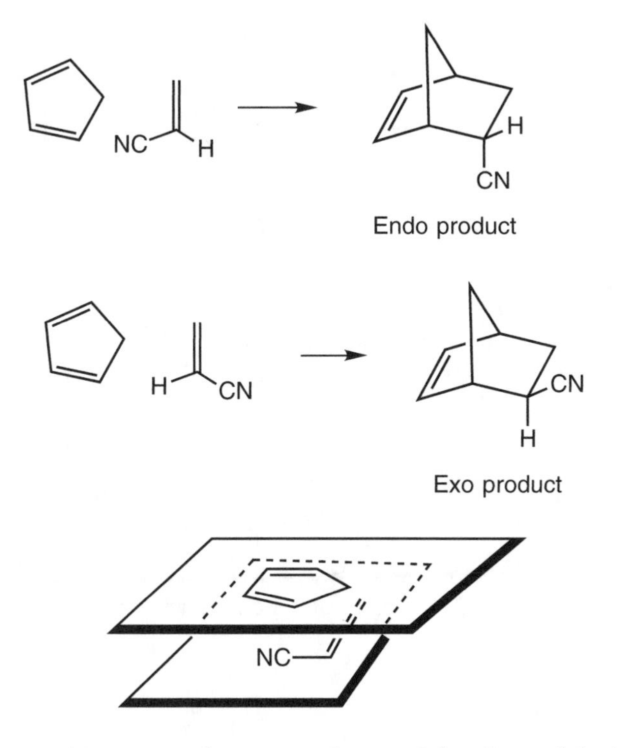

In the endo transition state, the two carbons of the dienophile lie under C-1 and C-4 of the diene. As new bonds form, the $—CH_2$ group is pushed up and the —CN group rotates down, giving the endo product.

The endo transition state, which is more compact, is the favored one. The endo product is the one with the electron-withdrawing group underneath the bicyclic ring. This kind of endo transition state also takes place with acyclic dienes.

EXAMPLES

Examples of endo and exo products are shown here. The endo product is the major product.

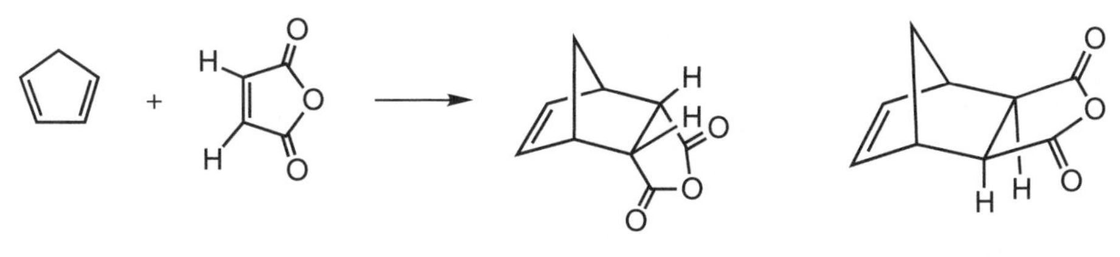

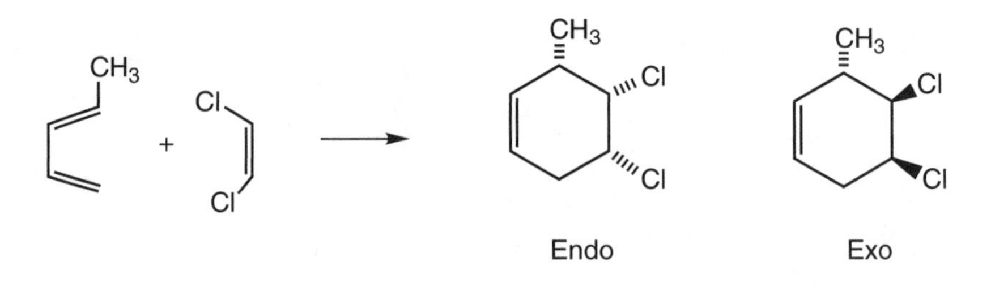

Recall that there is an enantiomer of the Diels-Alder product formed in an equal amount. The dienophile can approach from the top or bottom face of the diene.

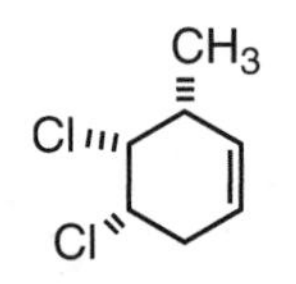

Enantiomer of endo product

13 RADICAL REACTIONS OF HYDROCARBONS

RADICAL STABILITY

Radicals are similar to carbocations in that they are electron-deficient. The unshared electron is in a *p* orbital, which gives the radical carbon sp^2 hybridization just as in carbocations. Thus, the order of stability of radicals is 3° > 2° > 1° > methyl. However, the extent of electron deficiency is less than in carbocations, so even methyl radicals can be made. Resonance stabilizes an unpaired electron just as it stabilizes charged species, so that simple allyl and benzyl radicals form more readily than 3° radicals.

Bond Dissociation Energies

When molecules are heated to very high temperatures in the gas phase, bonds can undergo **homolysis**. The bond is broken so that each half leaves with one electron. The energy (enthalpy) it takes to break a bond homolytically in the gas phase is called the **bond dissociation energy**. The same amount of energy is released when two radicals come together to form a bond.

From Table 13.1, one can see how the structure of a molecule affects the bond dissociation energy, or **bond strength**, of a given bond. Longer bonds tend to be weaker bonds. Thus, bonds to iodine are weaker than bonds to bromine or chlorine. In addition, bonds to groups that form stable radicals are weaker than other kinds of bonds. For example, breaking an H—Ph bond requires 112 kcal/mol, but breaking an H—CH_2Ph bond needs only 70 kcal/mol, less than that needed to break an H—I bond.

Most reactions in organic chemistry take place at moderate temperatures and are dominated by enthalpy, not entropy. The data on bond dissociation energies (BDEs) can be used to decide if a given reaction will take place. The enthalpy change of a given reaction, ΔH^0, equals the sum of the broken bonds' BDEs minus the sum of the BDEs of the bonds that form. A negative ΔH^0 means the reaction can take place as written.

TABLE 13.1. HOMOLYTIC BOND DISSOCIATION ENERGIES (kcal/mol)

Cl—Cl	Br—Br	I—I		CH_3—Cl	CH_3—Br	CH_3—I
58	46	36		84	70	56
H—Cl	H—Br	H—I		Et—Cl	Et—Br	Et—I
103	88	77		81	68	53
H—OH	H—NH_2	H—SCH_3	H—OCH_3	*i*—Pr—Cl	*i*—Pr—Br	*i*—Pr—I
119	103	88	91	80	68	51
H—CH_3	H—Et	H—*i*Pr	H—*t*—Bu	*t*—Bu—Cl	*t*—Bu—Br	*t*—Bu—I
104	98	95	91	79	65	50
H—CH=CH_2	H—C_6H_5			CH_3—CH_3	Et—CH_3	*i*—Pr—CH_3
104	112			88	85	84
H—CH_2CH=CH_2	H—CH_2—C_6H_5			*t*—Bu—CH_3	PhCH_2—CH_3	
87	70			81	72	

For example, if one wanted to invent a catalyst for a reaction between ethane and water to give methanol and methane, one would do a simple calculation:

$$HO\cdot + H_3C{-}CH_3 \longrightarrow HO{-}CH_3 + \cdot CH_3 \quad (13.1)$$

$$\cdot CH_3 + H_2O \longrightarrow CH_4 + HO\cdot \quad (13.2)$$

$$H_3C{-}CH_3 + H_2O \longrightarrow CH_4 + CH_3{-}OH \quad (13.3)$$

Breaking H_3C-§-CH_3: + 88 kcal/mol Forming HO—CH_3: −91 kcal/mol

Breaking H§OH: + 119 kcal/mol Forming H—CH_3: −104 kcal/mol

$$\Delta H^0(\text{reaction}) = 119 + 88 - 91 - 104 = +12 \text{ kcal/mol}$$

Adding Equations (13.1) and (13.2) and crossing out species that appear on both sides of the arrow give Equation (13.3) the desired equation. Breaking bonds requires energy, and forming bonds releases energy, so adding all the bond breaking and forming values gives the reaction's ΔH^0. In this example, it is +12 kcal/mol. The reaction takes place in the reverse direction. Thermodynamics depends on only the starting and ending compounds and not on the path taken. Therefore, any attempt to develop a catalyst to make methane and methanol from ethane and water would be fruitless.

EXERCISES

Calculate the ΔH^0 for direct chlorination, bromination, and iodination of methane with Cl_2, Br_2, and I_2 to give CH_3X and HX.

Solution

$$X_2 + CH_4 \longrightarrow CH_3X + HX$$

Cl_2: $\Delta H^0 = 58 + 104 - 84 - 103 = -25$ kcal/mol
Br_2: $\Delta H^0 = 46 + 104 - 70 - 88 = -8$ kcal/mol
I_2: $\Delta H^0 = 36 + 104 - 56 - 71 = +13$ kcal/mol

RADICAL REACTIONS

Free Radical Chain Reactions

If one mixes ethane and Br_2 together in the dark, no reaction takes place. However, if one shines light on the mixture, a reaction begins and ethyl bromide and HBr are formed. Alternatively, if one adds a compound with a very weak single bond, such as an organic peroxide, ROOR, to the ethane-Br_2 mixture in the dark and heats it, ethyl bromide and HBr are again produced. These reactions are examples of **free radical chain reactions**, where a radical (often called a **free radical**) is generated and then reacts to produce new radicals.

All chain reactions have three classes of reactions: **initiation**, where radicals form from nonradical precursors; **propagation**, where a radical reacts with a stable molecule to give a new radical and a product; and **termination**, where two radicals come together to form nonradical molecules. For two radicals to come together in a termination step, a third body must be present to carry away energy given off by bond formation. Often that is the wall of the reaction vessel. The bromination of ethane is a free radical chain process that can be initiated by light. Br_2 absorbs light, and the relatively weak Br—Br bond breaks to form two Br radicals. Normally, the intensity of light used is low enough that only a small fraction of bromine molecules form bromine radicals. The symbol $h\upsilon$ is used to indicate light.

Initiation:

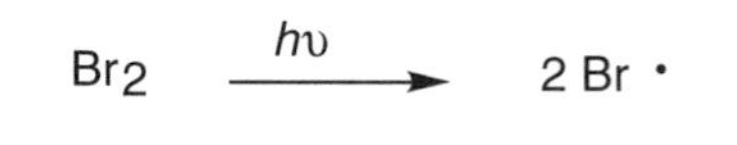

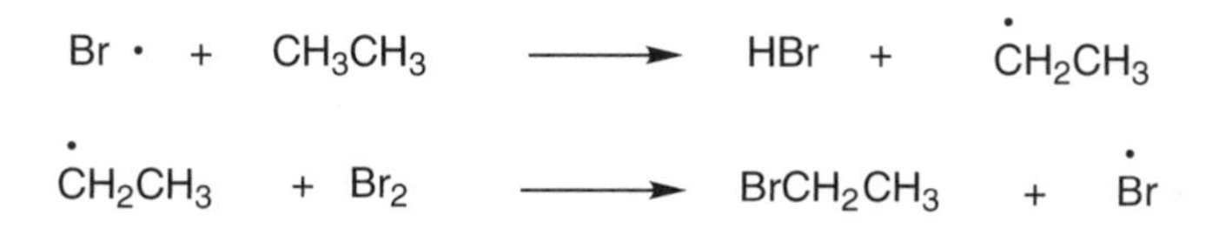

Termination:

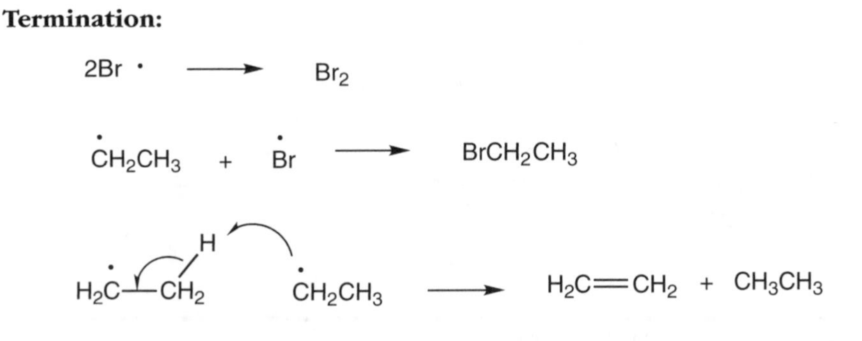

Normally, there are two or more propagation steps. Adding them together and crossing out species common to both sides gives the net reaction. A number of termination steps are possible, and usually one needs to write just one or two of the possible ones. The last termination step in the Br_2/ethane reaction is called a **disproportionation reaction**, in which two hydrocarbon radicals react to give an alkene and an alkane.

The important fact to remember about radical chain reactions is that radicals are always present in very low concentrations. The chance that two radicals will meet is extremely small since they are so rare. Thus, many propagation steps take place for each termination step. A common kind of mistake students make is to show the step $Br^{\bullet}$ + $CH_3CH_2^{\bullet}$ to give CH_3CH_2Br as a major product-forming step. It is the termination step because two radicals are coming together. The ethyl bromide product is made by reacting the rare ethyl radical with the abundant Br_2.

A number of compounds known as **radical initiators** decompose at moderate temperatures to produce radicals. With these compounds, initiation involves, first, the molecule breaking apart to generate radicals and, second, the radicals reacting with the substrate to start the chain reaction:

Di-*t*-butyl peroxide

Benzoyl peroxide

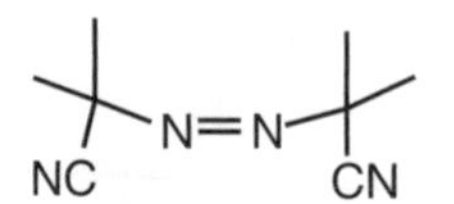

Azobis (isobutyronitrile) (AIBN)

O_2

Oxygen

$Bu_3Sn—SnBu_3$

Hexabutylditin

Recall that oxygen is a **triplet**, a molecule with two unpaired electrons, so it can act as a radical initiator at any temperature. The other molecules decompose between 40° and 100°C. AIBN breaks at the C—N bond to generate two carbon radicals and nitrogen.

EXERCISE

Write the free radical chain mechanism for the following reaction, identifying the initiation, propagation, and two possible termination steps.

$$CF_3H + Cl_2 + \text{di-}t\text{-butyl peroxide} \rightarrow CF_3Cl + HCl$$

Solution

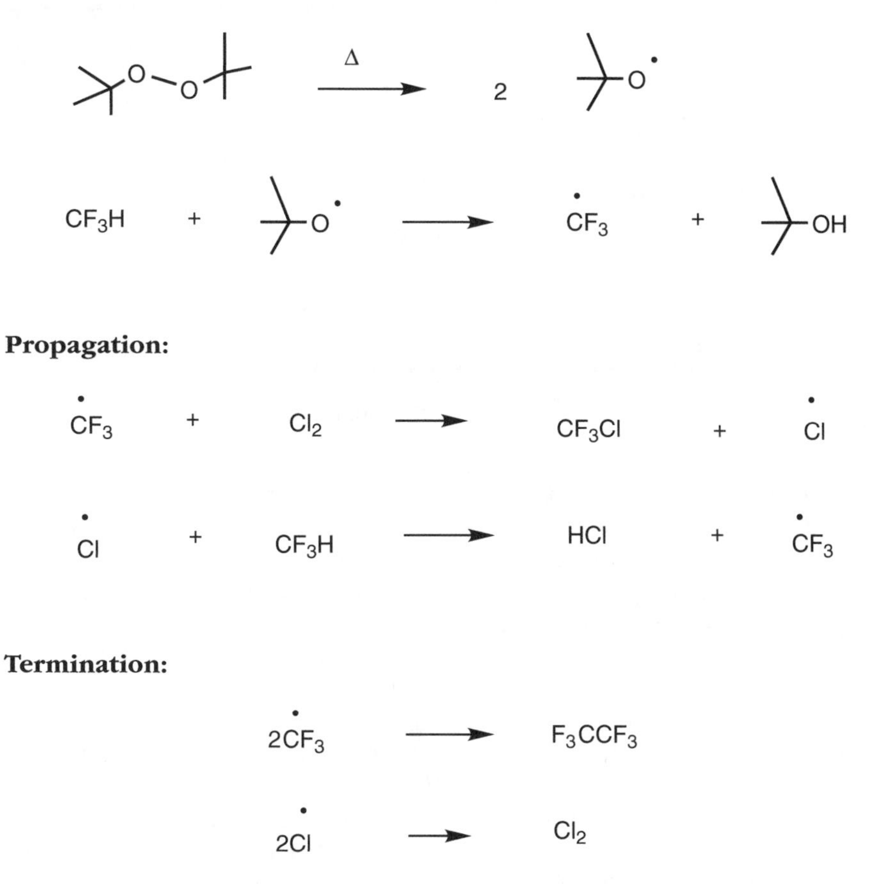

The *t*-butoxy radical is expected to react with the C—H bond to form an O—H bond rather than to react with Cl_2 to form an O—Cl bond because the O—Cl bond is very weak.

Selectivity in Radical Reactions

When a hydrocarbon reacts with Cl_2 under free radical conditions, a mixture of monochloro products is formed as long as there are different kinds of hydrogens to react. With propane, for example, there are six 1° C—H bonds and two 2° C—H bonds. If chlorination were purely statistical, one would expect a 3:1 ratio of $ClCH_2CH_2CH_3$ to $CH_3CHClCH_3$. In fact, the product ratio is 45:55. The 2° C—H bond is more reactive than the 1° C—H bond by 3.7:1 [calculated from the ratio (55/2):(45/6)]. The 1°:3° C—H bond reactivity ratio is 1:5. Thus, C—H bonds that are the weakest react the fastest in free radical chlorination.

In a radical chlorination reaction, the first product-forming step is the reaction of $Cl^{\bullet}$ with the hydrocarbon to give HCl and a 1°, 2° or 3° radical. The radical chlorination of 1° and 2° C—H bonds is shown here.

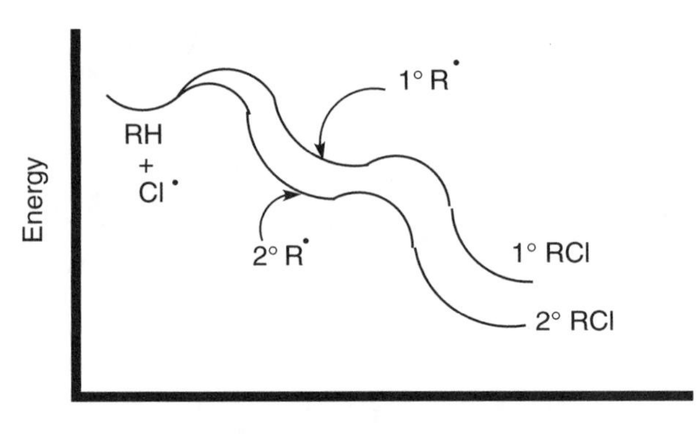

The activation energy needed to break the weaker 2° C—H bond is less than the activation energy needed to break the stronger 1° C—H bond. This leads to more of the 2° product being created since the reaction is irreversible.

Selectivity is much greater for free radical brominations of hydrocarbons. The 3°:2°:1° C—H bond reactivity ratio is 2000:100:1. The reason for this selectivity is that the first product-forming step is endergonic, +3 kcal/mol for a 3° C—H bond, +7 kcal/mol for a 2° C—H bond, and +10 kcal/mol for a 1° C—H bond. Thus, only the 3° C—H bond reacts to any extent. Since the overall reaction is exergonic for any kind of C—H bond, all the products are formed irreversibly.

In general, photochemical bromination and chlorination of hydrocarbons is seldom carried out in the laboratory. Chlorination, in particular, gives a mixture of products that is very difficult to separate. Selective oxidation of C—H bonds with halogens is usually confined to allylic or benzylic hydrogens, as described in the next section.

N-Bromosuccinimide Oxidations

N-Bromosuccinimide (**NBS**) is used to convert allylic and benzylic C—H bonds to C—Br bonds in a free radical chain process. One does not use molar concentrations of Br_2 for this kind of reaction because it would add to a double bond or engage in some other reaction. Light or a chemical initiator such as AIBN is needed to start the reaction:

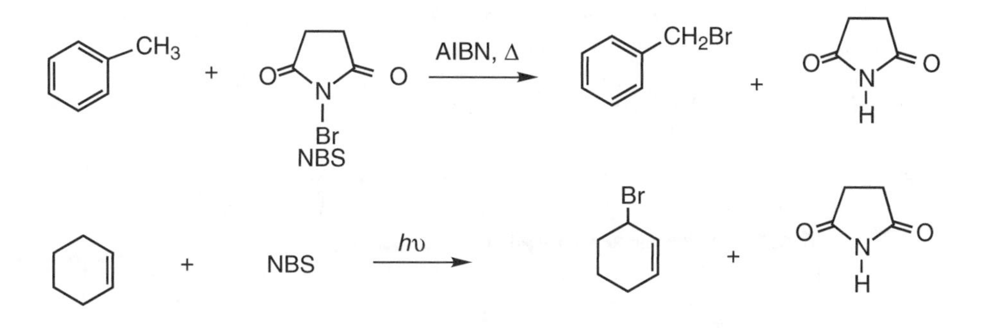

NBS acts as a source of Br_2 at very low concentrations, and the mechanism is a simple chain reaction:

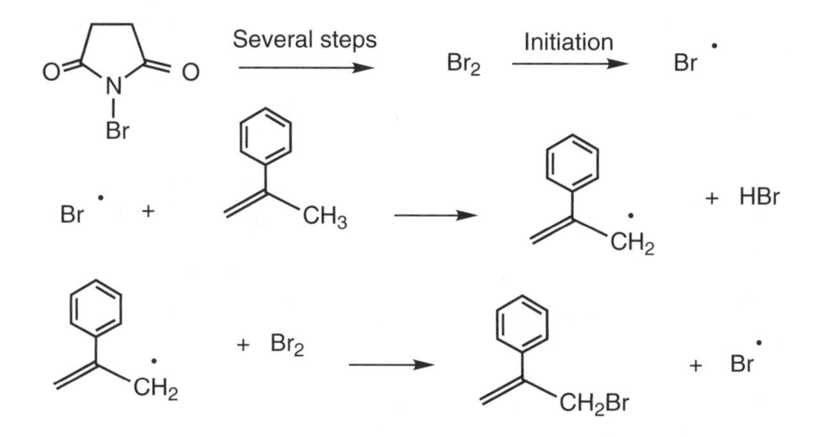

The reason the Br_2 NBS generated does not add to a double bond is that the addition reaction is a bimolecular reaction. The rate is the product of the alkene and Br_2 concentrations, and the Br_2 concentration is so close to zero that the rate is essentially zero. The $Br^•$ or hydrocarbon radical is sufficiently reactive that even at low concentrations it can engage in **allylic** or **benzylic brominations**.

Since allylic radicals are formed in NBS oxidations they can undergo allylic rearrangement and produce mixtures:

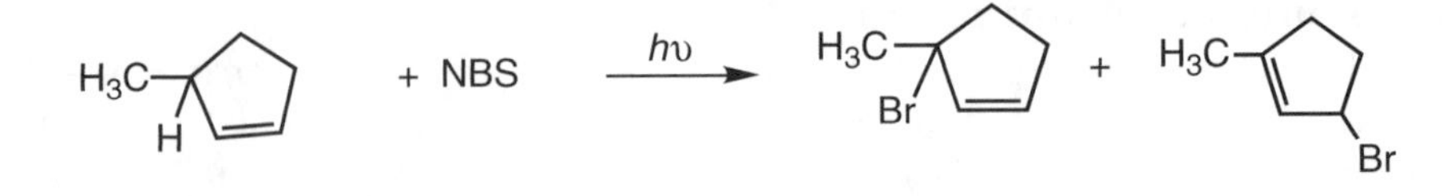

(*Note*: A C—H bond that is both allylic and 3° reacts in preference to one that is allylic and 2°.) Because allylic and benzylic bromides are useful substrates for S_N2 and elimination reactions, the NBS reaction is a useful way to convert hydrocarbons to other functional groups.

Reactions With Tributyltin Hydride

Tin is in the same group as carbon, and although it does not form stable Sn=C bonds, single bonds to tin are similar to single bonds to carbon. The major difference is that bonds to tin are longer, and therefore weaker, than bonds to carbon. The reagent tributyltin hydride, Bu_3SnH, is a versatile reagent for reducing carbon-halogen bonds to C—H bonds by means of a free radical chain process. The halides can be on sp^3 or sp^2 carbons. The relative rates of reactivity of C—X bonds are C—I > C—Br > C—Cl. C—F bonds are too strong to be reduced by this reagent. A chemical radical source such as AIBN is usually used to initiate the reaction:

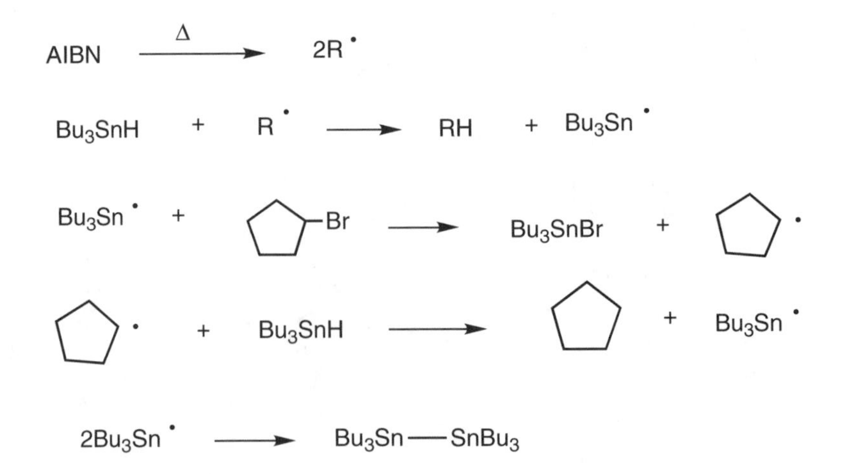

Notice that at no point is H• involved in this reaction. A hydrogen atom is a very high-energy species and is not normally part of any organic reaction mechanism. Instead, in this case, a weak Sn—H bond is broken and a strong C—H bond is formed in a radical propagation step.

Earlier it was mentioned that carbocations can react with double bonds to form new C—C bonds since a carbocation is an electrophile and a double bond is a nucleophile. Carbon radicals are likewise electrophiles, and they can react with double or triple bonds to form new C—C bonds. Carbon radicals have some advantages over carbocations because they do not undergo rearrangements and they are less likely to undergo eliminations than carbocations.

Carbon radicals are easily prepared by reacting an alkyl halide with Bu_3SnH, and if the radical can undergo intramolecular alkene or alkyne addition, it can avoid being directly reduced to a C—H group:

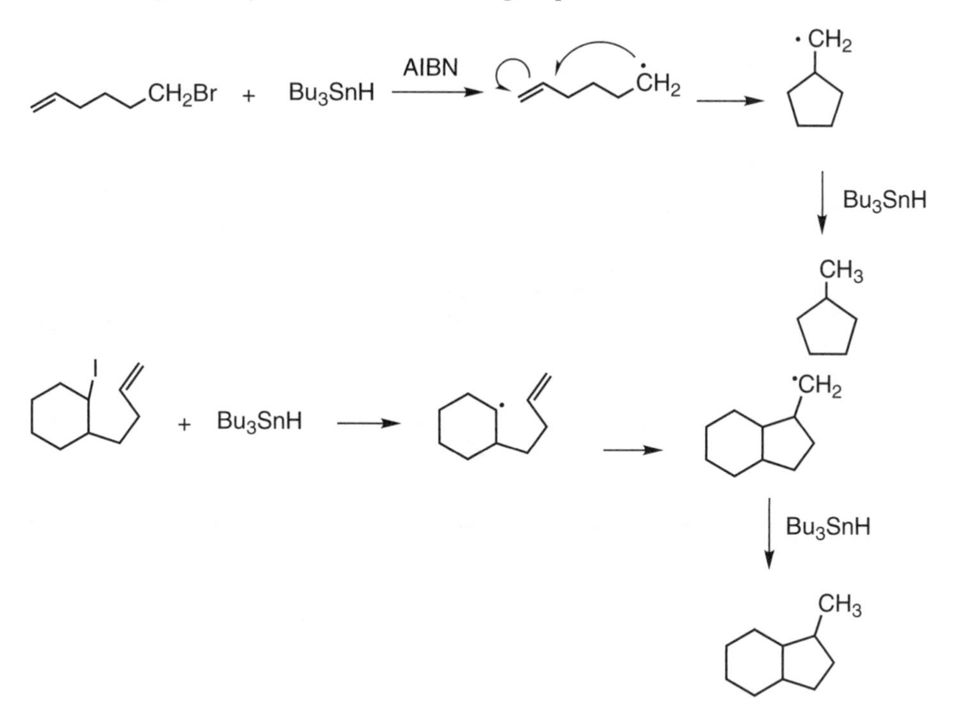

When there is a choice, the radical adds to the double bond to give a five-membered ring and not to the other carbon to give a six-membered ring. The reason for this preference has to do with how the radical carbon can approach the two carbons of the double bond. The constraints of the carbon chain's length and angles lead to a trajectory of approach that favors formation of a five-membered ring.

The Peroxide Effect in Reactions of HBr With Alkenes

HBr adds to alkenes following Markovnikov's rule, as described in Chapter 10. However, occasionally the reaction of HBr with an alkene gives the anti-Markovnikov addition product. Usually, this occurs with an impure sample of alkene. The reason for this regiochemistry change is a change in the mechanism of addition. The anti-Markovnikov addition product results from a free radical chain mechanism initiated by peroxides. When alkenes are stored in air, they can react with oxygen to produce small amounts of peroxide impurities. Since peroxides, ROOR, are efficient radical initiators, they cause HBr to add to a double bond through a radical and not through a polar mechanism:

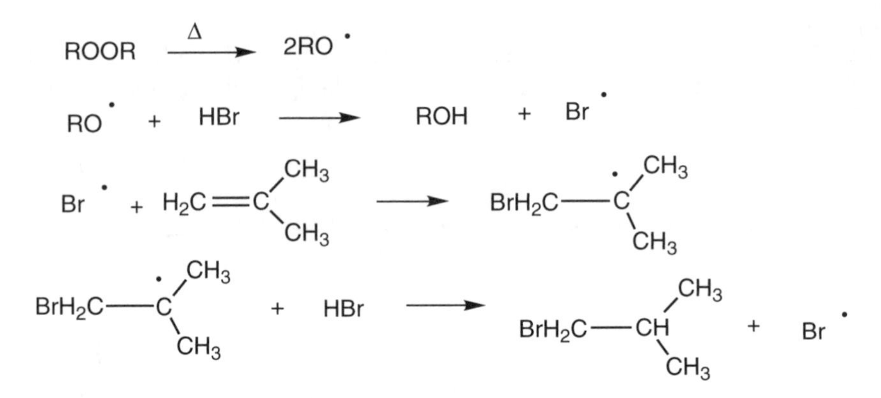

The reason the anti-Markovnikov addition product forms is because the bromine radical adds to the double bond first to give the most stable radical, which is 3° > 2° > 1°. Then the product carbon radical abstracts a hydrogen from HBr to give the observed product. Because Br• is so reactive, the radical addition of HBr to an alkene is much faster than the analogous polar addition.

EXERCISES

1. Fill in each blank with the appropriate compound based on the reactions introduced in this chapter.

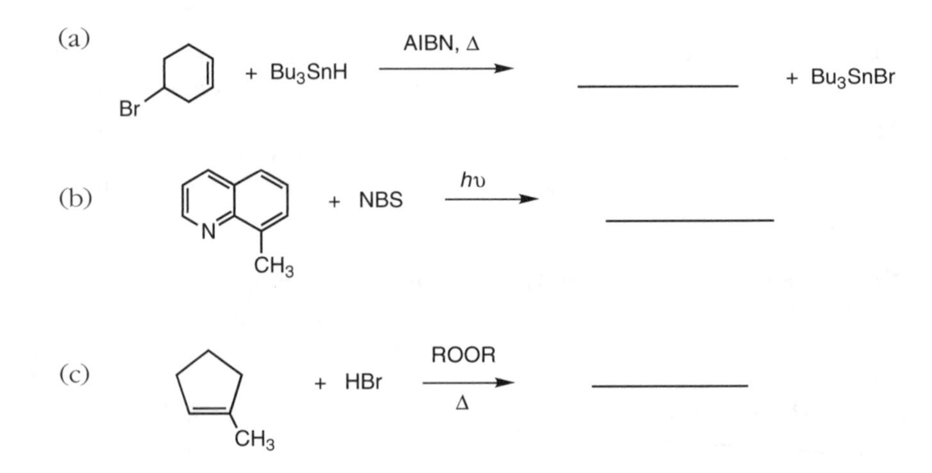

(d)

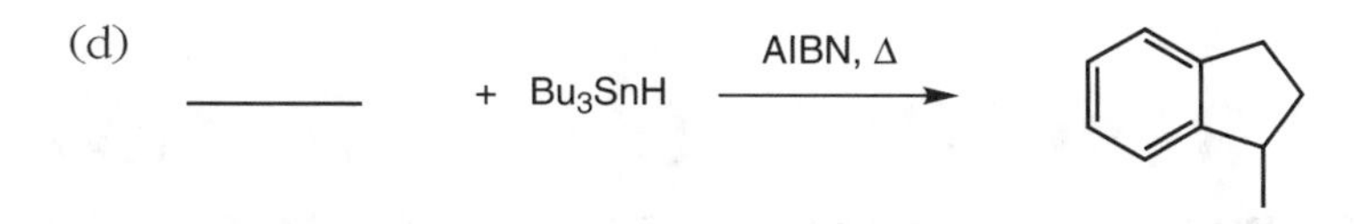

2. Starting with the indicated hydrocarbon, show how you would prepare the desired compound. More than one reaction step is required.

(a)

(b)

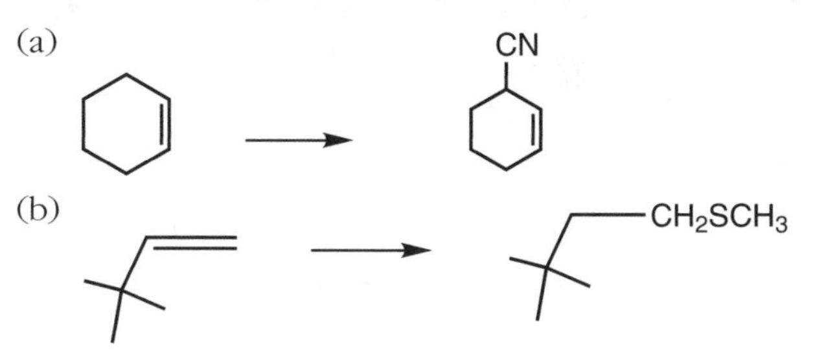

Solutions

1. (a)

(b)

(c)

(d)

2. (a)

(b)

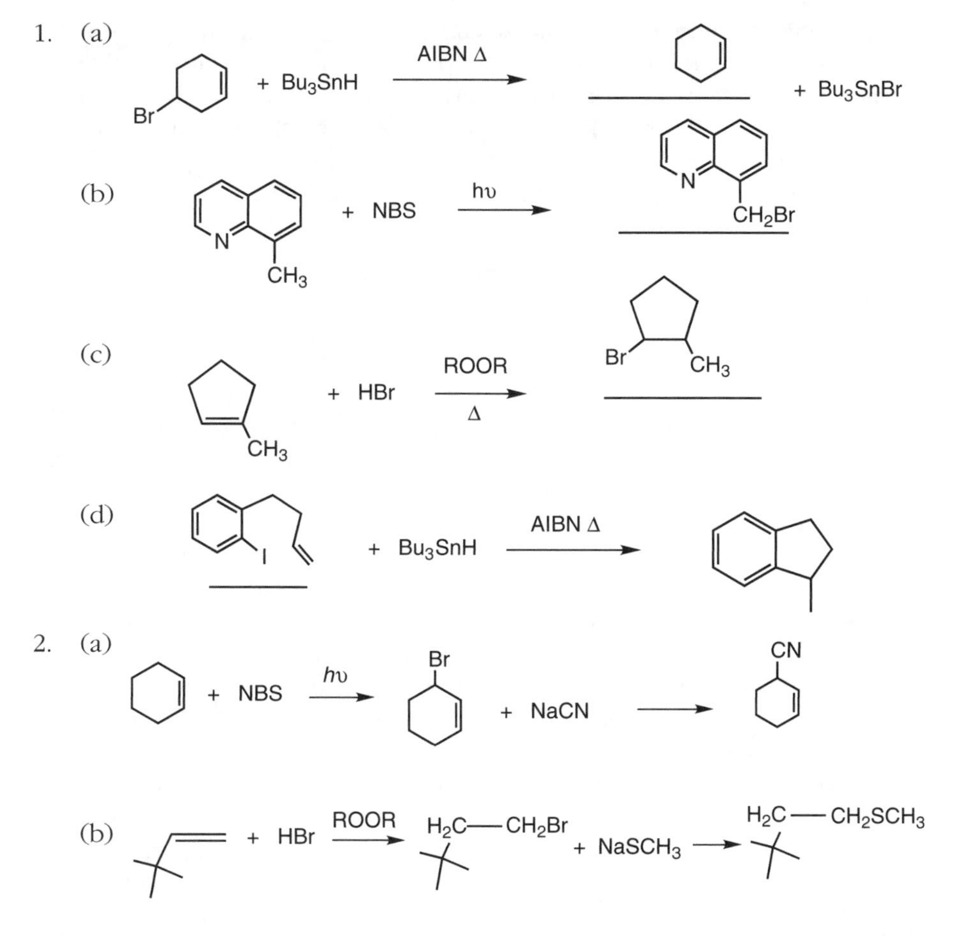

14
SPECTROSCOPY I: INFRARED SPECTROSCOPY, ULTRAVIOLET SPECTROSCOPY, AND MASS SPECTROMETRY

SPECTROSCOPY

Spectroscopy is the study of the interaction of matter with energy. It is a branch of physics, but it is also a vigorous area of chemical research. Using spectroscopic techniques, organic chemists can establish the structure of essentially any organic molecule. Most spectroscopic techniques involve the interaction of matter with **electromagnetic radiation**. Visible light, which has wave properties, is the best known example of such radiation. It travels at a constant velocity, designated $c = 3.0 \times 10^8$ m/s. Each color of light has a characteristic frequency and wavelength:

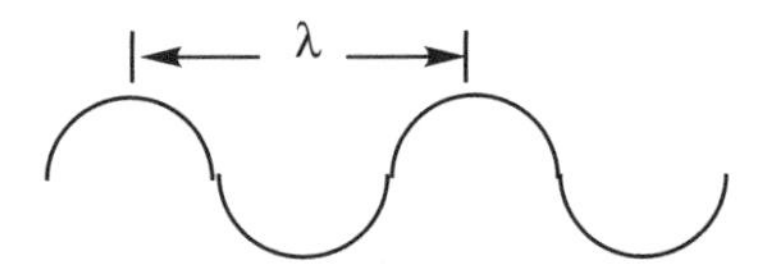

The wavelength λ is the distance between corresponding points on a wave. For red light, the wavelength is approximately 0.8×10^{-6} m, and violet light has a wavelength of 0.4×10^{-6} m. The frequency ν is the number of oscillations, or cycles per second, of the wave as it passes a fixed spot, and the units of frequency are hertz (Hz). There is a simple relationship among velocity c, λ, and ν: $\boldsymbol{\nu = c/\lambda}$. Since c is a constant, electromagnetic radiation of long wavelengths has small frequencies, and high-frequency radiation has short wavelengths. There is a second, very important relationship between energy and frequency, $\boldsymbol{E = h\nu}$. The constant h is known as Planck's constant and has the value 6.63×10^{-34} J s. The higher the frequency of the radiation, the more energy it possesses.

Electromagnetic radiation with different frequencies goes by different names. X rays are an extremely energetic form of electromagnetic radiation, exhibiting frequencies of approximately 10^{17} Hz. The energy corresponding to X rays is more than 300 kcal/mol, which is more than enough to break bonds in organic molecules if the energy is absorbed. Infrared radiation, which is felt as heat, has frequencies of approximately 10^{13} Hz. The amount of energy contained in infrared radiation is sufficient to cause the bonds in molecules to stretch and bend, assuming the molecule absorbs the radiation. Radiation is absorbed by molecules only when the radiation causes some process to take place. There has to be a match between the energy it takes to carry out some change and the energy (or frequency, since $E = h\nu$) of the radiation. With infrared radiation, for example, there has to be a match between the frequency at which a given bond vibrates and the frequency of the infrared radiation or no energy is absorbed. The technical term for this kind of energy match is **resonance**.

Any kind of spectroscopy involves beaming energy into a substance and seeing exactly which frequencies interact (normally, which frequencies are absorbed by the substance). If certain groups are present, energy will be absorbed at precise frequencies. If these groups are absent, no energy will be absorbed at that frequency. The results of a spectroscopic measurement are called **spectra**; the frequency or some related quantity is plotted along the x axis, and absorbance or some related quantity is plotted along the y axis.

INFRARED SPECTROSCOPY

Hooke's Law and Selection Rules

Infrared spectroscopy is the spectroscopy of bond stretching and bending because the amount of energy in infrared radiation is the energy it takes to stretch and bend bonds. The physics of molecular vibrations is very much like the physics of the vibration of springs. If one has a spring and two masses, m_1 and m_2, attached to each end of the spring, the spring will vibrate at a set frequency depending on the two masses and the **force constant** of the spring. Stiff springs, like garage door springs, have large force constants. Loose springs, like the children's toy Slinky, have small force constants. The equation for the vibration is derived from **Hooke's law**:

$$\nu' = \frac{1}{2\pi c}\sqrt{k\left(\frac{m_1 + m_2}{m_1 m_2}\right)}$$

In the case of a molecular vibration, such as the vibration of a C—H bond, k is the force constant of the bond, a measure of bond strength, m_1 and m_2 are the masses of carbon and hydrogen, and π and c are the usual constants. The frequency ν' normally has the units wavenumber, cm^{-1}, and is calculated by taking the reciprocal of the wavelength, $1/\lambda$ (cm). Wavenumbers, like frequency, are directly proportional to energy. Wavenumbers are used in infrared spectroscopy in part because they have convenient values. Most infrared absorptions occur between 4000 and 400 cm^{-1}.

The central idea of infrared spectroscopy is that different kinds of bonds absorb infrared energy at different energies (measured in wavenumbers). However, not all bonds absorb infrared energy as they stretch or bend. Quantum mechanics (whose study is needed for a thorough understanding of spectroscopy) imposes some rules on energy absorption. The most important of these **selection rules** is that a molecular vibration must result in a change in the bond dipole moment in order for vibrational energy absorption to take place. Therefore, a perfectly symmetric bond, such as the triple bond in dialkylacetylenes, RCCR, absorbs no infrared energy even when present in a molecule. However, a C=O double bond exhibits a large change in its dipole moment when it stretches, and so it exhibits a strong absorption band in an infrared spectrum.

General Absorption Values of Single, Double, and Triple Bonds

Infrared spectroscopy is the spectroscopy of functional groups. One can tell, by looking at an infrared (IR) spectrum if a molecule contains a carboxylic acid group, a nitrile, an aldehyde or ketone, and many other functional groups. A given functional group always absorbs IR radiation of approximately the same frequency in a similar environment. For example, an acyclic dialkyl ketone C=O group always absorbs at about 1720 cm^{-1} no matter what other functional groups are in the molecule.

From Hooke's law, it can be calculated that stretching vibrations involving bonds to the light atom hydrogen should have the highest vibrational frequencies. Triple bonds, because the force constant is the highest, should absorb energy at somewhat lower frequencies, followed by double-bonded groups and then single-bond stretches. Bending vibrations are lower in energy compared to stretching vibrations, and C—H bond bending vibrations overlap in frequency single-bond stretching vibrations. An outline of where different functional groups appear in the IR spectrum is given in Table 14.1.

The region from 1300 cm^{-1} to the limit of the spectrum, 600–400 cm^{-1}, depending on the instrument, is known as the **fingerprint region**. The spectrum normally has a number of peaks that arise from combination motions of the molecule's framework bonds in this region. It is different for each organic molecule (hence its name).

Normally, spectroscopy is used in the laboratory to identify molecules. If it is a known molecule, the spectra can be compared to data on file to establish its identity. If it is a new molecule, such as the product of a research project, the information in the spectra [using all the spectroscopic techniques, including mass spectrometry and nuclear magnetic resonance (NMR) which are discussed below] can be used to deduce the molecule's identity, normally with complete certainty. Spectroscopy is often covered in more detail in the laboratory part of an organic chemistry course, and students are expected to become competent in analyzing IR and NMR spectra. The only way to become truly proficient in this art is to look at lots of spectra. Most good libraries have

TABLE 14.1. INFRARED ABSORPTION FREQUENCIES OF COMMON ORGANIC STRUCTURAL GROUPS

4000–2700 cm^{-1}	2700–1300 cm^{-1}	1300–400 cm^{-1}
—OH (alcohol), 3550–3200, (hydrogen-bonded)	C=O (ketone), 1725–1705	Fingerprint region
—OH (alcohol), 3650–3590 (non-hydrogen-bonded)	C=O (aldehyde), 1740–1720	C—Cl, 800–600
—OH (carboxylic acid), 3500–2500 (very broad)	C=O (ester), 1750–1735	C—Br, 600–500
N—H, 3500–3310	C=O (carboxylic acid), 1725–1700	C—I, ~500
sp C—H, 3300	C=O (anhydride), 1850–1800 and 1790–1740	C—H (aromatic), 700–850
sp^2 C—H (alkene), 3095–3010	C=O (amide), 1700–1650	RCH=CHR (cis), ~690
Aldehyde C—H, 2900–2820 and 2775–2700	C=C (alkene), 1680–1620	RCH=CHR (trans), 970 and 1300
sp^2 C—H arene, ~3030	C=C (aromatic), ~1600	S=O, ~1100
sp^3 C—H, 2970–2850	C=N, 1690–1640	
	—C≡C—, 2250–2100	
	—C≡N, ~2250	

at least one IR and one NMR spectral atlas with hundreds of spectra with which one can become familiar.

To illustrate how IR spectroscopy can be used to identify the functional groups in a sample, look at the spectra of pivalic acid, $(CH_3)_3CCOOH$, and cyclooctanol shown here.

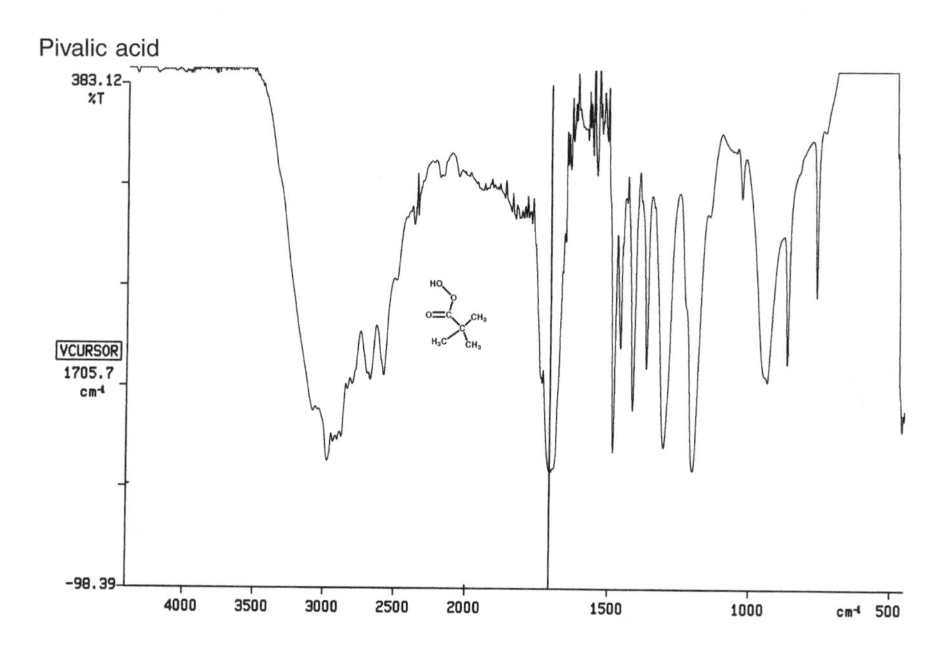

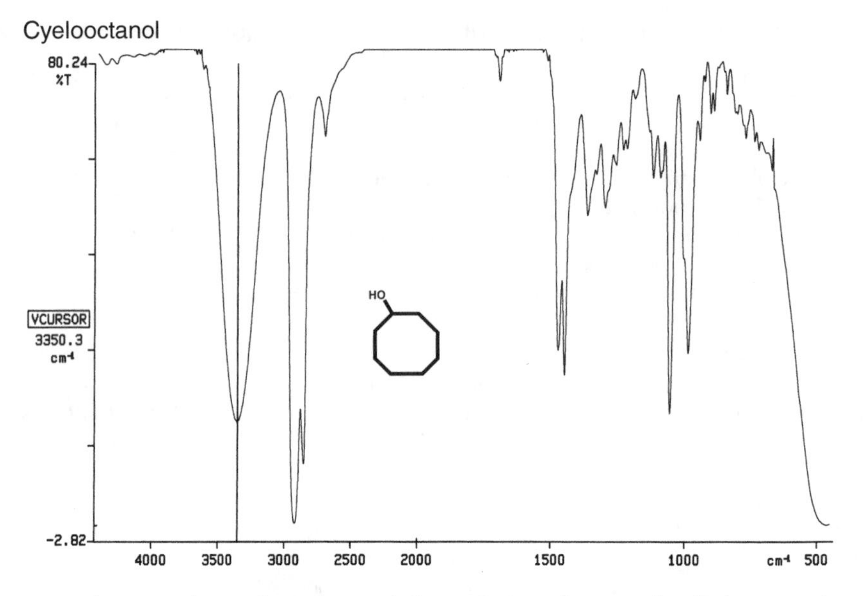

Notice how in the carboxylic acid the —OH peak extends all the way down to 2500 cm^{-1}, while in the alcohol the —OH peak is more symmetric and is centered at 3350 cm^{-1}. Furthermore in the alcohol spectrum there is no strong peak at about 1700 cm^{-1}, indicating the absence of any kind of C=O group. The very weak peak in the alcohol near 1700 cm^{-1} is probably what is called a **harmonic** of the strong —OH peak, somewhat similar to the echo of a loud bell. Only by looking at spectra can one know that such a peak is too weak to be any kind of carbonyl peak, such as the one seen in the acid spectrum at 1700 cm^{-1}.

Two further IR spectra, for *cis*-1,2-diphenylethene and 4-heptanone, are shown here.

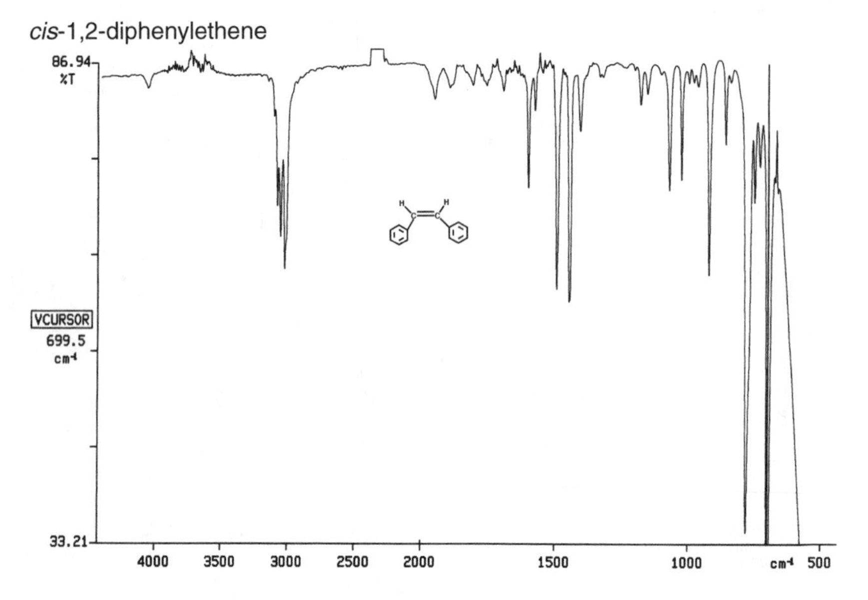

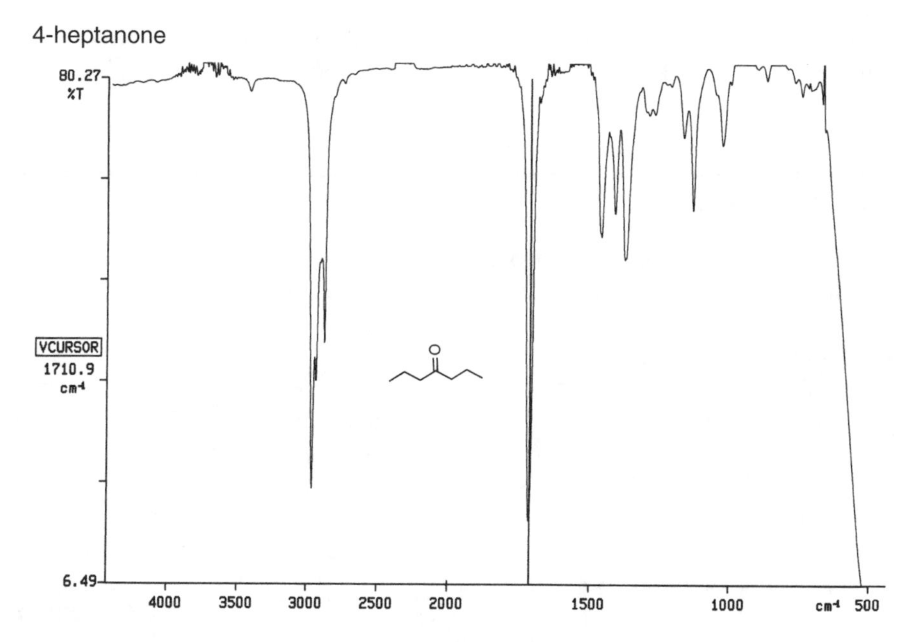

IR spectroscopy allows one to establish the presence or absence of C—H bonds to sp^2 and sp^3 carbons. If all the C—H stretches are below 3000 cm^{-1}, as in the case of 4-heptanone, the molecule contains only sp^3 C—H bonds. With *cis*-1,2-diphenylethene, all the C—H stretches are above 3000 cm^{-1} (here about 3080, 3055, and 3025 cm^{-1}), indicating that only sp^2 C—H bonds are present. The strong peak at 697 cm^{-1} suggests that there is a cis double bond (which we will later learn can be confirmed by NMR spectroscopy).

The environment of a functional group influences its absorption frequency to some extent. The most important example of this kind of effect is seen in the influence of conjugation on carbonyl group absorptions. When a carbonyl group (of any kind, ketone, ester, anhydride, and so on) is conjugated with an alkene or a phenyl ring, the frequency of the carbonyl stretch decreases by approximately 30 cm^{-1}. Thus, acetone, CH_3COCH_3, has a ketone stretch at about 1720 cm^{-1}, while acetophenone, CH_3COPh, exhibits its ketone stretch at 1689 cm^{-1}. When both groups attached to a carbonyl group are unsaturated, as in benzophenone, PhCOPh, the effect is additive and the carbonyl group absorbs at 1685 cm^{-1}.

A second effect of the environment is seen with alcohols. When an IR spectrum of an alcohol is taken in a dilute solution (too dilute to allow for hydrogen bonding with other alcohol groups), the —OH stretch occurs around 3600 cm^{-1}, and it is a very sharp peak. In a more concentrated solution, or as a **neat liquid** (which means that the IR is taken as a pure liquid sample), the —OH group is hydrogen-bonded with its neighbors. This leads to a family of slightly different —OH structures with slightly different geometries and bond lengths. The observed peak is a broad envelope resulting from many overlapping, unresolved peaks centered around 3350 cm^{-1}. This hydrogen bonding effect is even greater for the —OH group of carboxylic acids, resulting in the extremely broad peak seen for that functional group.

EXERCISE

How would you use IR spectroscopy to distinguish among samples of (a) benzaldehyde, (b) *N,N*-dimethyacetamide, (c) ethanol, and (d) formic acid?

Solution

Benzaldehyde, PhCHO, has a strong carbonyl stretch at around 1705 cm^{-1} and bands due to the aldehyde C—H bond at 2850 and 2750 cm^{-1}. *N,N*-Dimethyacetamide, $(CH_3)_2NCOCH_3$, has a strong carbonyl band below 1700 cm^{-1}, and its C—H stretches are between 2970 and 2850 cm^{-1}. Ethanol, CH_3CH_2OH, has no carbonyl stretch and a strong, broad —OH band centered around 3300 cm^{-1}. Formic acid, HCOOH, has a broad carboxylic acid —OH band between 3500 and 2500 cm^{-1} and a carbonyl band at a frequency higher than 1700 cm^{-1}. In solving these problems, take notice of what peaks will be absent as well as what peaks will be present.

ULTRAVIOLET SPECTROSCOPY

Ultraviolet (UV) spectroscopy uses high-energy electromagnetic radiation with wavelengths from 200×10^{-9} to 400×10^{-9} m (200–400 nm). This is the kind of radiation that reaches earth's surface and causes sunburn. Higher-energy ultraviolet light is blocked by the ozone layer, making life as we know it on land possible. Recall from the discussion of LUMO and HOMO orbitals in Chapter 3 that there are some energy levels in molecules that are unfilled, as well as energy levels that are filled. Ultraviolet light can cause a transition between electrons in a filled, bonding orbital to an empty, antibonding orbital. Electrons that are nonbonding, such as the lone pair on an oxygen atom, can also absorb energy and move into an antibonding orbital. The transitions excited by UV light are thus either from a π orbital to an antibonding π orbital, designated π to π^*, or from a nonbonding orbital (an *n* orbital) to a π^* orbital. Usually one needs two or more double bonds (C=C, C=O, or another type) in conjugation to obtain a UV spectrum. Isolated double bonds absorb such high-energy UV radiation that ordinary UV spectrometers cannot measure the absorption.

Ultraviolet spectra are quite broad, as shown in the accompanying diagram. Normally only one or two peaks are present in a UV spectrum. The maximum is reported as λ_{max}, and it changes slightly with the solvent, especially with a

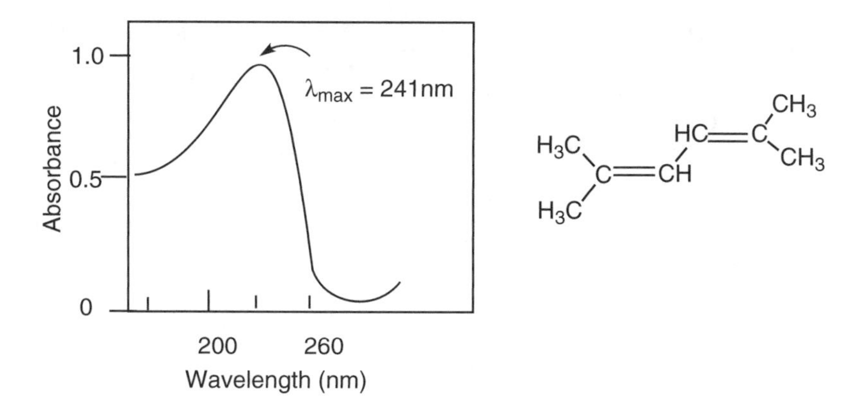

hydrogen-bonding solvent. Normally the solvent used for a given spectrum is noted.

Ultraviolet spectroscopy is very sensitive and is quite useful for determining the concentrations of organic and biological solutions. The relationship between the measured absorbance at λ_{max} and the concentration of a solution is given by the **Beer-Lambert law**, $\boldsymbol{A = cl\varepsilon}$, where A is the absorbance measured from the spectrum, c is the concentration of the solution (moles per liter), l is the path length of the cell (cm), normally equal to 1, and ε is the **molar absorptivity** (L/mol cm). Each compound has a different ε. It is a physical property like a melting point. One can think of molar absorptivity as how deeply colored the compound is in the UV. The higher the value of ε, the more UV radiation at λ_{max} is absorbed. Many compounds have molar absorptivity in the range 10,000 to more than 100,000.

A few examples of some UV-absorbing molecules, their λ_{max} and ε values are shown in Table 14.2. There are simple methods for calculating λ_{max} from a molecule's structure, but they lie outside the scope of this text.

As can be seen in the case of phenol and its conjugate base, species with more resonance structures tend to have lower energy λ_{max} values and higher molar absorptivities.

EXERCISE

The precursor of vitamin A, *trans*-β-carotene, contains a conjugated system of 11 double bonds. It absorbs in the visible region of the spectrum at λ_{max} 452 nm with an ε of 139,000. Suppose one has 20 ml of a solution of *trans*-β-carotene. One takes 0.01 ml of this solution, dilutes it to 1 ml, places it in a 1-cm-path-length cell, and measures the absorption at 452 nm as 0.4. What is

TABLE 14.2. ULTRAVIOLET ABSORPTION DATA FOR REPRESENTIVE ORGANIC MOLECULES

	λMAX	ε
$H_2C=CH-C\equiv CH$	219	7,600
$H_3C-CH=CH-C(=O)OH$	206	13,500
cyclohexenone (C=O)	225	10,300
$(H_3C)_2C=C(CH_3)-C(=O)CH_3$	246	5,300
$C_6H_5-CH_3$	206	7,000
C_6H_5-OH	210	6,200
$C_6H_5-O^{\ominus}$	235	9,400

the concentration of *trans*-β-carotene in the original sample? (Note that the Beer-Lambert law is valid for visible and UV spectroscopy.)

Solution

Since $A = \varepsilon lc$ and $A = 0.4$, $l = 1$ and $\varepsilon = 139{,}000$, the concentration (in mol/L) of the sample was 2.88×10^{-6} M. Since the sample was diluted 20/0.01 or 2000 times, the concentration in the original sample was 0.0576 M.

MASS SPECTROMETRY

Mass spectrometry is an extraordinarily sensitive method for determining the molecular weight and atomic composition of molecules. It can also give much useful structural information from a detailed analysis of the mass spectrum. Since submicrogram quantities of material can be analyzed, it is even useful for trace biochemical analysis of bodily fluids. It is used, for example, in the Olympics to ensure that no banned drug metabolites are present in an athlete's urine after a competition.

A mass spectrum is normally obtained in a high-vacuum chamber where a sample is volatized into the gas phase. Special techniques are available to volatize even very high-molecular-weight species. The individual molecules are bombarded with very high-energy electrons which cause the molecule to lose one electron and become an intermediate known as a **radical cation**. This is a species that has both an unpaired electron and a positive charge. Because electrons are so light, the mass of the molecule does not change significantly. The ion is subjected to a curved magnetic field, which causes its path to bend. At the same time a negatively charged field accelerates the positive ion in a specified direction through the magnet. The path of the ion depends on its mass-to-charge ratio, designated **m/z**, where *z* corresponds to charge. Normally the charge is 1. Neutral molecules are not accelerated and are not detected. Where the ion ends up on the detector is a function of its m/z ratio (and just the mass in the usual case where z is 1). High-resolution machines can distinguish ions differing in mass by 1 part in 10^5 parts.

Not all ions make it to the detector intact. Some can fragment into a neutral radical and a carbocation. In such cases the m/z of the carbocation fragment is detected (and the neutral radical is invisible to the detector). The intact radical cation is called the **molecular ion**, and its mass corresponds to the molecular weight of the molecule. The other ions are called **fragment ions** and give information about the structure of the molecule.

When one calculates the molecular weight of a compound, one normally uses the average atomic weights of the elements. For example, the atomic weight of carbon is listed as 12.01. This is because carbon is a mixture of two isotopes, 98.89 percent ^{12}C and 1.11 percent ^{13}C. However, any given small molecule likely has only ^{12}C isotopes, so when calculating the molecular weight of a molecule for mass spectrometry, one uses the masses of the most common isotopes. Some elements have several common isotopes. Bromine, for example, is 50.69 percent ^{79}Br and 49.31 percent ^{81}Br. Thus, for mass spectrometry, there are two more or less equal intensity CH_3Br molecular ions, one with a molecular weight of 94 (the ^{79}Br-containing molecule) and one with a molecular weight of 96 (containing ^{81}Br).

When one obtains a mass spectrum of an organic compound, there is normally a significant molecular ion and a smaller ion one mass unit higher, the ***M* + 1 peak**. This peak arises from the fact that carbon is a mixture of 98.89 percent ^{12}C and 1.11 percent ^{13}C. The $M + 1$ peaks contain a ^{13}C in place of a ^{12}C. The relative intensity of the $M + 1$ peak is a function of just how many carbon atoms are present in the molecule and can be calculated from the equation

$$\text{Intensity}_{M+1} = 0.011(\text{intensity}_{\text{M}}) \times \text{number of carbon atoms}$$

In addition to carbon and bromine, chlorine has two common isotopes, ^{35}Cl (75.77 percent natural abundance) and ^{37}Cl (24.23 percent natural abundance).

A typical mass spectrum, in this case of ethanol, is shown here. The m/z ratio, also called m/e, is along the x axis, and the relative intensity of each ion is plotted along the y axis.

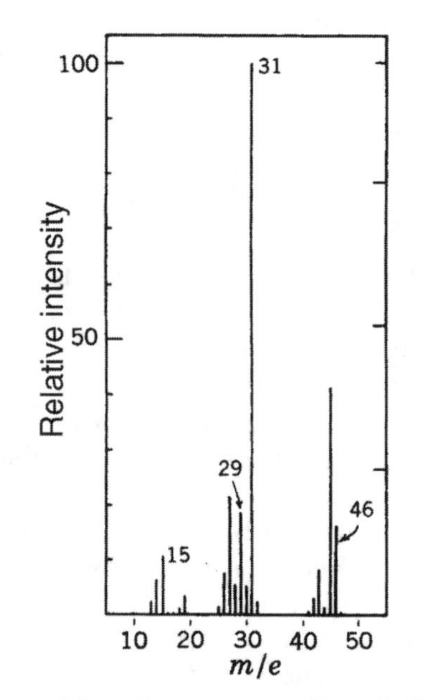

K. Biemann, Mass Spectrometry, Organic Chemical Applications, McGraw-Hill, New York, 1962, p. 73.

The molecular ion is seen at m/z 46. The peak with the greatest intensity is called the **base peak**, here m/z 31. In general, molecules fragment to break weaker bonds or to form more stable fragments. Recall that only positive ions are detected. Thus, the ion at $M - 1$, 45, must arise from a loss of $H^{\bullet}$, leaving behind $(C_2H_5O)^+$. Just from the mass of the fragment we do not know which H is lost. However, since the most stable carbocation is CH_3CH^+OH, because of resonance stabilization of the positive charge by the oxygen, this is the likely structure of m/z 45. Other fragmentation patterns are:

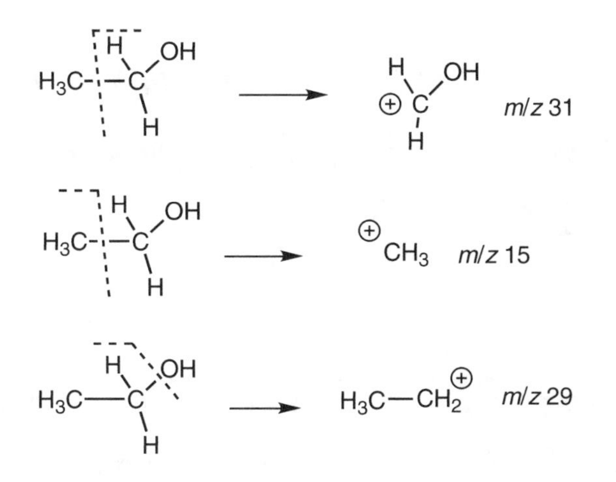

Because of the high energy used in ionizing molecules in mass spectrometry, a few radical cations undergo complex rearrangements, so not all the lower-intensity peaks can be assigned to simple fragmentation patterns.

Every functional group class has characteristic fragmentation patterns that can be used for structural identification. Furthermore, with high-resolution mass spectrometers, the identity of a given *m/z* peak can be unambiguously determined. For example, a peak with an *m/z* of 45 could be CH_2CH_2OH or CO_2H. However, the exact mass of the first ion is 45.0304 (using the masses of ^{12}C, ^{16}O, and ^{1}H), while the second ion has a mass of 44.9976. Thus, these ions could be easily distinguished. Furthermore, modern mass spectrometers come with computerized databases installed containing the mass spectra of thousands of known compounds, so if one has a known compound it can be quickly identified.

EXERCISE

Which kind of alcohol—primary, secondary, or tertiary—would be least likely to give a molecular ion?

Solution

Because tertiary carbocations are so stable, tertiary alcohols usually do not give a molecular ion in mass spectrometry unless special techniques are used.

15
SPECTROSCOPY II: NUCLEAR MAGNETIC RESONANCE SPECTROSCOPY

BASIC PRINCIPLES OF THE NUCLEAR MAGNETIC RESONANCE (NMR) MEASUREMENT

NMR spectroscopy is by far the most powerful spectroscopic tool that chemists have to identify and follow the reactions of organic molecules. It is based on the property that some nuclei have called **spin**. Some nuclei behave as if they are spinning tops, ^{1}H and ^{13}C nuclei, for example. In the absence of a magnetic field, these positively charged objects can have their axis of spin point in any direction. However, in the presence of a magnetic field, the spin axis can be aligned only with or against the field (quantum mechanics says the spin is quantized and can take only certain values). Aligned with the field is the low-energy state, and against the field, the high-energy state. The more powerful the magnetic field, the greater the energy difference between the two energy states.

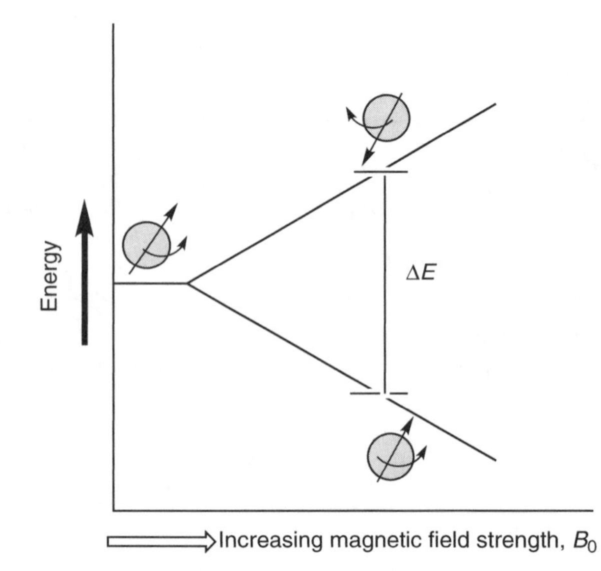

It is usual to call the higher-energy spin state the β state and the lower-energy spin state the α state. The amount of energy it takes to change a spin from the low-energy state to the high-energy state, even in the strongest magnetic field scientists can produce, is tiny, on the order of the energy in an FM radio signal (less than 10^{-5} kcal/mol). The energy can be calculated from

$$\Delta E = h\nu = h\gamma B_0/2\pi \tag{15.1}$$

Where h is Planck's constant, γ is the **gyromagnetic ratio**, and B_0 is the magnetic field strength (in tesla). The gyromagnetic ratio is different for each nucleus and is a measure of how much the nucleus couples with the magnetic field's energy. Some nuclei, such as ^{12}C and ^{16}O have no spin, and their gyromagnetic ratio is zero. The proton, ^{1}H, is one of the highest gyromagnetic ratios. That of ^{13}C is only one-fourth that of ^{1}H.

Most major research universities today use standard NMR machines with magnets with a field strength of approximately 7T. This corresponds to the frequency needed to change the spin state of ^{1}H by 300 MHz in what is called a **spin flip**. When the value of ΔE and the energy in the radiofrequency radiation match, energy is absorbed by the nucleus and one has what physicists call a **resonance** condition. With a magnet twice as strong, a 14-T magnet (about the size of a small motor home), the NMR machine will operate at 600 MHz to flip spins.

The heat energy at room temperature is much greater than the energy in 600-MHz electromagnetic radiation. Thus, nuclei in a magnetic field are at approximate thermal equilibrium in all types of magnets, and the ratio of nuclei in the low-energy state to nuclei in the high-energy state is about 10,001 : 10,000. The bottom line is that NMR is a very insensitive technique and that very high-quality electronics, radiofrequency transmitters, and receivers are needed to pick out a signal not much above the level of noise.

Modern NMR machines collect not one spectrum per sample but dozens or hundreds and then add them together. This has the effect of partly canceling out the noise (a technique called **signal averaging**) since the signals reinforce themselves but the noise sums to zero. The modern method for this collection uses a technique called **Fourier transformation**, and modern NMR machines are called FTNMRs. A description of how FTNMR works uses the analogy of a singing group. Suppose there are three singers, a baritone, a tenor, and a soprano. Each sings a note at a given frequency. One could ask each singer in turn to sing a note and record the frequency. Alternatively, one could have all the singers sing together and record the sonic waveform. There is a mathematical technique, called Fourier transformation, that can decompose the waveform into individual frequencies, giving the same information faster if the computer used for the Fourier transform is fast. This is how FTNMR works. All the nuclei are excited by a burst of radio frequency energy, and each nucleus relaxes back to the low-energy state in a process that depends on the energy difference between the α and β states. Everything is recorded in a few seconds, and the process is repeated until enough signal has been recorded to overcome the unavoidable electronic noise.

From what has been said so far, it seems that all protons in a sample have their spins flipped at the same frequency since they are all protons. Fortunately for chemists, this is not the case. The ΔE condition [Equation (15.1)] that defines the resonance condition (the frequency of radiation that is absorbed) involves the magnetic field that is felt at the nucleus. Different nuclei in a molecule experience slightly different magnetic fields. This difference arises because the electrons in the bonds around the nuclei, being charged, interact with the magnetic field and shield the nucleus slightly from the external field. Some kinds of bonds do a good job of shielding the nucleus from the external magnetic field, and other kinds do a poor job. The result is that almost every proton in a molecule experiences a slightly different magnetic field and thus undergoes a spin flip at a slightly different frequency. This frequency shift is called a **chemical shift**, and it allows one to determine what kinds of functional groups are near each proton.

STRUCTURAL DATA AVAILABLE FROM THE NMR MEASUREMENT

We will focus on four types of data that one can extract from a ^{1}H NMR measurement—the chemical shift, signal integration, the splitting pattern, and the magnitude of the coupling constant—and how this information can be used for structure determination. Later we will see how ^{13}C NMR spectroscopy is also valuable in structure determination.

Measurement of Chemical Shifts: The Parts per Million Scale

Classical physics teaches that a charged particle in a magnetic field moves to create a magnetic field opposing the external field. This is how electrons in bonds change the local field at the nucleus. The effective field at the nucleus is always less than the applied field of the external magnet. The greater the electron density around a proton, the more the electrons oppose the applied field and the smaller the net field.

From Equation (15.1), a smaller net magnetic field implies that a lower frequency will be needed to flip that spin. One of the most electron-rich compounds is the molecule **tetramethylsilane (TMS)**, $(CH_3)_4Si$. TMS is used as a chemical shift reference material and is assigned a value of 0. Other peaks in the NMR spectrum are given values relative to that of TMS. For example, in a 7-T spectrometer operating at 300 MHz, one can assign the TMS signal to come at 300,000,000 Hz. If a sample of CH_3OCH_3 is measured, its protons will be found to absorb energy at 300,000,972 Hz. The chemical shift of the dimethyl ether peak in the NMR spectrum is calculated as follows.

$$\text{Chemical shift}(\delta) = \frac{300{,}000{,}972 - 300{,}000{,}000 \times 10^6}{\text{Spectrometer frequency } (300{,}000{,}000)} = 3.24$$

If one runs the same sample in a 600-MHz NMR spectrometer, the dimethyl ether will absorb at 600,001,944 Hz, TMS will absorb at 600,000,000, and the spectrometer frequency will be 600,000,000, which gives the same δ value of 3.24. The advantage of the δ scale is that it gives the same values for all magnetic fields. The chemical shift has the units parts per million (ppm).

Regions of the NMR spectrum have names designating whether they represent high or low field strength: Normally TMS is on the right-hand end of the spectrum, which is the upfield end or the shielded region. The region to the left is called the deshielded region or downfield. The frequency of the peaks, or **resonances**, increases from right to left. Normally all proton resonances have values of 0–12 ppm.

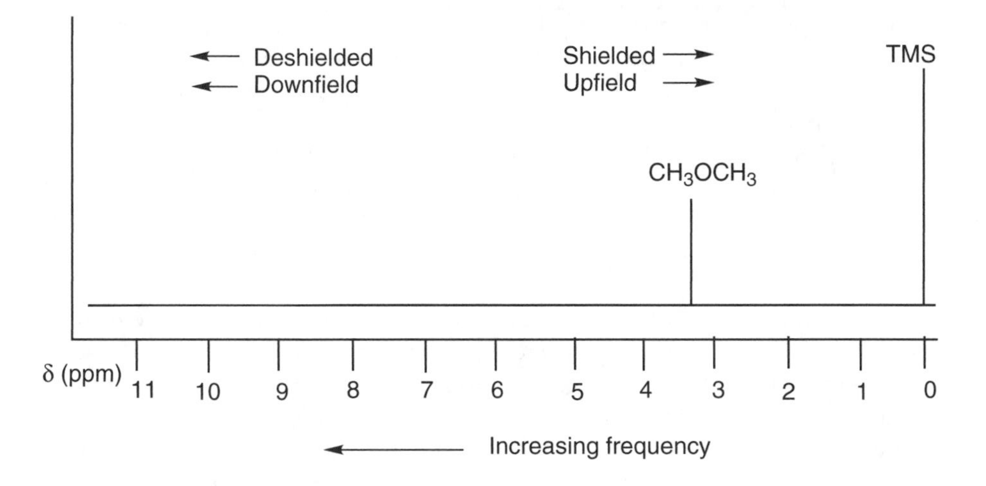

Equivalent Nuclei in the NMR Measurement

Some groups of protons in a molecule have the same chemical shift and are said to be **chemically equivalent**. For example, all the methyl protons in dimethyl ether have the same chemical shift. In general, if one has a group of protons, such as a CH_3 group, all the protons in the group are chemically equivalent if replacing any one of them with a different atom gives the identical compound. Invariably, all three protons in a methyl group are equivalent. Often both protons in a CH_2 group are equivalent, but not always. For example, the CH_2 protons in an ethyl ether group, $—OCH_2CH_3$, are equivalent because of free rotation around the single bonds. However, in a ring, the two CH_2 protons may be inequivalent depending on the ring geometry. Protons related by symmetry have the same chemical shifts. Protons that have similar but not identical chemical shifts are usually designated H_a, H_b, H_c, and so on.

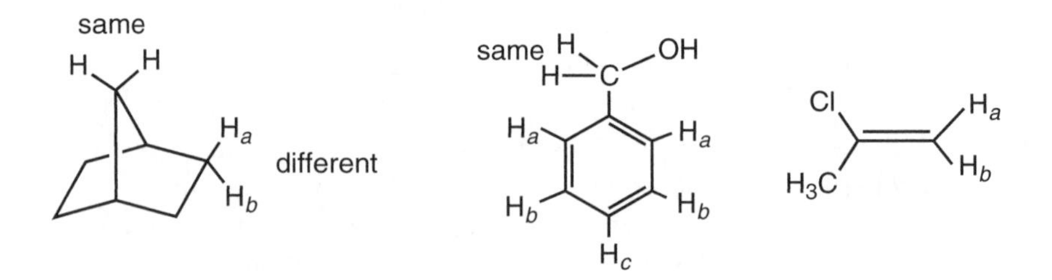

Factors That Influence the Magnitude of the Chemical Shift

Electronegativity, hybridization, and ring currents are the three major factors that influence proton chemical shifts. When electronegative atoms are bonded to a carbon, a proton also bonded to that carbon is deshielded. For example, the chemical shift of $CHCl_3$ is 7.3 ppm, that of CH_2Cl_2 5.3 ppm, and that of CH_3Cl 3.1 ppm. Because of the minor electronegativity difference between C and H, tertiary C—H groups on a hydrocarbon appear at about 1.5 ppm, hydrocarbon $—CH_2—$ protons resonate at 1.2–1.35 ppm, and hydrocarbon methyl groups appear at about 1.0 ppm.

The hybridization of the carbon makes a major difference in the chemical shift. The protons in ethane appear at 0.9 ppm, in ethylene at 5.3 ppm, in acetylene at 2.5 ppm, and in benzene at 7.3 ppm. The unsaurated compounds have their chemical shifts strongly influenced by what are known as **ring currents**.

In benzene and other aromatic molecules, the six aromatic ring electrons interact with the external field to create a ring magnetic current that opposes the external field. The field originates from the center of the ring and travels to the outside of the ring to close the magnetic field lines as shown opposite. The hydrogens on the outside of the aromatic ring have the external and ring current fields added together, resulting in strong deshielding and chemical shifts from 6.6 to 9 ppm, depending on the substituents on the ring. An analogous, but weaker, current is set up in an alkene, leading to protons exhibiting chemical shifts from 4.5 to 6.5 ppm. The current set up in an alkyne places *sp* C—H protons in a region of space where the induced field opposes the field, shielding the protons and giving them a chemical shift of 2.5 ppm.

A few general chemical shift values are shown in Table 15.1.

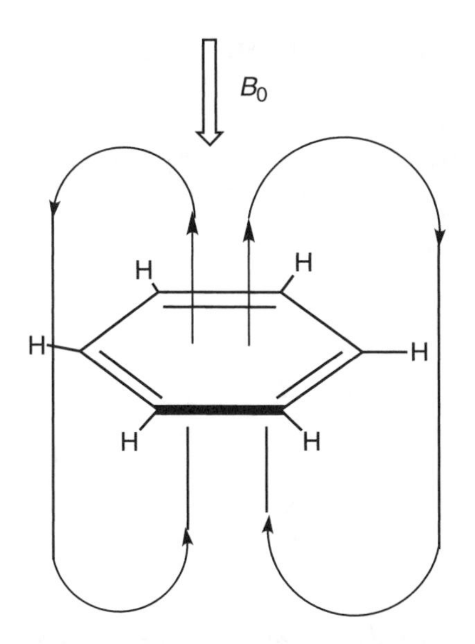

TABLE 15.1. CHEMICAL SHIFT VALUES FOR COMMON FUNCTIONAL GROUPS

Functional Group	Chemical Shift (ppm)	Functional Group	Chemical Shift (ppm)
CH_4	0.24	$CH_2{=}CR_2$	4.6–5
CH_3—C—C—X, where X = Cl, OH, C=O	0.9–1.1	H—C—Cl	3–4
CH_3—C=C	1.7	CH=C—COR	6–6.5
CH_3—C=O	2.1	Ar—H	6.5–9
CH_3—Ar	2.3	H—CO—N	8
CH_3—OR	3.3	H—N—R	1–3
CH_3—OCOR	3.7	H—OAr	6–8
H—C—Ar	2.3–2.9	H—OR	0.5–5
H—CO—R	9	H—OCOR	10–12

Integration of Peak Intensities in NMR Spectra

The size of an NMR peak is proportional to the number of protons present in the sample. For example, the NMR spectrum of acetaldehyde, CH_3CHO, has two peaks whose heights are in the approximate ratio 3 : 1. If one had a mixture of acetaldehyde and toluene, C_5H_5—CH_3, one could measure the ratios of the two methyl group areas and find the molar ratio of the components of the mixture. Earlier NMR spectrometers drew **integration** curves, steplike curves above each peak whose height for each step was proportional to the area under the curve. Newer NMR software prints out this integration data in table form. Not all peaks are of the same width, so the integration must calculate the area under the peak, not the relative peak height.

Coupling Constants and Molecular Connectivity Data

Looking at real NMR spectra, one finds out that not all peaks are single peaks, or **singlets**. Some are made up of two lines, or **doublets**, some are three lines, or **triplets**, and so on. For example, in diethyl ether, the methyl groups appear as three-line triplets and the methylenes as four-line quartets as shown here.

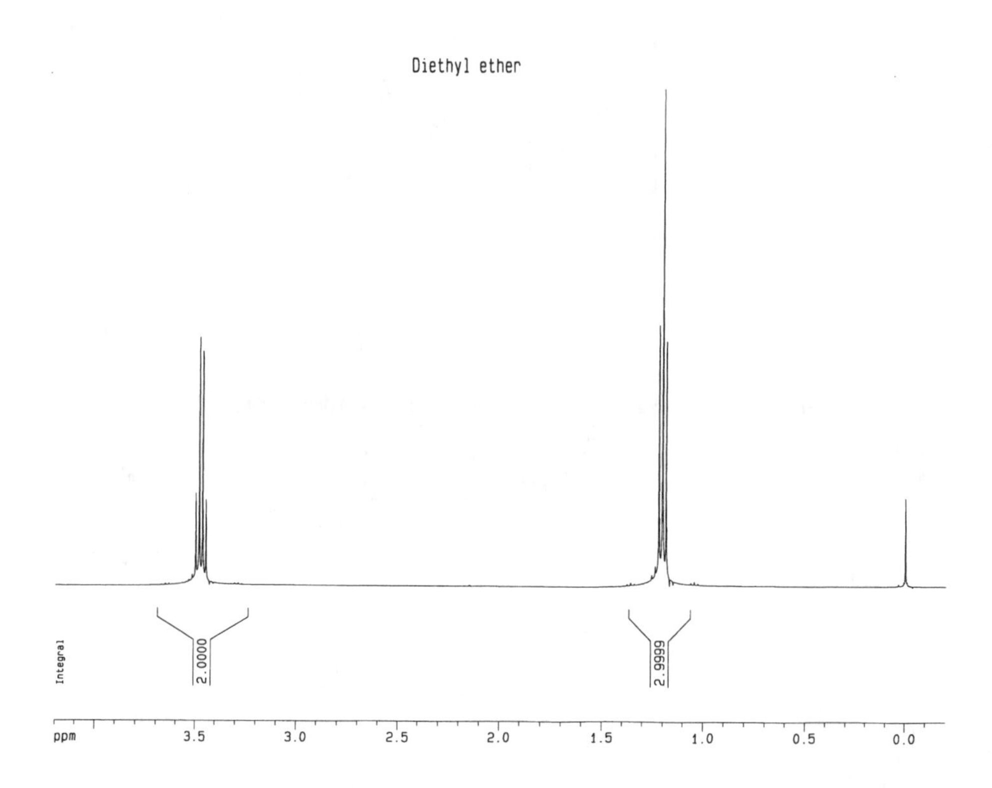

Since in an NMR spectrum every line corresponds to the energy of a spin flip, there must be three different kinds of proton energy states in the methyl group and four different kind of proton energy states in the methylene group. Furthermore, their energies are nearly the same because the distances between the lines are only about 7 Hz. Where do these energy differences come from?

These groups of lines, or **multiplets**, arise from a phenomenon called **spin-spin splitting**. Recall that a given proton can exist in the α or β spin state in a ratio of essentially 50:50. A —CH_2— proton in diethyl ether can be in a molecule whose methyl protons have eight possible orientations relative to the external magnetic field as shown at the top of the next page.

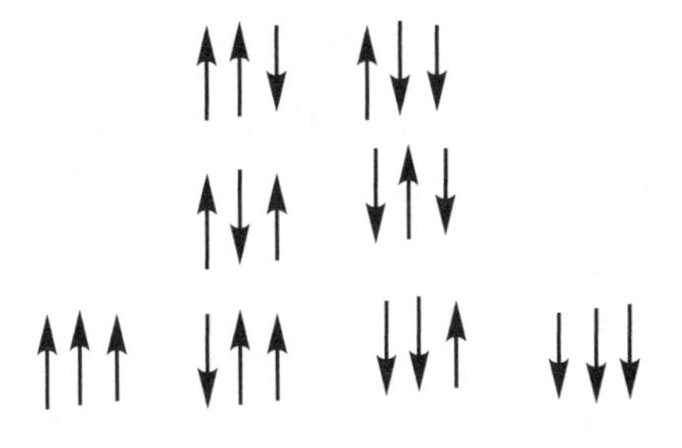

Any individual molecule has only one of these orientations, but they are all equally likely. Some are of unique energy, such as the ααα and βββ arrangements. Others have the same energy in a magnetic field, such as the ααβ, αβα, and βαα orientations. The net result is that the fields generated by the nuclear spins of the $—CH_3$ group affect the electrons in the bonds that connect the $—CH_3$ and $—CH_2—$ groups, and this in turn affects the field experienced by the $—CH_2—$ protons. The spins of one group of protons split the energy levels of nearby proton groups, which is why this effect is called spin-spin splitting.

For the left hand, ααα case, the nuclear spin field adds to the external field, and the $—CH_2—$ protons in such a molecule exhibit a resonant frequency downfield (at a higher frequency) from where they would resonate if this effect were absent. The next three spin arrangements also add to the external field but not as much as in the ααα state. The third arrangement opposes the external field, causing a slight upfield shift, and the final βββ state gives rise to the most pronounced upfield shift. The net result of this spin-spin splitting is to split the $—CH_2—$ resonance into four lines in the ratio 1:3:3:1, a **quartet**. The $—CH_2—$ protons are said to be **coupled** to the $—CH_3$ protons or **split** by the $—CH_3$ protons since the energy of one group affects the energy of the other.

EXERCISES

1. How many spin states can the protons in a $—CH_2—$ adopt, and how many of these states are different energies?
2. If you could take the NMR spectrum of a single diethyl ether molecule, would you see spin-spin splitting?

Solutions

1. The two protons can exist in αα, αβ, βα, and ββ spin states, and the αβ and βα states have the same energy. This causes any protons coupled to the $—CH_2—$ group to be split into a triplet, with relative intensities 1 : 2 : 1.
2. No. A given molecule would have the methyl protons in just one spin state, such as the ααα state. This would affect the energy of the $—CH_2—$ protons but lead to just one frequency and one line. If you then carried out the NMR measurement on a second single molecule, its methyl spin state might be ααβ, which would give a slightly different frequency for the $—CH_2—$ protons corresponding to another single line in the methylene quartet.

There is a simple rule for calculating the spin-spin splitting patterns in NMR spectroscopy, called the ***N* + 1 rule**. As long as the two groups have much different chemical shifts, N protons on adjacent atoms give $N + 1$ lines in the multiplet. The relative intensities of the peaks in the multiplet are given in Table 15.2.

TABLE 15.2. PATTERNS OF SPIN-SPIN SPLITTING

No of Adjacent Protons	No. of Lines	Name of Multiplet	Intensity Ratio of Lines
0	1	Singlet	1
1	2	Doublet	1 : 1
2	3	Triplet	1 : 2 : 1
3	4	Quartet	1 : 3 : 3 : 1
4	5	Quintet	1 : 4 : 6 : 4 : 1
5	6	Sextet	1 : 5 : 10 : 10 : 5 : 1
6	7	Septet	1 : 6 : 15 : 20 : 15 : 6 : 1

EXERCISE

Sketch the expected NMR spectrum of $BrCH_2CH_2CH_3$.

Solution

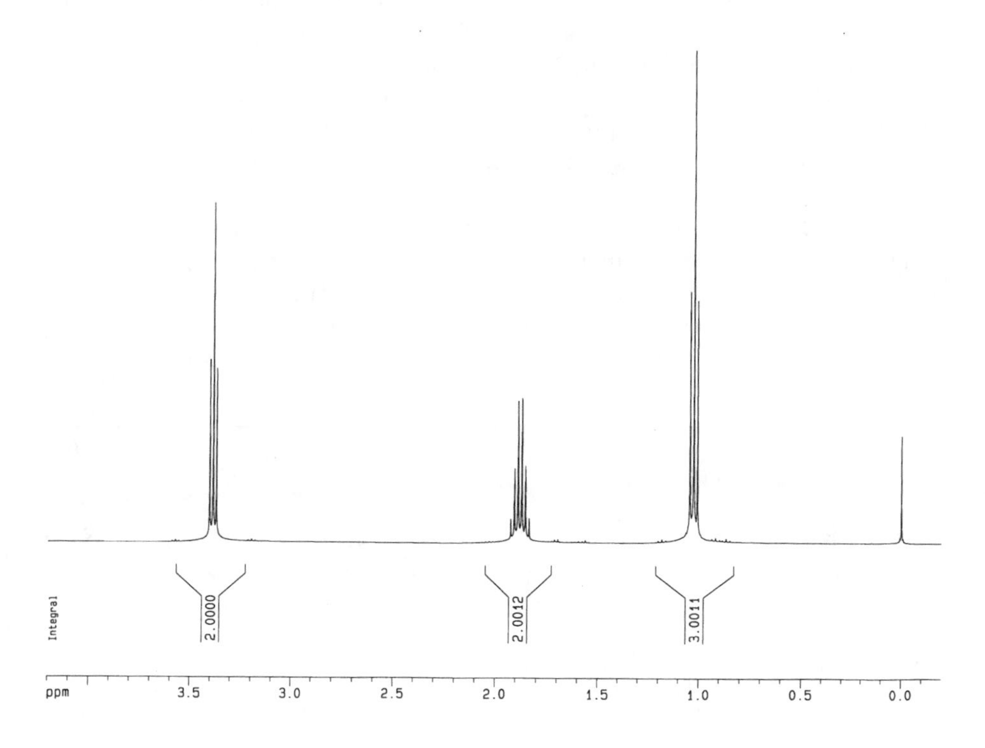

The triplets are not perfectly 1:2:1; they are said to *lean* toward the group that is splitting them. This is a common observation because the $N + 1$ rule is an approximation.

Coupling constants tell us about molecular connectivity. Each multiplet is like one piece of a jigsaw puzzle that one puts together to form the molecule. In the above case of 1-bromopropane, we see three groups. There is a group at about 3.5 ppm, two protons, that is a triplet. Triplets must be next to two protons. There is a second group at 1.8 ppm, two protons, that is a sextet, meaning it is next to five protons. Finally there is a three-carbon triplet at 1 ppm, showing that this group (clearly a $—CH_3$ group since it is three protons) is next to two protons. The only way one can put these pieces together is Br $—CH_2—CH_2—CH_3$. The isomer, isopropyl bromide, $(CH_3)_2CHBr$, would have a six-proton methyl doublet at approximately 1.2 ppm and a one-proton septet at about 3.8 ppm.

Coupling Constants as a Quantitative Measure of Spin-Spin Splitting

If one measures the distances between the individual lines in the ethyl ether spectrum, one finds that they are all 7 Hz apart. The distance between lines in a multiplet is called the **coupling constant**. It is measured in hertz and is designated ***J***. Since coupling constant effects are transmitted through bonds, the interaction between the protons in an ethyl group, $H_3C—CH_2—$, are called **three-bond coupling constants**, or $\boldsymbol{J_{H-C-C-H}}$. Coupling drops off rapidly as more bonds intervene between nuclei, so three-bond couplings are the most important. Because couplings arise from the nuclear spin itself, the external magnetic field does not influence the magnitude of the coupling constant. It is 7 Hz for diethyl ether in a 300- or 600-MHz NMR spectrum.

Coupling constants are affected by the types of bonds between protons and by the geometry. A collection of *J* values is shown in the accompanying table.

Group	*J* (Hz)	Group	*J* (Hz)
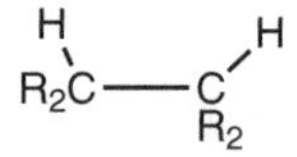 H, $R_2C—C$, H, R_2	~7	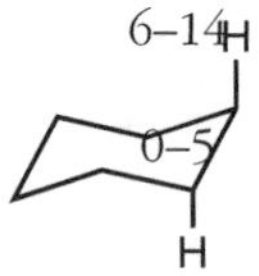 6–14 H, 0–5, H	6–14
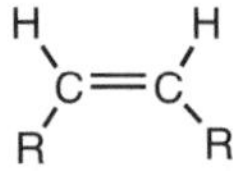 H, H, C=C, R, R	6–12	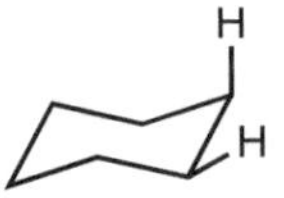 H, H	0–7
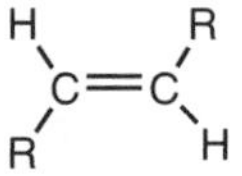 H, R, C=C, R, H	12–18	 H, H	0–5
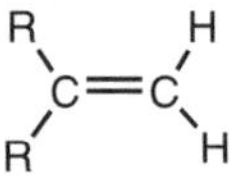 R, H, C=C, R, H	0–3	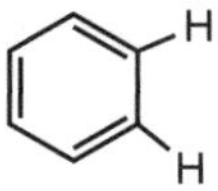 H, H	6–11

As can be seen, coupling constants are sensitive to the geometry of the carbon framework. They are very useful in telling cis and trans alkenes apart, for example.

Protons that are coupled to one another have the same coupling constant. For example, in an $—OCH_2CH_3$ group, the methyl and methylene protons split one another and so the coupling constant is the same for both groups, 7 Hz. Protons with the same chemical shift do not couple with one another. Thus, the protons in a methyl group, for example, do not split each other.

A factor that can influence coupling is chemical exchange. Normally one does not see coupling between an O—H proton and other protons in the molecule. This is because the —OH group is able to hydrogen-bond to water impurities in the solvent or on the surface of the sample tube. The H can rapidly exchange between oxygen atoms, and its spin becomes randomized, so there is no net coupling to any proton in the alcohol molecule.

EXERCISES

Describe the NMR spectrum of (a) 4-nitroethoxybenzene and of (b) methylacetate.

Solutions

(a) There are two 2-proton doublets in the aromatic region, one at about 8 ppm for the two equivalent protons near the electron-withdrawing nitro group and the other at about 7 ppm for the two equivalent protons near the ethoxy group. The coupling constants are equal and are between 6 and 11 Hz. There is a two-proton quartet at about 3.5 ppm and a three-proton triplet at about 1.8 ppm, both with coupling constants equal to 7 Hz.

(b) There are two singlets equal in height, one at about 2.2 ppm and one near 3.7 ppm for the $—OCH_3$.

APPLICATIONS OF ^{13}C NMR SPECTROSCOPY

The same spin flip process that results in a proton NMR spectrum can be used to produce a ^{13}C NMR spectrum. However, the sensitivity and abundance of ^{13}C makes the measurement more difficult. The magnetogyro ratio of ^{13}C is $\frac{1}{4}$ that of ^{1}H, and sensitivity scales as the third power of this quantity, so the same number of carbons gives $\frac{1}{64}$ of the signal of a group of protons. Second, the natural abundance of the ^{13}C isotope is only 1.1 percent, so this reduces the signal relative to ^{1}H by another factor of 100. Thus, ^{13}C NMR spectroscopy starts off being about 6400 times less sensitive than ^{1}H NMR spectroscopy.

A second diference in ^{13}C NMR is that the average organic molecule contains no ^{13}C nuclei, and only a few have even one ^{13}C. Thus, the ^{13}C nucleus is dilute, meaning that there is no significant ^{13}C—^{13}C coupling, although ^{13}C—^{1}H couplings exist in each organic molecule and can be observed if desired.

Although ^{13}C NMR has disadvantages, it was found to also have advantages, once instrument manufacturers were able to build machines sensitive enough to record ^{13}C spectra. The chemical shifts of ^{13}C groups range over more than

250 ppm. In ^{1}H NMR, different protons often have approximately the same chemical shift, and it is difficult to disentangle the overlapping peaks. With the huge chemical shift range of ^{13}C, normally every carbon in a molecule gives a nonoverlapping peak.

^{13}C spectra are usually recorded in the ^{1}H-decoupled mode. In this experiment, a broad radio frequency signal is sent that spin-flips all the ^{1}H nuclei. This has the effect of turning off all ^{13}C—^{1}H couplings, and all the ^{13}C signals appear as singlets. Since in a 300-MHz ^{1}H NMR spectrometer the ^{13}C signals are at 75 MHz, it is easy to build detectors that filter out the high-frequency ^{1}H noise and detect only the ^{13}C signal. Singlets concentrate the ^{13}C into a single peak that is easier to detect against the noise compared to a multiplet. Unfortunately, this **decoupling** affects the intensity of each peak somewhat differently, so the singlets cannot be integrated. One usually assumes that each peak represents a different carbon atom and there are no overlaps.

An example of a ^{13}C spectrum for the flavoring agent vanillin is shown here.

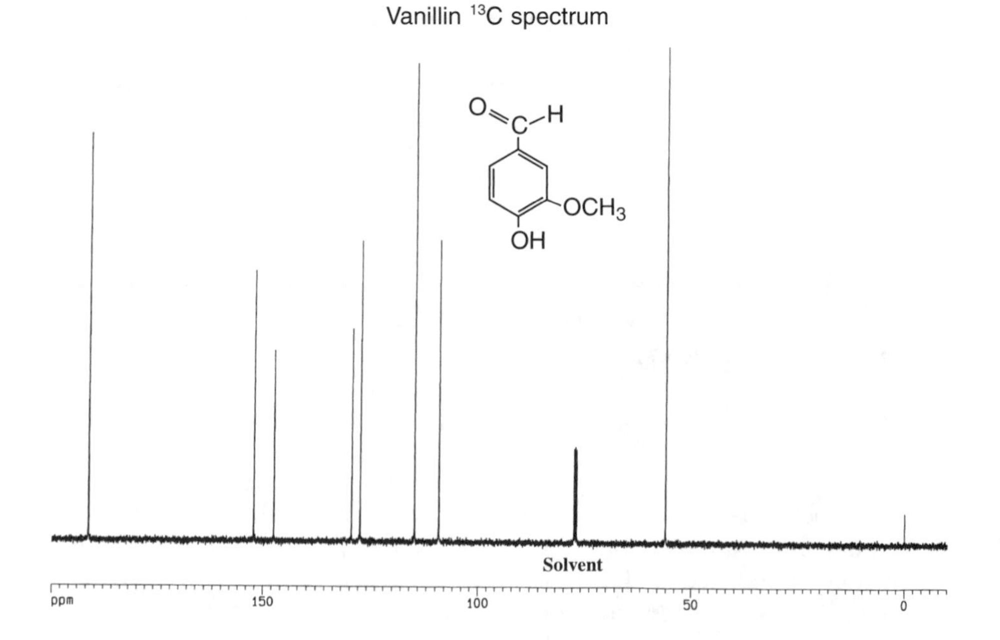

Carbons that do not have any hydrogens attached to them tend to have a low intensity in the NMR measurement, so the three lowest intensity lines are the three ring carbons attached to —OH, —OCH_3, and CHO.

The chemical shift for specific types of carbons has a wide range. Aliphatic carbons resonate between 0 and 65 ppm. The TMS molecule is used as the zero reference for ^{13}C spectra also. Carbons with *sp* hybridization appear between 25 and 105 ppm. Alkene carbons range from 55 to 165 ppm, depending on the substituents. Aromatic carbons range from 100 to 160 ppm. Carbonyl carbons are found between 140 and 220 ppm.

While ^{13}C NMR spectra are not normally measured with $^{13}C—^{1}H$ couplings, they can be detected by standard means. The J values are very large for J_{C-H} couplings, about 125 Hz for sp^3 C—H couplings, 160 Hz for sp^2 C—H couplings, and 250 Hz for sp C—H coupling constants. An advantage of recording the coupled ^{13}C NMR spectrum is that it allows one to easily distinguish $—CH_3$ groups, which are quartets, from methylenes, which are triplets, and from methines, which are doublets.

EXERCISE

The ^{1}H NMR spectrum of methane is a singlet. If one looks carefully, at approximately 63 Hz on either side of this singlet is a very small singlet. Why does it occur and about how high is each singlet relative to the main peak?

Solution

One percent of methane molecules are $H_4{}^{13}C$. There is a one-bond coupling constant of 125 Hz leading to a doublet centered at the $H_4{}^{12}C$ singlet. Each peak should be 0.011/2 times the height of the main peak. These small peaks are called ^{13}C satellites and are normally lost in the noise of the baseline.

16
REACTIONS AND SYNTHESES OF ALCOHOLS AND THIOLS

Alcohols can be considered organic analogs of water. Methanol, CH_3OH (known as wood alcohol), ethanol, CH_3CH_2OH (known as grain alcohol or just alcohol), and 1-propanol are completely **miscible** (soluble in all proportions) in water. In the case of alcohols with four carbon atoms, the more compact *t*-butanol is more soluble in water than the straight-chain 1-butanol. In general, for isomers with the same number of carbons, branching increases aqueous solubility.

Many important organic compounds are 1,2 diols. This functional group has the common name **glycol** and includes ethylene glycol, $HO{-}CH_2CH_2{-}OH$, and propylene glycol, $HO{-}CH_2CHOH{-}CH_3$, both used in automobile antifreeze. **Fats** and **oils** (such as olive oil and corn oil) are triesters of glycerol, $HOCH_2CHOHCH_2OH$.

Alcohols are about as acidic as water. Unhindered alcohols, such as ethanol, have a pK_a of 16. Hindered alcohols, such as *t*-butanol, are somewhat less acidic, with a pK_a of about around 18. The salt of an alcohol, an **alkoxide**, can be made by adding the alcohol to NaH or sodium metal, giving the sodium alkoxide and hydrogen.

SYNTHESES OF ALCOHOLS

Numerous procedures that give the alcohol functional group have been introduced in previous chapters. Markovnikov addition of water to alkenes, with or without Hg^{2+}, hydroboration/oxidation, S_N2 reactions of hydroxide with RX derivatives, and OsO_4 synthesis of glycols are important synthetic routes to alcohols that have already been covered. In this section, we introduce new procedures for alcohol synthesis, the first a functional group interchange process and the second a set of reactions that also form new carbon-carbon bonds.

Synthesis of Alcohols by Reduction of Carbonyl Compounds

Given the concept of carbon oxidation states, the conversion of an ester, aldehyde, or ketone to an alcohol is a reduction reaction. A large number of reagents exist that can carry out this reduction. The most useful ones act as a source of **hydride**, H^-. An extremely reactive hydride source is **lithium aluminum hydride** (LAH), $LiAlH_4$.

Because the hydride anion is a strong base, it can react with weak acids, such as water or alcohols, to give H_2. Thus, LAH cannot be used in protic solvents. Normally it is used in an ether solvent such as THF. All four of the Al—H groups are available for reduction, so that 1 mol of LAH reacts with 4 mol of an aldehyde or ketone:

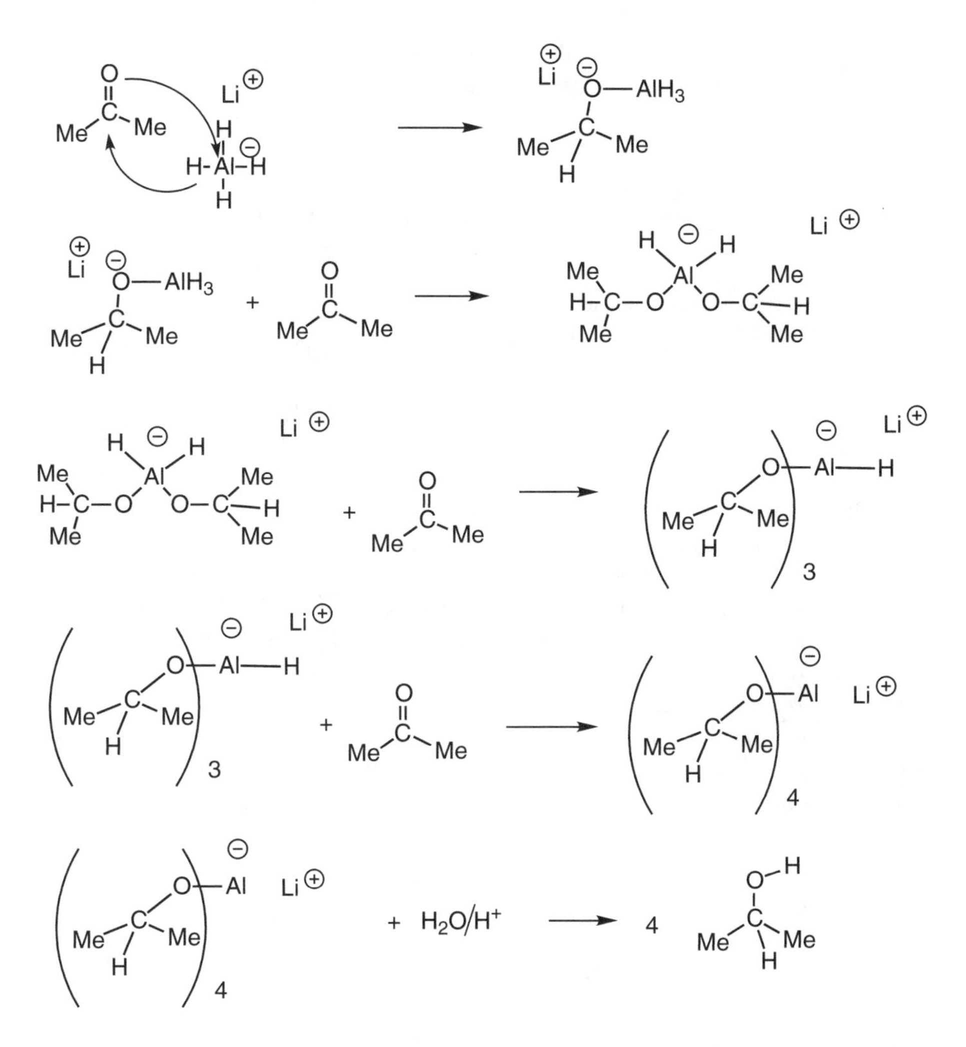

During the reduction, a hydride is transferred to the carbon of the carbonyl group and the oxygen bonds to the aluminum. The reaction continues until all four hydrides are transferred and there are four alkoxides bonded to aluminum. In a separate step, dilute acid is added to hydrolyze the O—Al bond and free the alcohols. An inorganic aluminum-oxygen polymer is a by-product.

Esters react with LAH to give, first, aldehydes: $RCO_2R' \rightarrow RCHO$. However, aldehydes are more reactive than esters in reduction reactions, so as soon as the aldehyde is formed, it is further reduced to give (after the reaction is hydrolyzed) a primary alcohol, RCH_2OH:

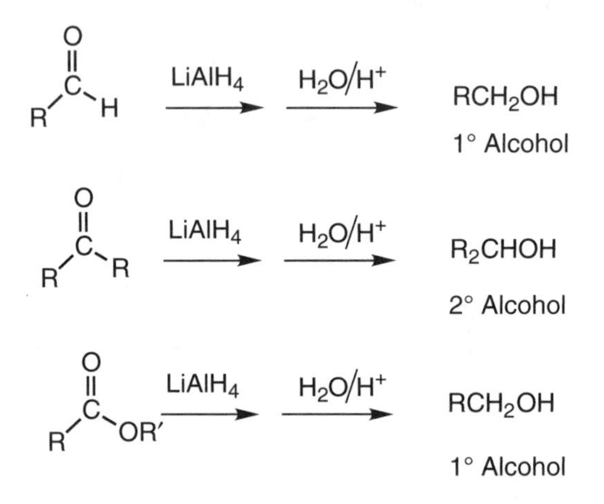

Sodium borohydride, $NaBH_4$, is a milder reducing agent than $LiAlH_4$. The B—H bond is more covalent than the Al—H bond. As a result, $NaBH_4$ reacts slowly with protic solvents, such as water and alcohols, and these solvents are used for $NaBH_4$ reductions. The mechanism for the $NaBH_4$ reduction of aldehydes and ketones is analogous to the $LiAlH_4$ mechanism. However, when protic solvents are employed, there is no need for a separate hydrolysis step to release the alcohol.

Both aldehydes and ketones are reduced to alcohols by $NaBH_4$. However, esters are generally more difficult to reduce, and they are recovered unchanged when treated with $NaBH_4$:

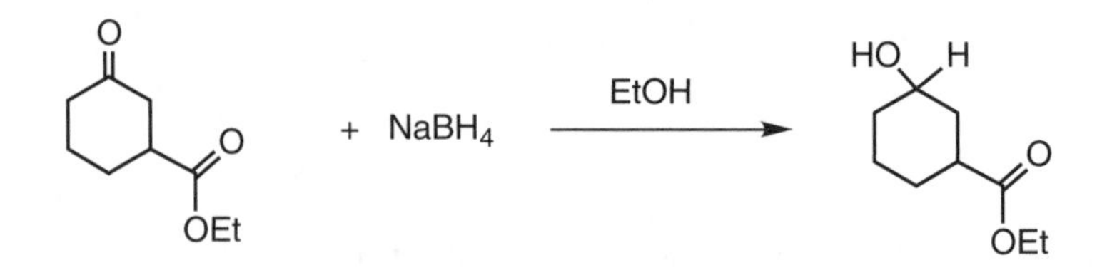

EXERCISE

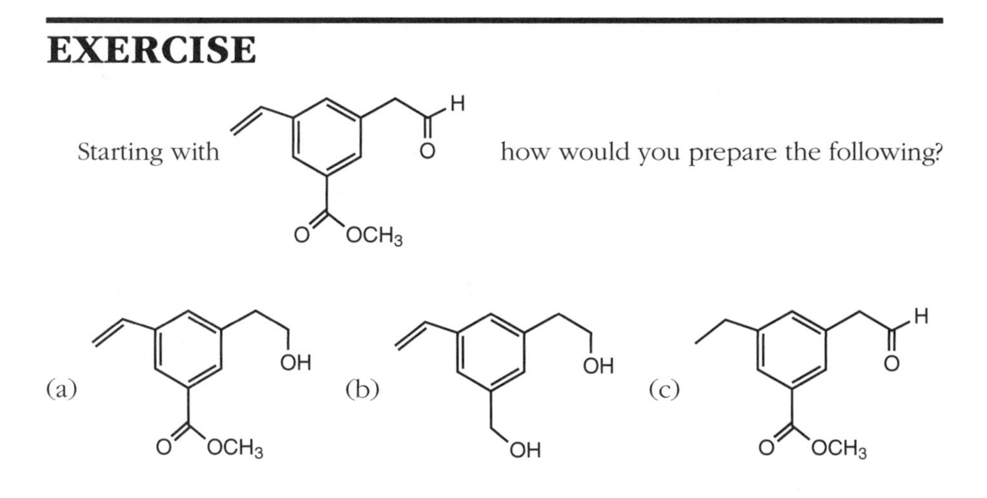

Solutions

(a) $NaBH_4/CH_3OH$; (b) $LiAlH_4$ followed by H_2O/H^+; (c) PtO_2/H_2. Notice that isolated carbon-carbon double bonds are not reduced by hydride reagents and that carbon-oxygen double bonds are reduced much more slowly than C=C bonds under catalytic hydrogenation conditions. With enough catalyst and at higher temperatures, catalytic hydrogenation can reduce aldehydes and ketones to alcohols. However, hydride reagents are more commonly used for this purpose.

Introduction to the Synthesis of Alcohols Using Organometallic Reagents

It is relatively easy to prepare molecules with C—Li, C—Mg, and C—Cu bonds. These are examples of **organometallic compounds**, compounds with carbon-metal bonds. Because of the electronegativity difference between carbon and metallic elements, the carbon-metal bond is polarized: $C\delta^-—M\delta^+$. As a result, these organometallic compounds are good sources of nucleophilic carbon. They are particularly reactive toward the electrophilic carbon of a carbonyl group, leading to the formation of a new carbon-carbon bond:

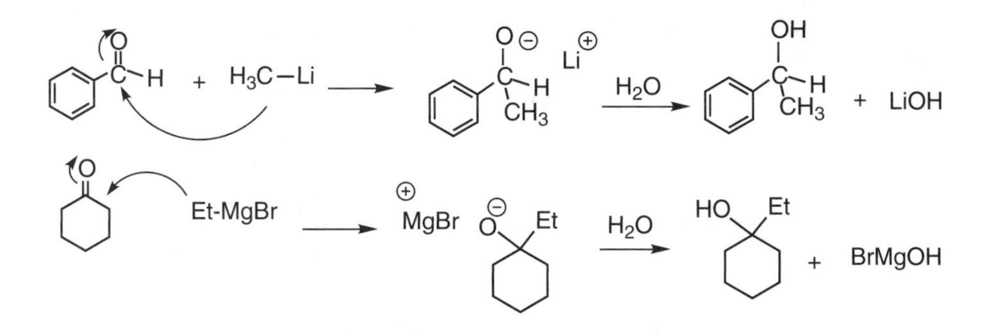

The initially formed compound in these reactions is a salt (an alkoxide) that is a reasonably strong base. Treatment with a dilute acidic solution or a buffer gives the desired alcohol product.

Synthesis of Organometallic Reagents

When organic halides, especially bromides or iodides, react with magnesium metal, they generate compounds of the type RMgX, known as **Grignard reagents** after the French chemist Victor Grignard. The use of an ether solvent, either diethyl ether or tetrahydrofuran (THF), is usually necessary for the reaction to take place. Alkyl, aryl, and vinyl halides are all reactive.

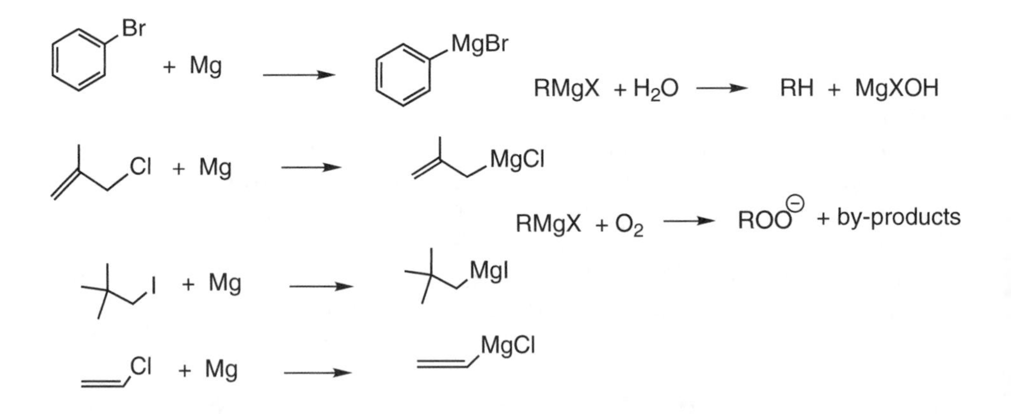

Grignard reagents are very strong bases and react with water or the oxygen in air. As a result, they must be prepared under **anhydrous** (water-free), **anerobic** (air-free) conditions with dry solvents.

Lithium reagents are made in the same way that Grignard reagents are made: $RX + 2Li \rightarrow RLi + LiX$. Organolithium reagents have essentially the same properties as Grignard reagents although they are somewhat more basic and can be easier to form. They often give somewhat better yields in additions to carbonyl groups, especially when the R group is sterically hindered.

The mechanism by which Grignard and lithium reagents form involves **single-electron transfer**. The metal surface donates an electron to the alkyl halide, creating an **anion radical**, $[RX]^{-\bullet}$, that, recalling the molecular orbital description of bonding, contains a single electron in an antibonding orbital. The anion radical is unstable and dissociates into an alkyl radical and a halide anion, $R^{\bullet}$ and X^-. The metal surface donates a second electron to $R^{\bullet}$ to give R^- and the appropriate counterion. A chiral alkyl halide produces a racemic organometallic reagent because radicals are involved in the mechanism:

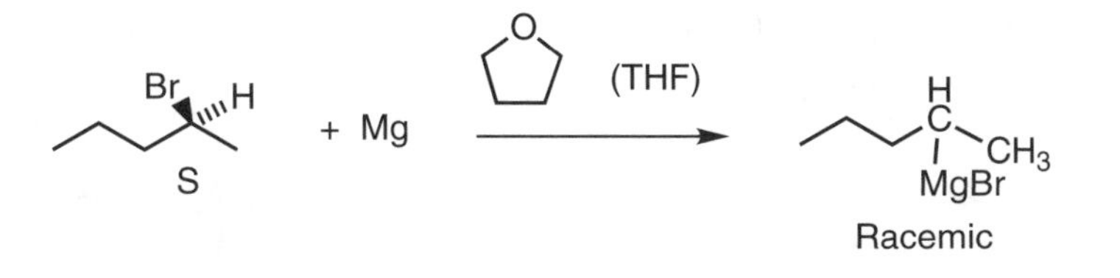

Reactions of RLi and RMgBr with Carbonyl Compounds and Epoxides

As shown above, lithium and Grignard reagents form new C—C bonds and generate the alcohol functional group when they react with aldehydes and ketones. Esters also react to give alcohols, but two new C—C bonds are generated at the carbonyl carbon. Esters, RCO_2R', first react with **R**Li or **R**MgBr, giving the ketone RCO**R** and displacing $R'O^-$. However, ketones are more reactive than esters, so as soon as one is formed it immediately reacts with a second **R**Li or **R**MgBr to give, after hydrolysis, the tertiary alcohol R**R**$_2$COH.

The reaction of epoxides with lithium and Grignard reagents gives another route to alcohols. The anionic carbon of the organometallic species reacts at the less hindered carbon of the epoxide in an S_N2 reaction where the ring C—O bond is the leaving group. The ring strain of the three-membered ring causes the ether bond to be a leaving group:

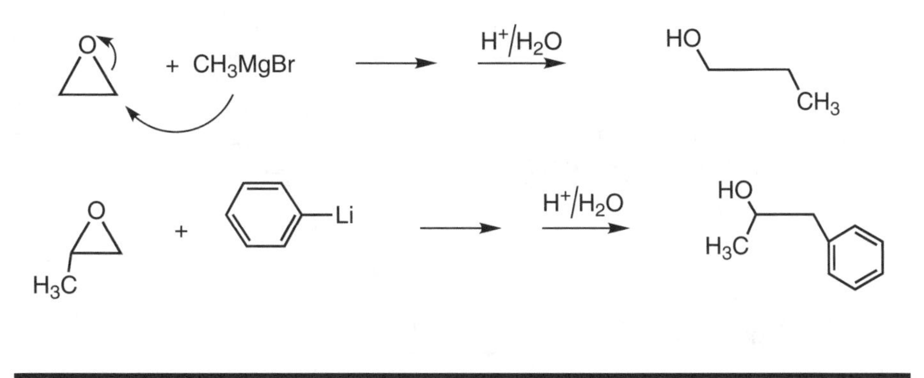

EXERCISES

1. Give three different combinations of carbonyl compound and lithium reagent that would produce 3-methyl-3-hexanol.

2. Show what reagents are needed to prepare the indicated products.

(a) 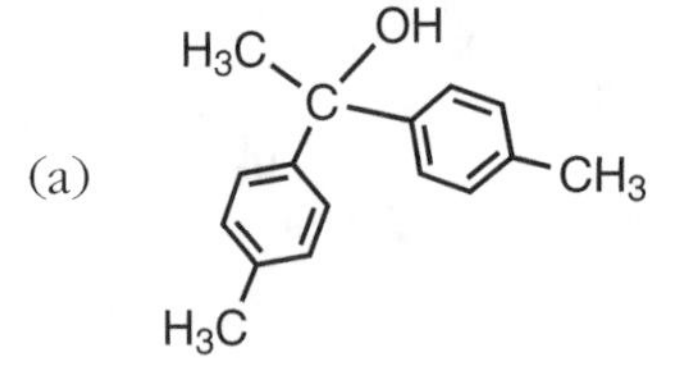

from an ester and a lithium compound

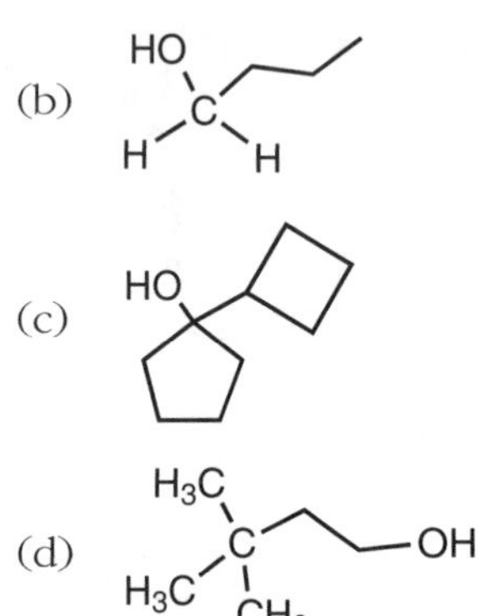

(b) from an aldehyde and a Grignard reagent

(c) from a ketone and a Li reagent, showing what halide you would use to propare the RLi compound

(d) from an RLi compound and an epoxide

Solutions

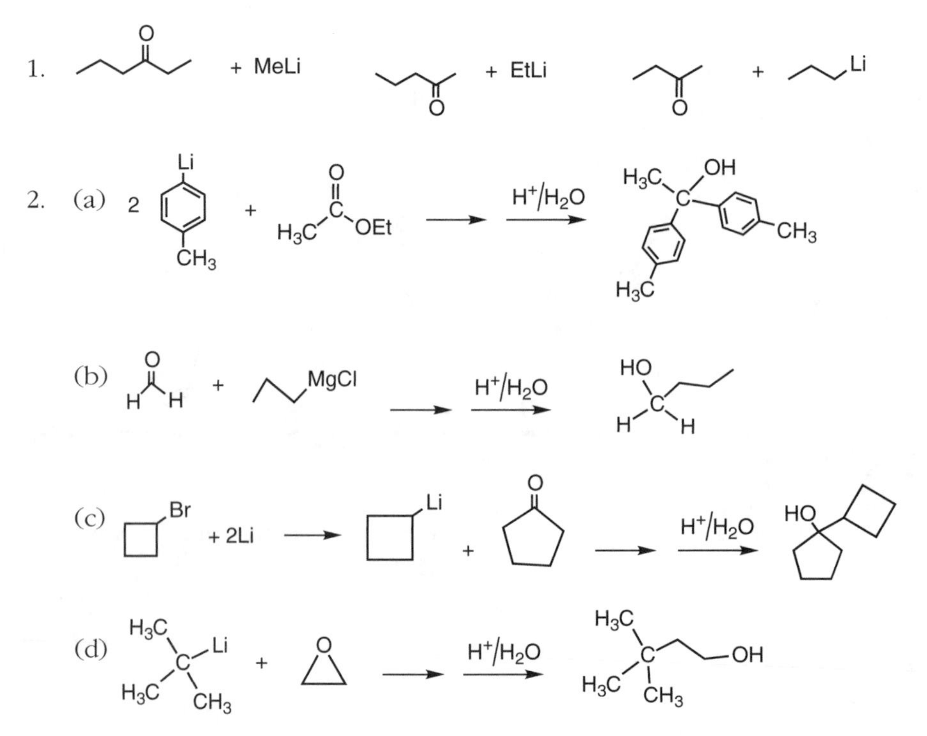

Limitations to the Preparation of RLi and RMgBr

Because lithium and Grignard reagents are strong bases, they react with many functional groups. Any protic group, including the —OH of an alcohol, phenol, or carboxylic acid, the —NH of an amine or amide, and the —SH of a thiol or the terminal C—H of an alkyne is deprotonated by these organometallic compounds. Furthermore, any carbonyl compound, including aldehydes, ketones, esters, anyhydrides, and acid chlorides, and compounds with carbon-nitrogen multiple bonds, such as nitriles, undergo additions with RLi and RMgX compounds. As a result, any halide, RX, with any one of these functional groups cannot be used to generate a lithium or Grignard reagent. Such a molecule reacts with itself or a second functionalized RX molecule. Normally, the only functional groups that can coexist in a Grignard or lithium reagent are carbon-carbon double or triple bonds, benzene rings, or ethers. Since alkyl and aryl fluorides are not very reactive toward Li or Mg metal, it is possible to form a Grignard or lithium reagent if the molecule also contains a C—Cl, C—Br, or C—I group.

Example The following molecules cannot be used to form Grignard or lithium reagents because of functional group incompatibilities.

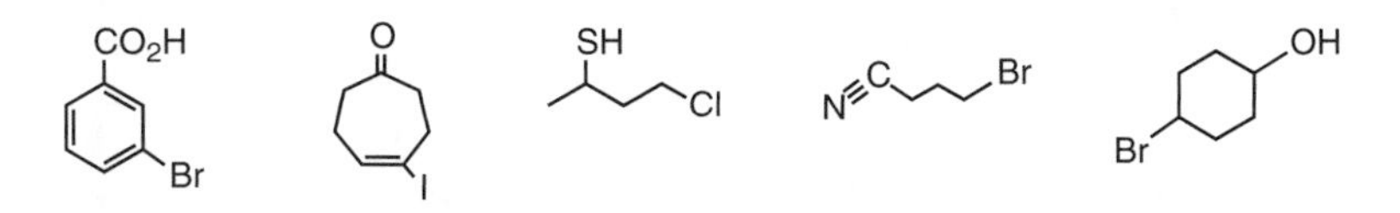

The following compounds form useful organometallic reagents.

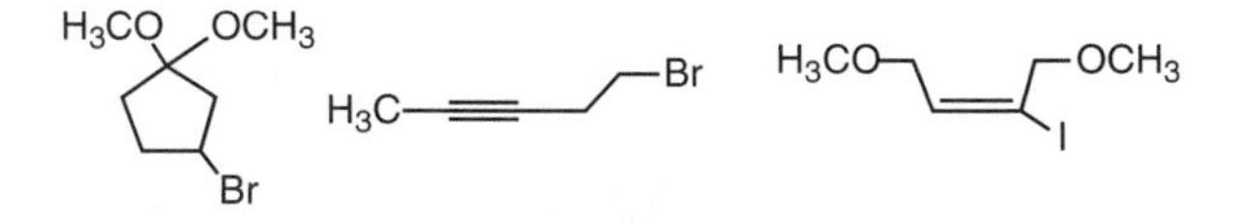

Lithium Diorganocopper Reagents

Lithium diorganocopper reagents, also known as **Gilman reagents** for Henry Gilman of Iowa State University who did the pioneering work on them, are an important class of organometallic reagents. While they are not normally used to prepare alcohols, they exhibit complementary reactivity to RLi and RMgX compounds. R_2CuLi reagents are prepared by treating CuI with two equivalents of RLi. These **cuprate** reagents have reduced basicity compared to RLi or RMgBr. While they react with protic groups, such as the OH of alcohols or acids, they do not react with the carbonyl groups of ketones or esters. However, R_2CuLi reacts with aldehydes to give 2° alcohols.

Gilman reagents are most useful in forming C—C bonds to alkyl, aryl, and vinyl halides (usually bromides or iodides): $\mathbf{R}_2CuLi + RX \rightarrow \mathbf{R}—R$. Neither RLi nor RMgX reagents react in a useful way with most alkyl halides. They do not undergo S_N2 reactions with all but the most reactive halides, such as CH_3I. Instead, R_2CuLi compounds are used for this transformation. The mechanism is not an S_N2 reaction (as can be seen by the fact that aryl halides react). Instead, it is complicated and probably involves some kind of single-electron transfer. Because RLi compounds are made from halides, RX, Gilman reagents provide a way to couple two RX molecules that are either the same or different.

Example Note the kinds of substrates that can be coupled using Gilman reagents.

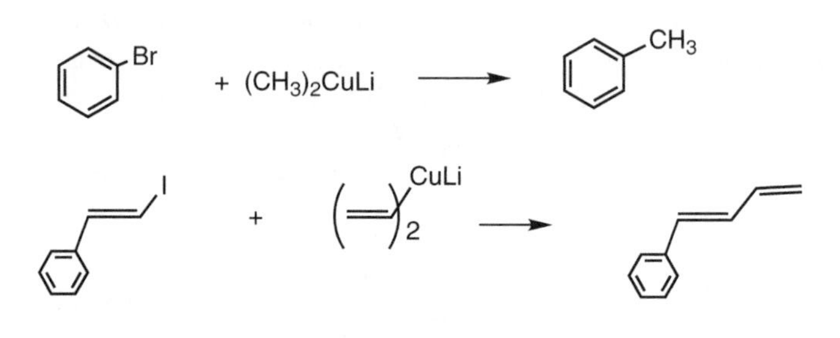

Notice that the E stereochemistry is maintained. The Z iodide gives the corresponding Z product. Thus, coupling to vinyl iodides is stereospecific.

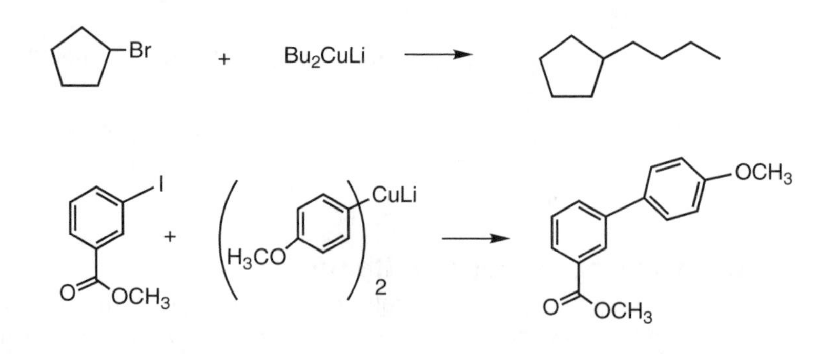

REACTIONS OF ALCOHOLS

Conversions of Alcohols to Alkyl Halides

In Chapter 8, the conversion of alcohols to alkyl halides using HX was discussed. A strong acid such as HI first protonates the —OH group to make it a leaving group, and then an S_N2 or S_N1 reaction takes place, depending on the structure of the alcohol. Because many molecules are destroyed by strong acids, other procedures are necessary to convert alcohols to halides.

Primary and secondary alcohols are usually converted to alkyl chlorides with the reagent **thionyl chloride**, $SOCl_2$. This reagent creates an intermediate, an **alkyl chlorosulfite**, which is a good leaving group:

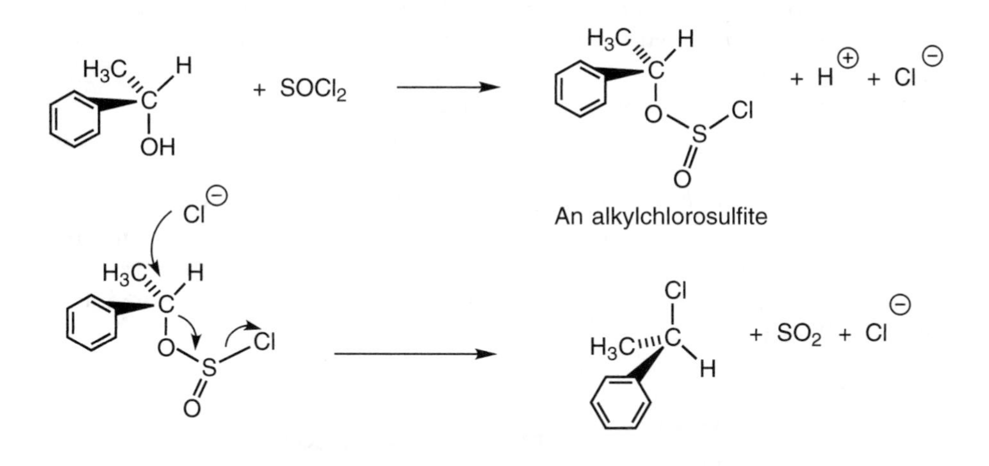

A chloride anion acts as a nucleophile to displace the chlorosulfite leaving group, which decomposes into sulfur dioxide and Cl^-. Because HCl is a by-product of this synthesis, it is common to add a weak organic base to this reaction to neutralize HCl. Pyridine is the usual choice:

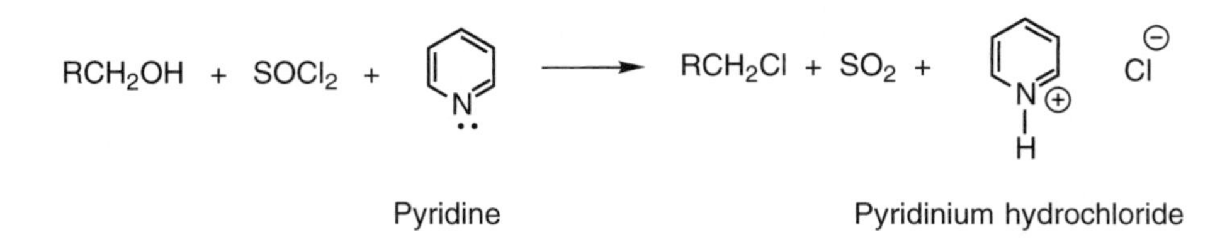

Since the thionyl chloride reaction is an S_N2 reaction, it is stereospecific and proceeds with inversion. Normally, 3° alcohols give elimination products with $SOCl_2$.

While $SOBr_2$ exists, it is more usual for organic chemists to use $\mathbf{PBr_3}$ to convert primary and secondary alcohols to alkyl bromides. This reagent first reacts with three equivalents of ROH to make $(RO)_3P$, a **trialkyl phosphite**. In the presence of H^+, the P atom is protonated and $(OR)_2(H)P^+{-}O$ becomes a leaving group. Three Br^- anions then sequentially displace the alkyl groups, finally giving three equivalents of RBr and H_3PO_3. Since the displacement reaction is an S_N2 reaction, the conversion of ROH to RBr is stereospecific and proceeds with inversion.

Oxidation of Alcohols to Carbonyl Compounds

Oxidation of the alcohol functional group is the most important method of preparing aldehydes, ketones, and carboxylic acids. Primary alcohols, RCH_2OH, are oxidized under mild conditions to aldehydes, RCHO, and conditions that are more vigorous give carboxylic acids, RCO_2H. Secondary alcohols, R_2CHOH, give ketones, R_2CO. Tertiary alcohols, R_3COH, are generally unreactive toward oxidizing reagents.

A number of chromium(VI) compounds are used in the laboratory for alcohol oxidation. Dissolving CrO_3 in water with H_2SO_4 gives **chromic acid**, H_2CrO_4. This combination is known as **Jones reagent**. It is excellent for the oxidation of 2° alcohols to ketones. Normally, 1° alcohols are oxidized to carboxylic acid with this reagent through an aldehyde intermediate. The mechanism of this chromium oxidation involves formation of a **chromate ester** from the alcohol and chromic acid, which decomposes to a carbonyl compound and a Cr(IV) species:

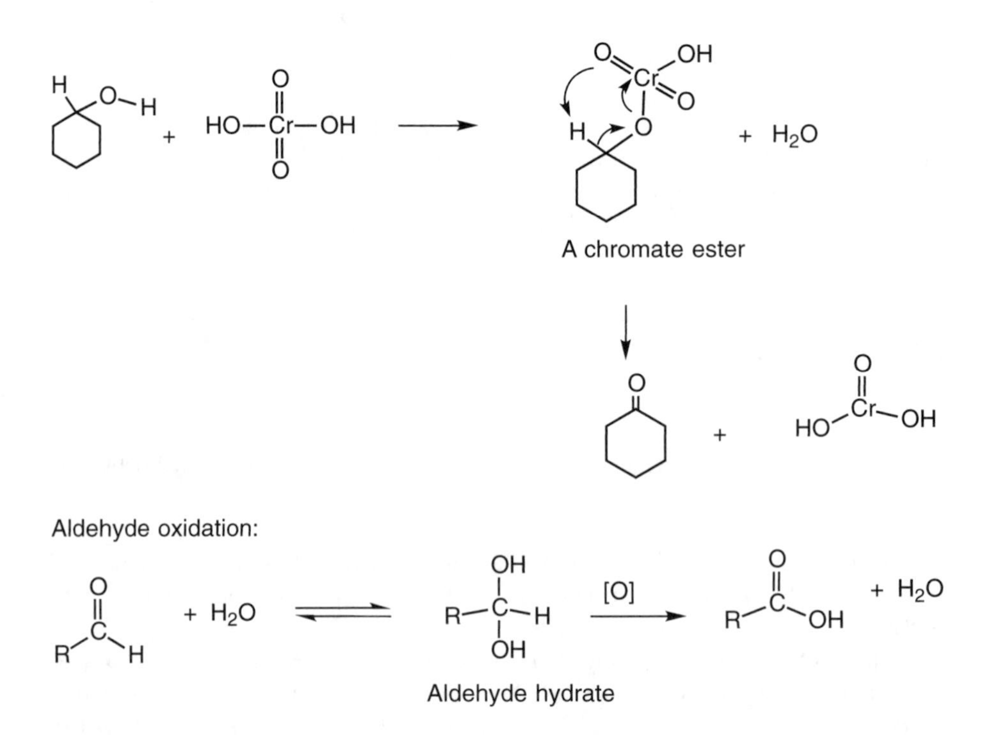

In the case of aldehydes, the presence of water creates an equilibrium amount of the unstable water addition product, called an **aldehyde hydrate**. Oxidation of this species (represented by a generalized oxidant [O]) produces a carboxylic acid. If one wishes to stop the oxidation of a 1° alcohol at the aldehyde stage, it is better to use a reagent that works under anhydrous conditions.

Two Cr(VI) reagents normally used in CH_2Cl_2 or other anhydrous solvents are **pyridinium chlorochromate (PCC)**, $C_5H_5NH^+ \ CrO_3Cl^-$, and **pyridinium dichromate (PDC)**, $(C_5H_5NH)_2^{+2} \ Cr_2O_7^{-2}$. Both of these reagents oxidize 1° alcohols to aldehydes and 2° alcohols to ketones. Most other functional groups are unreactive toward these reagents.

A highly selective reagent is the Mn(IV) compound **manganese dioxide**, MnO_2. It oxidizes only allylic and benzylic alcohols, C=C-CHROH or Ph-CHROH. The allylic and benzylic C—H bonds are weaker and more easily broken than other kinds of C—H bonds, which accounts for the ability of the weak oxidant MnO_2 to react with them:

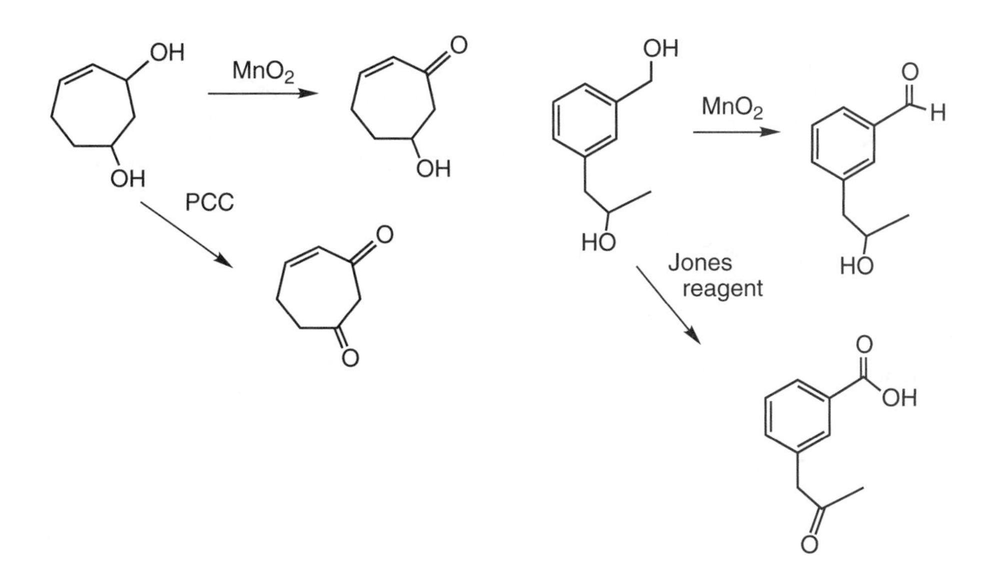

Notice how MnO_2 can selectively oxidize one alcohol in a **polyfunctional** molecule, that is, a molecule with several functional groups.

Aldehydes can be oxidized to carboxylic acids selectively with Ag^+ salts. **Tollens' reagent**, which is a solution of silver nitrate in basic ammonium hydroxide, $Ag(NH_3)_2^+NO_3^-$, is sometimes used for this oxidation. The Ag^+ ions are reduced to Ag metal, which can form a silver mirror on the inside surface of a glass flask.

EXERCISES

Give the correct reagent or product for each of the following reactions.

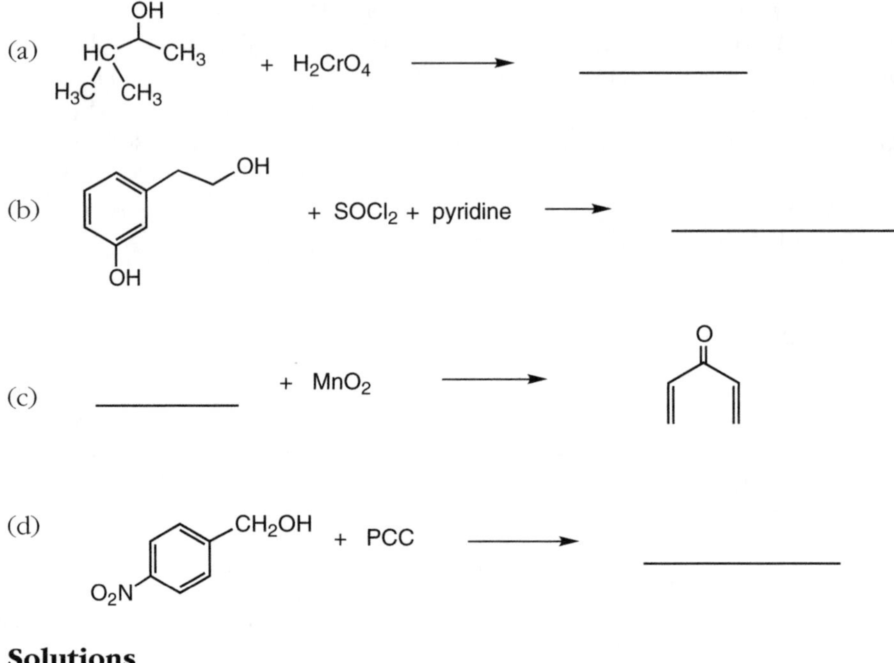

Solutions

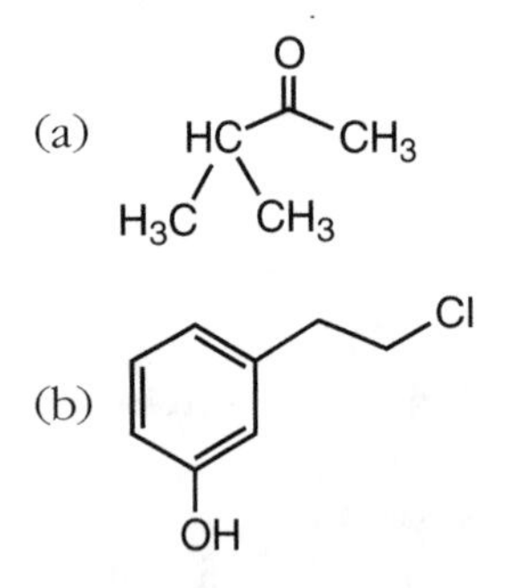

Because of the S_N2 mechanism, phenol —OH groups do not give —Cl.

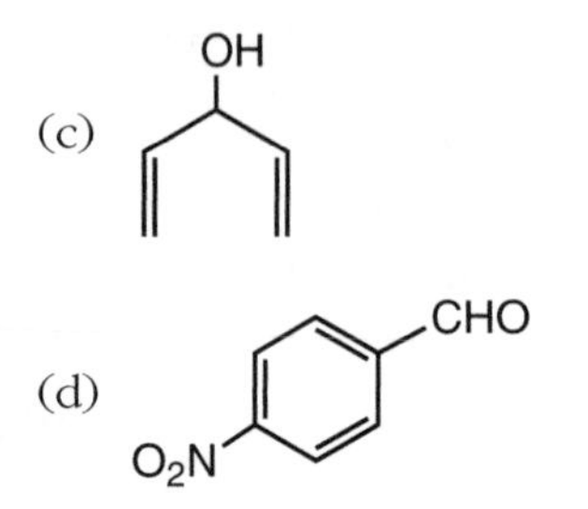

THIOLS AND RELATED SULFUR COMPOUNDS

Thiols are the sulfur analogs of alcohols. They are notorious for their bad smell. Ethanethiol, CH_3CH_2SH, is added in trace amounts to natural and bottled gas to give it an odor so that leaks can be detected. Although thiols are considered protic molecules, they are not good hydrogen bond donors or acceptors. Therefore, while alcohols form hydrogen-bonding networks in solution and have higher boiling points, thiols are monomeric in solution and have lower boiling points.

In the IUPAC system thiols are named by naming the parent alkane, retaining the final *-e*, and adding the suffix *-thiol*. The —SH group has priority only over halogens in numbering on a chain. Where the —SH group is a substituent in a molecule with a higher-priority functional group, it is indicated by the prefix *sulfanyl-*. Common names for thiols use the word *mercaptan* preceded by the name of the alkyl group. For example, in the IUPAC system $CH_3CH_2CH_2CH_2SH$ can be named 1-butanethiol or given the common name *n*-butyl mercaptan.

The most important feature of thiols is their acidity. While alcohol O—H groups have pK_a values of about 16 to 18, the pK_a of an alkyl thiol S—H is 7. Since sulfur is a heavy element, it is a better nucleophile than oxygen, and the **thiolate** anion is one of the most reactive nucleophiles in organic chemistry. Often, thiols are prepared using S_N2 reactions of NaSH with 1° or 2° alkyl halides. One must use an excess of NaSH in this reaction because the product thiol can react with a second molecule of alkyl halide to give a **thioether**, RSR.

A unique reaction of thiols is their oxidation to form **disulfides**:

$$2RSH + [O] \rightarrow RS{-}SR + H_2O$$

Many oxidants, including molecular oxygen or iodine, carry out this oxidation. This transformation is reversible, and an excess of certain thiols can reverse S—S bond formation. What is remarkable about this reaction is that a new σ bond is made (or broken) under exceptionally mild conditions. Such bond formation can cause a molecule to change its shape, as shown schematically here.

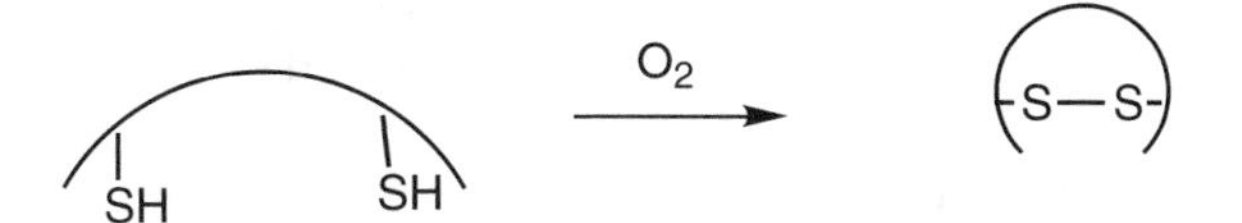

Nature uses disulfide bond formation to stabilize the structure of some proteins and hold certain parts of protein structures together.

Oxidation of thiols and thioethers can produce a variety of sulfur functional groups, depending on how powerful an oxidant is used:

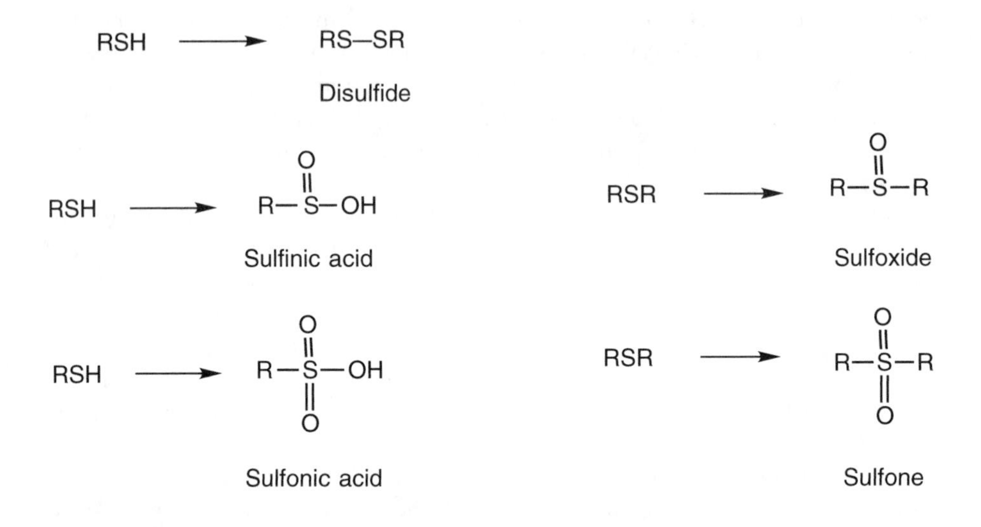

INTRODUCTION TO ORGANIC SYNTHESIS

In Chapter 7, reactions were broadly grouped into two categories: functional group interchange reactions and reactions that formed new C—C bonds. With the discussion of organometallic reactions in this chapter and introduction of the Diels-Alder reaction in Chapter 12, one has several powerful reactions for creating carbon frameworks. In addition, a number of reactions have been introduced that interconvert functional groups. These reactions make up the tools of organic synthesis, but one also needs a strategy for making molecules.

Most problems in organic synthesis are of the following type. Starting with X, synthesize Y. In a text, compound X is given, while in real life X is usually the compound that one can buy at a reasonable price. The best strategy for solving these synthesis problems involves **retrosynthetic analysis**, which is a way of working backward from the product, or **target**, to simpler molecules. When solving synthesis problems, it generally helps to focus on making the carbon framework first and then manipulating the functional groups to give the final product.

A typical synthesis problem would involve starting with a carbon compound of six carbon atoms or less and showing how one can prepare 1-phenylpropene:

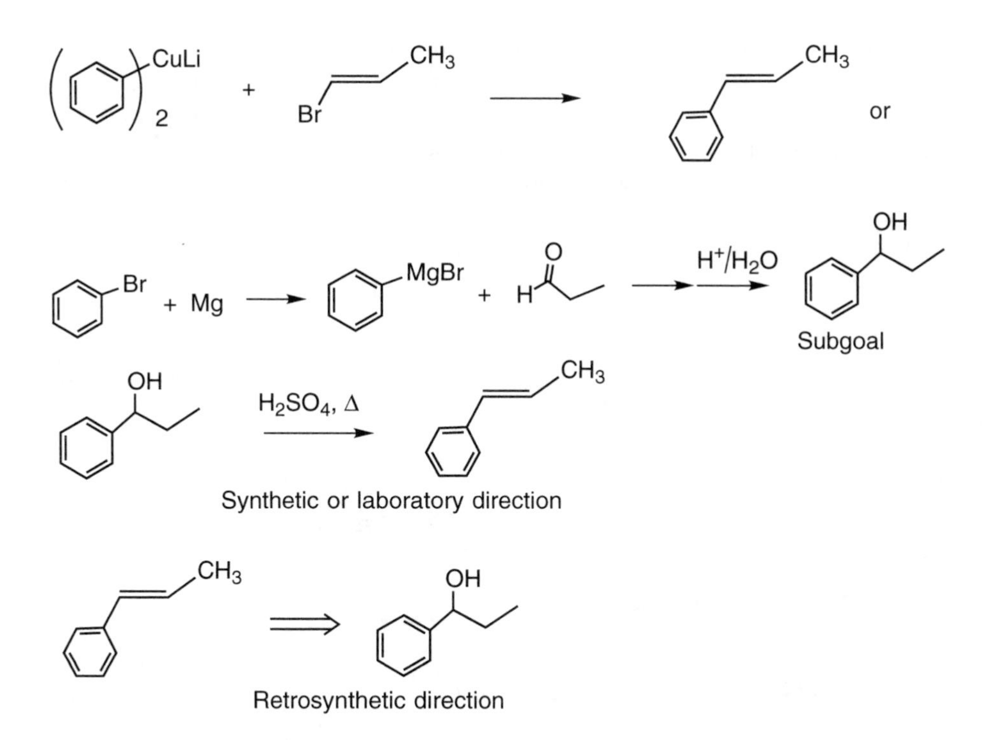

Sometimes one can see a direct way to carry out the synthesis. In this case, a cuprate reaction on a vinyl bromide will give the product in one step. Since each partner has six carbons or fewer, one does not need to show how to make the reagents. Normally, there is more than one way to carry out a synthesis. Here, one can look at the target and notice that it contains an alkene functional group. The student should review the reactions that produce double bonds. In this case, the double bond is conjugated with the phenyl ring, so it is easy to generate. If one had the molecule 1-phenylpropanol, it would undergo E1 elimination to give the desired product. The compound 1-phenylpropanol becomes a **subgoal** of the synthesis. If one had the subgoal molecule, it would be trivial to turn it into the product in the synthetic or laboratory direction.

Looking at the target, it is important to think of ways to simplify the structure in the retrosynthetic direction, which is indicated by the double-tailed arrow. In the synthetic direction, benzylic alcohols dehydrate to give alkenes. In the retrosynthetic direction, double bonds imply alcohols (among other functional groups). The reason for making an alcohol a subgoal is that carbon-carbon bond-forming reactions often generate an alcohol. Thus, reacting either PhMgBr with EtCHO or PhCHO with EtMgBr gives the subgoal.

There are several levels of complexity in synthesis problems. The easiest type of problem is a functional group manipulation problem. The carbon framework is the same in the starting material and product, and one must only convert some functional groups to other functional groups. The next level of difficulty involves creating both the carbon framework and the required functional groups. Finally, one may have to prepare one desired stereoisomer of a compound in addition to the carbon framework and functional groups.

For example, suppose one is asked to prepare (*Z*)-1-phenylpropene from compounds of eight carbons or less:

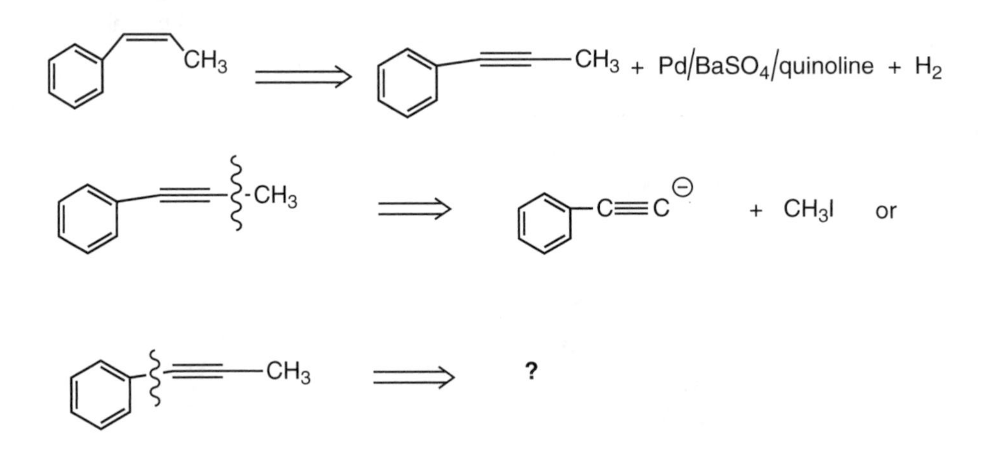

At this point, the only way one knows to make Z alkenes stereospecifically is by Lindlar hydrogenation of alkynes. Thus, in the retrosynthetic direction, the Z double bond implies a triple bond. 1-Phenylpropyne thus becomes a subgoal. We recall that it is easy to deprotonate terminal acetylenes and that these anions are good nucleophiles in S_N2 reactions. Thus, we can, in the retrosynthetic direction, imagine a **disconnection** that simplifies the carbon framework. Two disconnections can be imagined at either side of the triple bond. However, only one disconnection makes chemical sense. The phenylacetylide anion undergoes a facile S_N2 reaction with CH_3I, but the 1-propyne anion does not undergo an S_N2 reaction with phenyl iodide (and, although it was not stated above, acetylenic cuprates, unlike alkyl or aryl cuprates, are unreactive). In synthesis problems it is usually a good idea to consider several possible disconnections to simplify the carbon framework. Often, one disconnection is seen to be superior to the others.

EXERCISES

1. Fill in the blanks with subgoal molecules in the following examples of retrosynthetic analysis.

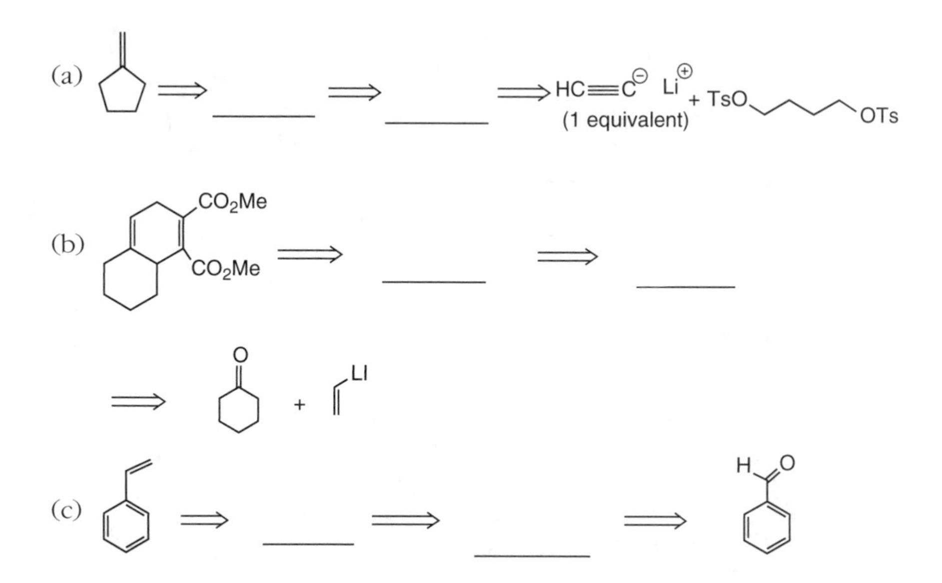

2. Synthesize each molecule as indicated. (*Note*: Once a molecule has been synthesized, it can be used in subsequent steps.)

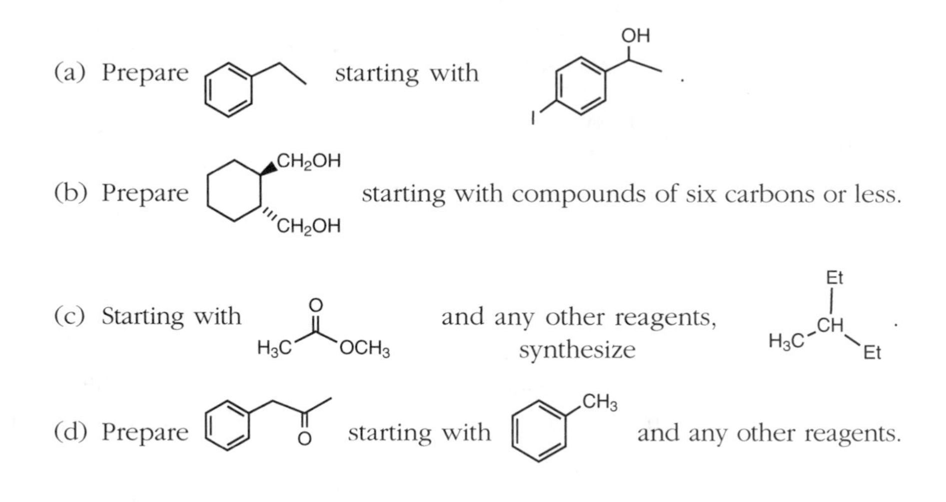

Solutions

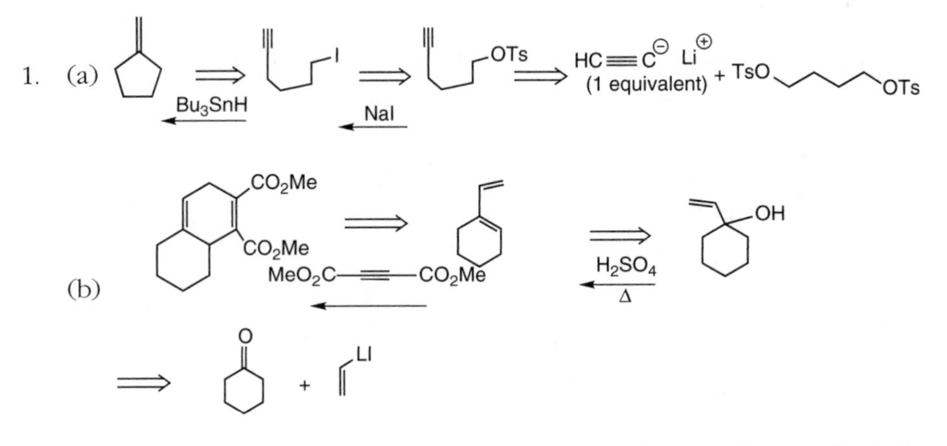

Whenever you see a six-membered ring with a double bond, think Diels-Alder reaction.

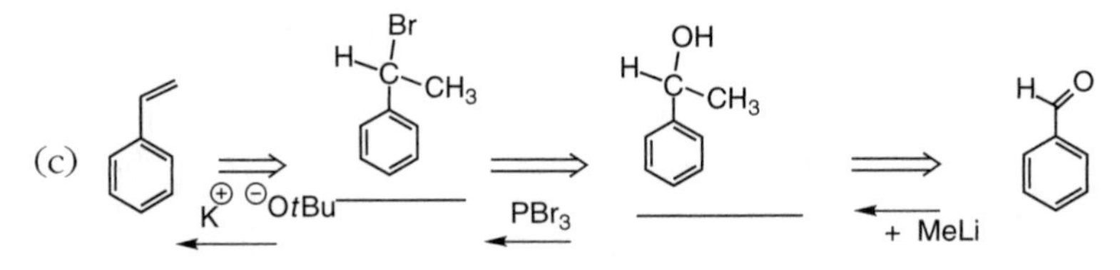

Alternatively, the alcohol could be eliminated directly under E1 conditions.

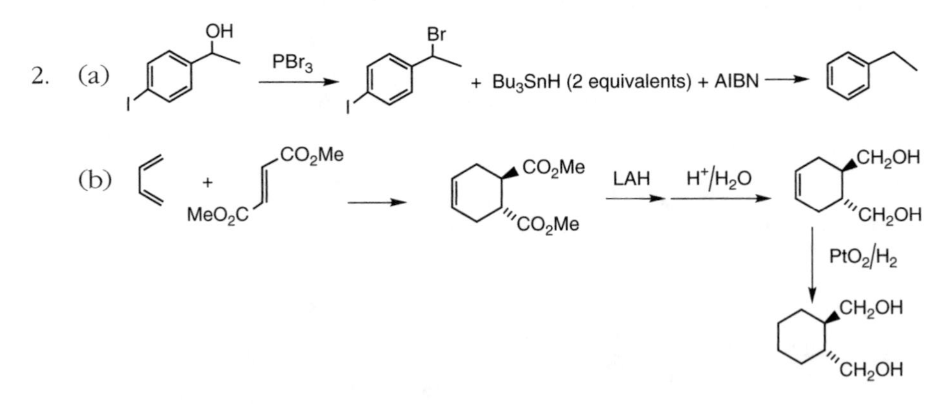

Because of the stereochemistry of the product, the dieneophile must be trans. Since dienophiles work best with electron-withdrawing groups, HOH_2C–CH=CH–CH_2OH would not be a good dienophile.

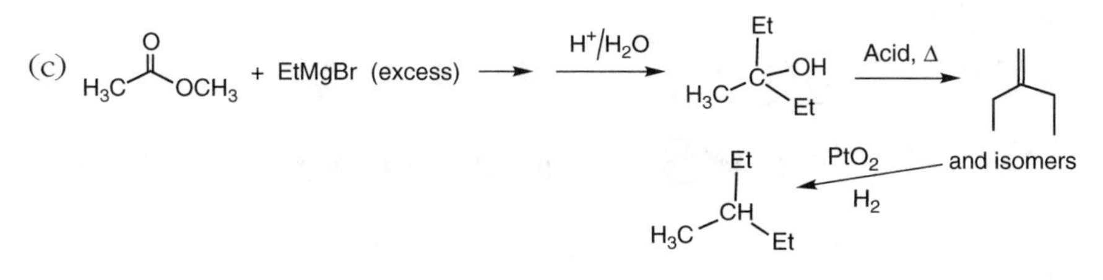

Both (a) and (c) require conversion of a C—OH to a C—H. In (c), the 3° alcohol would not be converted to a bromide by PBr_3, so the alternative dehydration-hydrogenation sequence is used. It would work in (a) too, but Bu_3SnH would still be needed to remove the I on the phenyl ring.

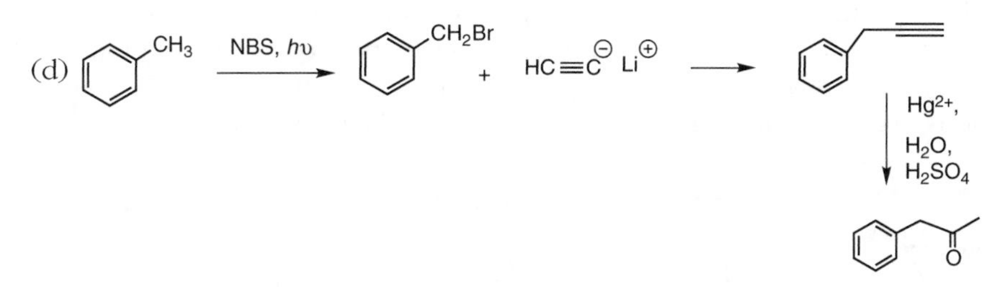

When starting with a hydrocarbon, usually the only way to introduce a functional group is by bromination or chlorination. Furthermore, the product contains a specific kind of ketone, a methyl ketone, which is made by hydration of a terminal acetylene. These two insights suggest this synthetic scheme.

17

SYNTHESIS AND REACTIONS OF ETHERS AND EPOXIDES

Ethers are among the least reactive of functional groups. Their main functions in organic chemistry are as solvents. An exception to this generalization are epoxides, three-membered ring cyclic ethers. Because of ring strain, epoxides undergo facile reactions with nucleophiles. Epoxides, as molecules with a heteroatom in a ring, are one example of **heterocyclic** compounds. The IUPAC names of common oxygen heterocycles are given below. The ring positions are numbered so that the oxygen atom is at position 1.

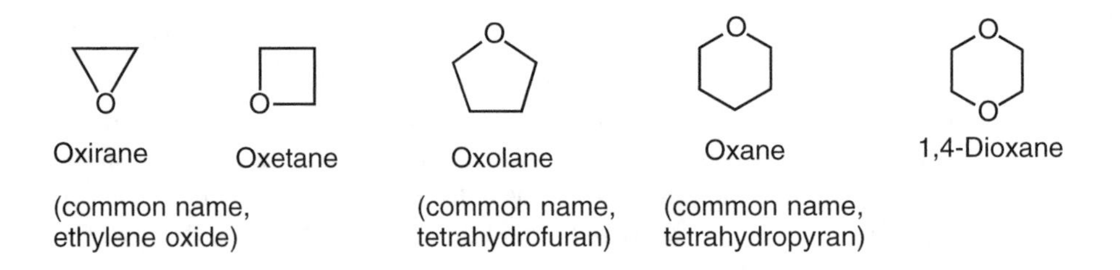

Because the oxygen atom of ethers is a Lewis base and a hydrogen bond acceptor, ethers and alcohols have similar solubilities in water. Furthermore, ethers, especially tetrahydrofuran, are able to dissolve small quantities of many inorganic salts. Lithium ions, in particular, interact favorably with ether oxygen lone pair electrons.

SYNTHESIS OF ETHERS AND EPOXIDES

Ether Synthesis

The traditional method of ether synthesis is **Williamson ether synthesis**. This reaction is just an S_N2 reaction between an alkoxide, RO^-, and an alkyl halide, R′X, giving an ether, ROR′. When the two ether groups are identical, the ether is a **symmetric ether**, and when the two R groups are different, the ether is an **unsymmetric ether**. Since alkoxides are strong bases, only CH_3X and 1° halides or sulfonates give good yields of ether products. Often, ethers can be made using two different combinations of nucleophile and electrophile.

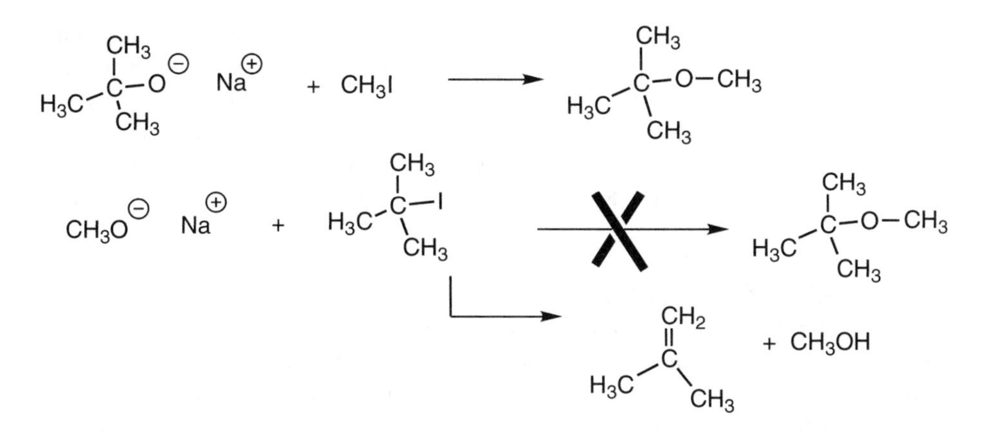

In many cases, one route is preferred because it minimizes elimination products. For example, any methyl ether, ROMe, should be prepared from RO^- and MeX since methyl halides cannot undergo elimination reactions.

Ethers can also be made by the acid-catalyzed addition of alcohols to alkenes. In the presence of a strong acid, such as H_2SO_4, ethanol reacts with $(CH_3)_2C{=}CH_2$ to give *t*-butyl ethyl ether. The mechanism involves protonation of the double bond to give the *t*-butyl carbocation which is then captured by ethanol.

More complicated ethers are made when ethylene glycol, $HOCH_2CH_2OH$, reacts with dihalides, such as 1,2-dichloroethane, under basic conditions:

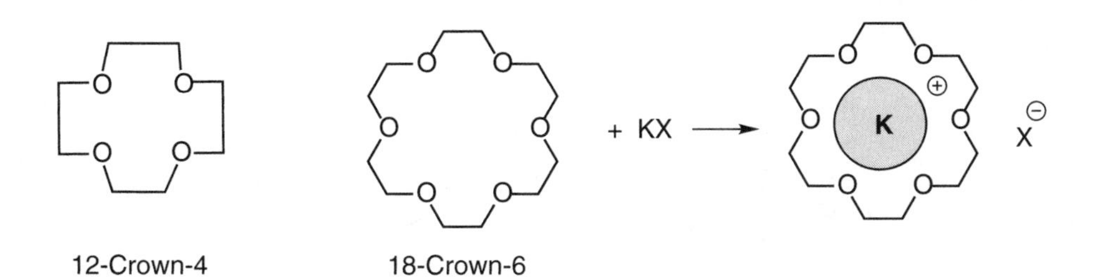

These cyclic polyethers were named **crown ethers** by their discover Charles Pedersen because their three-dimensional shape resembles a crown. The name *x*-crown-*y* indicates the overall ring size *x* and the number of oxygens in the ring *y*.

One can use different metal ions to promote the formation of a specific crown ether through a **template synthesis**. The interior of different crown ethers can accommodate alkali ions of different sizes. For example, K^+ ions fit snugly into the interior of 18-crown-6. Running the ether synthesis in the presence of K^+ ions leads to preferential formation of the crown ether best able to surround K^+.

Crown ethers are able to solubilize inorganic salts in nonpolar solvents. Since 18-crown-6 can encapsulate potassium ions, the K^+ becomes surrounded by hydrophobic groups and thus becomes soluble in hydrophobic solvents such as benzene. The anion also is solubilized because of the necessity of charge neutralization. A compound like the purple $KMnO_4$ is insoluble in benzene. However, when $KMnO_4$ is added to a solution of 18-crown-6 in benzene, some of it dissolves and the benzene turns purple.

Synthesis of Epoxides

The preparation of epoxides from bromohydrins was described in Chapter 10. It is more common to prepare epoxides directly from alkenes using **peroxycarboxylic acids**, RCO_3H, also known as **peracids**. Peracids react directly with double bonds in a concerted reaction to give an epoxide and the carboxylic acid:

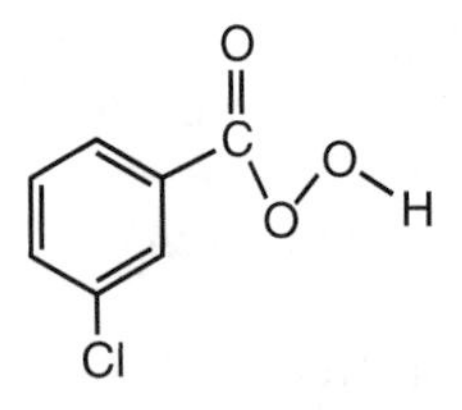

meta-Chloroperoxybenzoic acid (MCPBA)

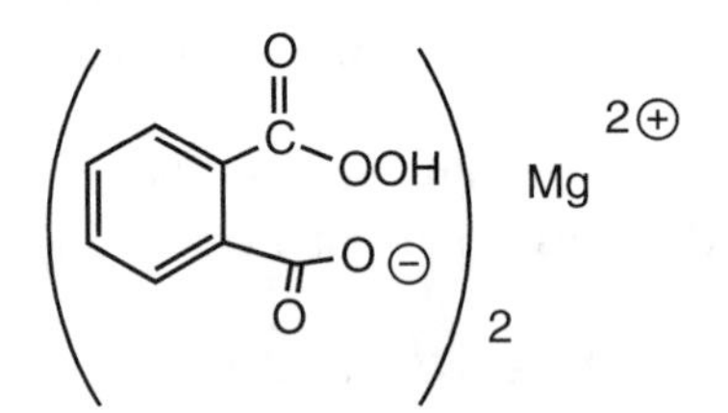

Magnesium monoperoxyphthalate (MMPP)

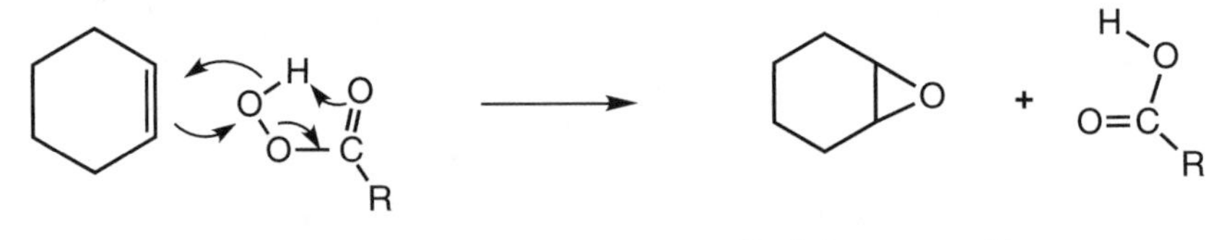

Cyclohexene oxide

Since epoxides are usually made from alkenes, common names for epoxides are derived by adding the word *oxide* after the name of the alkene. Note the structure of cyclohexene oxide shown above.

The peracid acts as the source of an electrophilic oxygen atom. It reacts faster with electron-rich double bonds, such as tetrasubustituted alkenes, than with monosubstituted alkenes. Double bonds substituted with electron-withdrawing groups, such as ester or ketone groups, do not react with most peracids. Because of the concerted mechanism of epoxidation, the stereochemistry of the alkene is maintained.

With a chiral catalyst, it is possible to prepare chiral epoxides from achiral alkenes. Barry Sharpless received the Noble Prize in chemistry in 2001 in part for inventing the reagent system *t*-butyl hydroperoxide/Ti(OiPr)$_4$/chiral diethyl tartrate, which is specific for forming epoxides of primary allylic alcohols:

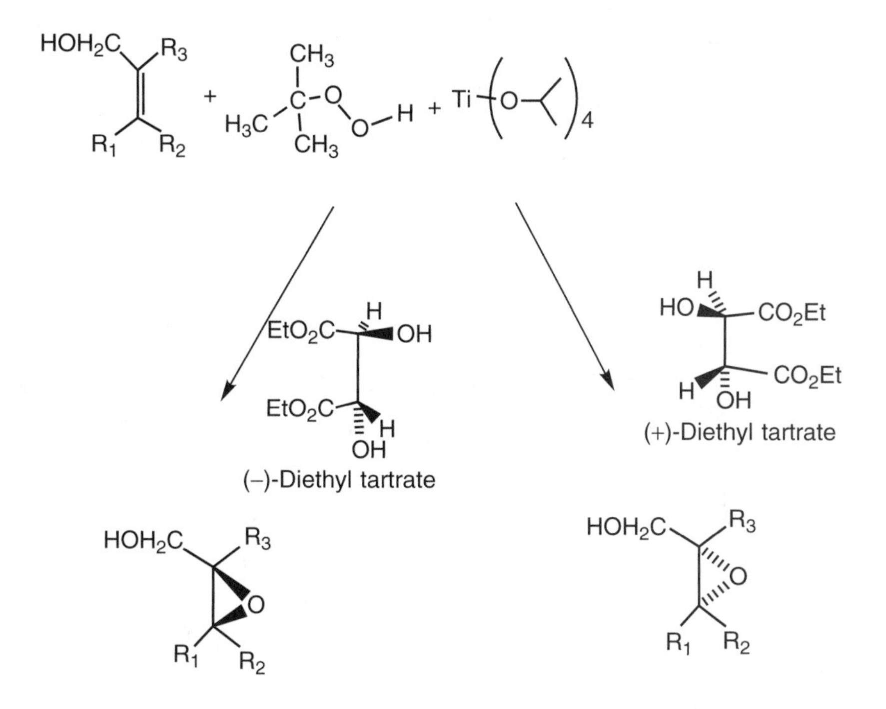

Diethyl tartrate is available as either enantiomer. In the presence of *t*-butyl hydroperoxide and titanium isopropoxide, it generates a chiral catalyst that stereospecifically adds an oxygen atom to the double bond from only one side of the double bond. For the allylic alcohol as shown above, with the alcohol group in the "northwest" quadrant, (–)-diethyl tartrate delivers the oxygen from the top face of the double bond and (+)-diethyl tartrate delivers the oxygen from the bottom face of the double bond. When all three R groups are different, this reaction generates two new stereogenic centers. Only the double bonds of allylic alcohols react with this reagent. Isolated double bonds are unreactive. This selectivity arises because the —OH of the allylic alcohol forms a complex with the Ti(O-*i*-Pr)$_4$, and this directs the reagent to the nearby double bond.

EXERCISES

Give the expected product or reactants for the following reactions.

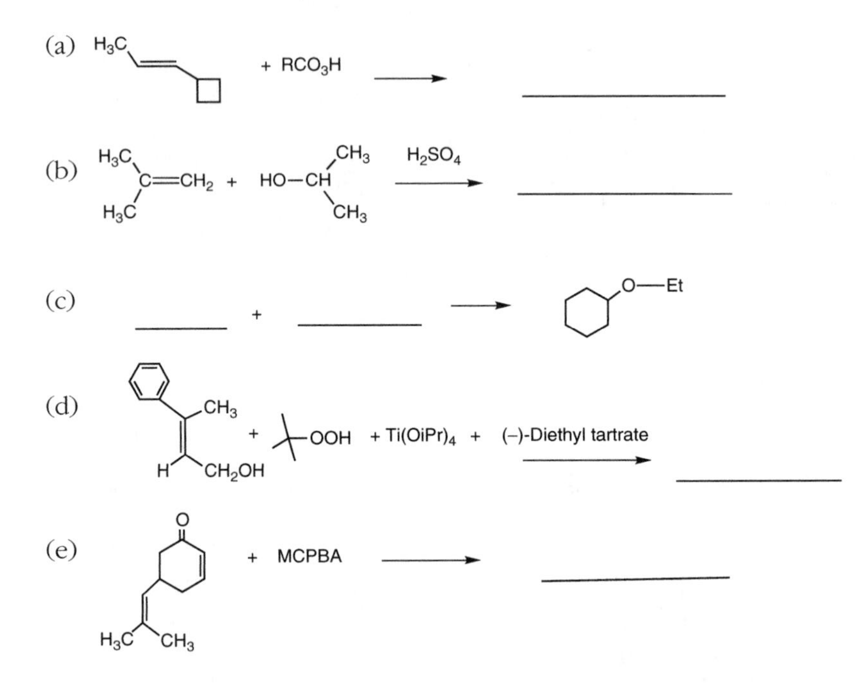

Solutions

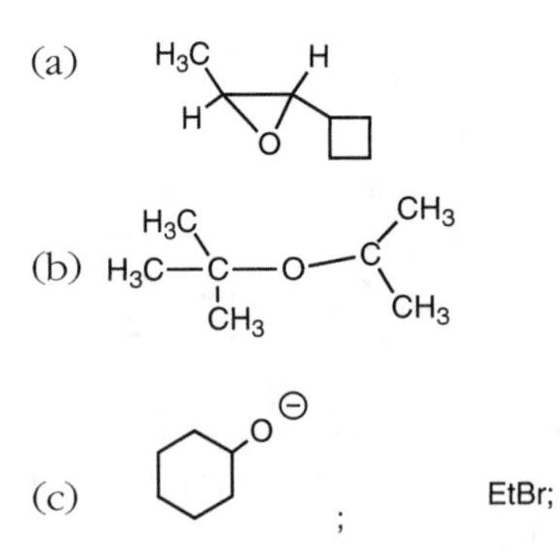

1° halides give less E2 product.

(d)

(e)

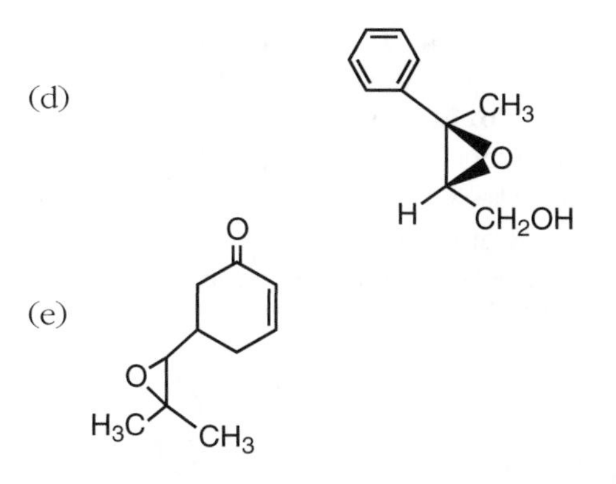

Ketone deactivates the conjugated double bond.

REACTIONS OF ETHERS AND EPOXIDES

Reactions of Ethers

There are very few useful reactions of ethers. The primary reaction is cleavage. When ethers are treated with a strong acid whose conjugate base is a good nucleophile, such as HI, they are cleaved to give an alcohol and an iodide:

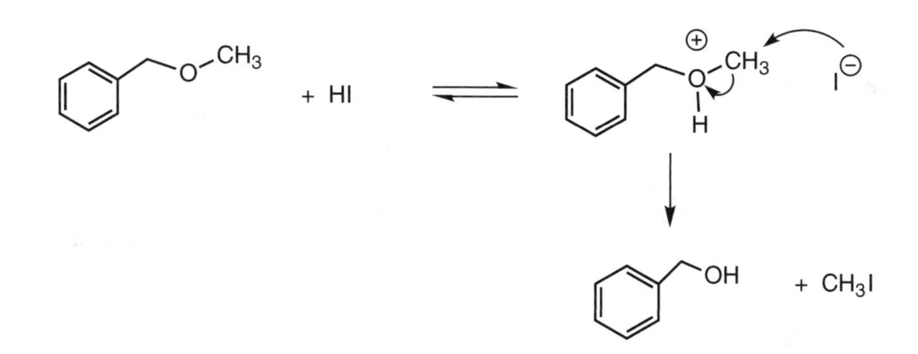

Notice that the nucleophile reacts with the less hindered alkyl group (Me > 1° > 2° > 3°).

If the reaction is allowed to proceed for an extended period of time, the alcohol ROH can react with a second molecule of HX by an S_N1 or S_N2 mechanism to give RX.

Another reagent that reacts similarly to HI is **trimethylsilyl iodide**, $(CH_3)_3SiI$. The trimethysilyl group acts like a large H^+, and this reagent cleaves ethers under nonacidic conditions. The reaction give an alkyl iodide and a **trimethylsilyl ether**, $ROSiMe_3$.

A potentially dangerous reaction of alkyl ethers is their reaction with atmospheric oxygen to give hydroperoxides. The reaction of a compound with oxygen without added reagents is called **autoxidation**. The reason autoxidation of ethers is a nuisance is that the products are shock-sensitive and can explode if the amount of ether hydroperoxide is sufficiently high. As a result, ethers, such as the common solvent diethyl ether, should be used only from freshly opened containers and should never be stored for extended periods of time exposed to air.

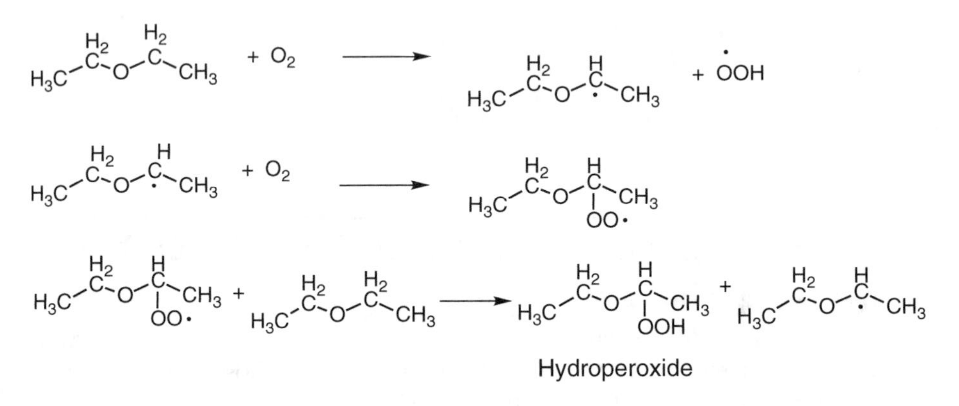

Carbon radicals next to an oxygen atom are stable because the half-filled, electron-deficient, radical orbital can interact with the lone pair of electrons on oxygen.

Reactions of Epoxides

The conversion of an alkene to an epoxide converts a nucleophile to a reactive electrophile. Epoxides can react with nucleophiles with or without added acid catalyst. Reactive nucleophiles, such as cyanide anion, RS^-, LiR, $RC{\equiv}C^-$, RMgBr, or cuprates react directly with epoxides through a standard S_N2 reaction. R_2CuLi reagents give particularly good yields with epoxides:

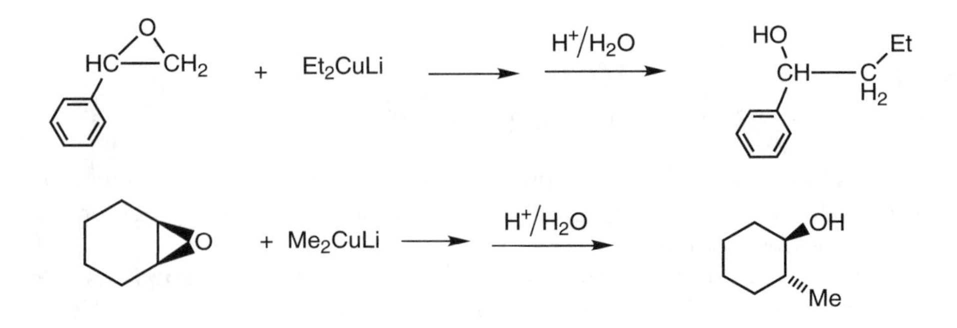

The nucleophile reacts at the less hindered carbon, and the reaction takes place with inversion of configuration.

Acids, either H^+ or Lewis acids such as BF_3, can interact with the Lewis base oxygen atom of an epoxide. This makes the oxygen a better leaving group and catalyzes the ring opening by a nucleophile:

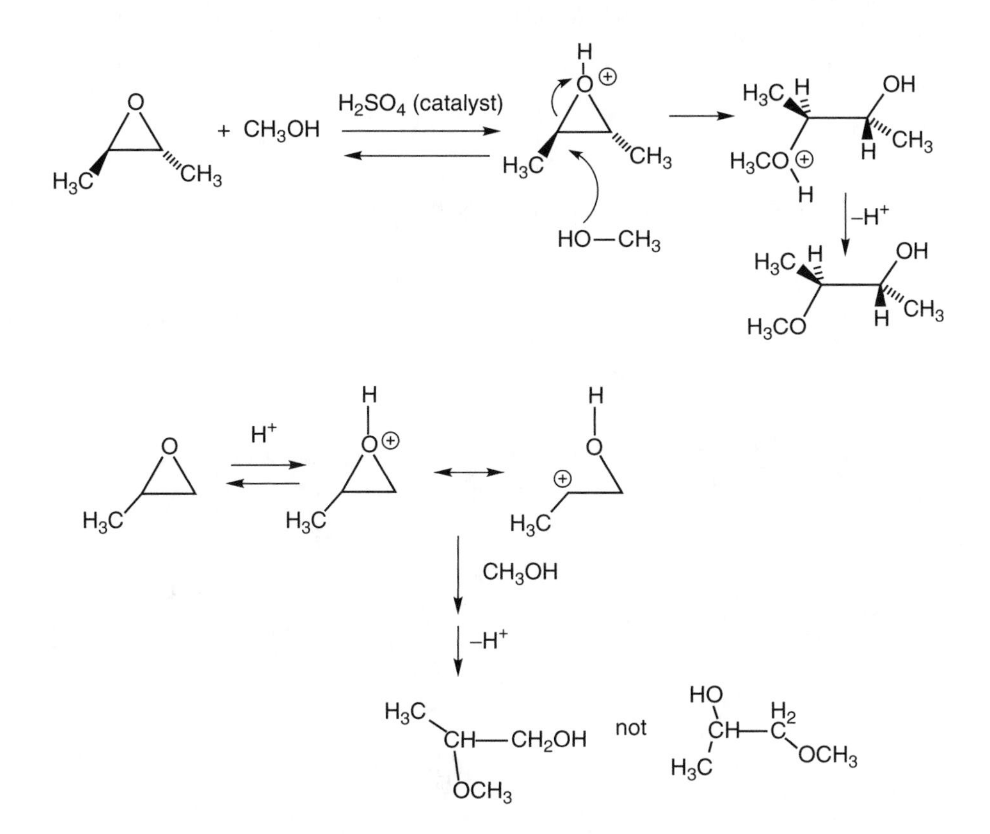

With an unsymmetric epoxide, the carbon best able to stabilize a positive charge becomes the electrophile. The nucleophile reacts at the more substituted carbon because of electronic effects rather than at the less substituted carbon because of steric effects. The same reactivity pattern was seen in halohydrin formation, discussed in Chapter 10.

EXERCISES

1. Starting with

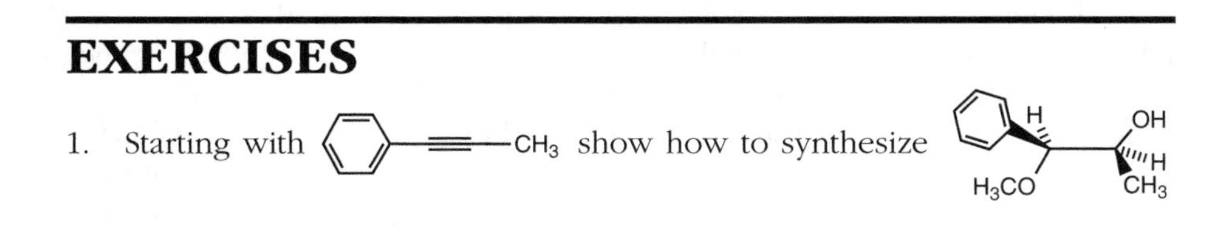

2. Propose a mechanism for the following reaction.

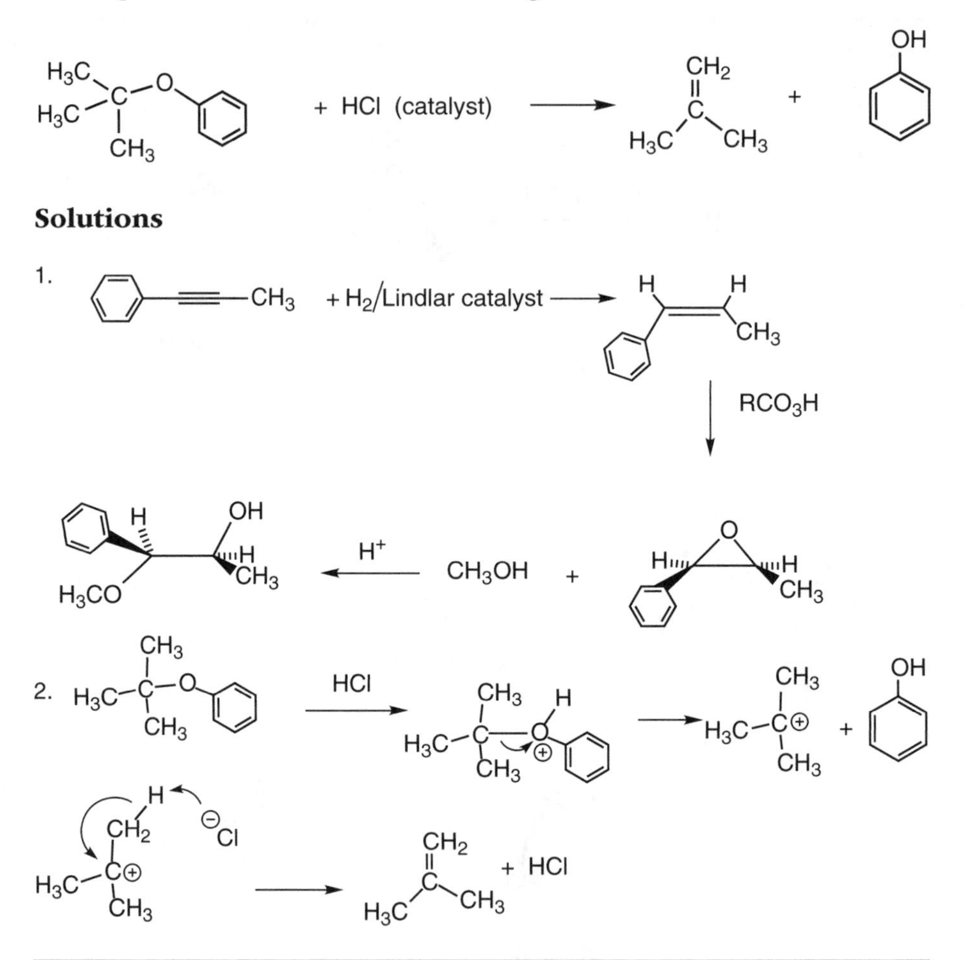

INTRODUCTION TO PROTECTING GROUPS

Certain combinations of reagents and functional groups undergo reaction. Sometimes the reaction is wanted; sometimes it is not. Organic chemists have developed ways to protect a given functional group so that it is unreactive toward almost all reagents. **Protecting groups** are groups that are easy to introduce onto a given functional group, are unreactive toward most reagents, and are easy to remove at the end of a synthesis.

Alcohols, being weak acids, often need to be protected when a molecule is exposed to a strong base and one wants to avoid reaction at the alcohol site. Several protecting groups based on ethers are used to protect the alcohol functional group:

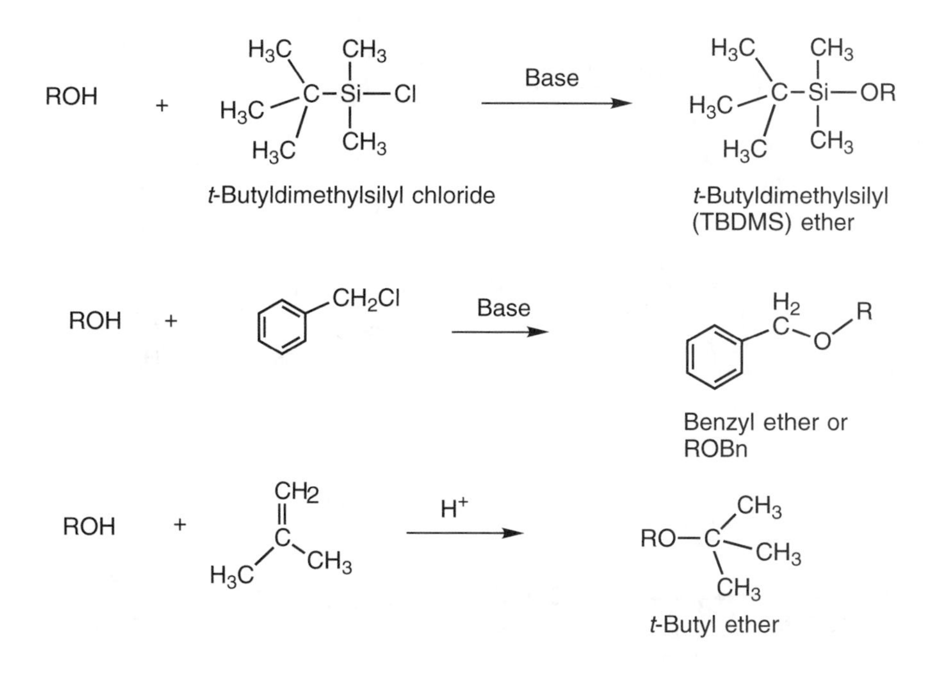

t-Butyldimethylsilyl chloride

t-Butyldimethylsilyl (TBDMS) ether

Benzyl ether or ROBn

t-Butyl ether

While scores of protecting groups are reported for alcohols, the *t*-butydimethysilyl ether, benzyl ether, and *t*-butyl ether groups are among the most commonly used. They are all stable to base, RLi, and related organometallic reagents, and halogenating reagents such as PBr_3 and $SOCl_2$. The *t*-butydimethysilyl ether group is removed with a soluble source of F^- such as $Bu_4N^+F^-$ since the Si—F bond is exceptionally strong. Benzyl ethers are cleaved to ROH and Ph—CH_3 using Pd/C—H_2. In this kind of catalytic hydrogenation, called **hydrogenolysis**, a single bond is cleaved by H_2. It is mainly limited to benzylic bonds. Finally, *t*-butyl ether groups are cleaved to alcohols and 2-methylpropene by strong acid (see Exercise 2 above for the mechanism). There are protecting groups for virtually every functional group, and other examples will be introduced in subsequent chapters.

Example These examples show how protecting groups allow one to carry out syntheses that otherwise would be difficult or impossible.

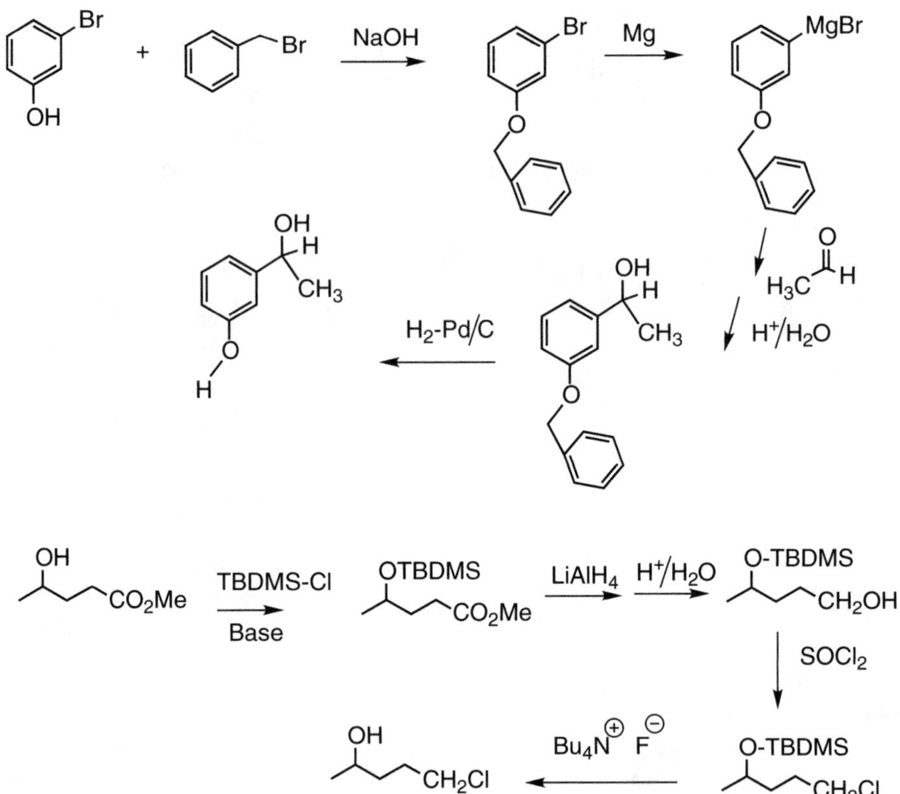
Br
OH
+
Br
NaOH
Br
Mg
MgBr
O
O
OH
H
CH3
H2-Pd/C
OH
H
CH3
O
H
O
H3C
H
H+/H2O
O
OH
CO2Me
TBDMS-Cl
Base
OTBDMS
CO2Me
LiAlH4
H+/H2O
O-TBDMS
CH2OH
SOCl2
OH
CH2Cl
Bu4N F

O-TBDMS
CH2Cl

18
ALDEHYDES AND KETONES I: SYNTHESIS AND ADDITION REACTIONS

We have already seen that the carbonyl group of aldehydes and ketones undergoes addition reactions with carbon nucleophiles and hydrides. In fact, addition reactions are a general feature of carbonyl group chemistry. Because of the electronegativity difference between carbon and oxygen, there is a significant degree of charge separation in a carbonyl group. The carbon is at the + end of the carbonyl dipole, and the oxygen atom is at the – end of the carbonyl dipole. Virtually all nucleophiles add to the carbon of the carbonyl group, some reversibly and some irreversibly.

There is a significant difference in the rates of addition to the carbonyl group in going from formaldehyde, H_2CO, to RCHO, and to R_2CO, with formaldehyde being the most reactive. Alkyl groups are electron-donating relative to H. Since the C of a carbonyl group has a partial positive charge, alkyl groups stabilize the C=O group. Because the formaldehyde and aldehyde carbonyl groups are less stable than the ketone carbonyl group, they are more reactive in addition reactions. Steric effects also influence carbonyl group reactivity and act in the same direction as electronic effects. Because H is smaller than any alkyl group, steric effects reinforce the reactivity order, $H_2CO > RCHO > R_2CO$.

The nomenclature of aldehydes and ketones was covered in Chapter 3. It is sometimes useful to use Greek letters to denote the position of a group relative to a carbonyl CO group. The carbon next to the CO group is the α carbon, the next one is the β carbon, and so on:

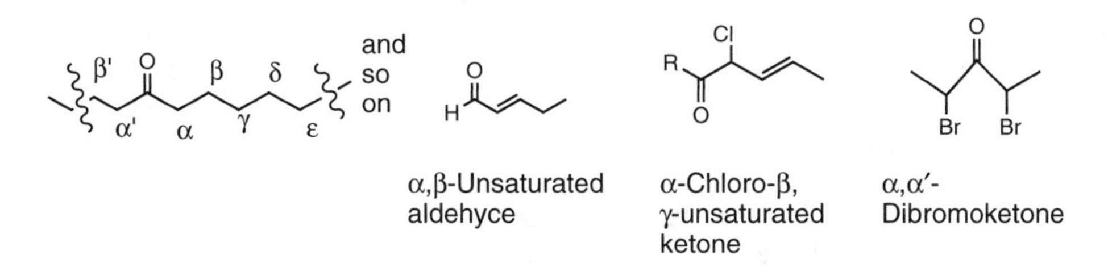

With this nomenclature, it is convenient to talk about classes of compounds. For example, α,β-unsaturated carbonyl compounds have different properties from β,γ-unsaturated carbonyl compounds, as will be discussed in Chapter 19.

SYNTHESIS OF ALDEHYDES AND KETONES

A number of important synthetic routes to ketones and aldehydes have already been introduced. Oxidation of 1° and 2° alcohols, hydration of alkynes, and hydroboration-oxidation of alkynes all provide useful synthetic pathways to aldehydes and ketones. Several oxidation or reduction reactions are frequently used.

Ozone, O_3, is a reactive form of oxygen, generated by passing an electric discharge through O_2. It is a powerful electrophile that reacts rapidly and selectively with alkenes:

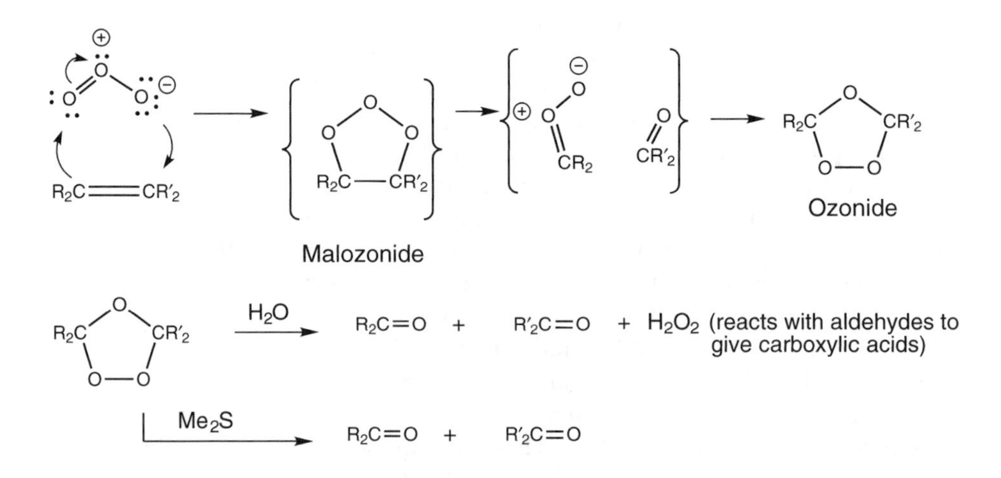

Ozone gives several unstable intermediates before forming the relatively stable (but shock-sensitive) **ozonide**. Note that the carbon-carbon double bond is cleaved in this product. Hydrolysis of the ozonide gives two carbonyl compounds, either ketones or aldehydes, depending on the nature of the R groups. If any aldehydes form, the hydrogen peroxide will normally oxidize them to carboxylic acids. Treating the ozonide with the reducing agent Me_2S (or Zn metal) gives only the carbonyl compounds and no hydrogen peroxide, so aldehydes, even formaldehyde, can be recovered. The overall cleavage reaction of alkenes with ozone is called **ozonolysis**.

A second oxidative route to carbonyl compounds involves the oxidation of 1,2 diols by potassium periodate, KIO_4. Recall that 1,2 diols are made in excellent yield from alkenes by OsO_4 oxidation. The mechanism of oxidation involves the formation of a five-membered cyclic periodate intermediate that decomposes in a concerted process to two carbonyl compounds:

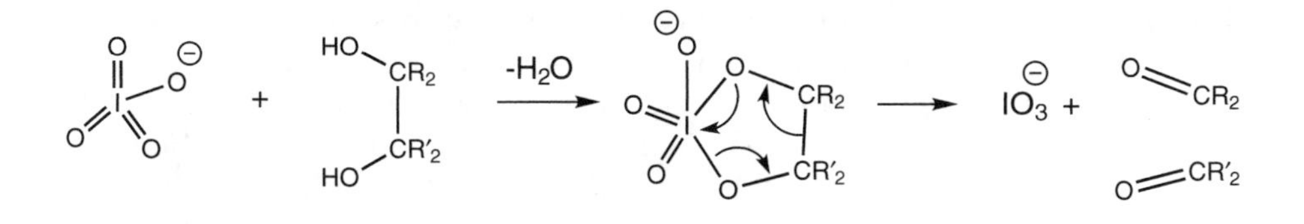

The **Lemieux-Johnson reaction** combines OsO_4 and KIO_4 to cleave alkenes directly to carbonyl compounds. This mild process avoids the inconvenience of generating ozone:

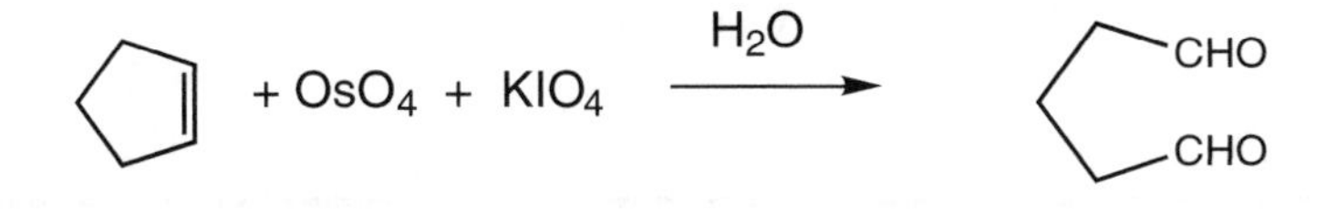

Aldehydes and ketones can be prepared by acidic hydrolysis of **vinyl ethers**, which are alkenes with an OR substituent:

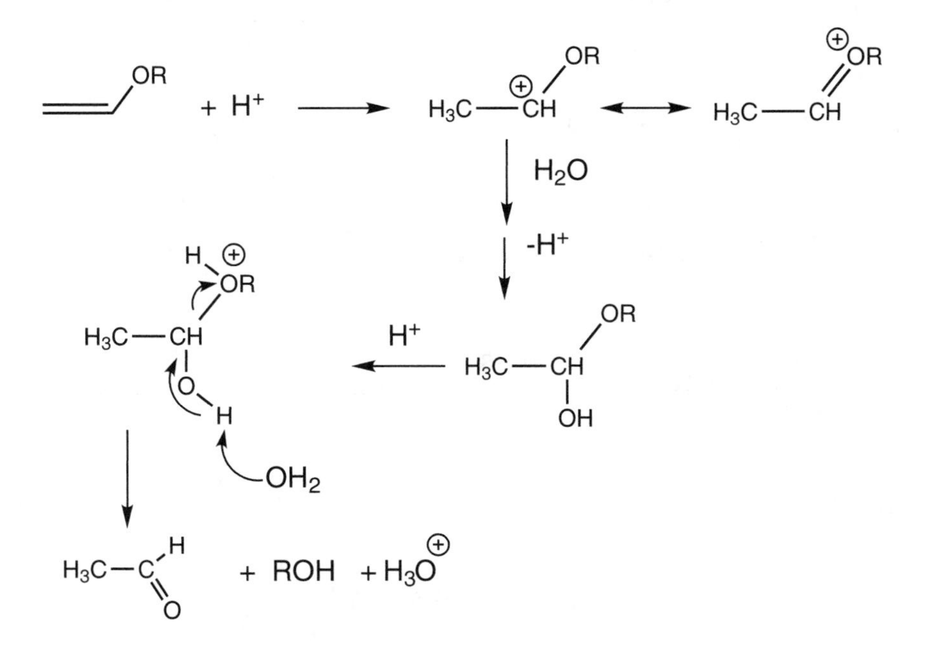

The double bond is protonated since the—OR group strongly stabilizes an adjacent carbocation through resonance. Water adds to the carbocation and, after the loss of a proton, a compound is formed that has an OR and an OH attached to the same carbon atom. This functional group is called a **hemiacetal**. Hemiacetals are quite unstable and decompose to a carbonyl compound and an alcohol. Much more will be presented about hemiacetals in the next section on addition reactions.

Aldehydes can be made by hydride reduction of acid chlorides, RCOCl. However, the hydride reagent must be strongly deactivated so that it does not further react with the product aldehyde. Experimentally, it is found that **lithium tri(*t*-butoxy)aluminum hydride**, LiAlH(O-*t*-Bu)$_3$, is such a reagent. It is made by adding three equivalents of *t*-butyl alcohol to $LiAlH_4$. As long as the reaction is cooled to −78°C (the temperature of frozen carbon dioxide, dry ice), good yields of aldehydes are obtained.

EXERCISES

Fill in each blank with the appropriate substrate or reagent.

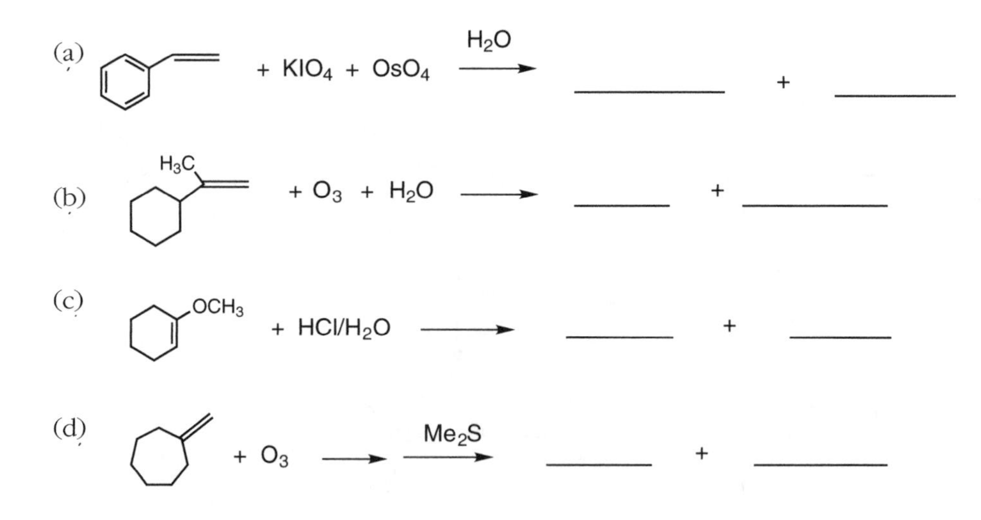

Solutions

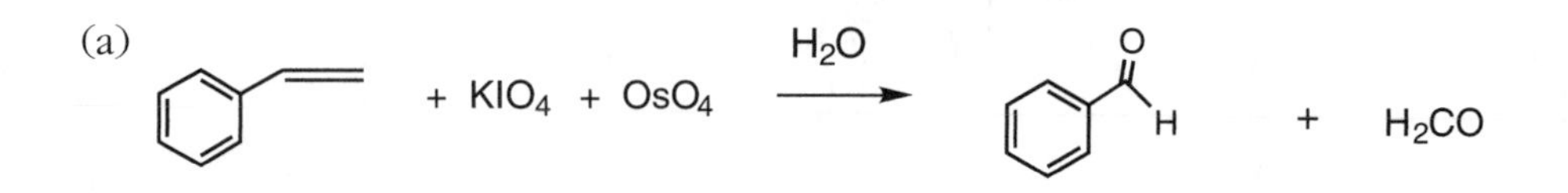

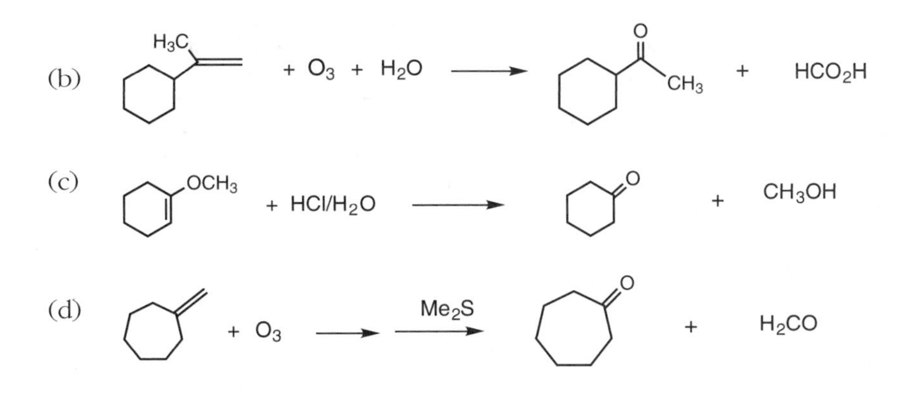

ADDITION REACTIONS OF ALDEHYDES AND KETONES

Addition of Water and Alcohols

Some carbonyl groups are so unstable because of the presence of electron-withdrawing groups that they react with water to give addition compounds called carbonyl **hydrates**:

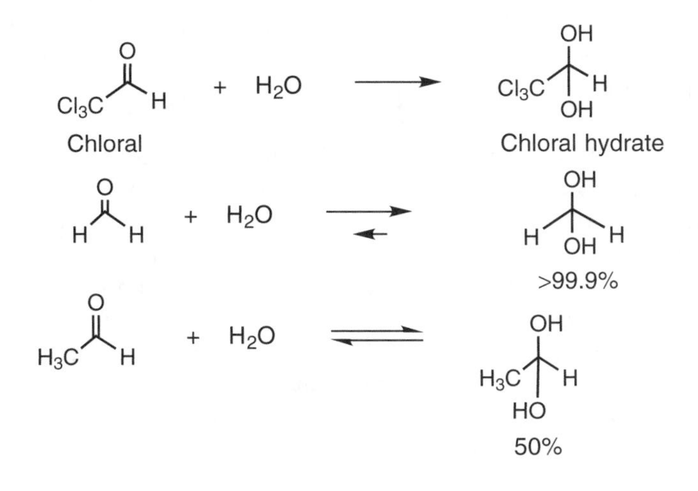

Ketones in water have equilibrium hydrate concentrations well below 1% because the two alkyl groups stabilize the carbonyl group.

Hydrate formation can be catalyzed by acid or by base. In base-catalyzed hydrate formation, the first step is nucleophilic addition of HO^- to the carbonyl carbon as shown here.

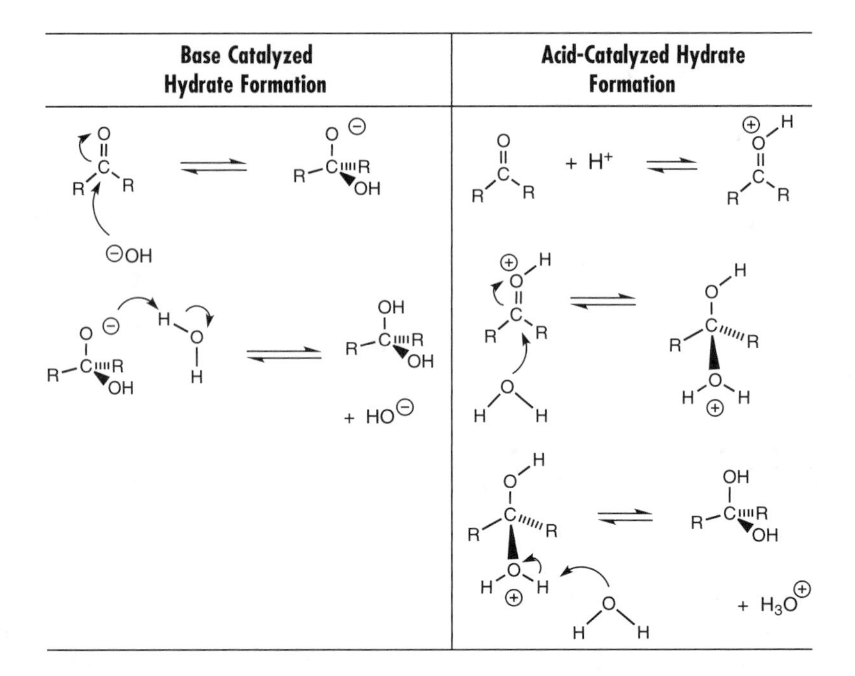

The addition of HO^- gives a **tetrahedral intermediate** which is protonated by water to produce the hydrate. In the acid-catalyzed pathway, protonation of a lone electron pair on oxygen gives an intermediate that is highly activated toward nucleophilic attack. Water attacks the protonated carbonyl group, and finally proton transfer to water or another base in the system produces the hydrate.

There are a number of important points to notice about these mechanisms that will apply to other addition reactions in this chapter. First, hydrate formation is reversible, so all the steps in the mechanism are reversible. Second, in a *base*-catalyzed reaction in water, the *strongest acid* one can use is *water*. Any stronger acid would immediately react with HO^- and give water. Third, in the *acid*-catalyzed reaction in water, the *strongest base* one can use is *water*. Any stronger base would react with H_3O^+ to give water. Using H^+ in an aqueous base-catalyzed reaction or HO^- in an aqueous acid-catalyzed reaction is one of the most common errors made by students in introductory organic chemistry. Additionally, notice that in acid-catalyzed reactions of carbonyl groups, the function of the acid (which can be H^+ or a Lewis acid) is to interact with the carbonyl oxygen and activate the carbonyl group to nucleophilic addition. Finally, notice that all the intermediates have no more than one charge. It is rare to have intermediates in organic reactions that have multiple charges.

Alcohols react with carbonyl compounds in a manner that initially is similar to the reaction of water with these compounds. However, the ultimate result is quite different for base-catalyzed and acid-catalyzed reactions as shown here.

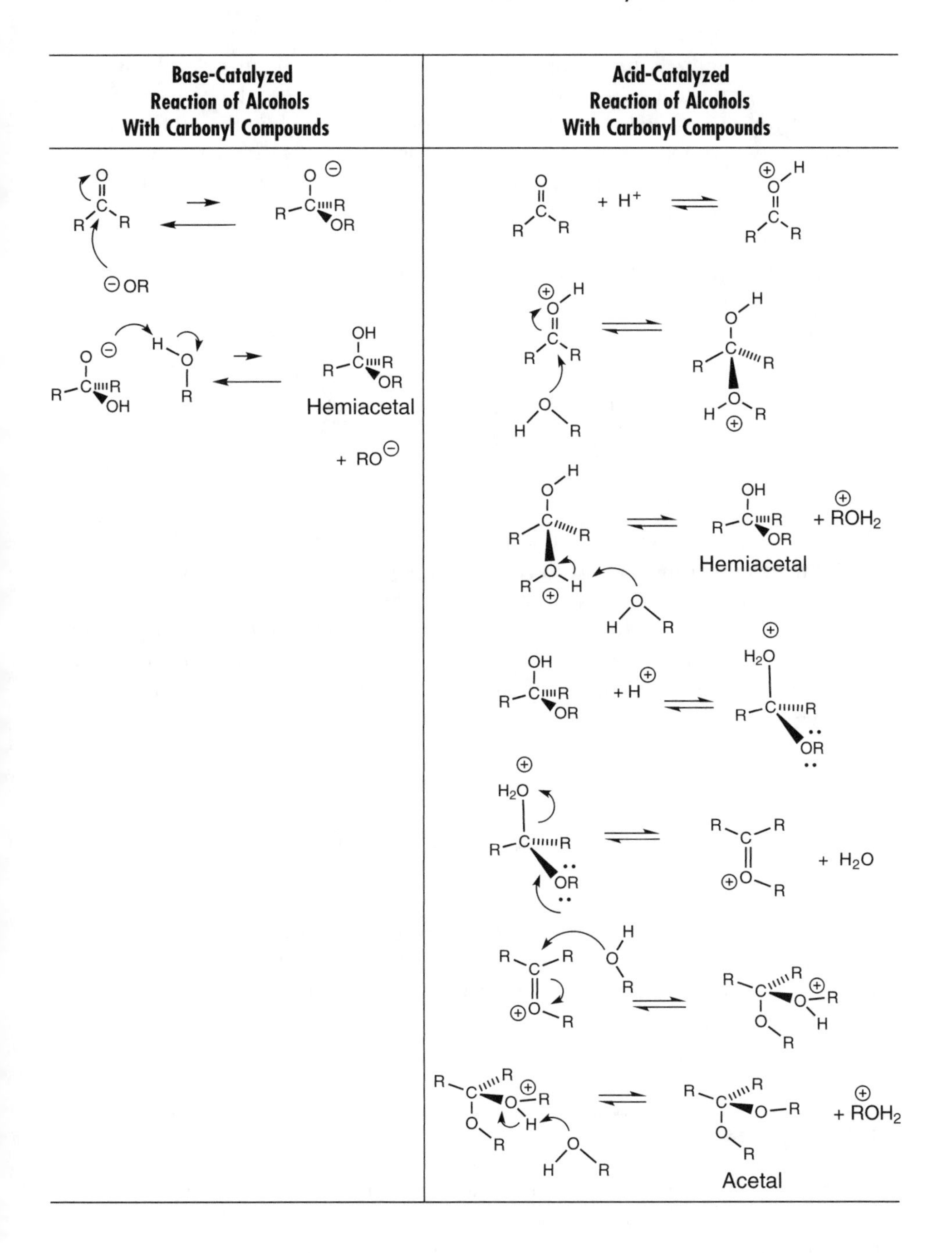
Base-Catalyzed
Reaction of Alcohols
With Carbonyl Compounds
Acid-Catalyzed
Reaction of Alcohols
With Carbonyl Compounds
Hemiacetal
Hemiacetal
Acetal

In a base-catalyzed reaction, it is possible to form a hemiacetal. However, since HO^- is not a leaving group, no further reaction can take place. Since the aldehyde or ketone is favored at equilibrium over the hemiacetal for almost all molecules, no net reaction takes place. Conversely, the acid-catalyzed addition of alcohols to aldehydes and ketones forms the hemiacetal intermediate, but when the—OH group is protonated, it becomes a leaving group. The lone pair electrons on the alcohol oxygen are able to displace water, giving an O-alkyl carbonyl intermediate that is stabilized by resonance:

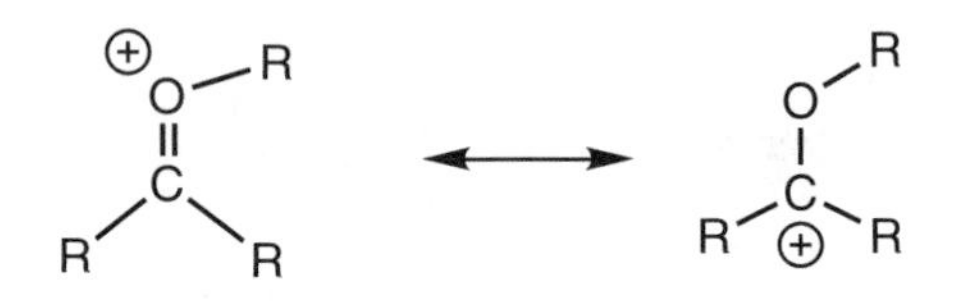

This intermediate reacts with a second alcohol molecule which, after deprotonation, gives a compound with two OR bonds to the same carbon, an **acetal**.

Acetals are not favored at equilibrium for most alcohol-carbonyl group combinations. However, the reaction can be forced to the give the acetal product by removing water (some specialized reaction conditions make this possible). By the law of mass action, the equilibrium is forced to the side of the acetal product:

$$R_2C{=}O \quad + \quad 2R'OH \quad \xrightleftharpoons{H^+} \quad R_2C(OR')_2 \quad + \quad H_2O \uparrow$$

At this point, the student may want to return to the mechanism of vinyl ether hydrolysis outlined above. It will be seen that hemiacetals can be formed from vinyl ethers, as well as carbonyl compounds. However they are formed, their ultimate fate is the same.

EXERCISE

Ethylene glycol, $HOCH_2CH_2OH$, reacts with aldehydes or ketones to give an acetal. Since one molecule of ethylene glycol and one molecule of a carbonyl compound give one acetal molecule and one molecule of water, this reaction is not disfavored entropically, unlike normal acetal formation, where three molecules give two molecules of product. Give the mechanism for acetal formation with ethylene glycol.

Solution

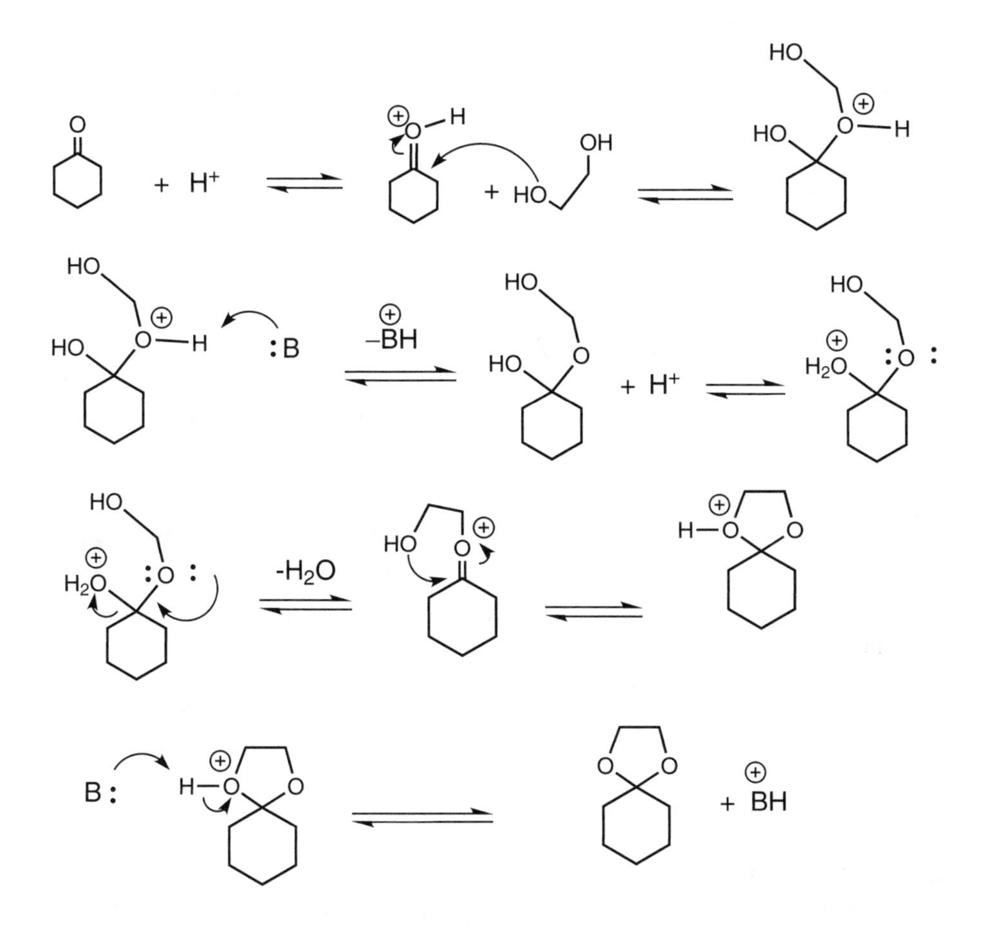

One feature of the above mechanism can confuse students. In the fourth step, there are two alcohol—OH groups and one ether oxygen. The tertiary alcohol is shown as being protonated. In fact, all the oxygens can be protonated in solution at one time or other. However, only the protonation step shown is on the pathway to products. Other protonation steps take place but are irrelevant. In any mechanism problem, always be aware of one's objective. Each step should lead to the formation of one of the bonds one needs to form or to the breaking of one of the bonds one needs to break.

Acetals as Protecting Groups

The acetal functional group is predominately used as a protecting group for aldehyde and ketone carbonyl groups. For example, it allows one to prepare lithium or Grignard reagents from carbonyl compounds:

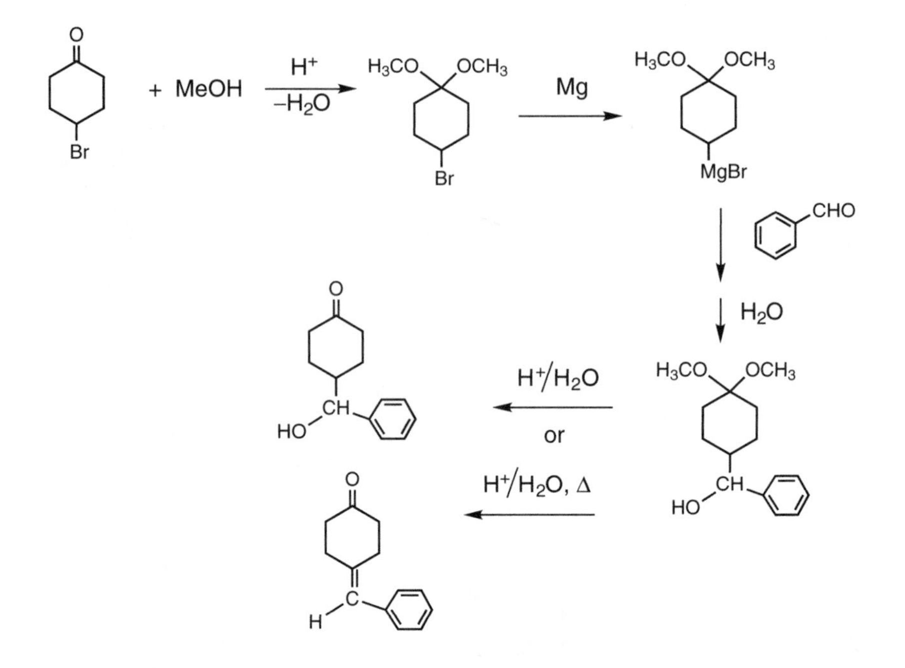

The acetal functional group survives the mild hydrolysis step used in Grignard reactions. It can usually be removed without causing elimination reactions of alcohols. If desired, vigorous conditions can hydrolyze an acetal and cause E1 elimination of an alcohol in the same flask. Acetals form only from aldehydes and ketones. Carboxylic acids, esters, and other carbonyl compounds give different products when heated with alcohols under acidic conditions, as will be described in Chapters 20 and 21.

Dithioacetals

Thiols react with aldehydes and ketones in a manner similar to the reaction of alcohols to give the sulfur analog of acetals, **dithioacetals**. These compounds are more resistant to acidic hydrolysis than acetals, but the addition of mercury salts to the hydrolysis reaction facilitates the hydrolysis since sulfur and mercury form strong bonds. Dithioacetals are frequently used with the reduction reagent **Raney nickel** (RaNi) to **deoxygenate** aldehydes and ketones:

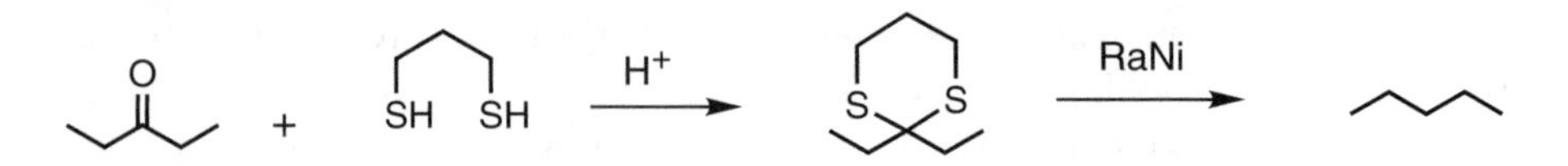

Raney nickel is prepared by treating an aluminum-nickel alloy with NaOH solution. The base dissolves the aluminum and generates H_2 that absorbs on the nickel surface. The heterogeneous nickel reagent cleaves C—S bonds and replaces them with C—H bonds.

EXERCISE

How many different methods are available for removing oxygen from an organic molecule?

Solution

Three useful ways to carry out this transformation have been described so far:

1. Convert an aldehyde or ketone to a dithioacetal and reduce with Raney nickel.
2. Convert an alcohol to a bromide with PBr_3 and reduce with Bu_3SnH.
3. Eliminate an alcohol to an alkene either through an E1 process or by forming a tosylate and carrying out an E2 elimination; then reduce the alkene with PtO_2/H_2.

Formation of Imines from Carbonyl Compounds and Primary Amines

Primary amines react with aldehydes and ketones to form compounds with a carbon-nitrogen double bond, **imines** (known as **Schiff bases** in earlier chemistry texts):

$$R_2C{=}O + R'NH_2 \overset{H^+}{\rightleftharpoons} R_2C{=}NR' + H_2O$$

This reaction is catalyzed by acid, but the pH-vs-rate profile has a maximum at about pH 7 as shown here. The reason for this pH dependence is explained by the mechanism for imine formation:

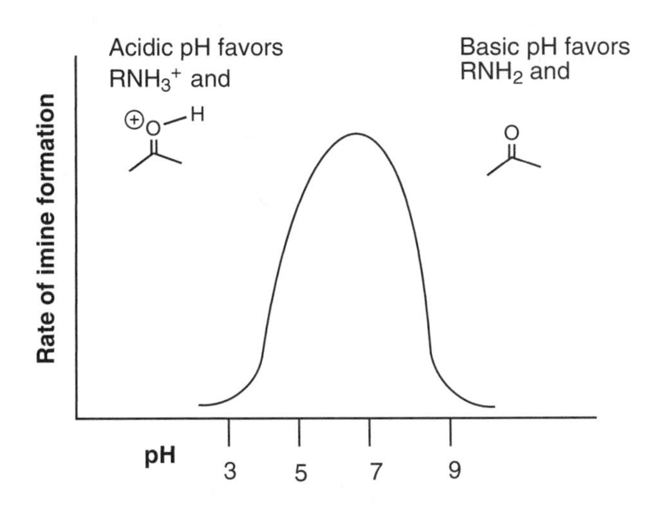

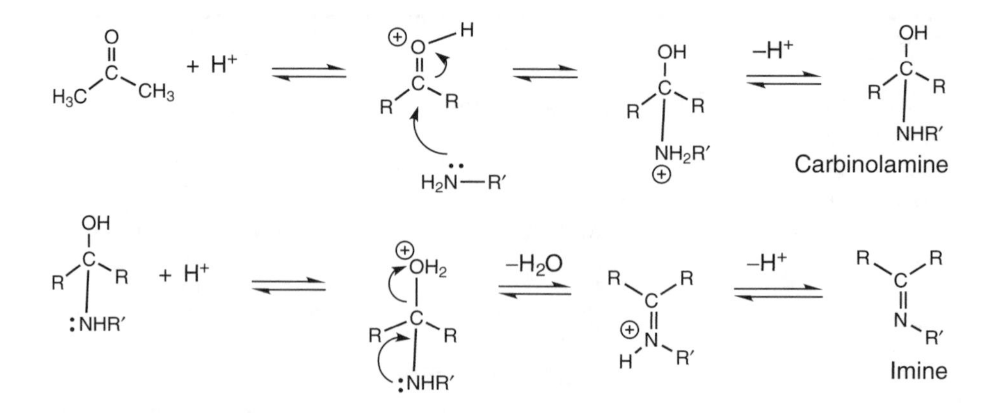

Amines are organic bases, and they are much more basic than carbonyl compounds. Thus, in an acidic solution, there is a mixture of protonated amine and protonated carbonyl compound, with RNH_3^+ predominating. If the solution is too acidic, all the amine will be tied up as RNH_3^+ and there will be essentially no RNH_2 nucleophile to react with the low-equilibrium concentration of protonated carbonyl. At higher pH values, there is ample RNH_2, but no electrophilic protonated carbonyl compound. At about pH 7 there is an optimum concentration of electrophile and nucleophile. This pH optimum has extremely important consequences for biochemistry since imine formation takes place at biological pH. A study of biochemical reactions can reveal scores of processes that involve imine formation.

The **carbinolamine** intermediate in imine formation is analogous to the hemiacetal intermediate in acetal formation. It is normally too unstable to be isolated. Imines, like acetals, do not form under basic conditions because acid is needed to turn the—OH group into a leaving group. Carbonyl compounds other than aldehydes or ketones do not give carbon-nitrogen double bonds when reacted with primary amines.

A number of amine derivatives form compounds with carbon-nitrogen double bonds that are analogous to imines:

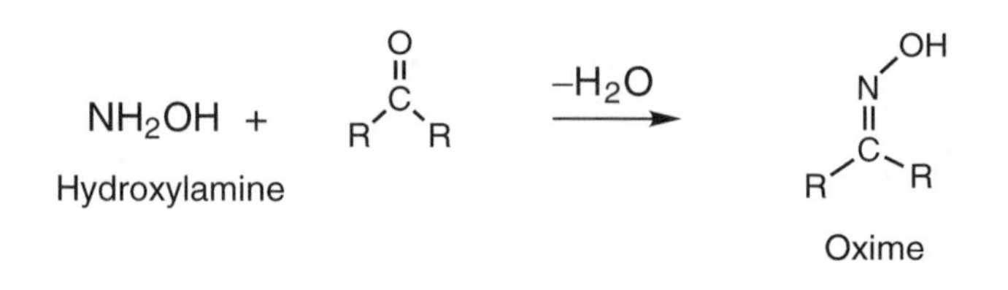

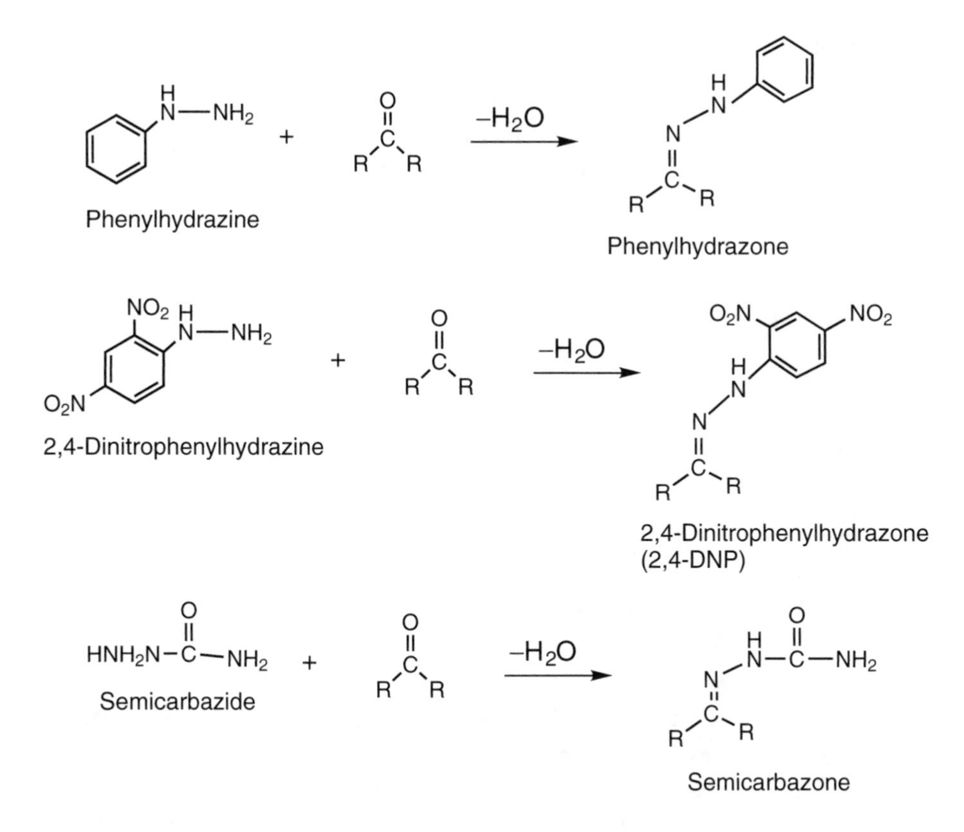

All these reactions are acid-catalyzed, and their mechanisms are the same as those described for imine formation. The **oxime**, **phenylhydrazone**, **2,4-dinitrophenylhydrazone**, and **semicarbazone** adducts are stable derivatives of aldehydes and ketones that are often crystalline solids. Before IR and NMR spectroscopy were invented, organic chemists made chemical derivatives of molecules and used the melting point of the derivative to identify the molecule.

EXERCISE

Propose a mechanism for the **transamination** reaction that converts one imine into another.

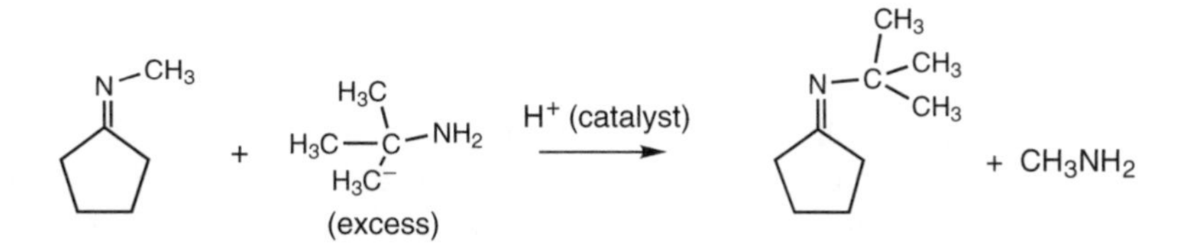

Solution

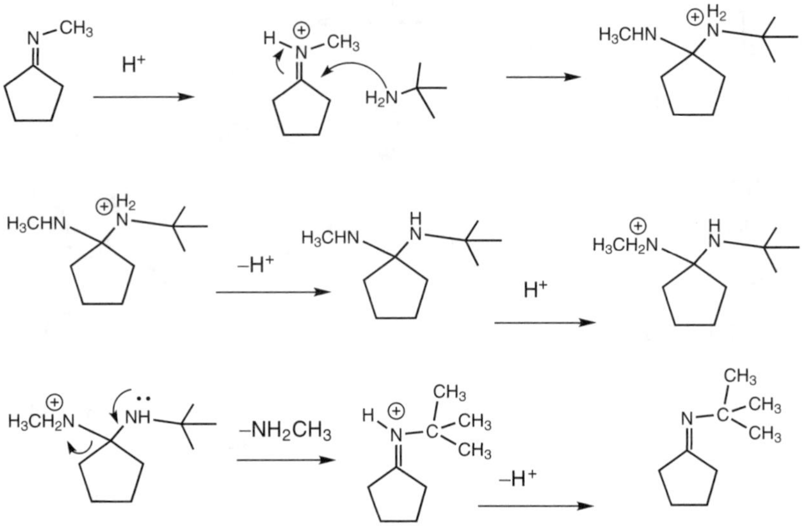

Formation of Enamines From Aldehydes or Ketones and From Secondary Amines

Secondary amines cannot form imines with neutral carbon-nitrogen double bonds. Instead, they react to give **enamines**, which are analogous to vinyl ethers:

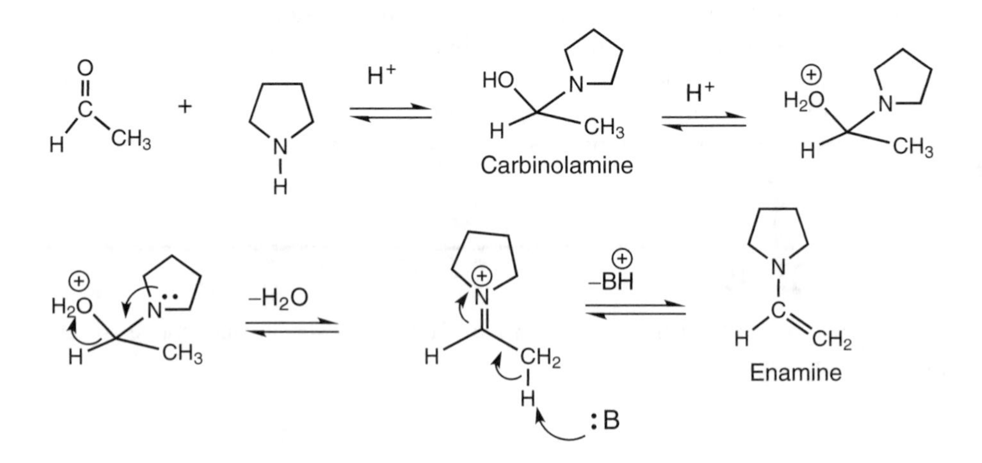

The mechanism of carbinolamine formation is the same as that followed by primary amines. However, once water is lost, the charged intermediate cannot lose a proton attached to nitrogen. Instead, an adjacent carbon is deprotonated, generating a new carbon-carbon double bond. Enamine formation is reversible, so water must be removed to drive the equilibrium to the enamine side of the equation. Enamines are useful reagents for creating new carbon-carbon bonds, as will be described in Chapter 19.

Base-Catalyzed Addition of HCN to Aldehydes and Ketones

The additions of amines and alcohols to aldehyde and ketone carbonyl groups are acid-catalyzed reactions. The addition of HCN to a carbonyl group is a rare example of a base-catalyzed carbonyl addition reaction. HCN is a weak acid (pK_a = 9.1), and $^{-}$CN is an excellent nucleophile. This combination allows for the creation of a HCN addition complex with carbonyl groups, called a **cyanohydrin**:

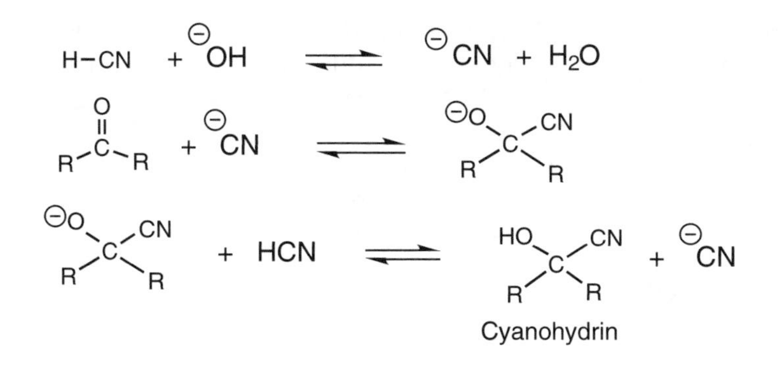

Cyanohydrins are stable to weak acids. Since the —CN group is hydrolyzed to —COOH under vigorous conditions (Chapter 21), cyanohydrins can be used to make α-hydroxy acids. Strong bases cause cyanohydrins to decompose back to the starting materials:

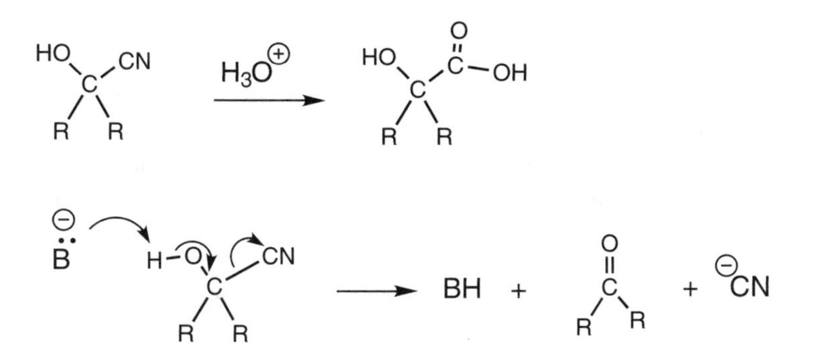

Formation of Carbon-Carbon Double Bonds by the Wittig Reaction

Up to this point, only elimination reactions that form double bonds have been described. In fact, mixtures of regioisomers are frequently produced when either E1 or E2 reactions are used to form double bonds:

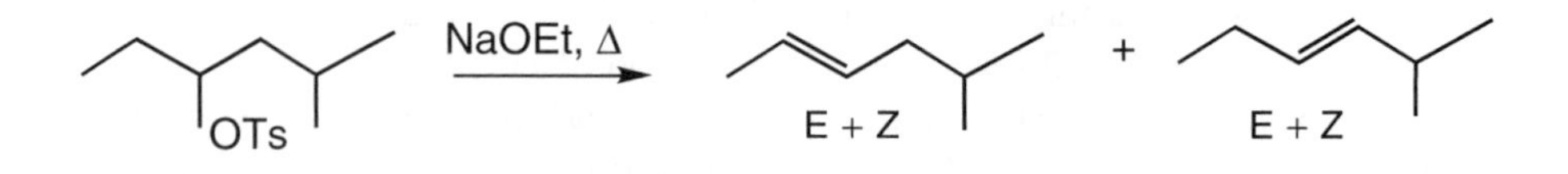

An alternative reaction, the **Wittig reaction**, named for its discoverer, the German chemist Georg Wittig, forms double bonds in a completely regiospecific manner. The Wittig reaction involves the addition of a nucleophilic phosphorus species, a **phosphonium ylide**, to an aldehyde or ketone carbonyl and decomposition of the initial adduct to an alkene and a **phosphine oxide**:

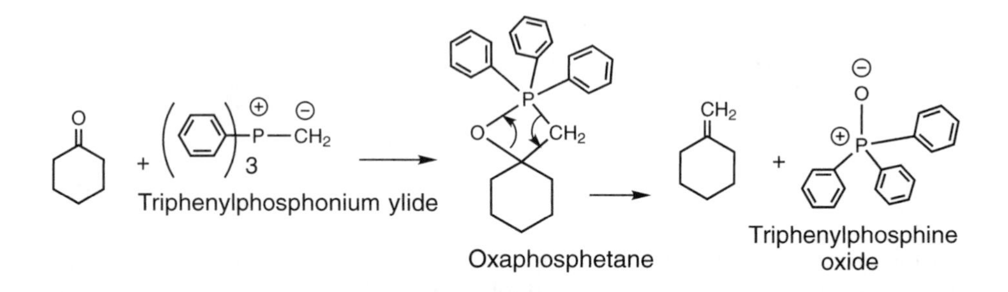

Ylides are compounds with a negatively charged carbon next to a positively charged heteroatom. They exist for sulfur, nitrogen, and other atoms, but we will consider only phosphorus ylides. Since phosphorus can expand its octet, chemists sometimes draw phosphonium ylides as $Ph_3P{=}CR_2$. However, the dipolar representation, $Ph_3P^+{-}CR_2^-$, gives a better description of the charge distribution in an ylide. The stability of triphenylphosphine oxide drives this reaction to the right, just as it did in the Mitsunobu reaction.

Ylides are made by a two-step process. First, a phosphine, normally triphenylphosphine, Ph_3P, reacts with CH_3I or a primary or secondary halide to give a **phosphonium salt**. This is a normal S_N2 reaction. The salt then reacts with a strong base, such as BuLi, to deprotonate and produces the ylide. Since the phenyl group has no protons that can be removed, PPh_3 is the phosphine of choice:

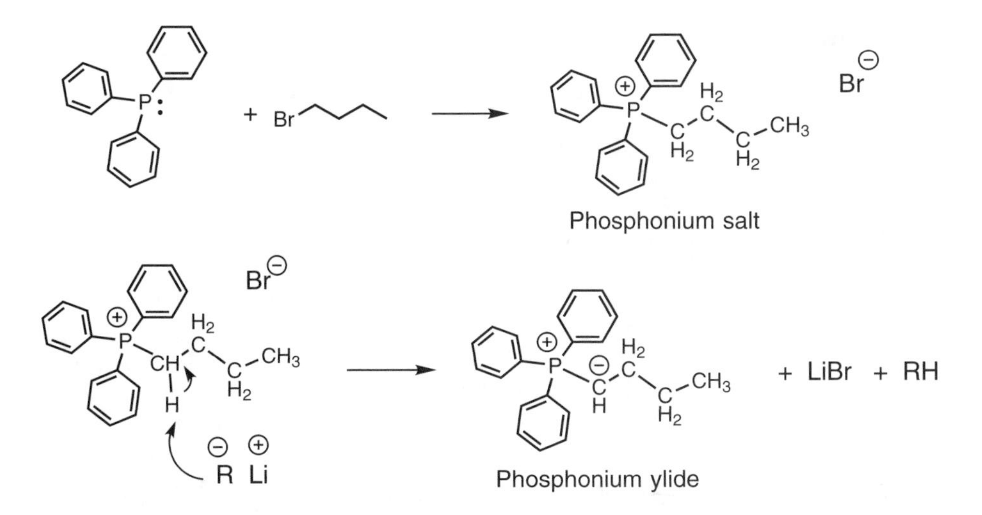

Because the newly formed double bond consists of the carbon of the carbonyl group and the anionic ylide carbon, its structure is unambiguous:

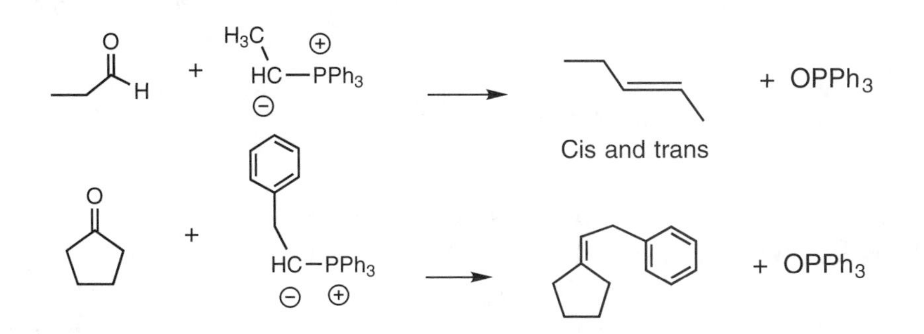

While the Wittig reaction is regiospecific, it is not stereospecific. A mixture of E and Z isomers forms whenever possible, although the Z isomer is often the major product with alkyl group substituents.

EXERCISES

Show how the following alkenes could be prepared using the Wittig reaction. Start with the appropriate alkyl halide and carbonyl compound.

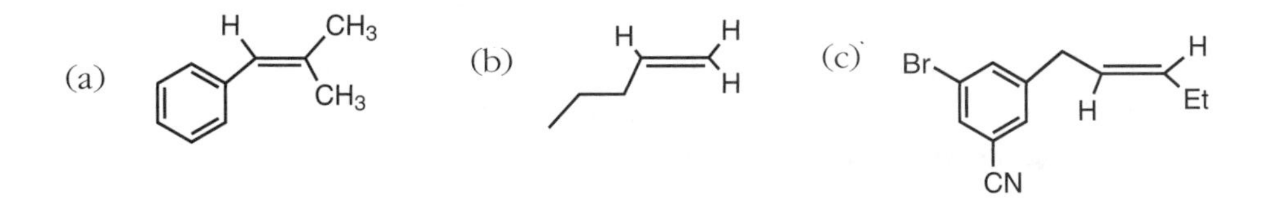

Solutions

(a)

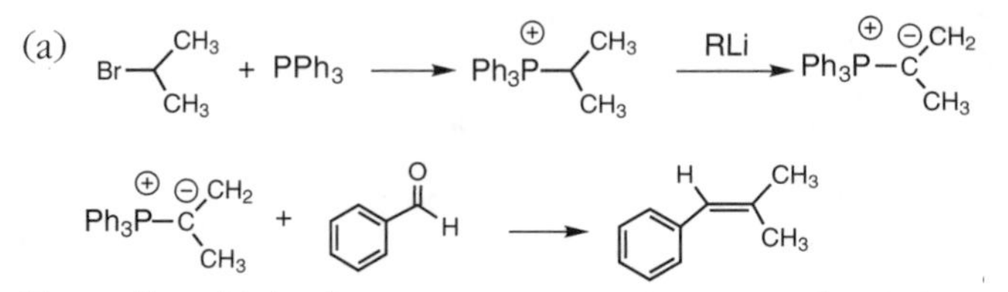

Normally, aldehydes are more reactive toward Wittig reagents than ketones, so this pair should work better than

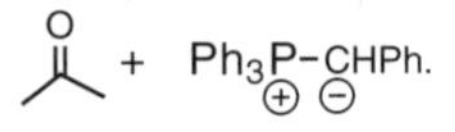

(b) $CH_3I + PPh_3 \longrightarrow \xrightarrow{RLi} Ph_3\overset{+}{P}\text{—}\overset{-}{C}H_2$ + (butanal) $\longrightarrow$ (1-pentene)

Another pair using butyl bromide and formaldehyde would also work.

(c)

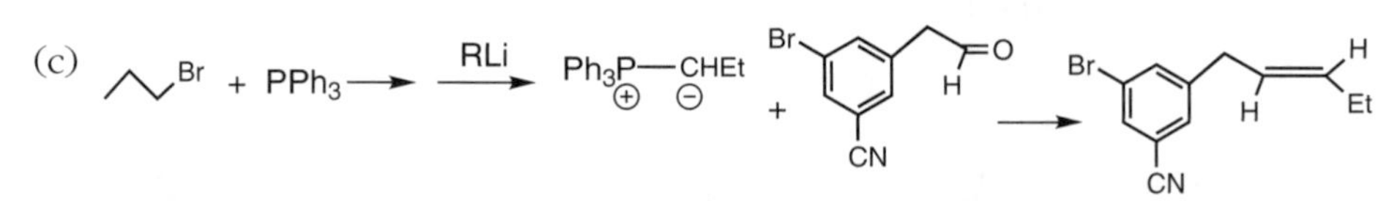

Cis and trans: either pair of reagents would work.

Oxidation of Aldehydes and Ketones With Peroxycarboxylic Acids: The Baeyer-Villiger Reaction

In addition to oxidizing alkenes to epoxides, peroxycarboxylic acids can oxidize aldehydes and ketones to carboxylic acids and esters, via the **Baeyer-Villiger** reaction:

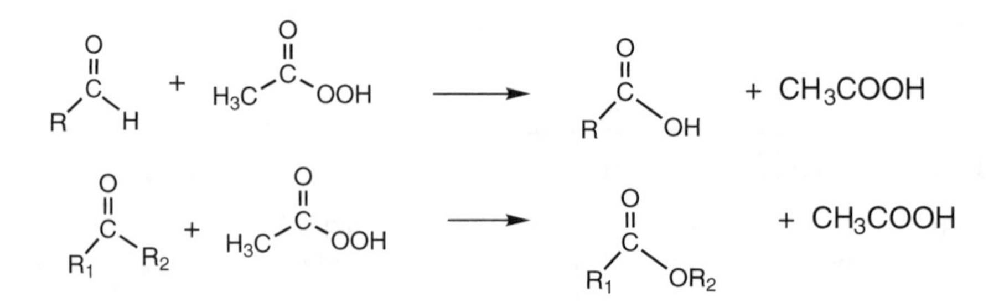

This is an unusual reaction since it appears that an oxygen atom is inserted into a C—H or a C—C bond. In fact, this is a rearrangement reaction driven by breaking the very weak O—O bond of the peracid:

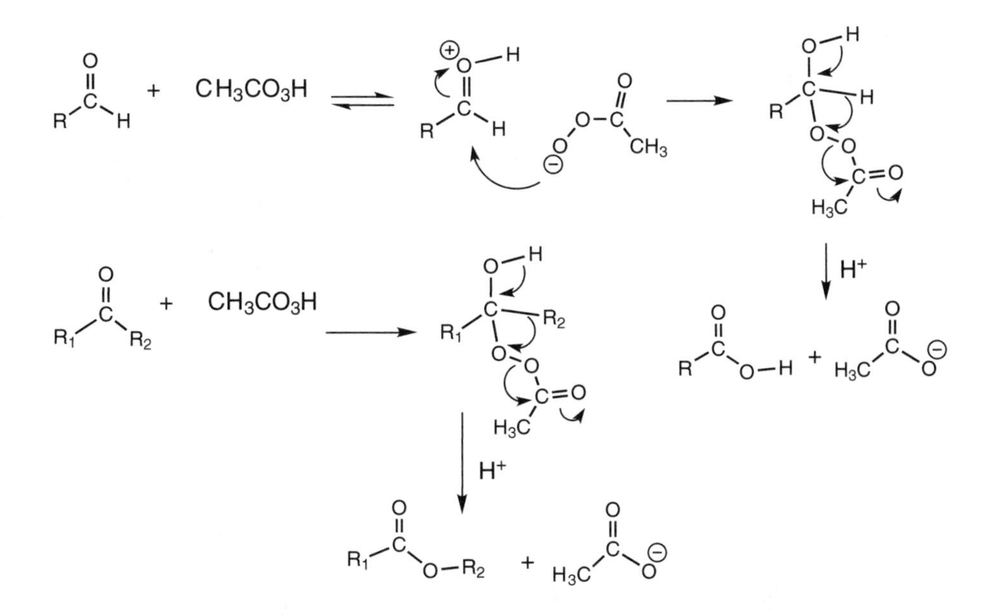

The formation of an adduct between the carbonyl compound and the peracid is similar to formation of the carbonyl adducts seen with alcohols or amines. However, this adduct breaks down in such a way as to cleave the weak O—O bond and to form a strong C=O bond. In aldehydes, the H migrates to oxygen, giving a carboxylic acid. In ketones, the alkyl group that migrates is the one that forms the more stable carbocation, so the migration preference is 3° > 2° > 1°.

EXERCISES

Fill in each blank with the appropriate substrate or reagent.

(a)

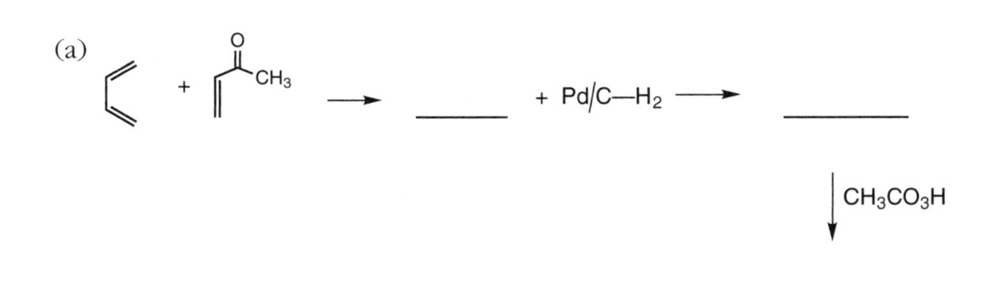

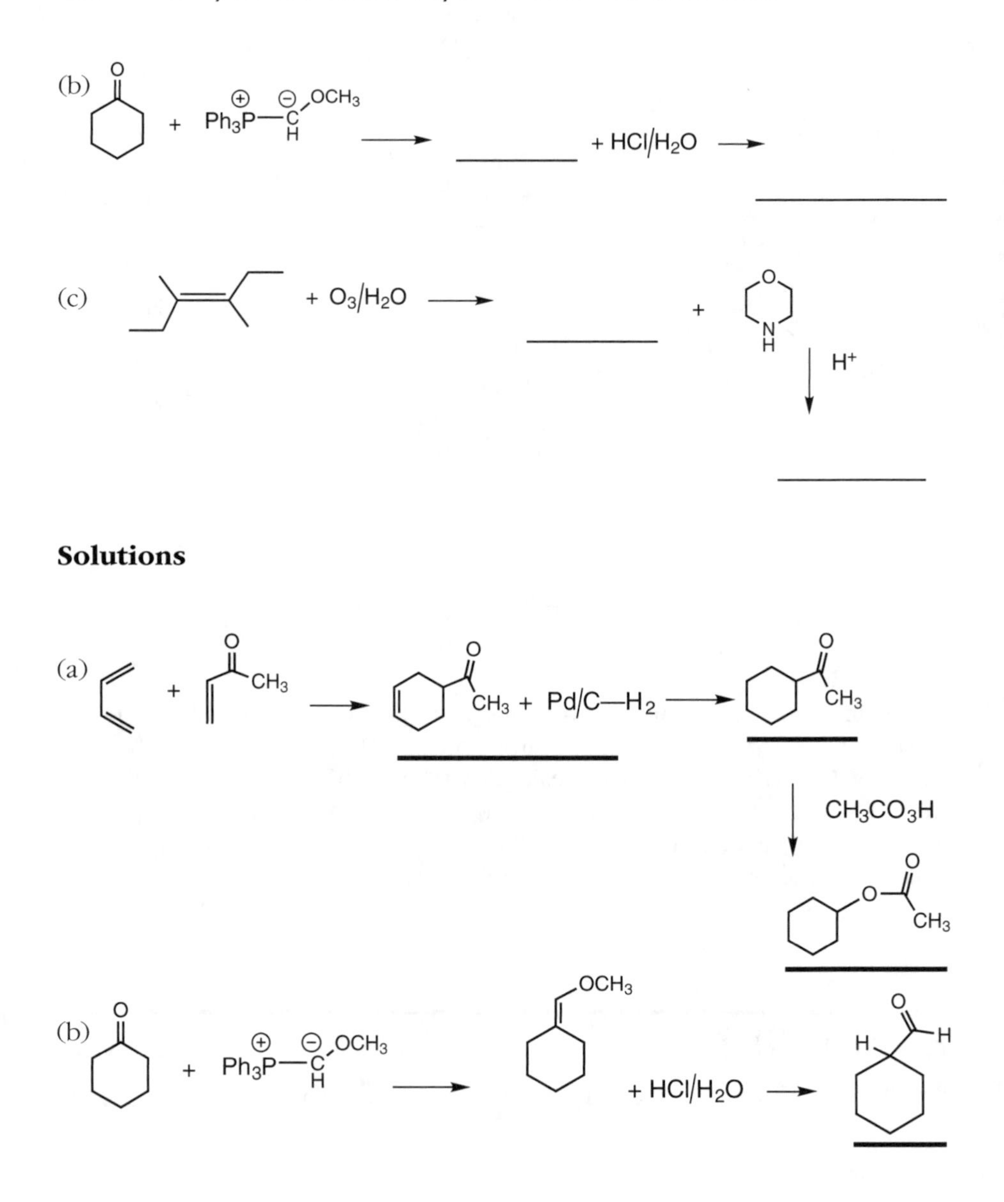
(b)
O
+
Ph3P—C
OCH3
H
+ HCl/H2O
(c)
+ O3/H2O
+
O
N
H
H+
Solutions
(a)
O
+
CH3
O
CH3 + Pd/C—H2
O
CH3
CH3CO3H
O
O
CH3
(b)
O
+
Ph3P—C
OCH3
H
OCH3
+ HCl/H2O
O
H
H

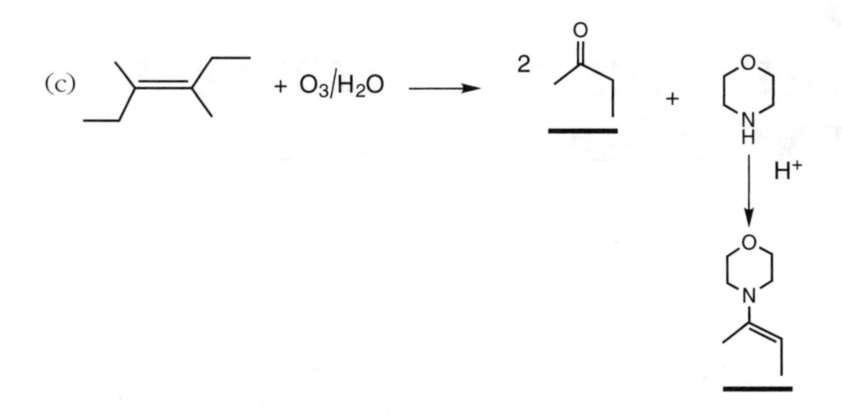

The enamine with the more stable, trisubstituted double bond is formed in preference to: since the reaction is reversible.

19
ALDEHYDES AND KETONES II: ENOLS AND ENOLATES

In Chapter 18, the electrophilic character of carbonyl groups was described. However, the reason the carbonyl group is so central to understanding organic chemistry is that it can also be converted to a reactive nucleophile. The C—H bond next to an aldehyde or ketone carbonyl group is only slightly less acidic than an alcohol —OH. The pK_a of the ethanol —OH is 16.0, that of the —CH_3 of CH_3CHO is 17.0, and that of the —CH_3 of acetone, CH_3COCH_3, is 20.0. The reason for this acidity is the resonance stabilization of the resulting anion, called an **enolate**:

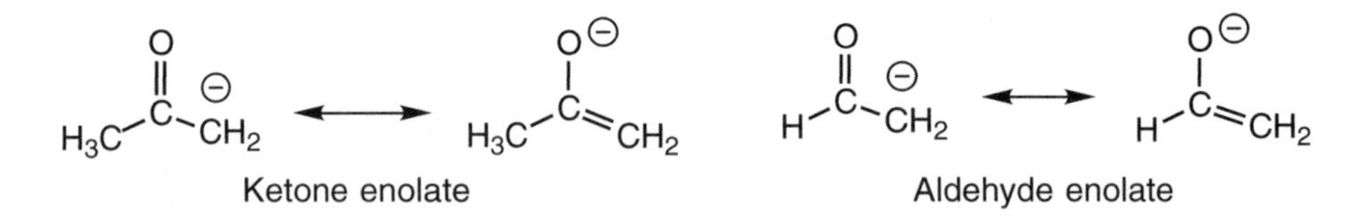

The aldehyde C—H of the CHO group is not unusually acidic since it is perpendicular to the carbonyl π system. Only a C—H bond that overlaps the C=O π bond is acidic. In a rigid carbonyl system, such as is found in a *trans*-Decalin, only the C—H bond that overlaps the carbonyl π system is deprotonated to form an enolate:

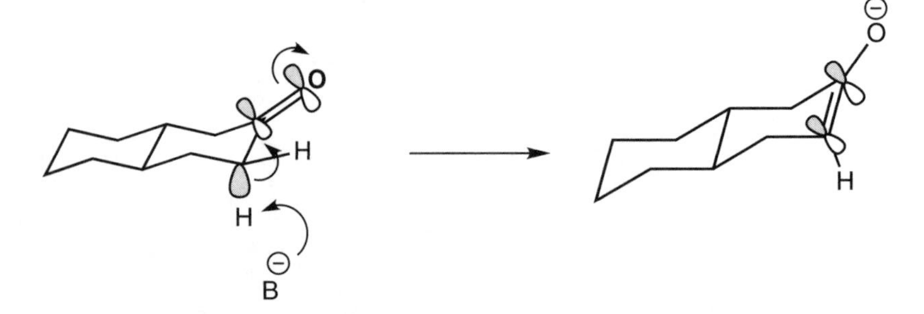

In Chapter 11, it was shown that the addition of water to alkynes gives an —OH-substituted alkene called an **enol**. Protonation of an enolate on oxygen gives an enol, $R_2C{=}CROH$. It was pointed out in Chapter 11 that enols and carbonyl compounds are in equilibrium through a process called **tautomerization**:

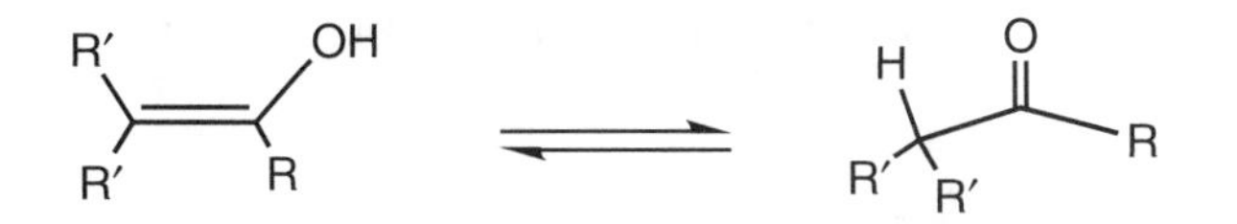

For simple aldehydes and ketones, tautomerization favors the carbonyl compound over the enol. Acetone, for example, contains less than 0.1 percent of the enol form at equilibrium. If the enol form can be stabilized by a hydrogen bond, as is the case in β diketones, significant amounts of the enol form will exist at equilibrium:

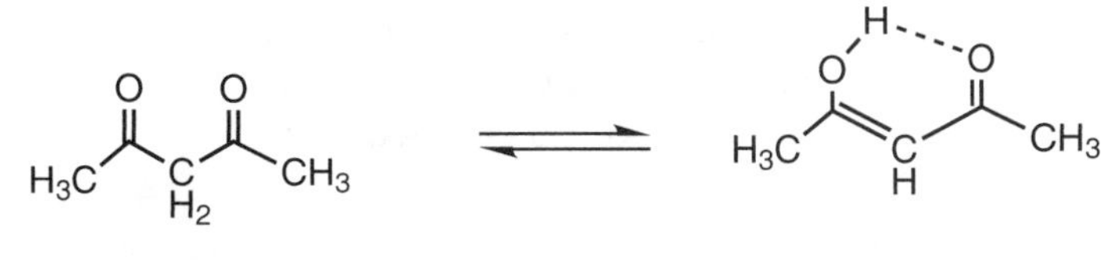

Enol double bonds are like alkene double bonds in that they are stabilized by alkyl groups and conjugation.

REACTIONS OF ENOLS AND ENOLATES

Mechanism of Acid- and Base-Catalyzed Enolization

Enols are formed and converted to carbonyl compounds under acid- or base-catalyzed conditions, which we have already learned is called tautomerization. In the presence of acid, the first step is, as usual, protonation of the carbonyl oxygen:

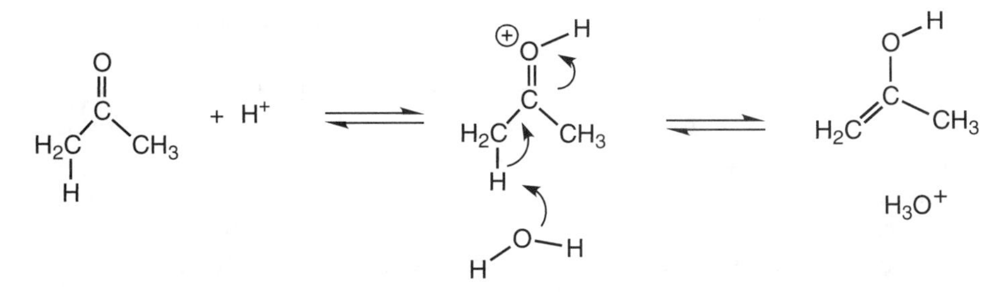

A base, such as water, abstracts the α proton, giving the enol. The **principle of microscopic reversibility** says that for all reactions, the mechanism of the reverse reaction follows the same path as the forward reaction, but in reverse. Thus, under acidic conditions, enols are protonated on carbon and loss of the —OH proton results in the carbonyl compound.

Strong bases directly deprotonate carbonyl compounds to give the enolate anion:

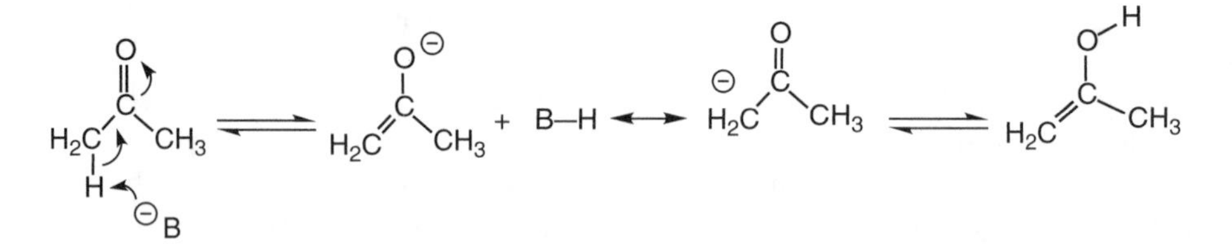

The enolate anion is an example of an **ambident** ion: It can react at different sites, at carbon or at oxygen. Protonation at oxygen gives the enol, and protonation at carbon gives the carbonyl compound. While protonation at oxygen is favored, most reactions with other electrophiles involve reaction at the carbon atom of the enolate anion.

Kinetic and Thermodynamic Enolates

An unsymmetric ketone can be deprotonated to give two different enolates:

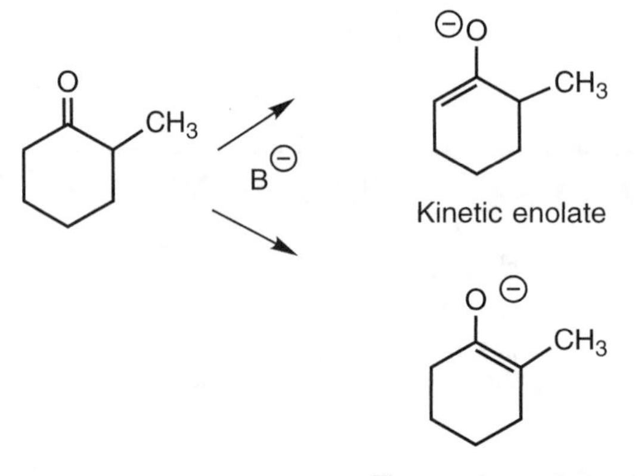

The less substituted double-bond isomer is called the **kinetic enolate**, and the more substituted double bond isomer is called the **thermodynamic enolate**. Different reaction conditions result in one or the other enolate. The thermodynamic enolate has the most stable double bond, so it is the most stable and forms under reversible conditions. Reversible conditions involve the presence of a weak acid, such as water, an alcohol, or an unreacted carbonyl compound. The kinetic enolate is formed irreversibly, and one needs to use a strong base, such as LDA [$LiN(i\text{-}Pr)_2$] or KH. The base abstracts the secondary CH_2 hydrogens in preference to the hindered tertiary CH hydrogen, and the conjugate acid, $HN(i\text{-}Pr)_2$ or H_2 in the above cases, is not acidic enough to protonate an enolate.

In order to prepare a kinetic enolate, one must add the ketone to a solution of LDA or an other strong hindered base. Since a ketone is a weak acid, any excess (which would be the case if the LDA were added to the ketone solution) produces thermodynamic enolate conditions:

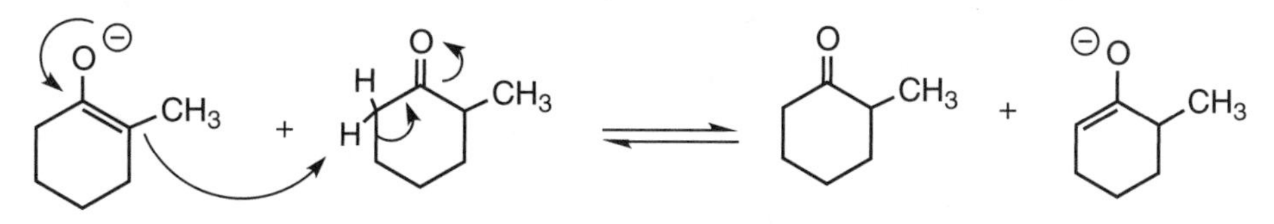

Example Forming an enolate using an alcoholic base gives mostly the thermodynamic enolate, and LDA gives the kinetic enolate:

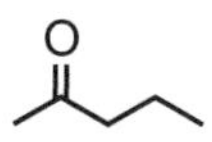

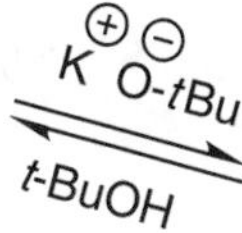

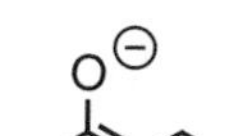

Major isomer

Major isomer

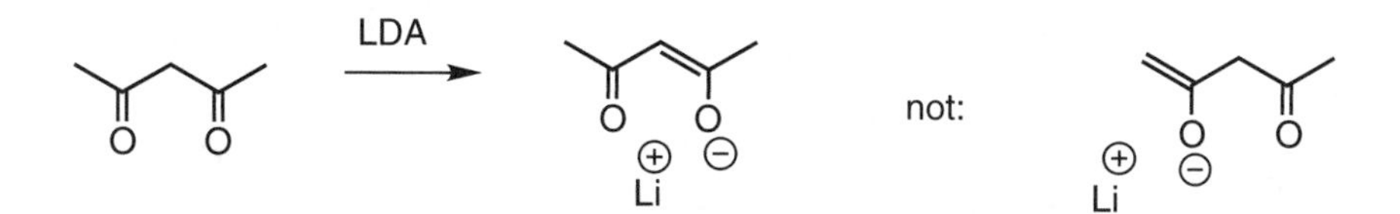

The acidity difference of the CH_2 protons, pK_a = 9, compared to the CH_3 protons, pK_a = 20, is so great that any base deprotonates the more acidic site. LDA gives the kinetic enolate only when the pK_a of the more hindered and less hindered protons are approximately the same.

Halogenation of Carbonyl Compounds via Enols and Enolates

If an aldehyde or ketone is placed in aqueous acid with Cl_2, Br_2, or I_2, the carbonyl compound undergoes halogenation at the α carbon. All three halogenation reactions take place at the same rate, which implies that the carbon-halogen bond-forming step is not the rate-limiting step. Instead, the slow step is enol formation. The enol, which is nucleophilic, then rapidly reacts with the halogen electrophile:

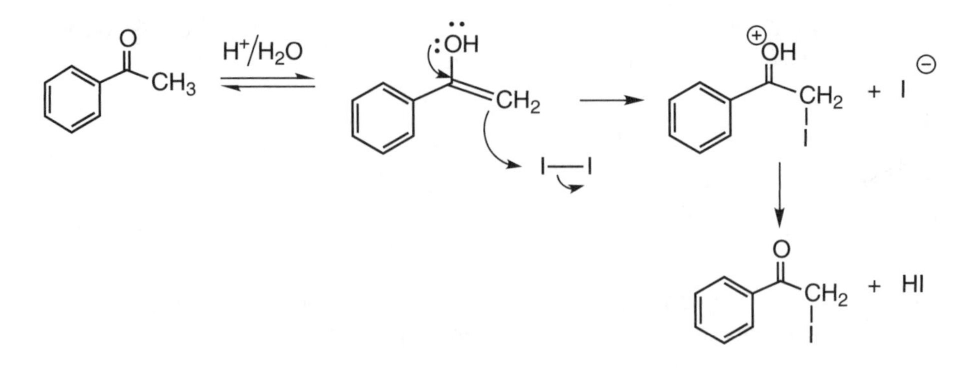

Because of the electron-withdrawing nature of a halogen, the product is not as reactive as the starting carbonyl compound, so the reaction can be stopped after the addition of one halogen.

The reaction of ketones under basic conditions follows a slightly different pathway. Instead of an enol, an enolate ion is produced, which also reacts with a halogen to form an α-halo carbonyl compound:

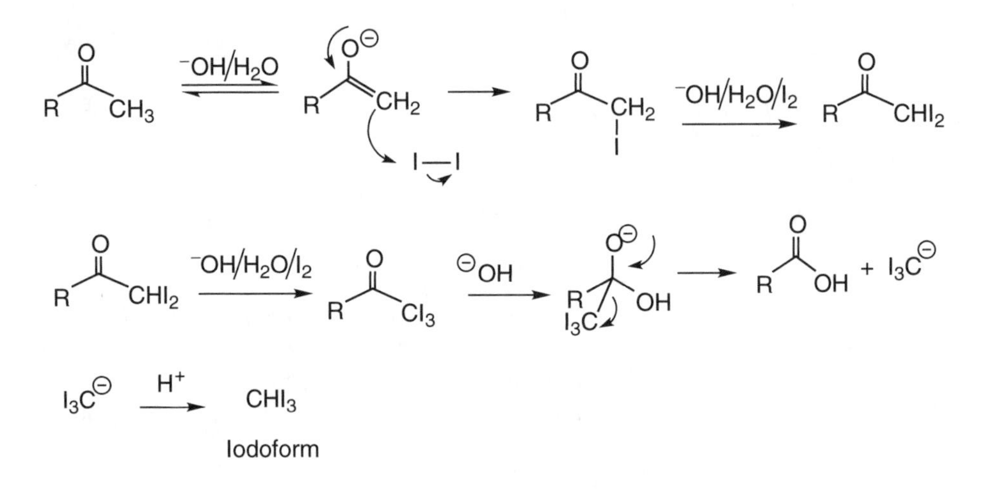

The α-halo carbonyl compound is more reactive in forming an enolate than the original ketone because the electron-withdrawing halogen stabilizes the enolate anion. It reacts a second and third time, if one has a methyl ketone, to give a trihalomethyl ketone. The hydroxide anion can add to the carbonyl group, forming a tetrahedral intermediate. Because an X_3C^- anion is stabilized by inductive effects, it serves as a leaving group, forming a carboxylic acid and X_3C^-. The anion is protonated with acid to form a **haloform** ($HCCl_3$ is chloroform, $HCBr_3$ is bromoform, and HCI_3 is iodoform). The oxidation of a methyl ketone to a carboxylic acid and a haloform is called the **haloform reaction**.

Epimerization and Transposition

A carbonyl compound with a stereogenic center at the α position can undergo an acid- or base-catalyzed process called **epimerization** in which the stereocenter is interconverted between R and S forms. The name ***epimerization*** comes from an earlier term for enantiomers, **epimers**. Epimerization results from enol or enolate formation, which converts a tetrahedral carbon alpha to a carbonyl group to a flat sp^2 carbon:

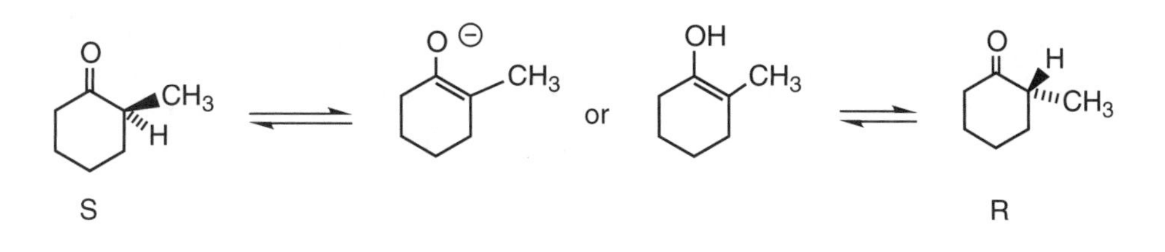

The enol or enolate can be protonated from the top or bottom face, which leads to a racemic mixture when starting from a single enantiomer.

In the case of diastereomers, one stereogenic center can interconvert if it is next to a carbonyl group, while the other centers are unchanged. For example, in fused ring systems, the presence of a ketone can facilitate the interconversion of *cis*-Decalins and *trans*-Decalins:

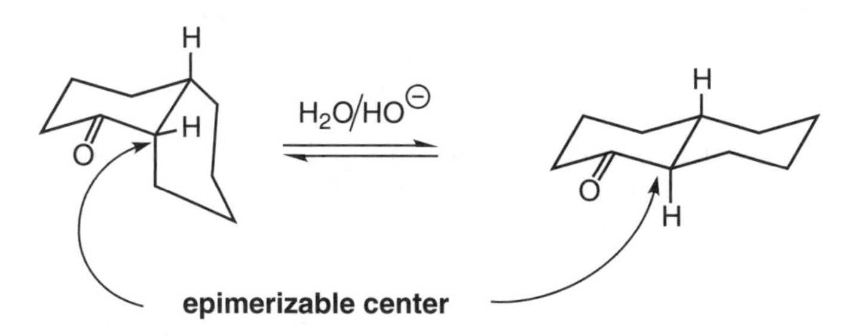

A stereogenic center with one C—H bond next to a carbonyl group is called an **epimerizable center**. In the presence of acid or base, its configuration can be changed through reversible enolization.

Enolization can also lead to a chemical reaction, **carbonyl transposition**, when a carbonyl group is next to a hydroxyl group:

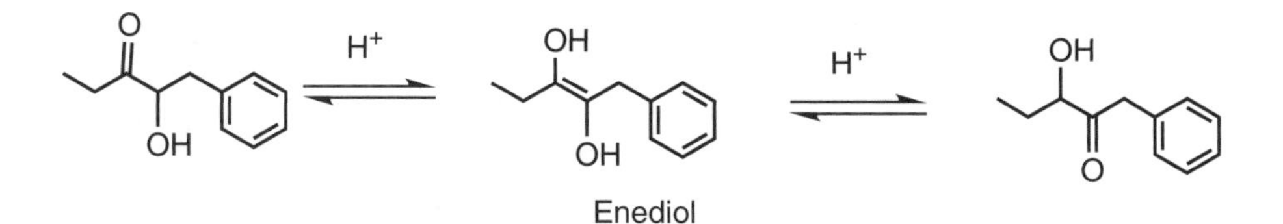

Either acid- or base-catalyzed enolization produces an **enediol** (a double bond bonded to two —OH groups) or an **enediolate**. This species can enolize back to the starting compound or enolize in the opposite direction, producing a transposed ketone. In the prior example, the two isomers are of approximately equal stability, so the equilibrium constant is 1. For an α-hydroxy aldehyde, carbonyl transposition favors the ketone isomer since ketones are more stable than aldehydes:

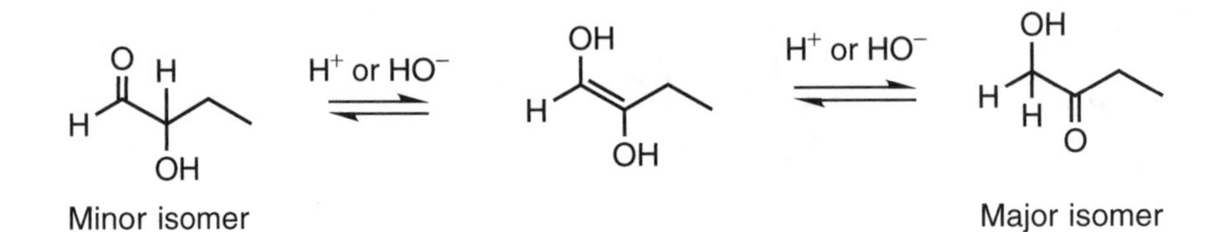

EXERCISES

1. Circle all epimerizable centers in each molecule.

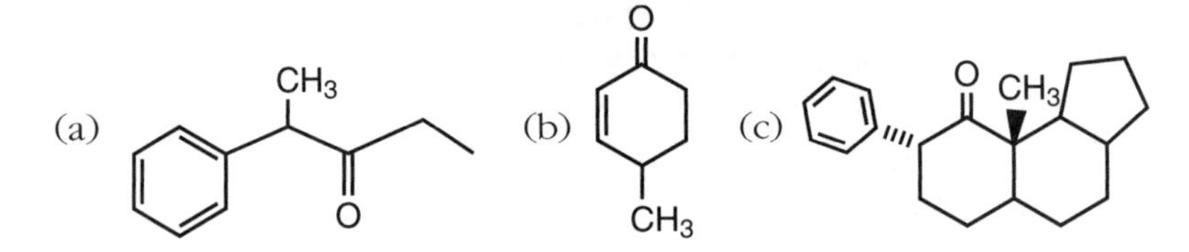

2. Write the mechanism for the aqueous acid-catalyzed carbonyl transposition of 2-hydroxybutanal to 1-hydroxy-2-butanone.

Solutions

1.

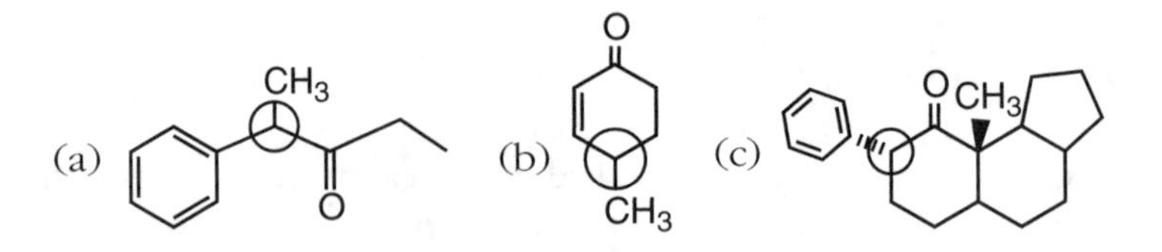

In the case of (b), enolization gives a dieneolate anion:

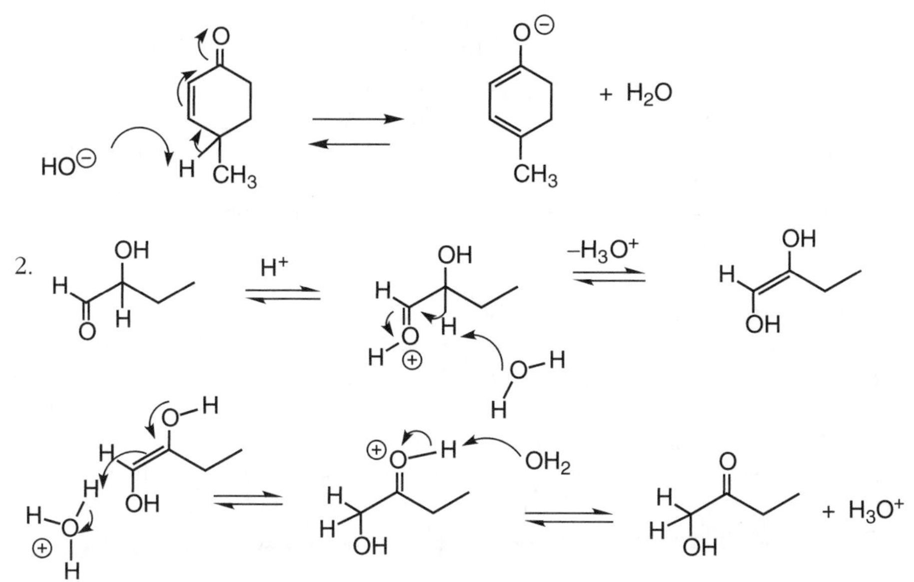

Alkylation of Enolates

The feature of aldehydes and ketones that makes them so useful in organic synthesis is that they are reactive electrophiles (reacting with RLi and RMgX) and also can be converted to good nucleophiles by forming enolates. Like all nucleophiles, enolate anions react best with unhindered electrophiles such as MeI, 1° halides, epoxides, and allylic and benzylic systems. Hindered 2° or 3° electrophiles usually react with enolates to give E2 elimination products:

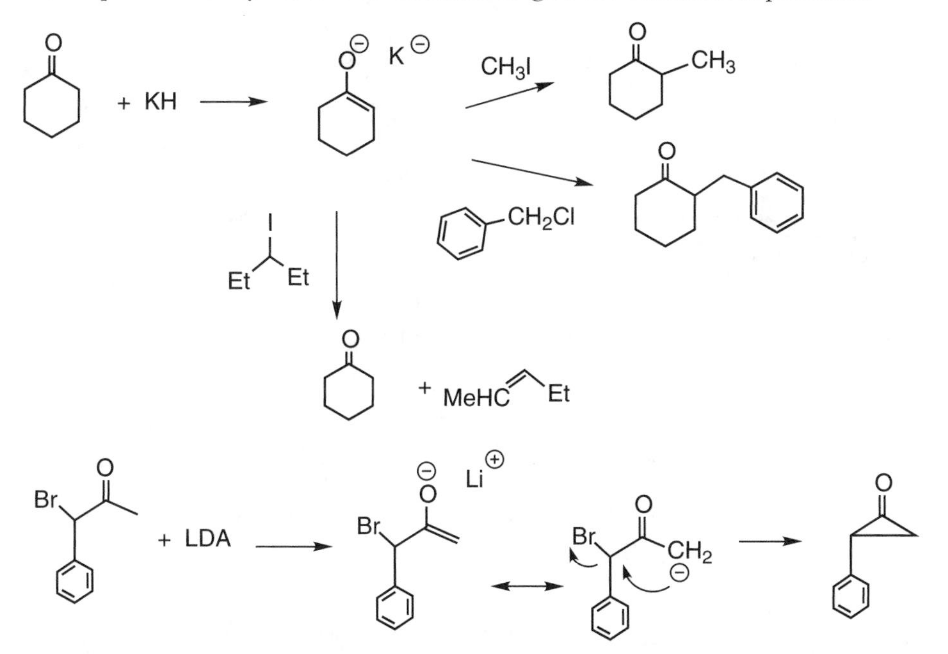

Enolates can react in an **intermolecular** reaction with external electrophiles, or they can react in an **intramolecular** reaction with an electrophile that is part of the enolate molecule. Intramolecular reactions form new rings. The rate at which rings form varies with ring size, normally in the order 6,5 > 3 > 4 > 7. Six- and five-membered rings are relatively unstrained. Cyclopropanes, while strained, have an entropic advantage because the nucleophilic and electrophiles sites are held so close together. Four-membered rings are difficult to form because they are strained and lack the entropic advantage of cyclopropanes. The formation of larger rings is difficult because of the low probability of the ends finding each other, although formation of seven- and eight-membered rings is not impossible.

When a CH_2 group is attached to two electron-withdrawing groups, as in β diketones and related compounds, it is quite acidic (pK_a ~9, similar to RCOO**H**). It is easily deprotonated by NaOH or even R_3N, and the resulting delocalized enolate undergoes alkylation. With two equivalents of base, dialkylation products can form with reactive electrophiles. While alkylation on oxygen can be a problem in these systems, specialized procedures can be used to favor alkylation on carbon:

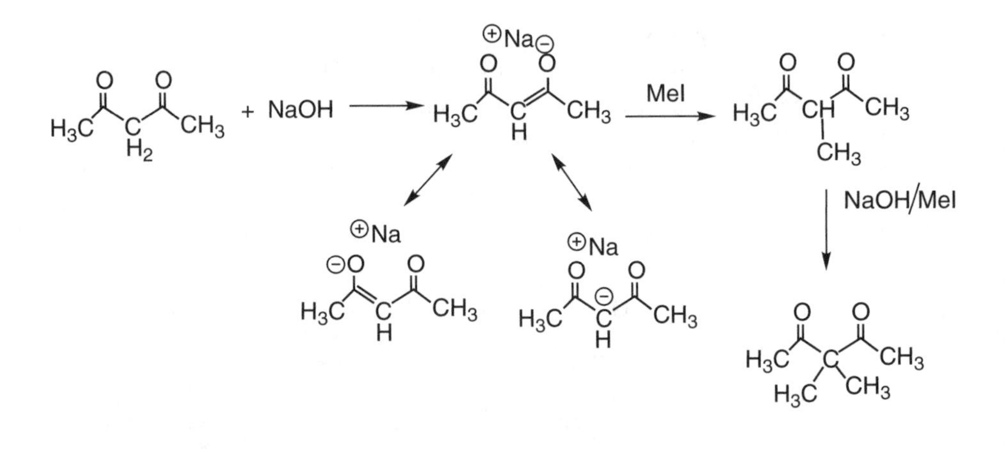

EXERCISES

Propose a synthesis for each of the following molecules using at least one enolate-based carbon-carbon bond-forming reaction.

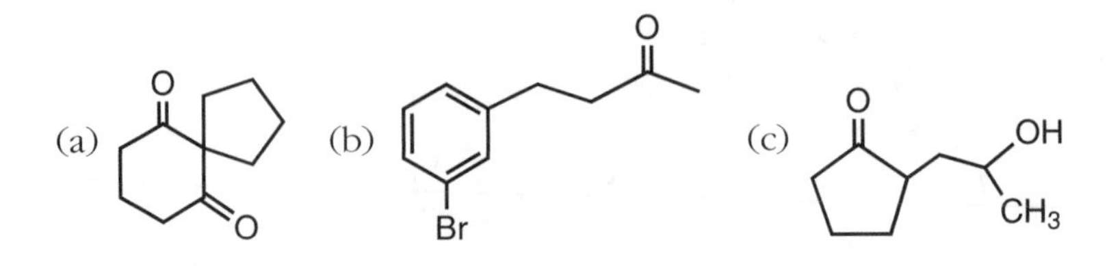

Solutions

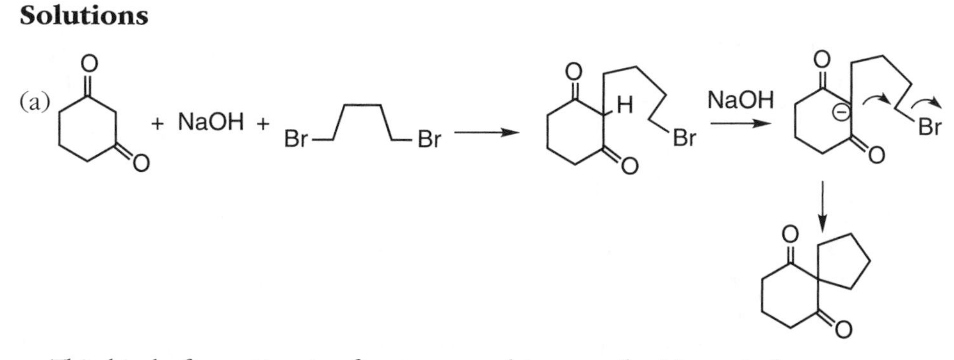

This kind of question is often answered incorrectly. Many students answer

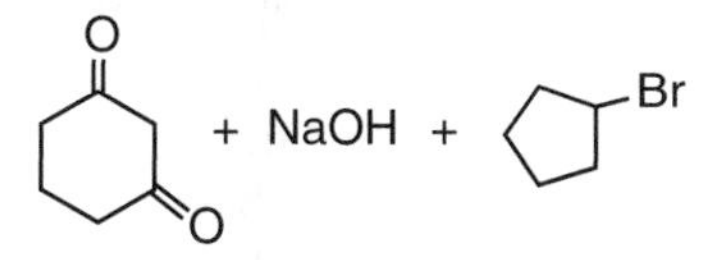

but that would give a mixture of elimination and

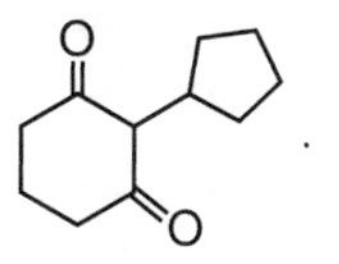

When doing a synthesis, one needs to look at the changes in going from the starting material to the product. In this case, the starting ketone has a CO—CH_2—CO group going to a CO—CR_2—CO group, so two C—H bonds must be broken and two C—C bonds formed.

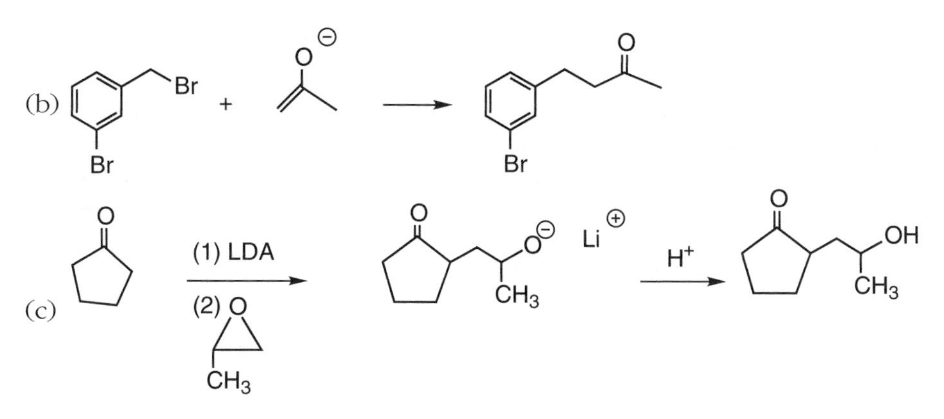

The Aldol Reaction

When an aldehyde or ketone enolate reacts with the carbonyl group of an aldehyde or ketone in a nucleophile-electrophile reaction, it is called an **aldol reaction**. Aldol is the name of the aldehyde-alcohol product formed by the self-reaction of acetaldehyde:

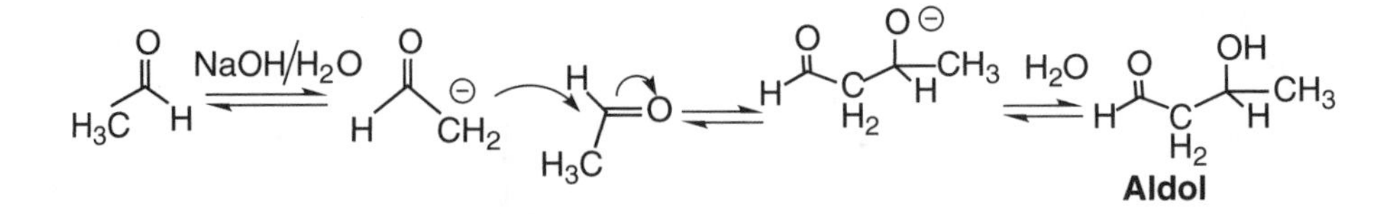

The aldol reaction initially gives a β-hydroxy aldehyde or ketone. Notice that all the steps in the aldol reaction are reversible, including the carbon-carbon bond-forming step.

The aldehyde carbonyl group is more electrophilic than the ketone carbonyl group. As a result, when an aldehyde enolate is formed, it tends to react with as yet unreacted aldehyde to give aldol products. Because of the possibility of aldol reactions, the alkylation of aldehyde enolates is an inefficient reaction.

Some carbonyl compounds are only electrophiles (sometimes called acceptors in the aldol reaction) because they have no enolizable protons. These include formaldehyde; benzaldehyde, PhCHO; 2,2-dimethypropanal; and benzophenone, PhCOPh. Treating a mixture of a ketone and one of the nonenolizable carbonyl compounds gives a single aldol product since the ketone enolate reacts with the more reactive aldehyde carbonyl rather than the less reactive ketone carbonyl:

When two different carbonyl compounds react under aldol conditions, it is called a **mixed aldol reaction**. While, as in the above case, only one product can form in a mixed aldol reaction, a mixture of products normally results. Both carbonyl compounds can form enolates, and both can serve as acceptors:

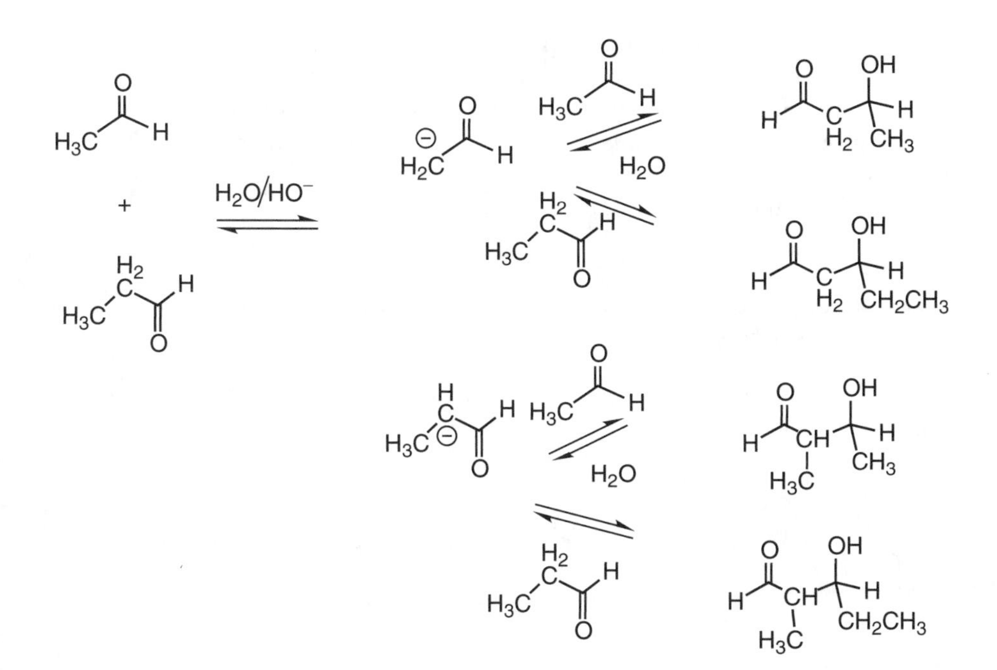

Notice that one, and often two, new stereogenic centers are formed in the aldol reaction. A special nomenclature is used to describe the stereochemistry of the compounds that are diastereomers, using the term **erythro** and **threo**. Erythro and threo are also used to describe the stereochemistry of sugars, so they are not restricted to aldol products. To decide if a diastereomer with two adjacent stereogenic centers is erythro or threo, first draw the molecule using wedge and dashed bonds. Draw similar or identical groups eclipsed. Similar means alkyl groups and C—H groups in this case:

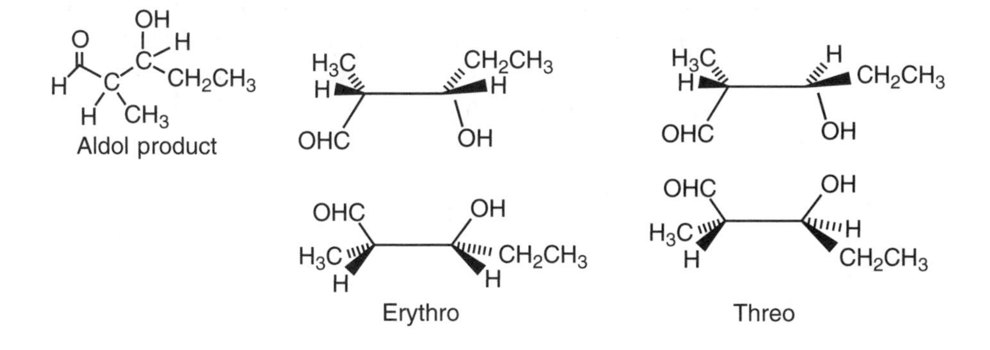

When all the C—H and C—R groups on adjacent carbons are located in mirror image positions, the diastereomer is the erythro isomer. When the groups are not in mirror image positions, the diastereomer is the threo isomer. In other words, groups are in the *same* positions in the *erythro* isomer and in *different* positions in the *threo* isomer. If the groups on adjacent carbons are too different, one cannot use this nomenclature.

EXERCISES

Show what carbonyl compounds would react to give each of the following aldol products.

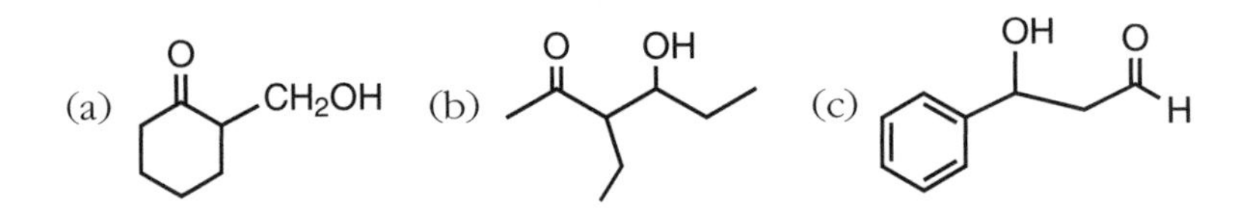

Solutions

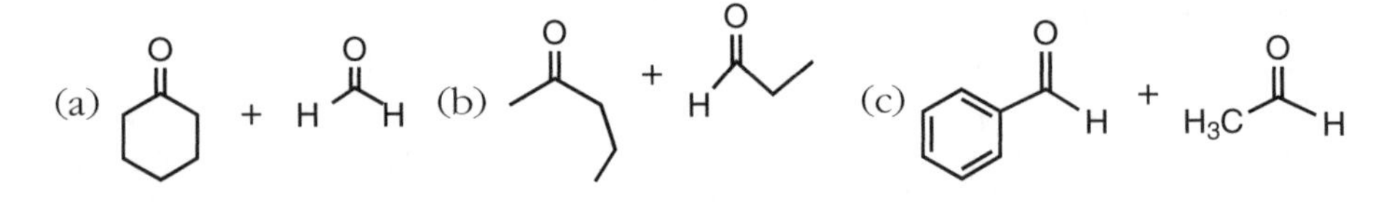

Dehydration of Aldol Products

The aldol reaction, as pointed out above, is reversible. For aldehydes, the equilibrium often lies on the side of the aldol product, but when ketones react, the equilibrium often lies on the side of the ketone starting materials. However, the aldol reaction can be driven to completion if the β-hydroxy carbonyl compound undergoes dehydration, forming an α,β-unsaturated carbonyl compound. Elimination takes place at higher temperatures through a two-step process, called an **E1cb elimination**, which stands for unimolecular elimination, conjugate base:

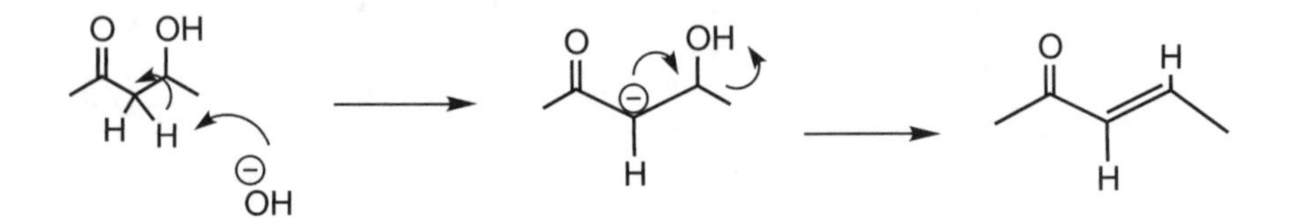

A base forms the enolate (the conjugate base) of the aldol product that then undergoes elimination of HO^- to form the conjugated double bond. This reaction is irreversible, thus driving the aldol reaction to completion. Unlike the S_N2 reaction, in this case HO^- is a leaving group. Whenever the hydroxyl group is next to a negatively charged atom, an elimination reaction occurs to form a double bond. The more stable trans double bond is the predominate product.

The aldol reaction can be catalyzed by acid, in which case an enol reacts with a protonated carbonyl compound to give a β-hydroxy carbonyl compound. In acid, the product can undergo an E1 or E2 elimination to give the α,β-unsaturated carbonyl compound. Normally, the elimination reaction is so facile in acidic media that the β-hydroxy carbonyl compound cannot be isolated:

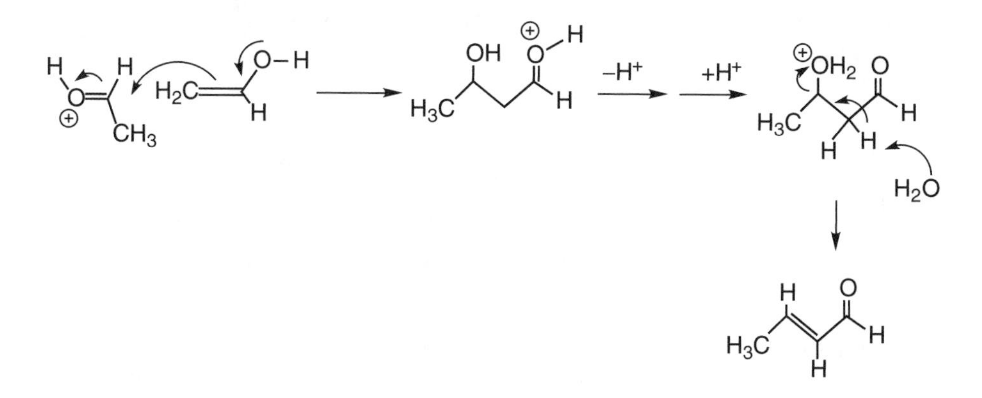

The acid- and base-catalyzed aldol reactions are quite similar. In the base-catalyzed reaction, a reactive nucleophile, the enolate, reacts with a moderately reactive electrophile, the carbonyl compound. In the acid-catalyzed reaction, a moderately reactive nucleophile, the enol, reacts with the very reactive protonated carbonyl group.

It is not difficult to decide which two carbonyl compounds have undergone an aldol/dehydration reaction. In the enone product, $O_1{=}C_2{-}C_3{=}C_4$, numbering the atoms as shown, the new carbon-carbon bond is formed between C-3 and C-4, and C-4 was originally the acceptor carbonyl group.

Examples Make sure that, given the products, you can write the starting materials for aldol reactions.

(a)

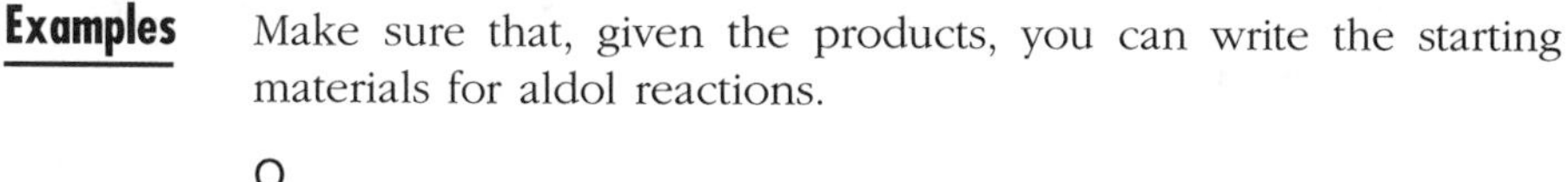

(b)

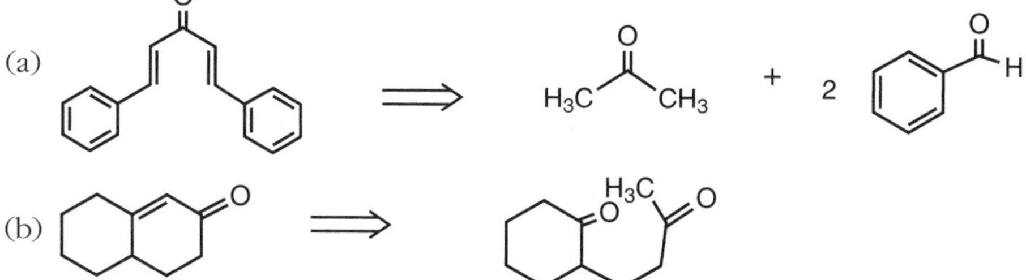

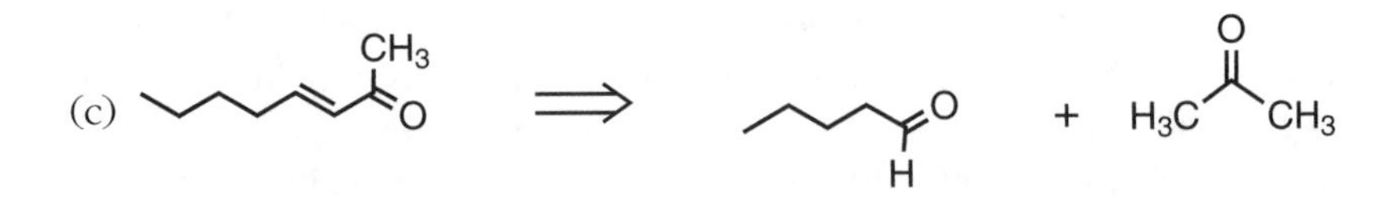

Reactions of Enamines

The formation of enamines from aldehydes or ketones and secondary amines was described in Chapter 18. Enamines are similar to enols except that they can be isolated and purified. They undergo alkylation reactions with methyl and reactive 1° halides, giving **iminium salts** that can be hydrolyzed to carbonyl compounds:

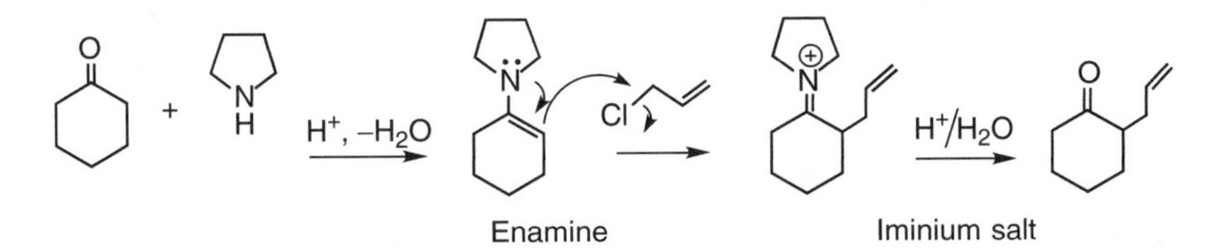

Enamines are useful in **acylation** reactions, in which an acyl group, RCO—, is introduced into a molecule. For example, enamimes react with acid chlorides to give, after hydrolysis, β-dicarbonyl compounds:

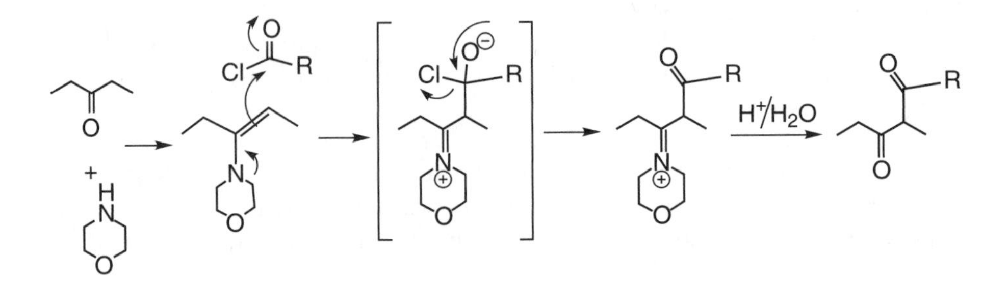

The enamine reacts with an acid chloride to give a tetrahedral intermediate which collapses to form an acyl group.

EXERCISES

Show how each of the following compounds could be made using enamine alkylation or acylation.

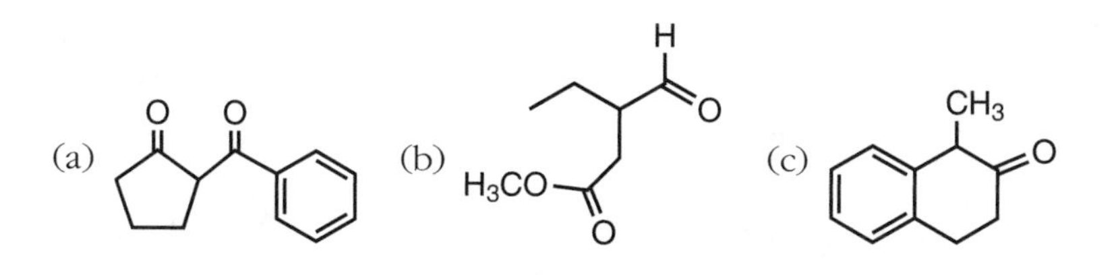

Solutions

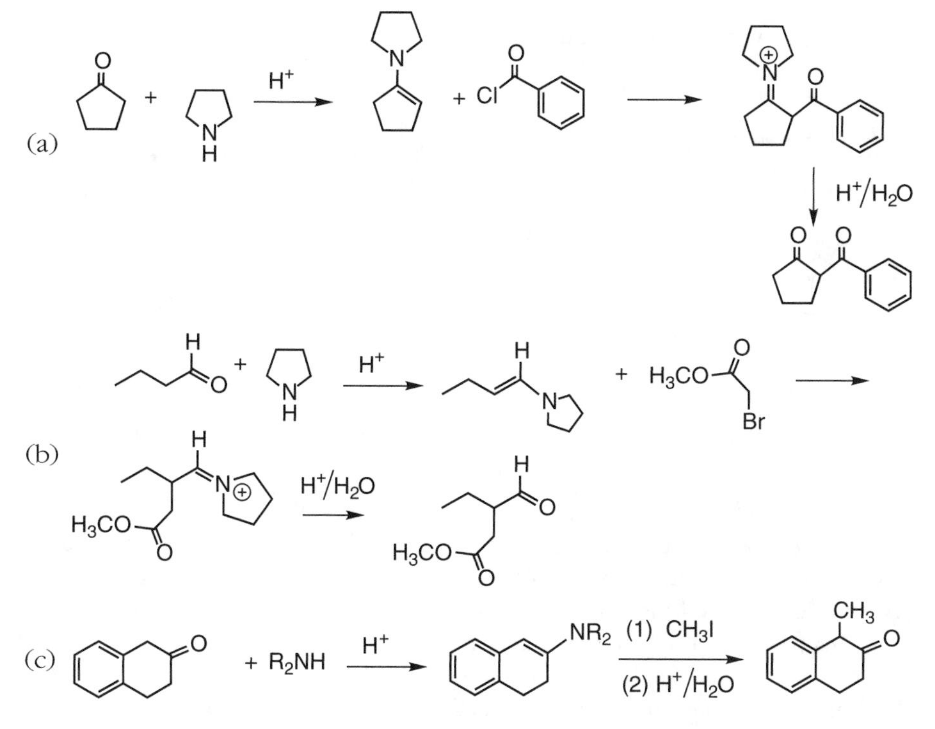

Note: The double bond forms the more stable conjugated regioisomer

CONJUGATE ADDITION REACTIONS

Conjugate addition is the reaction of a nucleophile with an electron-deficient carbon-carbon double bond or triple bond. A number of different substituents are sufficiently electron-withdrawing to make an unsaturated group reactive:

W = $-C(=O)R'$, $-C(=O)H$, $-C(=O)OR'$, $-C(=O)NR_2'$, $-CN$, $-NO_2$

W = electron-withdrawing group

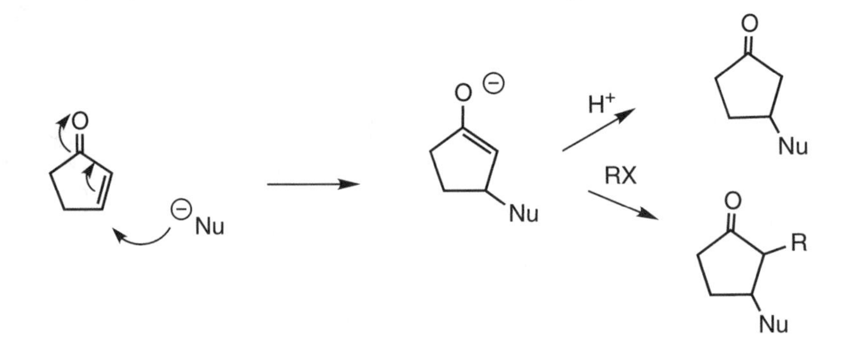

α,β-Unsaturated ketones can undergo two different kinds of reactions with nucleophiles. Some nucleophiles add to the C of the C=O group in what is called a **1,2-addition** reaction. Other nucleophiles react at the end of an electron-deficient double bond in a **conjugate addition** or **1,4-addition** reaction. This reaction also is known as a **Michael** reaction, after Arthur Michael, who was one of the first internationally recognized American organic chemists. Note that the Michael reaction of an α,β-unsaturated ketone gives an enolate anion. The nucleophile always reacts with the electron-deficient double or triple bond to place the negative charge next to the **W group** (electron-withdrawing group) where it can be stabilized:

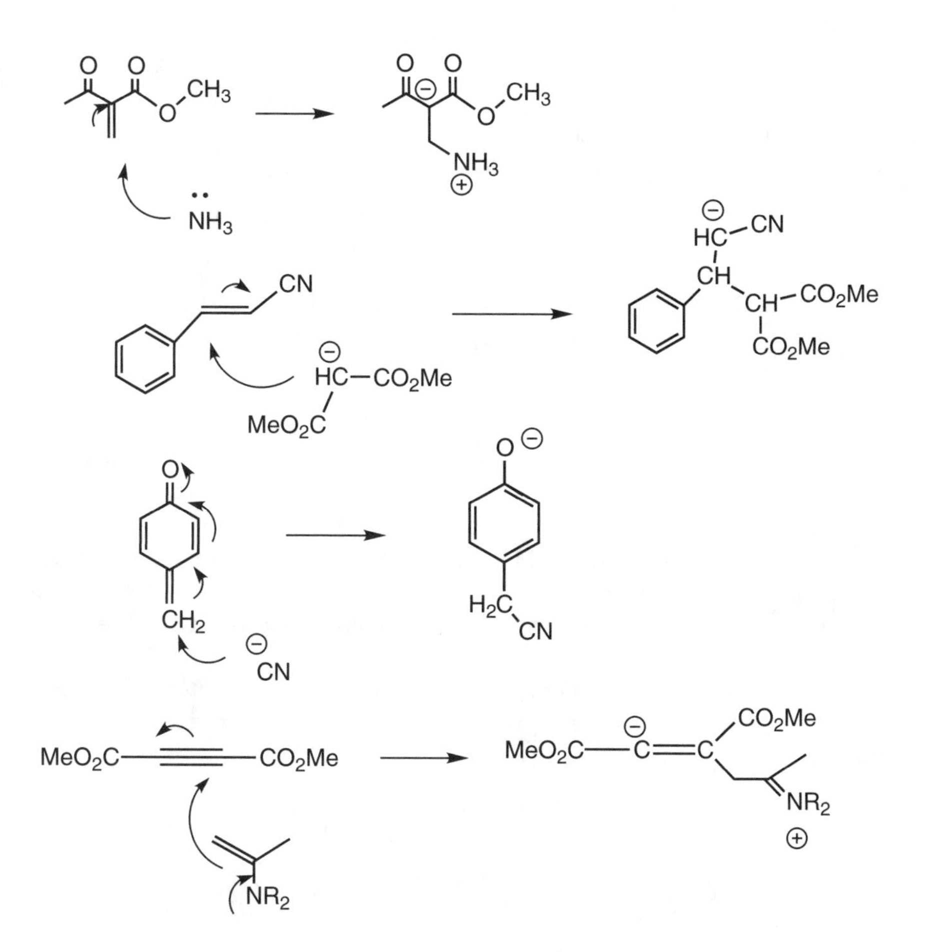

All the examples above generate a stabilized anion. The initially formed product can undergo protonation, alkylation or, in the case of the enamine reaction, hydrolysis. Thus, the Michael reaction is capable of forming a bewildering array of products. Normally, the beginning organic chemistry student needs to know only what compound can be the electrophile, or **Michael acceptor**, in a Michael reaction and what kinds of nucleophiles add to α,β-unsaturated substrates.

Very basic anions, such as RLi, RMgX, and the H^- derived from $LiAlH_4$, add 1,2 to an α,β-unsaturated aldehyde or ketone. Stabilized, less basic nucleophiles and anions tend to add 1,4 to an α,β-unsaturated aldehyde or ketone. Compounds of the type W_2CH^-, where W is an electron-withdrawing group, are most often used in Michael reactions. Enamines, amines, RS^- anions, and acids with a pK_a of less than about 14 all undergo the Michael reaction:

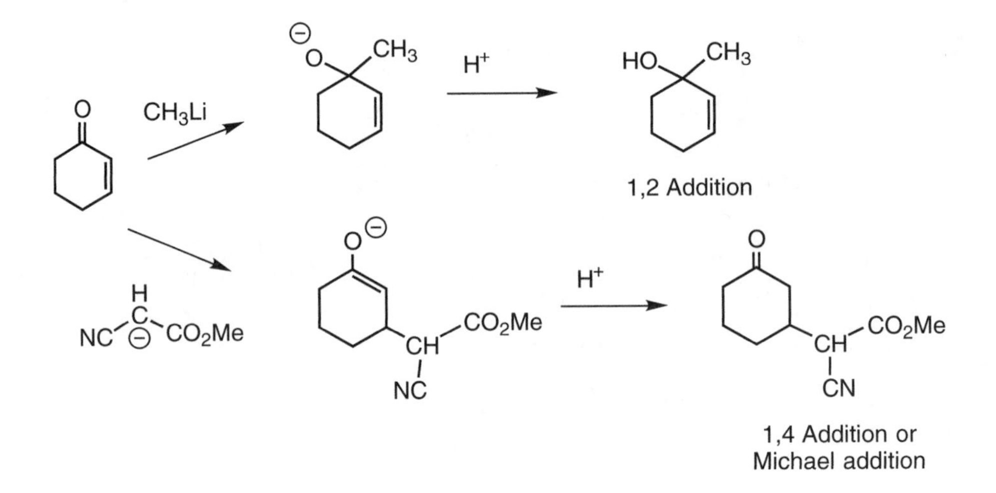

Many nucleophiles used in the Michael reaction are stabilized carbonyl enolates, as in the above example. When they react with α,β-unsaturated carbonyl compounds, the product is a 1,5-dicarbonyl compound, —CO—C—C—C—CO—.

While RLi compounds undergo 1,2 addition with α,β-unsaturated carbonyl compounds, cuprates, R_2CuLi, are excellent reagents for performing conjugate addition of alkyl and aryl groups:

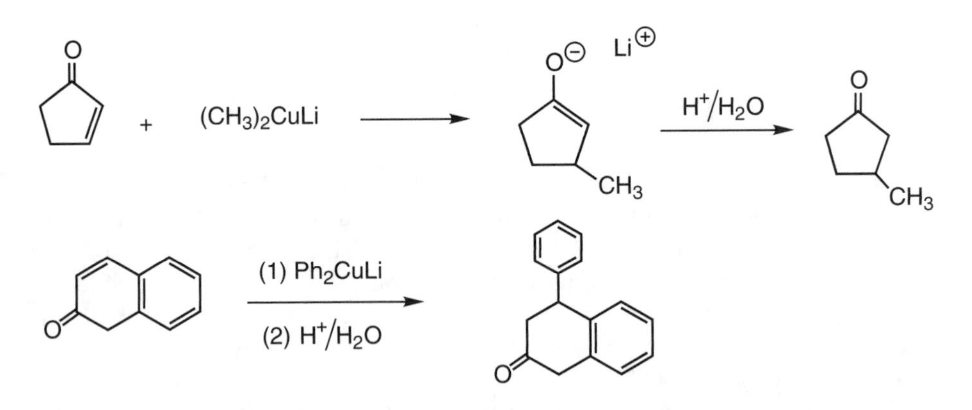

Unlike the Michael reaction, this reaction is not a simple nucleophilic addition reaction. Its mechanism is not completely understood.

EXERCISES

1. Show how the following compounds could be made using conjugate addition reactions. An addition reaction, such as alkylation, reduction, and so on, may be needed.

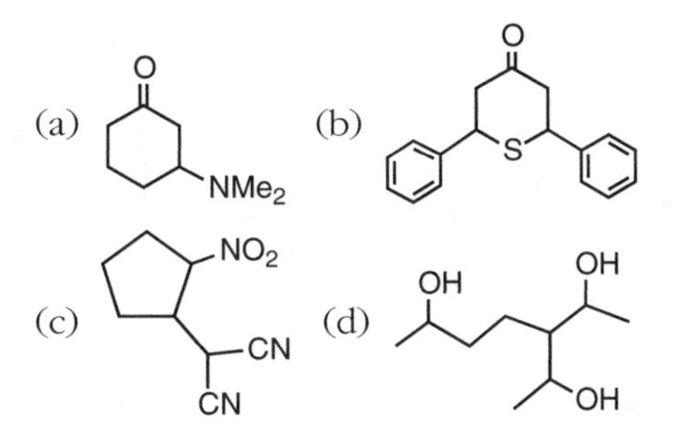

2. Fill in each of the blanks with the proper product.

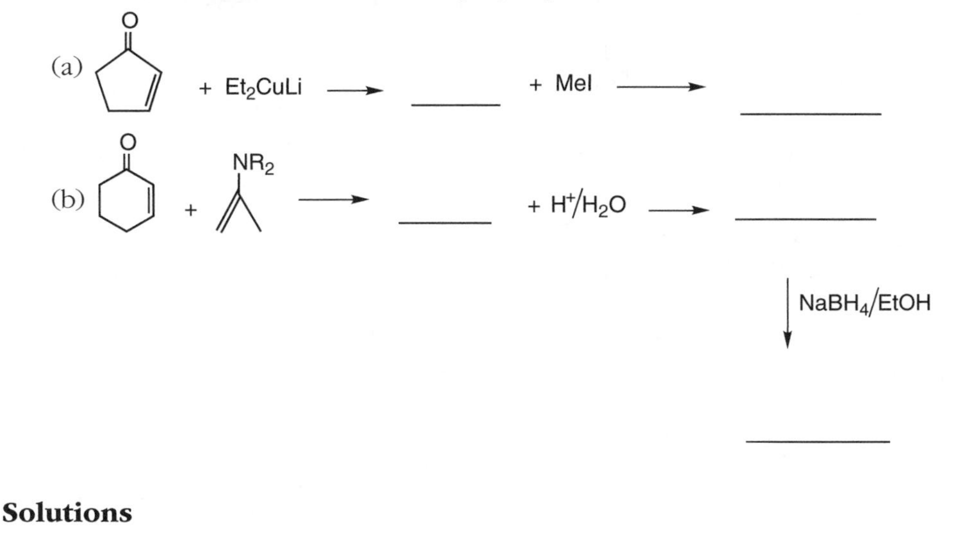

Solutions

1.

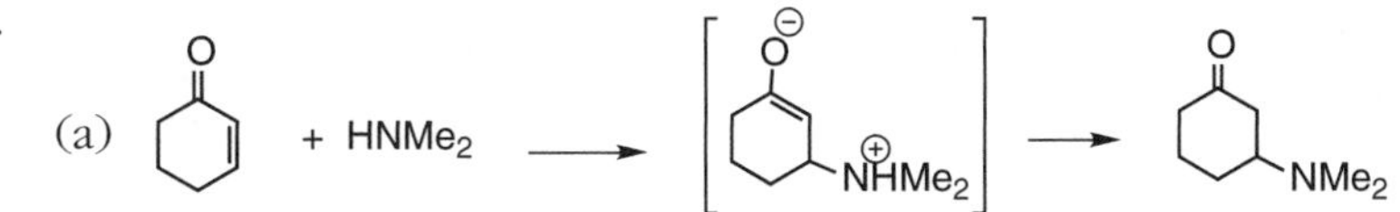

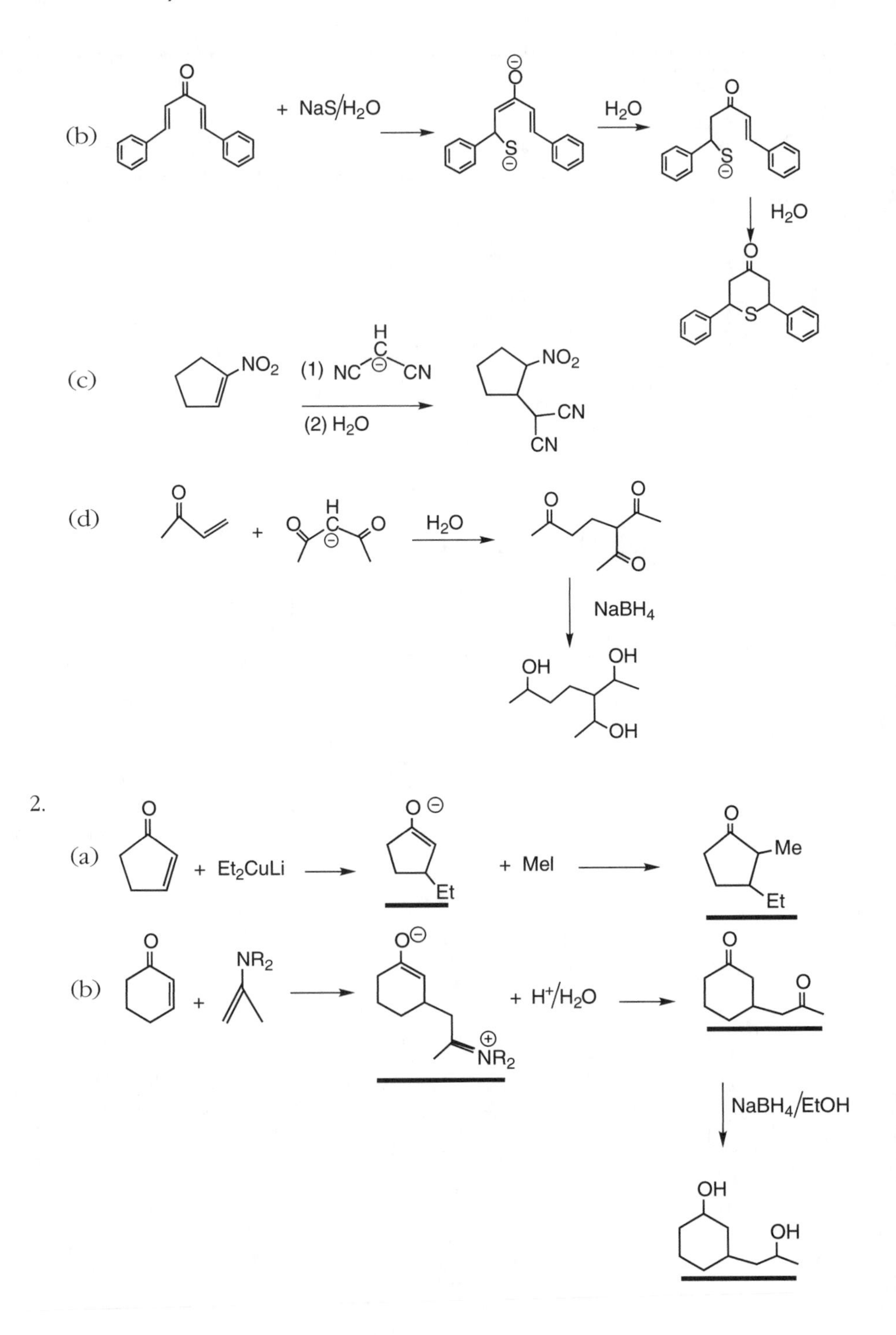
(b)
+ NaS/H_2O
H_2O
H_2O
(c)
NO_2
(1) NC C CN
(2) H_2O
CN
(d)
+
H_2O
$NaBH_4$
OH
2.
(a)
+ Et_2CuLi
Et
+ MeI
Me
(b)
NR_2
+ H^+/H_2O
$NaBH_4$/EtOH
OH

The Robinson Annulation Reaction

Annulation refers to ring formation, and the Robinson annulation reaction is a two-step ring-forming reaction involving a Michael reaction followed by an aldol condensation. It is a powerful method for creating new six-membered rings and was invented by the British chemist Robert Robinson:

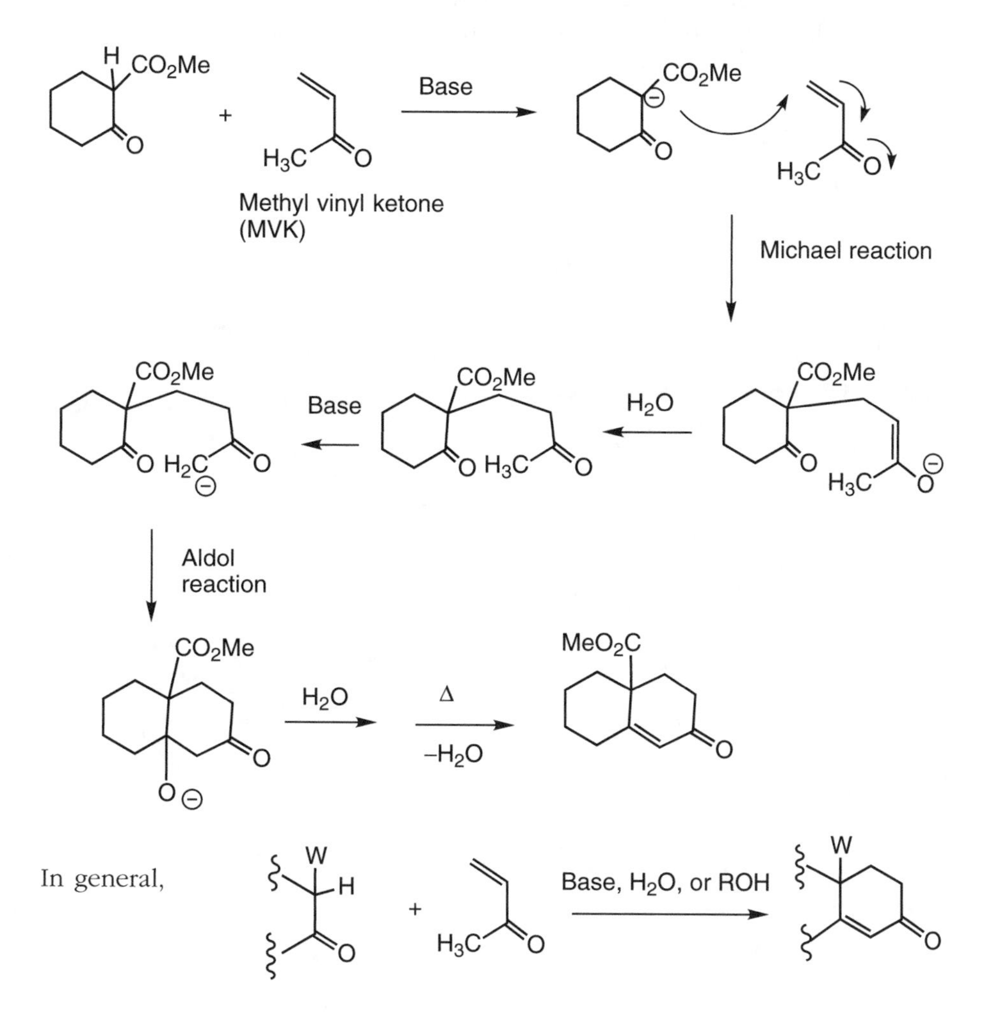

The Michael acceptor is methyl vinyl ketone or a related compound. The ketone reacting with MVK need not be cyclic. After the first carbon-carbon bond is formed, an aldol reaction forms the second ring bond. While one can imagine other aldol reactions taking place (e.g., using an enolate formed from the cyclohexanone reacting with the ketone of the methyl ketone substituent), the one shown leads to a stable six-membered ring. The student should compare this ring-forming reaction to the Diels-Alder reaction, which also forms six-membered rings.

20
REACTIONS AND SYNTHESIS OF CARBOXYLIC ACIDS

The central feature of the carboxylic acid functional group, RCO_2H, is its acidity. Different R groups enhance or lessen the acidity of the carboxylic acid group, giving organic chemists a chance to study how structure influences physical properties. Another reason to study carboxylic acids is that many acids are important natural products, playing central roles in biochemistry. Because so many acids have been known for hundreds of years, many common carboxylic acids have trivial names. Some of these names and the corresponding compounds are listed in Table 20.1.

Carboxylic acids are unusual in having relatively high boiling points compared to other polar molecules. For example, propanoic acid has a boiling point of 141°C, while 2-butnanol boils at 99°C. The reason for this difference is that carboxylic acids can form hydrogen-bonded dimers (and, in concentrated solution, trimers and higher aggregates):

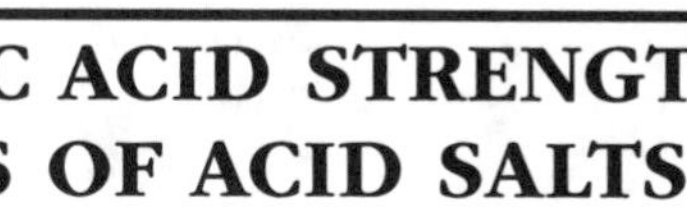

Hydrogen-bonded carboxylic acid dimer

CARBOXYLIC ACID STRENGTHS AND PROPERTIES OF ACID SALTS

Acidity of Acids With One —CO_2 Group

Acetic acid has a pK_a of 4.8, while ethanol has a pK_a of 16. Using the free energy equation $\Delta G^0 = -RT \ln K_a$, the **free energy of ionization** of ethanol is 21.7 kcal/mol, while for acetic acid the free energy of ionization is 6.5 kcal/mol. In other words, it takes approximately 15 kcal/mol less energy to ionize acetic acid than to ionize ethanol. As discussed in Chapter 4, stronger acids have more stable conjugate bases.

Two factors are at work in stabilizing the carboxylate anion: resonance and the inductive effect of oxygen. Resonance in a carboxylate anion was

TABLE 20.1. COMMON NAMES FOR SOME CARBOXYLIC ACIDS

Structural Formula	IUPAC Name	Common Name
$H-C(=O)-OH$	Methanoic acid	Formic acid
$H_3C-C(=O)-OH$	Ethanoic acid	Acetic acid
$CH_3CH_2CH_2-C(=O)-OH$	Butanoic acid	Butyric acid
$CH_3CH_2CH_2CH_2-C(=O)-OH$	Pentanoic acid	Valeric acid
$CH_2{=}CH-C(=O)-OH$	Propenoic acid	Acrylic acid
$H_3C-CH(OH)-C(=O)-OH$	2-Hydroxypropanoic acid	Lactic acid
$CH_3(CH_2)_{14}CH_2-C(=O)-OH$	Tetradecanoic acid	Palmitic acid
$CH_3(CH_2)_{16}CH_2-C(=O)-OH$	Octadecanoic acid	Stearic acid
$HO-C(=O)-CH_2-C(=O)-OH$	Propanedioic acid	Malonic acid
$HO-C(=O)-CH_2-CH_2-C(=O)-OH$	Butanedioic acid	Succinic acid

illustrated in Chapter 4. The fact that the C—O bond lengths in the carboxylate anion are equal provides experimental support for the resonance description:

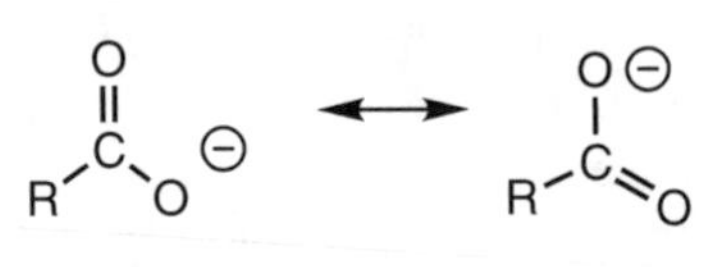

Resonance

Inductive effect

Notice that whatever the structure of the R group, it cannot delocalize the charge by way of resonance. Only the carboxylate oxygens can delocalize the charge.

The inductive effect of the carbonyl group also acts to stabilize the carboxylate anion. The positive end of the carbonyl dipole lowers the energy of the negative charge on oxygen. In addition, the nature of the R group can enhance or oppose the stabilizing effect of the carbonyl group's inductive effect. An electron-withdrawing R group (such as a —CF_3 group) stabilizes the carboxylate anion and makes the corresponding acid a stronger acid. A selection of pK_a values for different carboxylic acids, illustrating the influence of inductive effects on acidity, is given in Table 20.2.

TABLE 20.2. INFLUENCE OF INDUCTIVE EFFECTS ON THE ACIDITY OF CARBOXYLIC ACIDS

Carboxylic Acid	pK_a
$H_3C{-}C({=}O){-}OH$	4.7
$ClH_2C{-}C({=}O){-}OH$	2.9
$Cl_2HC{-}C({=}O){-}OH$	1.3
$Cl_3C{-}C({=}O){-}OH$	0.9
$F_3C{-}C({=}O){-}OH$	0.2
$C_6H_5{-}C({=}O){-}OH$	4.2
$H_3CH_2C{-}C({=}O){-}OH$	4.9

Inductive effects are additive, as illustrated by the acidities of chloroacetic acids. Measuring the pK_a values of different acids gives information on a group's electron-withdrawing ability. Benzoic acid, C_6H_5COOH, is more acidic than acetic acid, and this implies that an sp^2 carbon is more electron-withdrawing than an *sp*^{3} carbon. Finally, the inductive effect drops off rapidly with distance. When four or more σ bonds separate two groups, the inductive effect is almost undetectable. For example, $ClCH_2CH_2CH_2COOH$ has a pK_a of 4.5, which is very close to that of $CH_3CH_2CH_2COOH$, 4.9.

Acidities of Dicarboxylic Acids

When a molecule contains two carboxylic acids close together, the ionization of one group hinders the ionization of a nearby group because of charge-charge repulsion. This leads to two different ionization constants even when the two carboxylic acid groups are identical by symmetry. For example, in oxalic acid, HOOC—COOH, ionization of either of the groups gives HOOC—COO$^-$, the **monoanion** of the diacid. The first ionization constant is 1.2 because of the electron-withdrawing nature of the other —COOH group. The second ionization constant, which produces the **dianion** $^-$OOC—COO$^-$, is 4.3. As the two carboxylic acid groups become separated by more and more —CH_2 groups in individual dicarboxylic acids, the two ionization constants become similar. For adipic acid, HOOC—$(CH_2)_4$—COOH, pK_{a1} is 4.43 and pK_{a2} is 5.41. Although the inductive effect drops to near zero in the case of the adipic acid monoanion, recall that ionization constants are measured in water. The **field effect** of the negative charge (its ability to influence nearby charges through space and through polarization of the solvent water molecules) results in the second ionization constant being larger than the first even for adipic acid.

EXERCISES

Indicate which acid in each pair is the stronger acid.

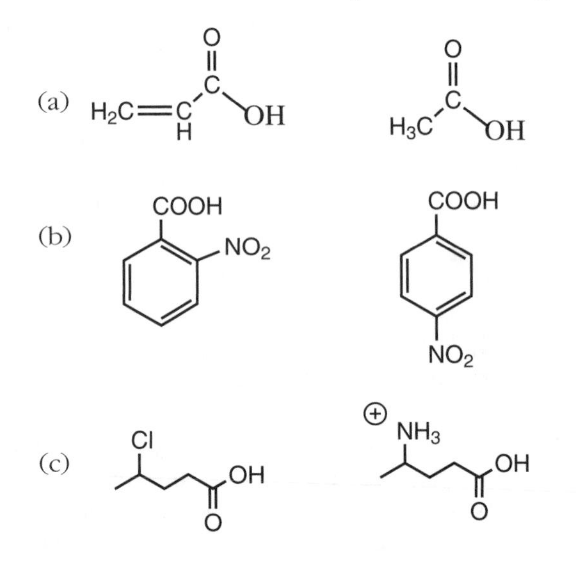

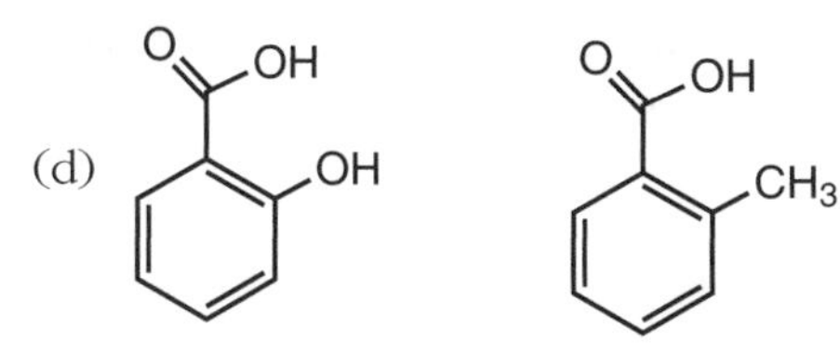

Solutions

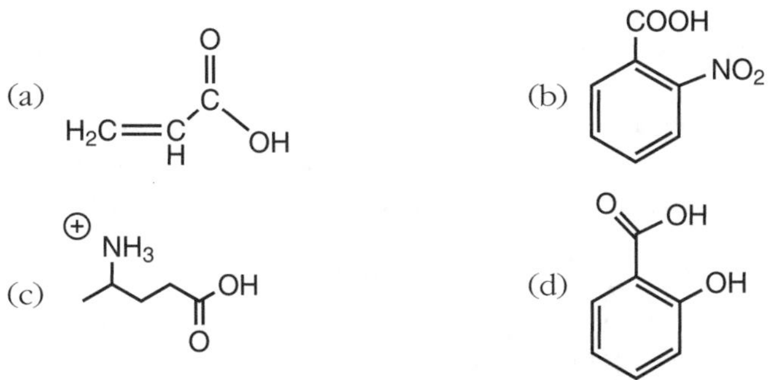

Strong Organic Acids Related to Carboxylic Acids

In addition to carboxylic acids, a few other types of strong acid derivatives are encountered in organic chemistry and biochemistry. Organic derivatives of sulfuric acid with a C—O bond are named **sulfates**. Since sulfates are good leaving groups, these compounds are reactive electrophiles in S_N2 and S_N1 reactions. Organic derivatives of sulfuric acid where an —OH group is replaced by a carbon group are called **sulfonic acids**. They are often used as acid catalysts in organic reactions, especially *para*-toluenesulfonic acid. Organic derivatives of phosphoric acid with a C—O bond are named **phosphates**. They were mentioned in Chapter 8 since they are precursors in biological cation-olefin cyclizations.

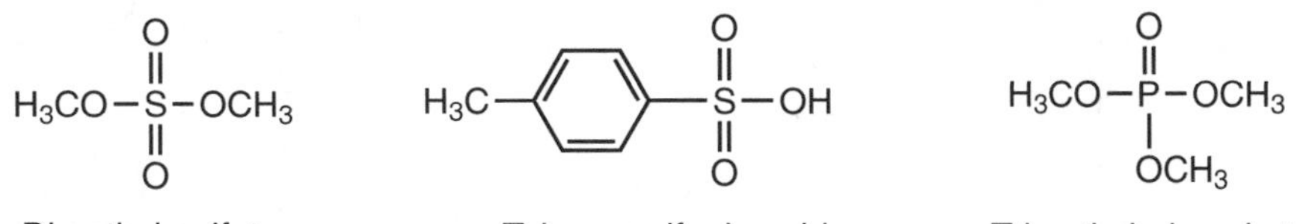

Dimethyl sulfate *para*-Toluenesulfonic acid Trimethyl phosphate

When carbon dioxide dissolves in water, carbonic acid, H_2CO_3, is formed reversibly. Carbonic acid is a weaker acid than carboxylic acids, with a first ionization constant of 6.36. The product bicarbonate ion, HCO_3^-, is a weak acid, with a pK_a of 10.3. Many minerals are carbonate or bicarbonate salts; for example, limestone is $CaCO_3$.

Carboxylic Acid Salts and Micelles

Since proton transfer between an acid and a base is normally the fastest type of chemical reaction, the treatment of a carboxylic acid with a base results in salt formation. Carboxylic acid salts are named using two words. The first word of the name is the cation (such as sodium, potassium, or ammonium) and the second word is derived from the name of the acid by replacing the *-ic acid* ending with *-ate*. For example, ammonium hydroxide reacts with benzoic acid to form ammonium benzoate, NH_4^+ $C_6H_5CO_2^-$, along with one molecule of water.

Carboxylic acid salts are examples of **amphiphiles**. These kinds of molecules have one region that is **hydrophilic** (water-loving) and one region that is **hydrophobic** (water-avoiding). An example of an amphiphile is sodium laurate, Na^+ $^-O_2C(CH_2)_{10}CH_3$. The charged end is hydrophobic, and the **lipophilic** (fat-loving) hydrocarbon chain is hydrophobic. When an amphiphile is added to water, the molecule has to solve the problem that part of it is attracted to water and the other part is repelled by water. If the concentration of the amphiphile is sufficiently high, it solves this problem by forming **micelles**. Micelles are molecular assemblies of 50 or more molecules, roughly spherical, with the polar end, the **head group**, on the outside surface and the hydrocarbon tails jumbled together on the inside.

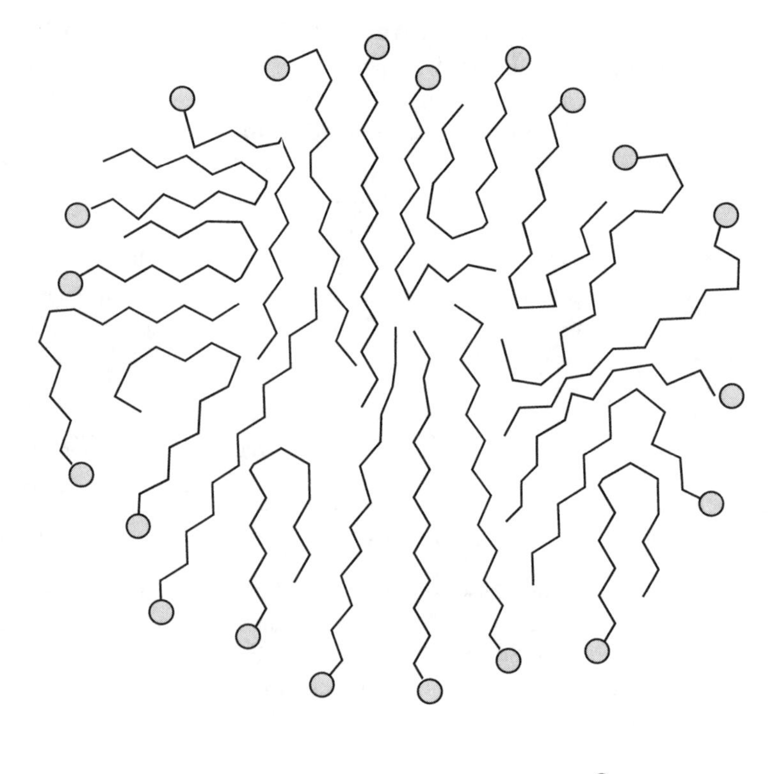

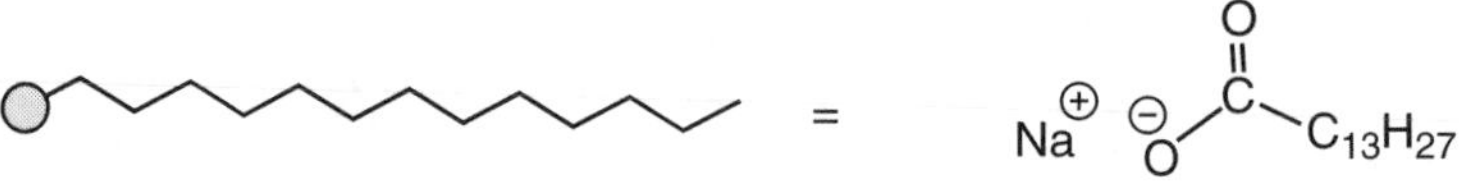

Greasy, hydrophobic molecules are solubilized in water by micelles. The hydrophobic interior of a micelle accommodates nonpolar solutes. This solubilization accounts for the cleaning action of soaps. When one does laundry, any water-soluble dirt dissolves immediately. The detergent, an amphiphile, forms micelles that encapsulate oil and grease. In the past detergents were salts of **fatty acids** (long-chain carboxylic acids). However, if any calcium or magnesium ions were present in the water, they formed insoluble salts with carboxylates, leading to soap scum. Modern detergents use sodium alkyl sulfates that are not precipitated by divalent ions.

SYNTHESIS OF CARBOXYLIC ACIDS AND ESTERS

There are three major ways to prepare the carboxylic acid functional group. Oxidation methods, as described in Chapter 16, acting on primary alcohols or aldehydes are versatile procedures. Alternatively, hydrolysis reactions of esters, amides, or nitriles, which will be covered in Chapter 21, work if the starting functional groups are available. Finally, carboxylation of RLi or RMgX reagents gives carboxylic acids with an increase in the number of carbon atoms.

Synthesis of Carboxylic Acids by Carboxylation of Grignard Reagents

The carbon of carbon dioxide is electrophilic. CO_2 reacts rapidly and in good yield with Grignard reagents to form the magnesium salt of a carboxylic acid, with creation of a new C—C bond. The addition of dilute acid forms the free carboxylic acid:

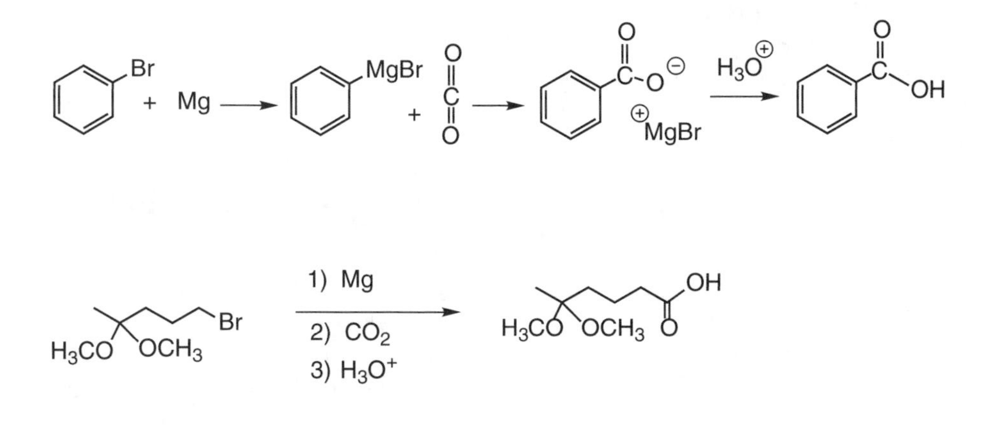

Example A typical synthesis problem requires one to start with 2-butanone and any other reagents and prepare 2,2-dimethylbutanoic acid. To solve this problem, note that one needs to add two carbon atoms and change the functional group from a ketone in the starting material to a carboxylic acid in the product. The following route is the most direct solution, and notice how choosing an alkyl iodide subgoal simplifies the problem:

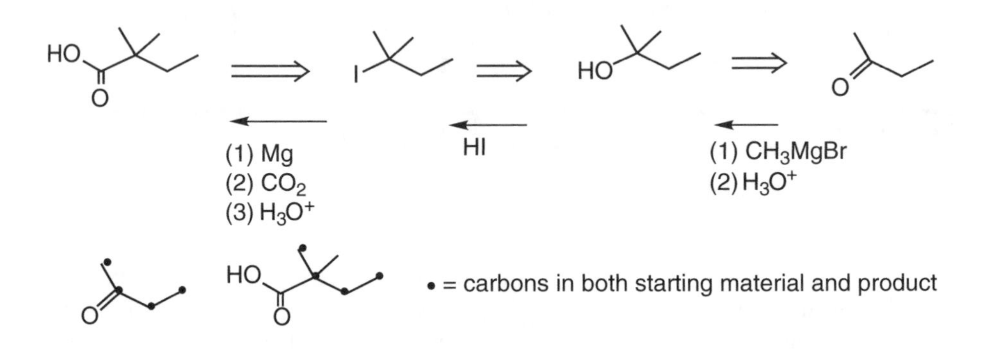

Taking an inventory of carbons present in both the starting material and the product helps to focus on the region of the molecule undergoing change.

SYNTHESIS OF CARBOXYLIC ACID ESTERS

Acid-Catalyzed Ester Synthesis

When a carboxylic acid and alcohol are mixed in the presence of an acid catalyst, an ester is formed reversibly:

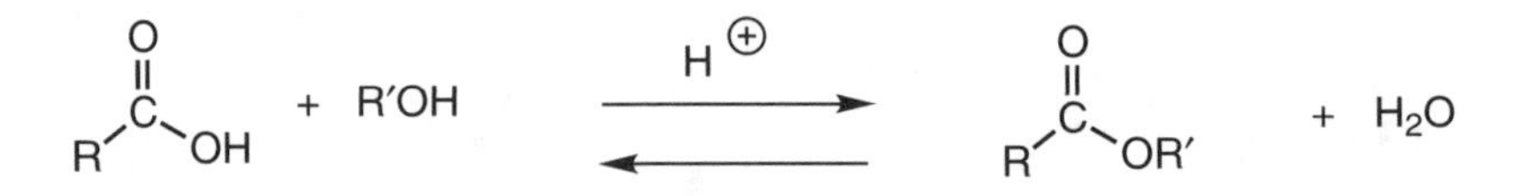

The mechanism of this organic reaction was in part established by the use of oxygen isotopes to determine which bonds were broken during the course of the reaction. When an ^{18}O-labeled alcohol, such as $CH_3{}^{18}OH$, was reacted with an acid, the ester product retained the isotope label, $RCO{-}^{18}OCH_3$, and no ^{18}O was found in the water. In addition, when an ^{18}O-labeled carboxylic acid, $RC^{18}O_2H$, was reacted with an alcohol, some of the ^{18}O appeared in the product water. Because of hydrogen atom transfer between carboxylic acids, any ^{18}O-labeled acid has the ^{18}O distributed equally between the two oxygens:

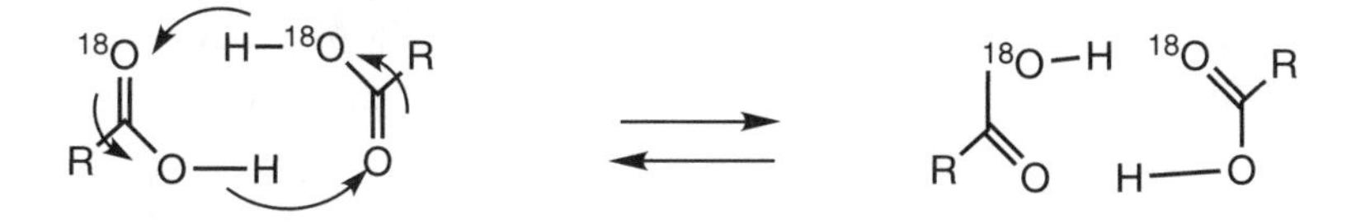

The oxygen isotope results can be explained by a mechanism that involves a tetrahedral intermediate similar to the intermediate that forms during acid-catalyzed hydrate formation between an aldehyde and water (Chapter 18). Notice that in acid-catalyzed ester formation, the acid serves first to make the acid carbonyl group more electrophilic and then to make an —OH group a leaving group:

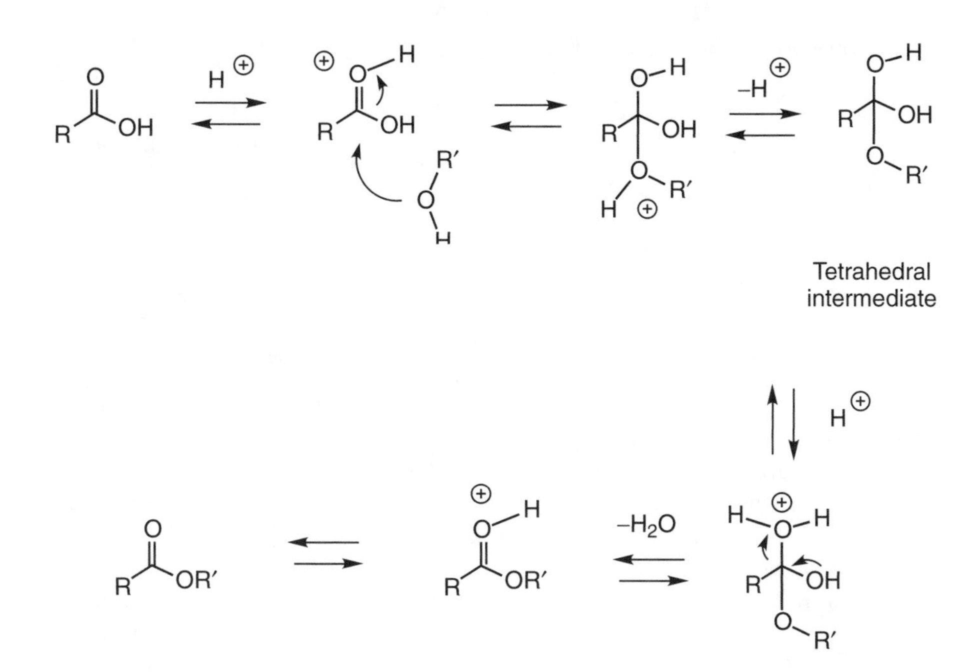

This mechanism is consistent with the ^{18}O isotope studies because the oxygen associated with the alcohol ends up in the ester and one of the oxygens of the carboxylic acid group leaves as water.

Concentration effects determine the equilibrium position of the ester formation reaction. Reacting a carboxylic acid in an alcohol solvent (especially methanol) with an acid catalyst, such as H_2SO_4, leads to ester formation in a reaction often called **Fischer esterification**. Without an excess of one of the reaction partners, one must separate the water by-product from the organic materials, normally by distilling the water away.

An acid catalyst is needed for ester formation from an alcohol and carboxylic acid. In the presence of base, a carboxylate salt forms. Since both the carboxylate and the oxygen of an alcohol are nucleophiles, no reaction takes place.

Nucleophilic Ester Synthesis

Carboxylate anions are moderate nucleophiles that react with strong electrophiles, such as CH_3I or $PhCH_2Br$, to give esters through an S_N2 reaction. This is not a base-catalyzed reaction because one full equivalent of base is needed to form the carboxylate nucleophile. Often the silver salt of the carboxylic acid (prepared from RCOOH and Ag_2O) is used since formation of the insoluble silver halide by-product drives the reaction to completion.

Diazomethane for Methyl Ester Synthesis

Diazomethane, CH_2N_2, is a thermally unstable, potentially explosive, carcinogenic compound. However, if one needs to convert a valuable acid to a methyl ester, no reagent is more likely to provide a quantitative yield. Diazomethane is protonated by a carboxylic acid to form a methyldiazonium salt which undergoes an S_N2 reaction to give the ester and nitrogen:

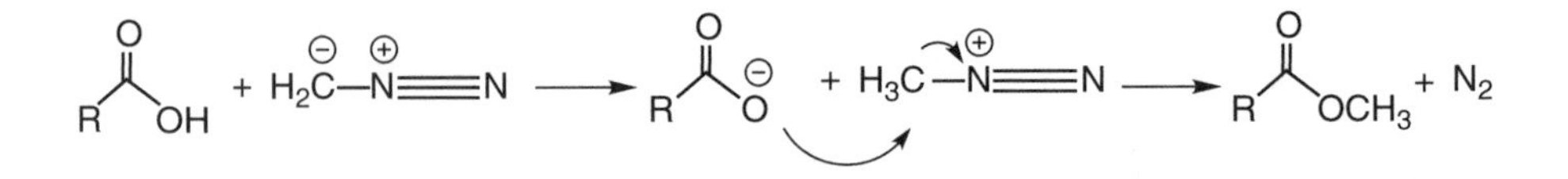

Synthesis of Lactones

Lactones are cyclic esters in which both the carboxylic acid group and the alcohol are in the same molecule. In lactone formation, one molecule, the hydroxyacid, forms two molecules, the lactone and water, so that lactone formation is preferred entropically. A strain-free five- or six-membered lactone ring is normally favored at equilibrium over the open-chain hydroxyacid:

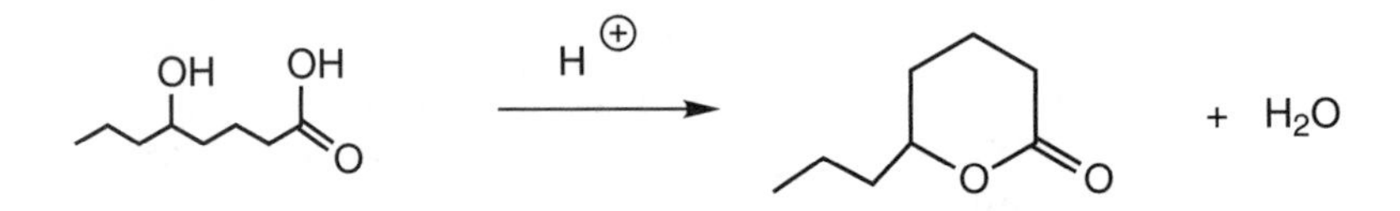

Lactones can be prepared intramolecularly in a reaction known as **halolactonization**. In this process, a halogen forms an intermediate cyclic halonium ion (such as a bromonium ion) with a double bond. Then a nearby carboxylate anion attacks the ion, forming new C—X and C—O bonds:

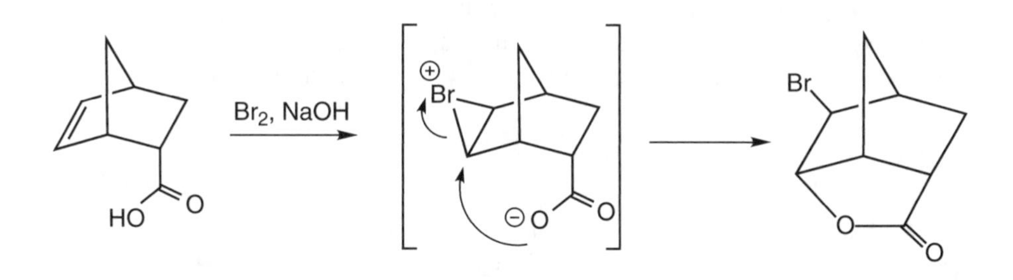

EXERCISES

1. What is the expected product from the acid-catalyzed reaction of CH_3CH_2COOH with $HOCH_2CH_2OH$?
2. Write a mechanism for the acid-catalyzed formation of 4-butanolactone from 4-hydroxybutanoic acid.

Solutions

1. Either or , depending on the concentrations of alcohol and carboxylic acid. Acids do not form acetals of the type since such a hemiacetal derivative would be unstable, as hemiacetals generally are.
2.

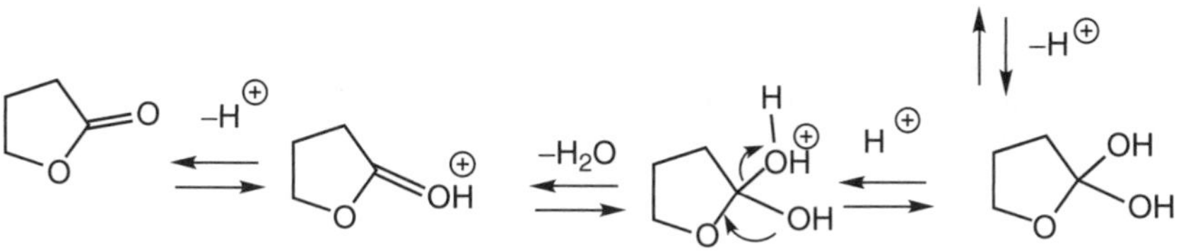

DECARBOXYLATION

Certain polyfunctional carboxylic acids can undergo a loss of CO_2, **decarboxylation**, under mild conditions. For this reaction to take place, the carboxylic acid must have a carbonyl group at the carbon beta to the carboxylic acid carbon. The reaction occurs through a cyclic transition state:

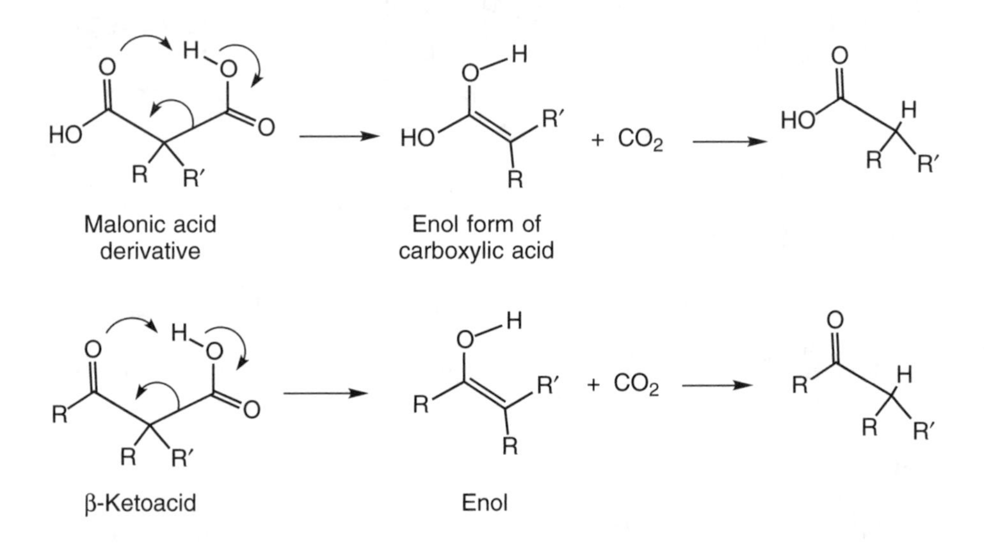

The usual substrates for decarboxylation are **β-ketoacids** or derivatives of **malonic acid**. This reaction, which is reversible in biological systems, is an important step in many biochemical reactions. Decarboxylation is the source of the CO_2 that is exhaled in every breath, and the reverse reaction, **carboxylation**, is responsible for the first step in CO_2 fixation by green plants in photosynthesis.

21
DERIVATIVES OF CARBOXYLIC ACIDS

Five functional groups are normally grouped together as derivatives of carboxylic acids: carboxylic acid chlorides, carboxylic acid anhydrides, carboxylic acid esters, carboxamides, and nitriles. All these functional groups share the property that they can be hydrolyzed to carboxylic acids. In addition, carboxylic acids are readily converted to any of these groups. The factors that cause some of these carboxylic acid derivatives to be much more reactive toward nucleophiles than others are one of the most important topics presented in this chapter.

REACTION PATTERNS OF CARBOXYLIC ACID DERIVATIVES

Each of the groups covered in this chapter that contain a carbonyl group reacts with nucleophiles to generate a tetrahedral intermediate:

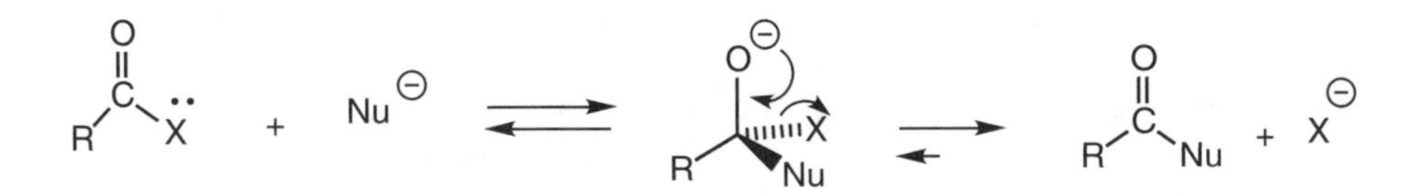

However, each functional group differs in the rate at which it reacts with a nucleophile. The approximate relative rates at which acid chlorides, anhydrides, esters, and amides react with an oxygen nucleophile at pH 7 are 10^{11}, 10^{7}, 10^{1}, and 10^{-3}, respectively. The resonance stabilization of the carbonyl group largely explains these reactivity differences.

The more a carbonyl group is stabilized by resonance, the less reactive it is:

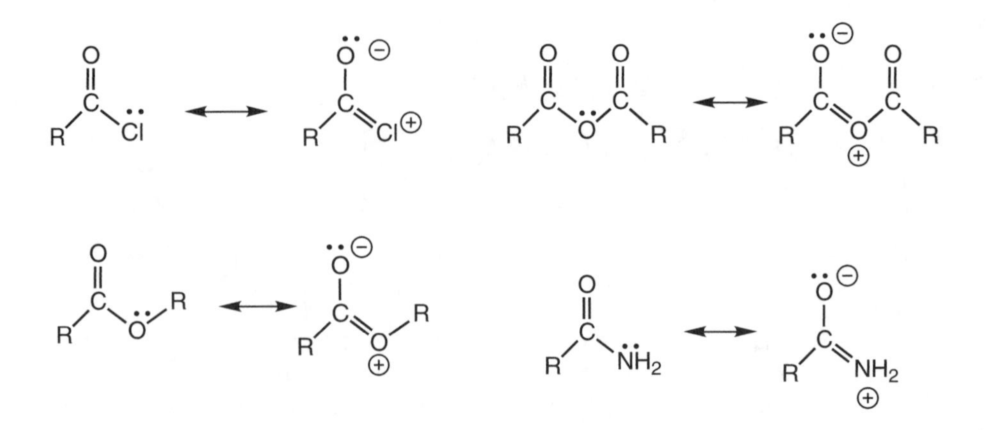

Since chlorine is a third-row element, its lone electron pair orbitals are larger than the orbitals on the second-row carbon atom. This size difference leads to poor overlap and poor resonance interactions in acid chlorides. Also, the inductive effect of chlorine promotes the addition of a nucleophile to an acid chloride's carbonyl. In the case of anhydrides, the lone pairs on oxygen can engage in resonance interactions with two separate carbonyl groups. Compared to those of esters, the lone pairs are less available for resonance delocalization. Finally, in comparing esters to amides, a nitrogen is less electronegative than an oxygen. Therefore, nitrogen's lone pair in an amide is better able to engage in resonance interactions with a carbonyl group than the oxygen lone pair in an ester.

Amides show, as clearly as anything can, the consequences of resonance. Because the two amide resonance structures contribute approximately equally to the ground state, the C—N bond in amides has properties intermediate between those of a single and a double bond. First, the C—N bond exhibits **hindered rotation**. Normal single bonds have rotation barriers of only a few kcal/mol, while rotation around a double bond requires breaking the π component of the bond (at least 55 kcal/mol). The amide C—N bond possesses a rotation barrier of approximately 20 kcal/mol, which means that rotation takes place slowly at room temperature. Second, the length of the C—N bond in amides (1.35 Å) is significantly shorter than a normal C—N single bond (1.47 Å). Third, while amine nitrogens are tetrahedral (with the lone electron pair occupying one tetrahedral position) the geometry around an amide nitrogen is planar, with all the atoms in the group C—CO—NR_2 lying in the same plane. Finally, while simple amines are weak bases, the nitrogen of an amide is not basic. If an amide is treated with a strong acid, protonation takes place on oxygen.

Within the group acid chloride, anhydride, ester, and amide, it is easy to convert a more reactive functional group to a less reactive one. Thus, acid chlorides are versatile starting materials for the synthesis of anhydrides, esters, and amides. However, amides with substantial resonance stabilization are unreactive toward most weak nucleophiles.

SYNTHESIS AND REACTIONS OF ACID CHLORIDES

Synthesis of Acid Chlorides

Thionyl chloride, $SOCl_2$, was introduced in Chapter 16 as a reagent for the conversion of alcohols to alkyl chlorides. It is also a valuable reagent for the preparation of acid chlorides from acids. Merely heating an acid, RCOOH, with $SOCl_2$ generates RCOCl, sulfur dioxide, SO_2, and HCl. Excess thionyl chloride and any dissolved SO_2 and HCl can be removed by distillation. Not all carboxylic acids are stable in the presence of HCl. In such cases, the sodium salt of the carboxylic acid can react with thionyl chloride (leading to the generation of NaCl), or the reaction can be carried out in the presence of a weak organic base, such as pyridine. Acid fluorides, RCOF, and acid bromides, RCOBr, can be synthesized, but they seldom exhibit any advantages over acid chlorides.

Reactions of Acid Chlorides With Nucleophiles

Oxygen and nitrogen nucleophiles react with acid chlorides to form other carboxylic acid derivatives:

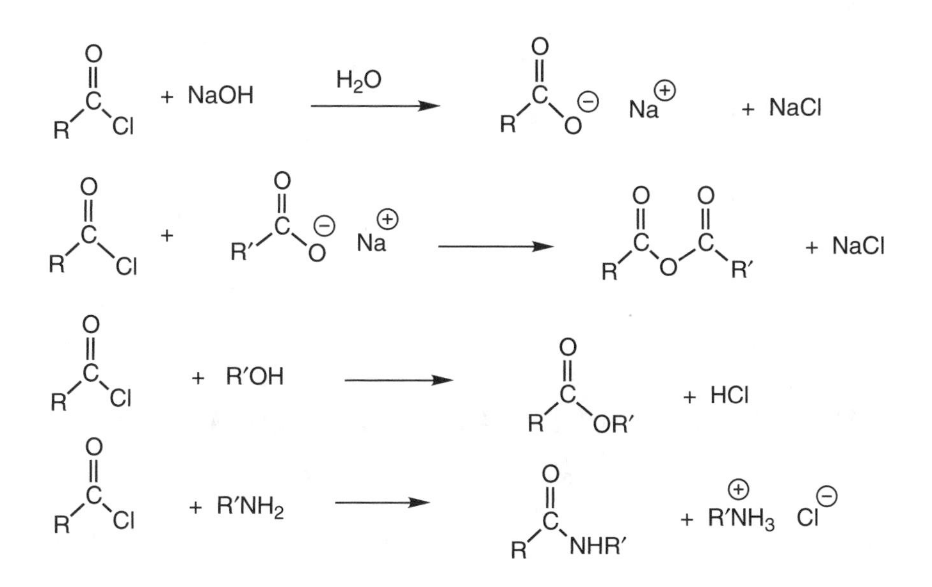

All these reactions take place rapidly in good yields in most cases. The mechanism for all of them proceeds through nucleophilic attack on the carbonyl carbon to form a tetrahedral intermediate, followed by the expulsion of Cl^- and generation of the carbonyl compound:

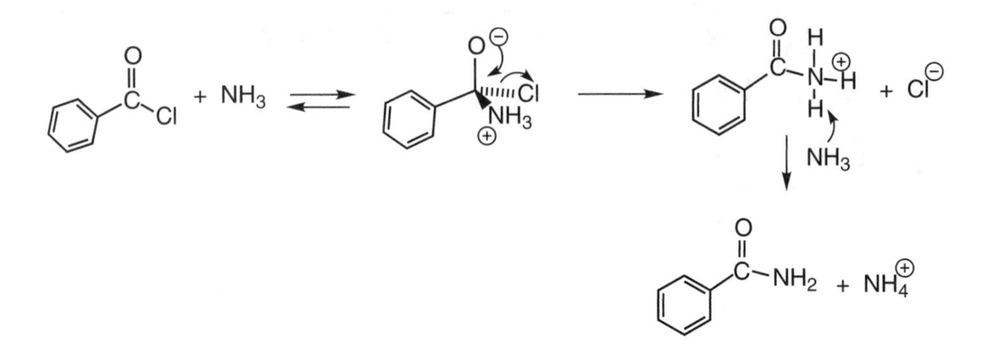

In general, the rate of nucleophilic attack on an acid chloride is much faster than the S_N2 reaction at a tetrahedral carbon. Since the carbonyl group is flat, it is sterically less crowded for a nucleophile to approach it than to approach even a primary tetrahedral carbon.

Acid chlorides react with various carbon nucleophiles to give new C—C bonds. With Grignard (or lithium) reagents, normally two equivalents of R′MgX react with one equivalent of acid chloride to give, after hydrolysis, a 3° alcohol:

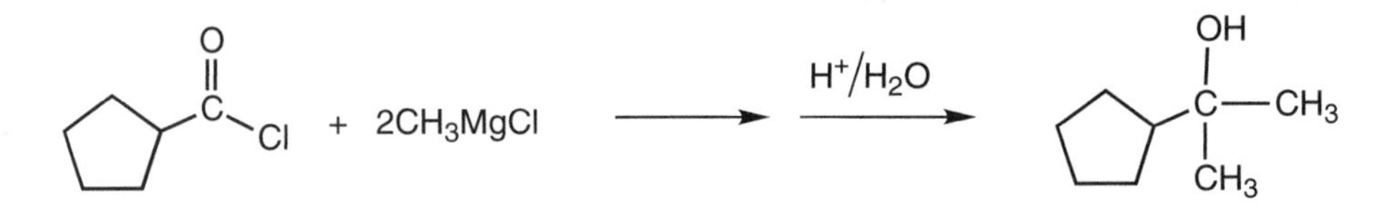

Acid chlorides are converted to ketones by reaction with cuprates:

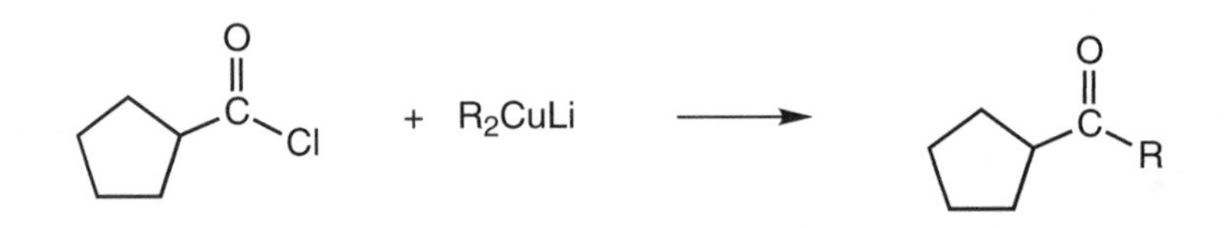

SYNTHESIS AND REACTIONS OF CARBOXYLIC ACID ANHYDRIDES

Most anhydrides are **symmetric anhydrides**, RCO—O—COR, although **unsymmetric anhydrides**, RCO—O—COR′ can be prepared by the reaction of an acid chloride, RCOCl, with an acid salt, $R'CO_2Na$. Simple anhydrides can be prepared by vigorous heating of carboxylic acids to distill out water, leaving behind the anhydride. Of course, only symmetric anhydrides can be made this way.

A commercially important anhydride is the cyclic anhydride maleic anhydride. This compound is an excellent dienophile in Diels-Alder reactions and can be used to make structurally complex anhydrides:

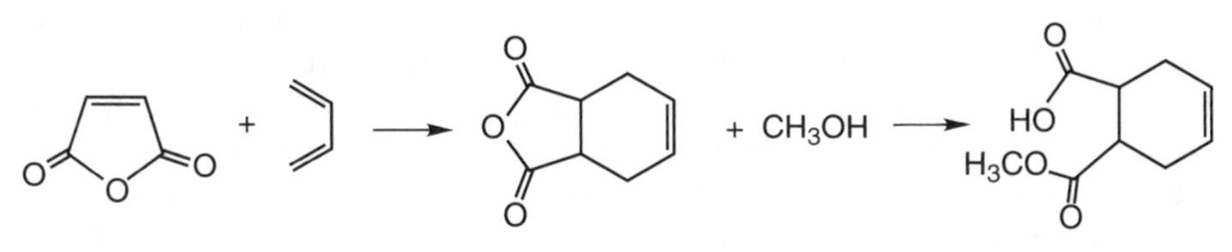

Both acid chlorides and anhydrides are examples of **acylating agents**. These compounds transfer an acyl group (RCO—) to nucleophiles, especially oxygen, nitrogen, or carbon nucleophiles. Anhydrides are often used as acylating agents for acid-sensitive molecules since the by-product of acylation is a carboxylic acid, which is less acidic than the HCl by-product formed when acid chlorides are used:

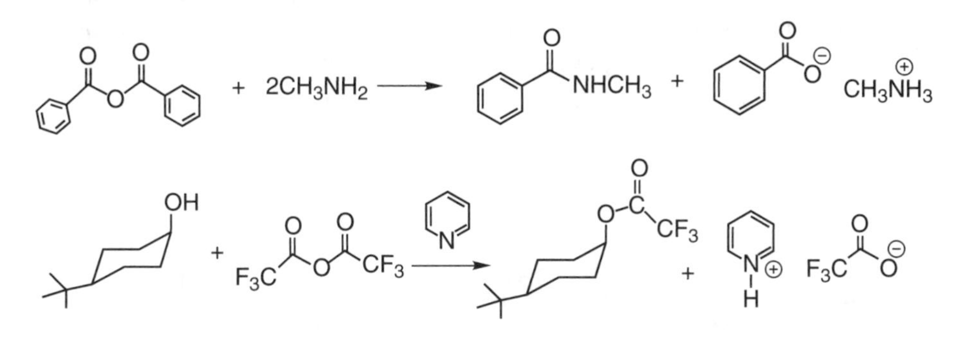

As is the case for acylations with acid chlorides, a weak base such as pyridine is added to neutralize the acid by-product.

Anhydrides with enolizable hydrogens, such as acetic anhydride, can undergo aldollike condensations with aromatic aldehydes. This reaction, **Perkin condensation**, is catalyzed by RCO_2^-:

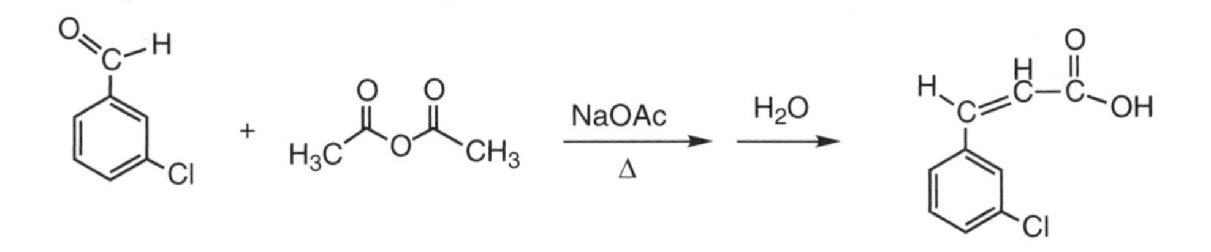

SYNTHESIS AND REACTIONS OF CARBOXYLIC ACID ESTERS

In organic chemistry, whenever an acid of any kind has an —OH group replaced by an —OR group, the compound is an ester. For example, $(HO)_2PO{-}OCH_3$ is a monoester of phosphoric acid. However, unless there is a reason to suspect otherwise, the term *ester* is assumed to refer to a carboxylic acid ester. Common methods for ester synthesis were covered in previous chapters (Fischer esterification, nucleophilic synthesis with RCO_2^-, the CH_2N_2 reaction, and the Baeyer-Villiger reaction) and in the sections above (the reaction of acid chlorides and anhydrides with alcohols). When one has a reasonably complex carboxylic acid, it is common to use the Mitsunobo reaction (Chapter 8) to prepare an ester. When the Mitsunobu conditions are used, one does not have to prepare an acid chloride or anhydride in a separate chemical step.

In biology, nature must use acylating agents that are reasonably stable in the presence of water yet react rapidly with nucleophiles at 25°C. Thioesters, RCO—SR, provide the solution to this reactivity dilemma. Thioesters are similar to acid chlorides because the sulfur atom is a large, third-row element that does not overlap well with second-row elements to produce resonance stabilization. However, because sulfur is not as electronegative as chlorine, a thioester is not as electrophilic as an acid chloride. An important, naturally occurring thioester acylating agent is **acetyl coenzyme A**:

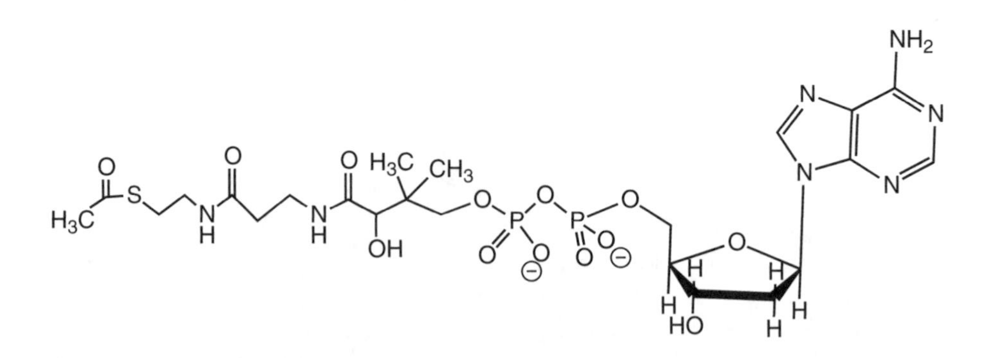

Reactions of Carboxylic Acid Esters

The acid-catalyzed synthesis of an ester and one molecule of water from a carboxylic acid and an alcohol (Chapter 20) is reversible, leading to ester hydrolysis:

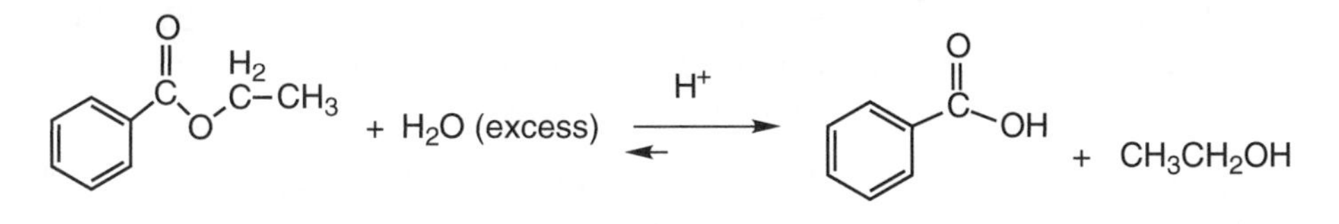

The mechanism of acid-catalyzed ester hydrolysis is the exact reverse of ester formation. An exception is found for *t*-butyl esters, which undergo hydrolysis by an S_N1 mechanism:

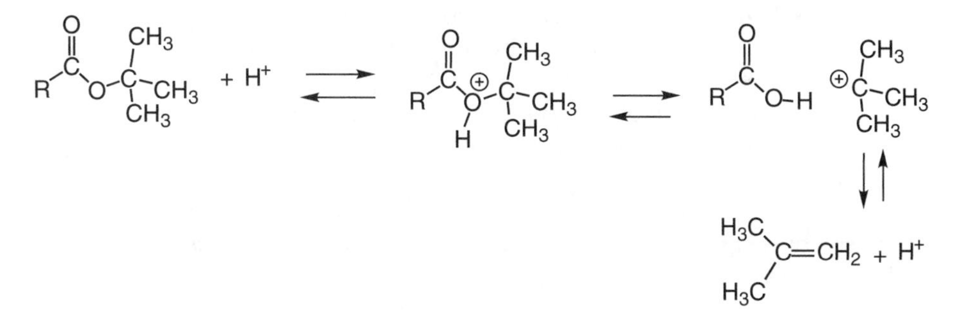

The reason for this unusual mechanism is the stability of the *t*-butyl cation.

Unlike acid-catalyzed ester hydrolysis, the **saponification reaction**, which is the hydrolysis of esters under basic conditions, is irreversible. This irreversibility is attributable to the final step of the mechanism, the reaction of the newly generated carboxylic acid with base:

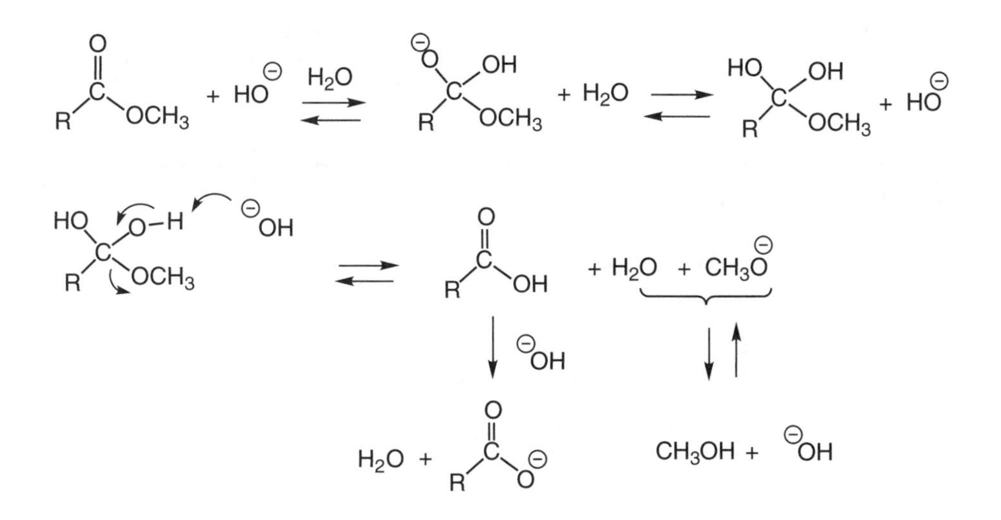

Related to the acid-catalyzed hydrolysis of esters is the **transesterification reaction**. In this reaction, an ester and an alcohol are heated with an acid catalyst and an exchange of alcohol groups takes place:

$$R\overset{O}{\overset{\|}{C}}OCH_3 + HO\text{~~~~} \underset{}{\overset{H^+}{\rightleftharpoons}} R\overset{O}{\overset{\|}{C}}OCH_2CH_2CH_2CH_2CH_3 + CH_3OH$$

The mechanism for this reaction is the same as for acid-catalyzed ester hydrolysis, with ROH replacing HOH. Transesterification is an excellent way to make high-boiling esters by distilling away low-boiling alcohol products such as methanol and ethanol.

From our discussion of carbonyl group stability, one would expect that heating an ester with ammonia or a primary or secondary amine would give an amide (tertiary amines, R_3N, have no NH proton to be replaced by an acyl group). While heating an ester with an excess of an amine can give an amide, the yields of this reaction are often very low and the reaction is slow. Nevertheless, it does take place.

$$\overset{O}{\|}\text{—}OCH_3 + \text{(Ph)~~}NH_2 \xrightarrow{\Delta} \overset{O}{\|}\text{—}\underset{H}{N}\text{~~(Ph)} + CH_3OH$$

Esters are unreactive toward mild nucleophiles, such as $NaBH_4$. However, the more reactive hydride source, $LiAlH_4$, reduces esters down to primary alcohols, $RCO_2R' \rightarrow RCH_2OH$. The highly reactive carbon nucleophiles RLi and RMgX react with esters to give tertiary alcohols, as discussed in Chapter 16.

EXERCISES

1. Write a mechanism for the acid-catalyzed transesterification of ethyl benzoate with 1-butanol.

2. Give a resonance-based explanation of why the α positions of thioesters are more acidic than the α positions of esters.

3. Fill in each of the blanks with the appropriate reagent or product.

(a) 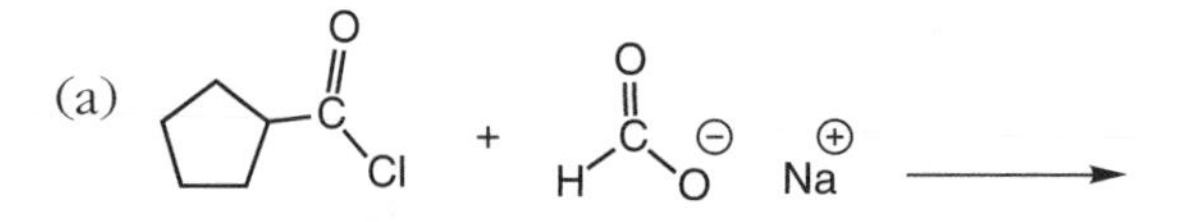

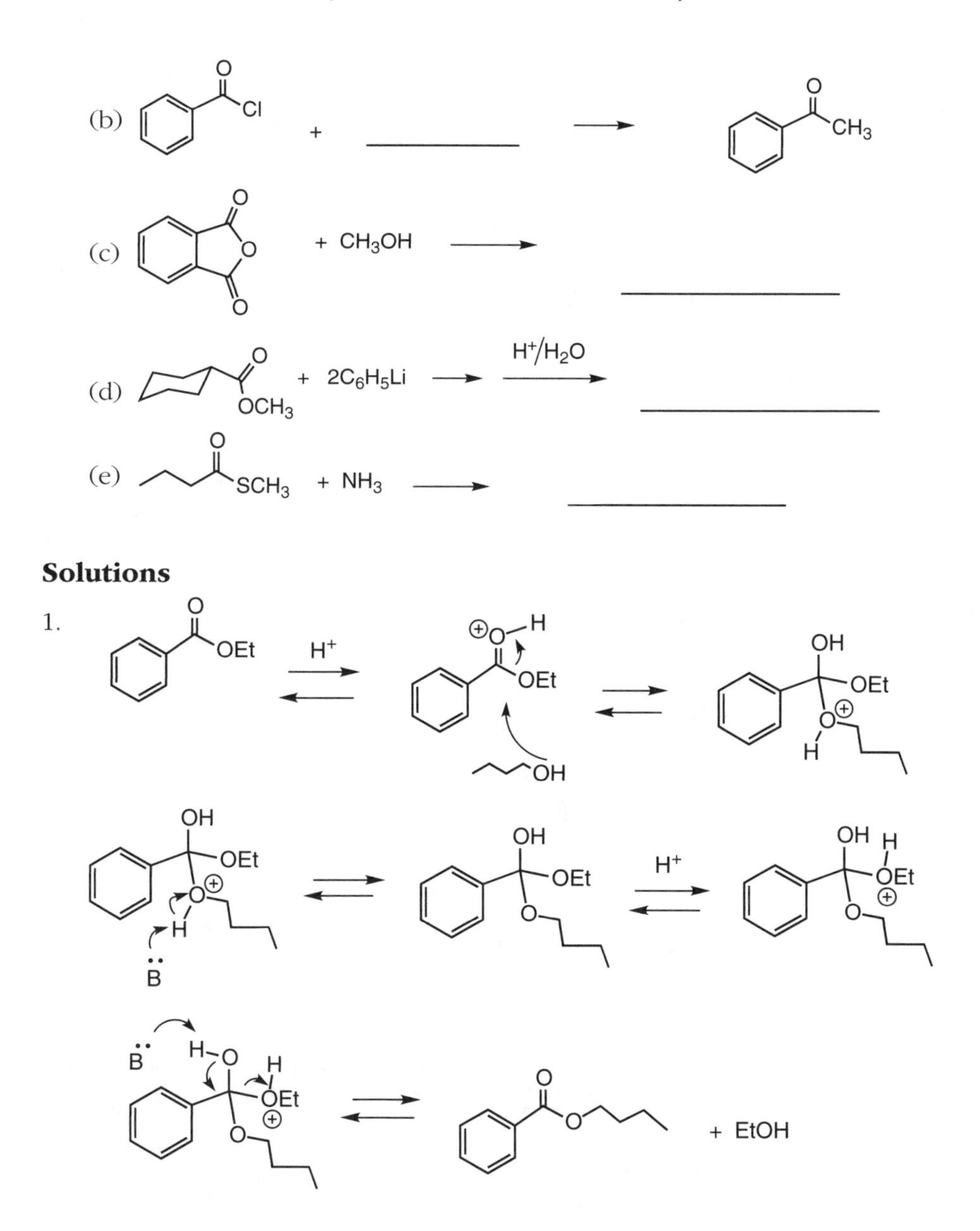

2. The oxygen lone electon pairs in an ester engage in resonance with the carbonyl group. In an ester enolate, the electron pair of the carbanion must compete with the oxygen lone pairs for delocalization into the carbonyl group. In the case of the enolate of a thioester, the lone pairs on sulfur are not involved with resonance interactions with the carbonyl group. As a result, the carbanion lone pair has no competition for delocalization.

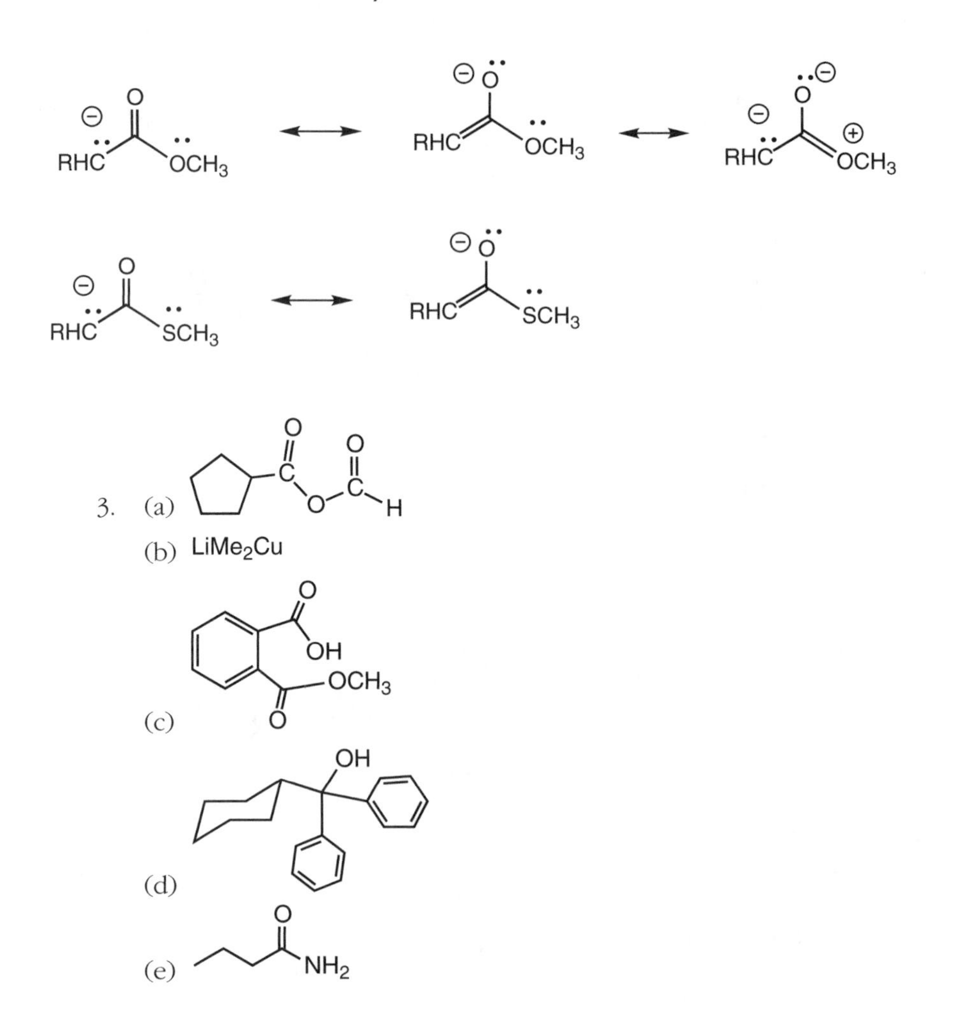

SYNTHESIS AND REACTION OF AMIDES

Synthesis of Amides

Each type of amide—primary, $RCONH_2$; secondary, RCONHR′; and tertiary, $RCONR_2'$—can be made by reaction of an amine with an acid chloride or anhydride. However, some amides are sensitive to even traces of acid and need to be prepared under neutral conditions. For these reactions, **coupling agents** have been invented that allow an amine and a carboxylic acid to react directly, generating an amide. A common coupling agent is dicyclohexylcarbodiimide (**DCC**), which can be used to make both normal acyclic amides and cyclic amides, which are called **lactams**. Even strained, four-membered ring lactams (known as β lactams) can be made with DCC:

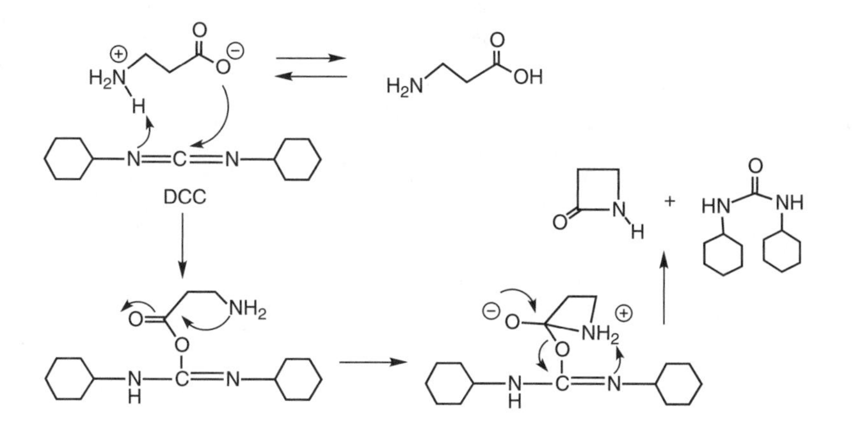

The intermediate in this reaction is chemically similar to an anhydride in that it has a O=C—O—C=N unit instead of the O=C—O—C=O group seen in anhydrides. Thus, the function of DCC is to generate an acylating agent directly from the carboxylic acid, which reacts with the nitrogen nucleophile to form the amide bond. The by-product is dicyclohexylurea (an adduct of water and DCC) whose insolubility helps to drive the reaction to completion.

When an amine reacts with an acid chloride, one equivalent of HCl is released. Since an amine is a base, an additional equivalent of amine is needed to neutralize the HCl. However, if the amine being used is expensive or rare, another base can be added to neutralize the HCl. A common practice is to react an aqueous solution of an amine and dilute NaOH with an acid chloride. Even though in such a reaction there is a large molar excess of water, an amine is a better nucleophile than water. Thus, the faster reaction, amide formation, wins out. Recall that in our discussion of nucleophilic substitution reactions we pointed out that protic solvents lower nucleophilicity through hydrogen bonding. Both water molecules and HO^- are more extensively hydrogen-bonded in water than amines, leading to the observed reactivity.

Reactions of Amides

The most important reaction of amides is hydrolysis, which can be catalyzed by an acid or a base. However, because amides are so stable, the hydrolysis must be carried out at an elevated temperature. Since skin, hair, and nails contain amide bonds, it is a good thing that hydrolysis is so slow; otherwise swimming and bathing would be impossible.

Basic hydrolysis of amides is initiated by a nucleophilic attack of HO^- on the amide carbonyl, followed by collapse of the tetrahedral intermediate and reaction of the product carboxylic acid with a base to form a salt. Thus, basic hydrolysis is irreversible:

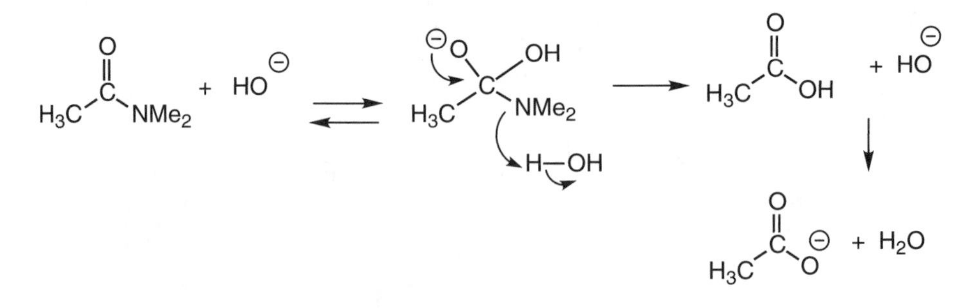

Acid hydrolysis is also irreversible because the product amine is irreversibly protonated under the reaction conditions:

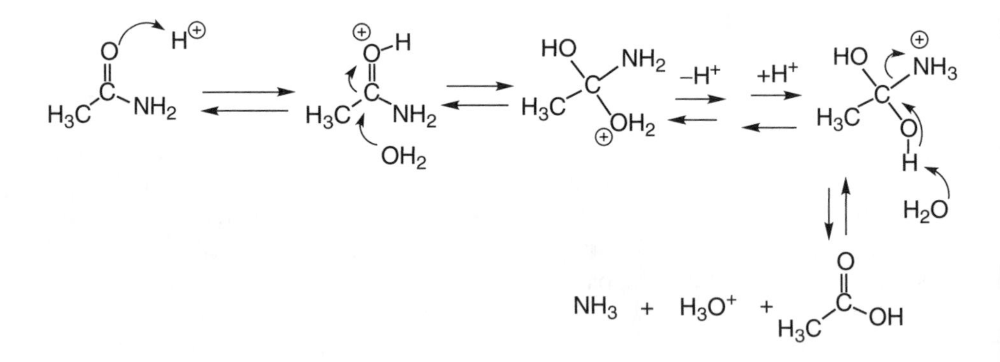

In general, acid hydrolysis of amides is a faster reaction than basic hydrolysis. Protonation of the amide activates it to nucleophilic attack, and the acid medium protonates the nitrogen in the tetrahedral intermediate, making it a better leaving group.

An amide N—H is about as acidic as an alcohol O—H, so amides can be deprotonated by strong bases. The resulting nucleophile can react with an alkyl halide or tosylate in a normal S_N2 reaction, giving rise to 2° or 3° amides. For example, the reaction of acetamide with NaH, followed by treatment with benzyl bromide, yields *N*-benzylacetamine, $CH_3CONH(CH_2C_6H_5)$.

Amides undergo an unexpected reaction on reduction with $LiAlH_4$. The product is the amine, that is, $RCO—NR_2' \rightarrow RCH_2—NR_2'$. The mechanism for this reduction is

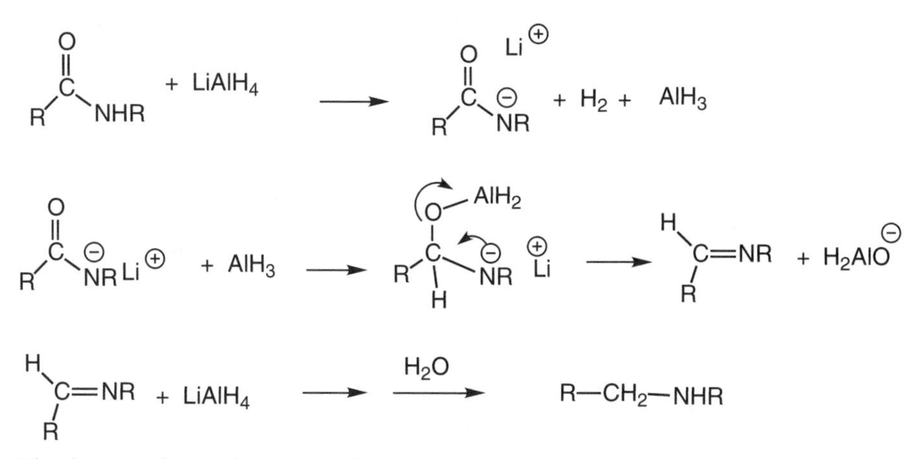

The key to this reduction is the strength of the aluminum-oxygen bond. It is eliminated from the intermediate reduction product to form an imine, which is rapidly reduced by a second equivalent of $LiAlH_4$, leading, after hydrolysis, to the amine product. The first step in the reaction is an acid-base reaction. It is not required, however, because tertiary amides are reduced to amines also.

Amides are useful acylating agents for RLi and Grignard reagents as long as the reaction is carried out at low temperatures (usually –78°C, dry ice temperature). The initially formed tetrahedral intermediate is stable at low temperatures. It is protonated to give a carbinolamine (also called a **hemiaminal**), which, as described in Chapter 18, decomposes to a carbonyl compound and an amine:

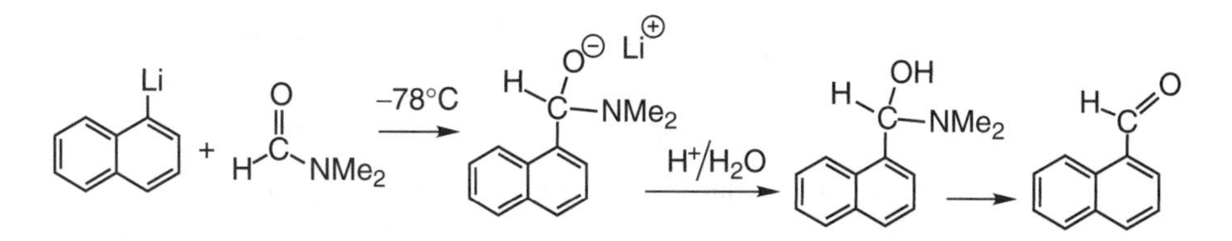

SYNTHESIS AND REACTIONS OF NITRILES

Since nitriles are hydrolyzed to carboxylic acids, they are normally considered carboxylic acid derivatives. Most nitriles are prepared through S_N2 reactions because cyanide anion is an excellent nucleophile. However, nitriles can be synthesized, as well, by the dehydration of primary amides. Many dehydrating agents, such as P_2O_5, carry out this reaction. For example, heating benzamide, $C_6H_5CONH_2$, with P_2O_5 gives benzonitrile, C_6H_5CN.

Nitriles can be hydrolyzed under acid or basic conditions. In both instances, an amide is formed as an intermediate. Usually, the hydrolysis can be stopped at the amide stage or continued on to the carboxylic acid:

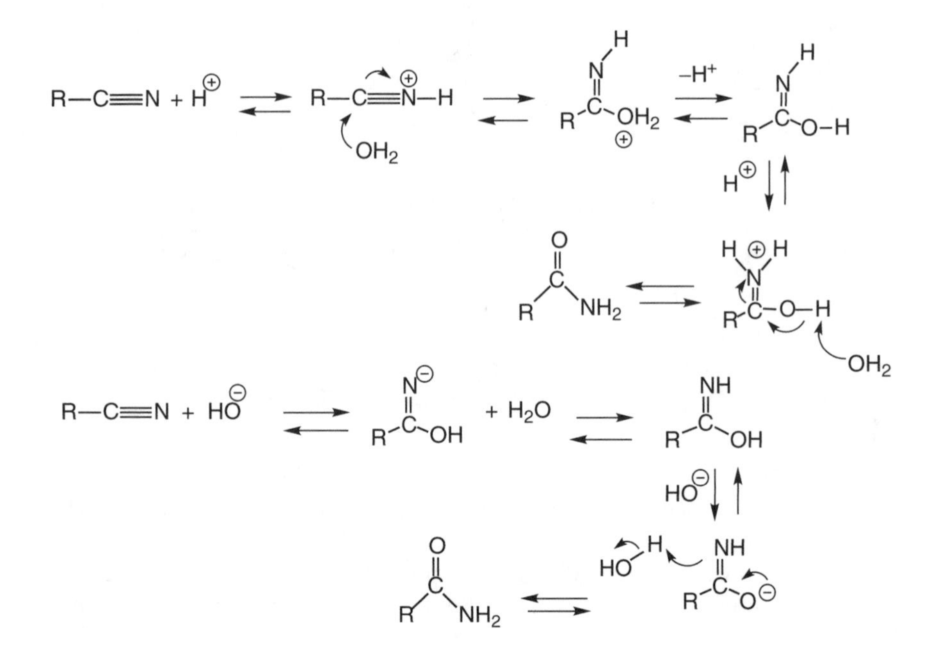

Nitriles react with $LiAlH_4$, but not with $NaBH_4$, to give, after hydrolysis, primary amines: $RCN \rightarrow RCH_2NH_2$. Nitriles also react with RLi or RMgX to give, after hydrolysis, ketones:

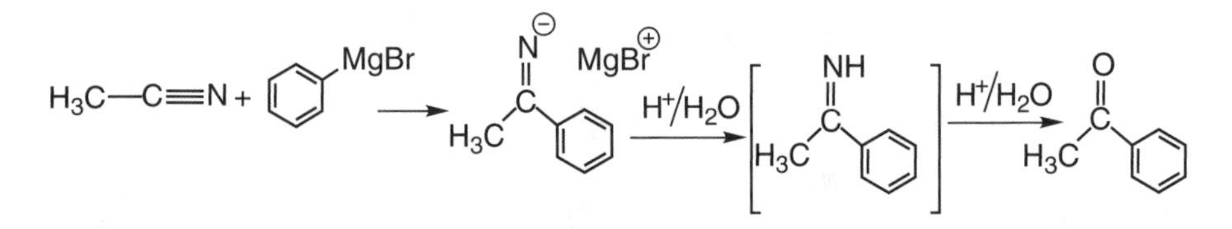

Since the initial addition product is an anion, it does not react further with a second anion. Protonation gives an imine, which undergoes rapid hydrolysis to a ketone, as described in Chapter 18.

EXERCISES

1. Starting with $PhCH_2CH_2Br$, show how you would synthesize
   (a) $PhCH_2CH_2CHO$ (b) $PhCH_2CH_2CH_2NH_2$ (3) $PhCH_2CH_2COCH_3$

2. When an amide, $RCONH_2$, is heated in acidic isotopically labeled water, $H_2{}^{18}O$, some of the ^{18}O is becomes incorporated into the amide, $RC^{18}ONH_2$. What is the mechanism for this isotopic enrichment?

Solutions

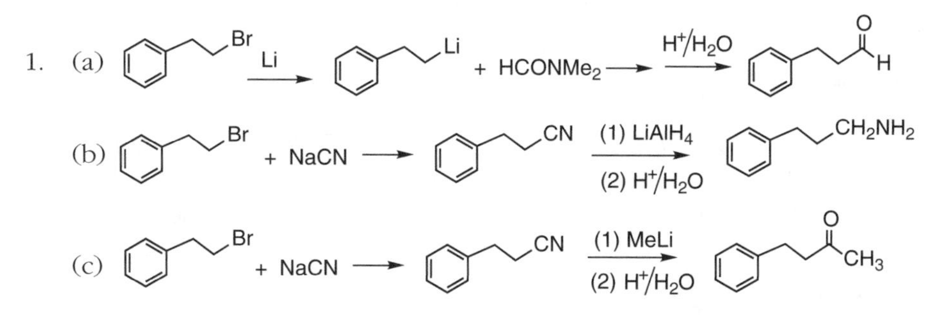

2. In the mechanism for amide hydrolysis, water adds reversibly to the protonated amide intermediate to generate a tetrahedral intermediate:

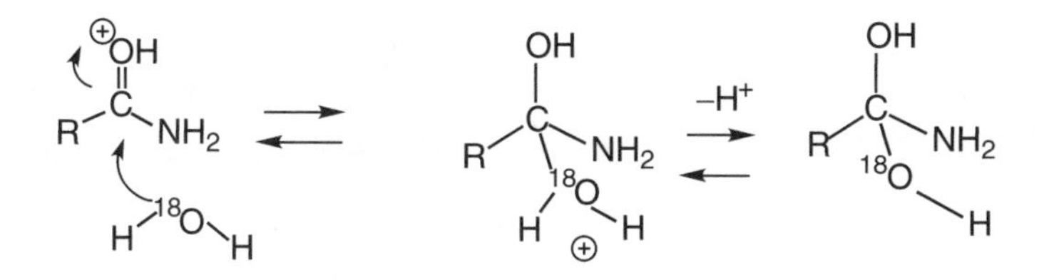

When this intermediate is deprotonated, the two —OH groups are equivalent. To the extent that the tetrahedral intermediate reverts to the starting amide, there is a 50% chance that the amide will contain ^{18}O.

POLYAMIDES AND POLYESTERS

Polymers are high-molecular-weight molecules made up of small molecular units called **monomers**. Hundreds, thousands, or even more monomer units can be linked together to form a polymer. Much of the chemical industry is based on the design and synthesis of synthetic polymers with desirable properties. Initially, synthetic polymers were created to replace natural polymeric materials such as silk, wood, cotton, wool, and rubber. Today, new materials are designed to have specific properties that exceed anything available from biological sources.

Some polymers are made up of repeating units of a single monomer. These **homopolymers** include polyethylene (whose **repeating unit** $—CH_2CH_2—$ arises from polymerizing ethylene) and poly(vinyl chloride) which is made from polymerizing vinyl chloride:

Monomer or Monomers	Polymer Repeating Unit
$H_2C{=}CH_2$ Ethylene	$-CH_2CH_2-$ Polyethylene
$H_2C{=}CH-Cl$ Vinyl chloride	$-CH_2CH(Cl)-$ Poly(vinyl chloride)
$H_2C{=}CH-C_6H_5$ Styrene	$-CH_2CH(C_6H_5)-$ Polystyrene
$H_2C{=}CH-C(=O)OCH_3$ Vinyl acetate	$-CH_2CH(C(=O)OCH_3)-$ Poly(vinyl acetate)
$^{\oplus}NH_3(CH_2)_5COO^{\ominus}$ 6-Aminohexanoic acid	$[-HN-(CH_2)_5-C(=O)-]$ NH $(CH_2)_5$ Nylon 6
$C_6H_4(CO_2Me)_2$ + $HOCH_2CH_2OH$ Dimethyl terphthalate, 1,2-Ethanediol	$[-OCH_2CH_2O-C(=O)-C_6H_4-C(=O)-]$ OCH_2CH_2O- Poly(ethyleneterphthalate) (Dacron)

Other polymers are made up of two or more different monomers known as **copolymers**, such as the **polyester** Dacron.

Polyesters are usually made through a transesterification reaction in which a volatile alcohol such as methanol is distilled away to drive the synthesis of a polymer. **Polyamides**, such as nylon 6, can be made by heating the starting monomer, distilling away a molecule of water for each amide bond formed. Some polyamides are copolymers. Nylon 66, for example, is made by reacting adipic acid, $HO_2C(CH_2)_4CO_2H$, with 1,6-hexanediamine, $H_2N(CH_2)_6NH_2$.

EXERCISES

Indicate what monomer or monomers were polymerized to generate the following polymers.

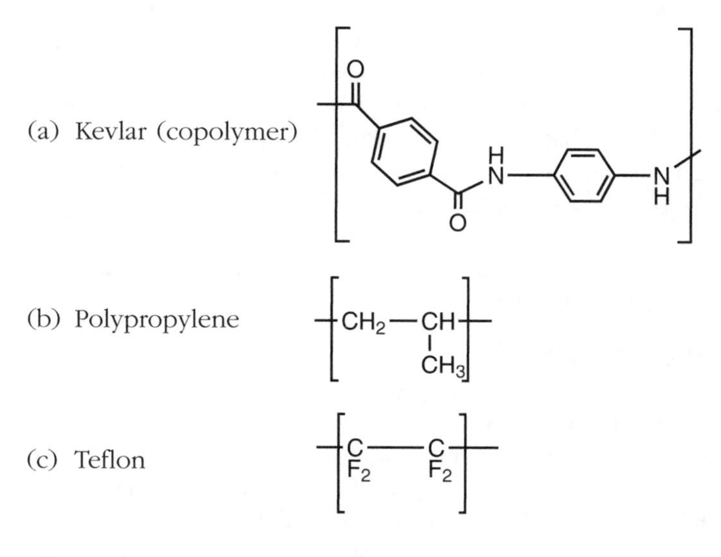

Solutions

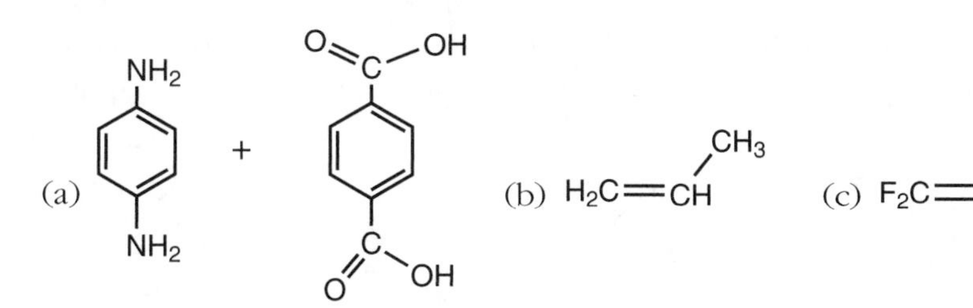

22
REACTIONS OF ESTER ENOLATES

Although esters are not as acidic as aldehydes or ketones, one can use common bases, such as $NaOCH_3$, to generate usable amounts of ester enolates. In addition, there are several commonly available ester derivatives that are quite acidic, so enolate derivatives can be generated using NaOH. Because ester enolates can be used to make desirable classes of compounds such as ketones and acids, they are important synthetic intermediates. Furthermore, a study of biochemistry finds that many important classes of compounds are made in living systems using ester enolates and their derivatives.

REACTIONS OF ACIDIC ESTER DERIVATIVES

The Malonic Ester Synthesis

Diethyl malonate, $EtO_2CCH_2CO_2Et$, contains a $—CH_2$ group activated by two electron-withdrawing groups. Consequently, its pK_a is 13. The resulting enolate is a reasonably reactive nucleophile that reacts with allylic, benzylic, and 1° alkyl halides. The resulting $EtO_2CCHRCO_2Et$ diester can be deprotonated in a subsequent step, and the enolate can react with a different alkyl halide, R′X, giving $EtO_2CCRR'CO_2Et$. Heating a malonic ester in acidic water hydrolyzes the ester groups to carboxylic acids. As discussed in Chapter 20, β diacids undergo decarboxylation, producing acetic acid derivatives:

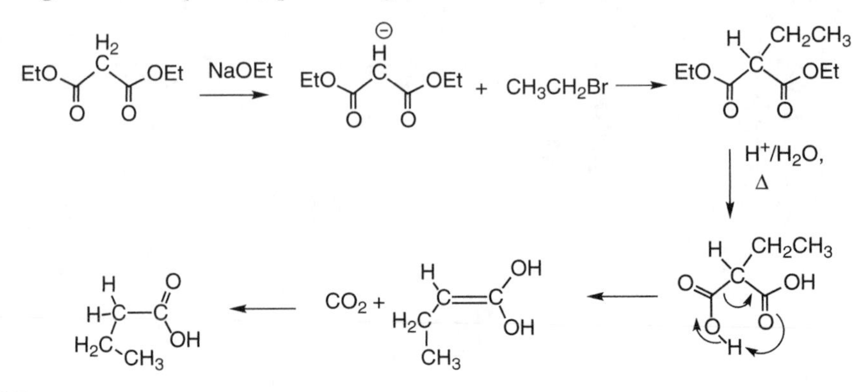

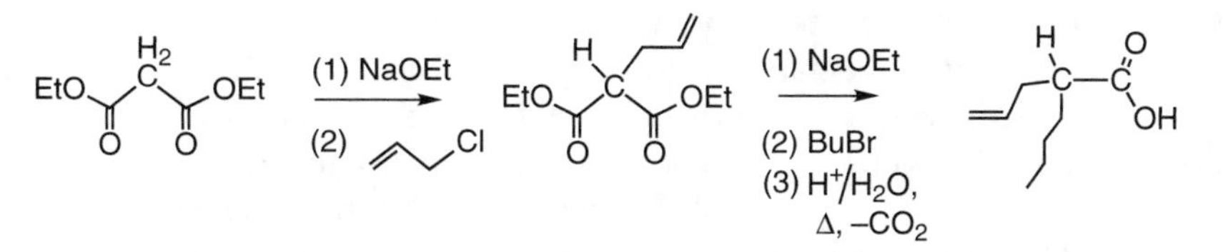

The **malonic ester synthesis** is an excellent way to prepare carboxylic acids of the type RCH_2CO_2H or $RR'CHCO_2H$.

EXERCISES

1. Indicate which acids could be made using the malonic ester synthesis and which cannot. For those that can be made this way, what alkyl halides would one use?

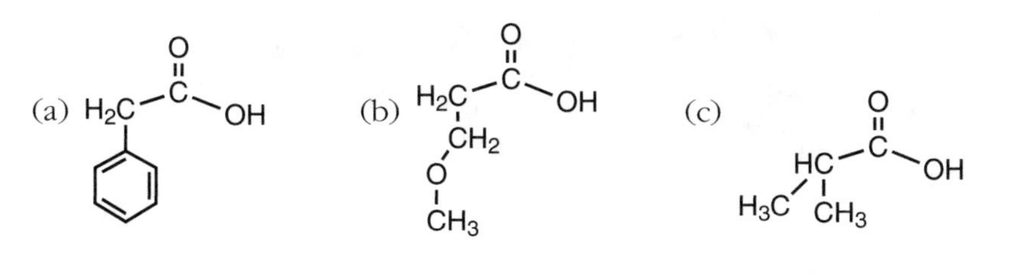

2. How would you use malonic ester syntheis to synthesize ?

Solutions

1. (a) No. PhX do not undergo nucleophilic substitution reactions.
   (b) Yes, using CH_3OCH_2Cl.
   (c) Yes, using CH_3I, followed by a second deprotonation and the addition of more CH_3I.
2. One must use Br—$CH_2CH_2CH_2CH_2$—Br and not cyclopentylbromide.

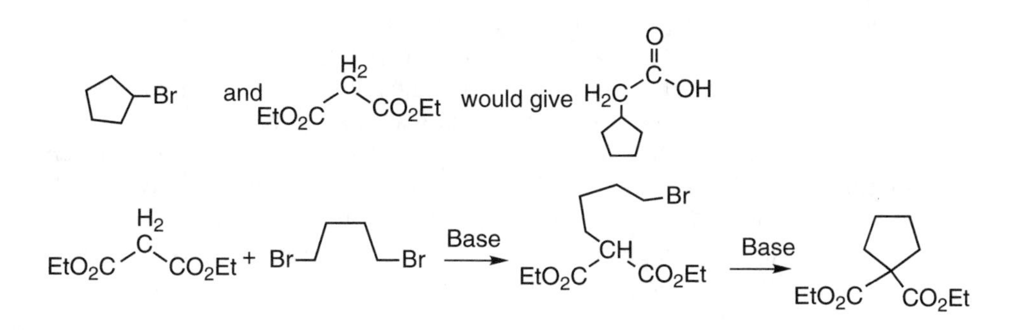

The Acetoacetic Ester Synthesis

A reaction that is mechanistically very similar to the malonic ester synthesis is the **acetoacetic ester synthesis**. Ethyl acetoacetate, $CH_3COCH_2CO_2Et$, also has a doubly activated, acidic $—CH_2$ group. It can undergo one or two alkylation reactions to give β-ketoesters of the type $CH_3COCHRCO_2Et$ or $CH_3COCRR'CO_2Et$, which after hydrolysis of the ester group and decarboxylation of the β-ketoacid gives methyl ketones of the type CH_3COCH_2R or $CH_3COCHRR'$:

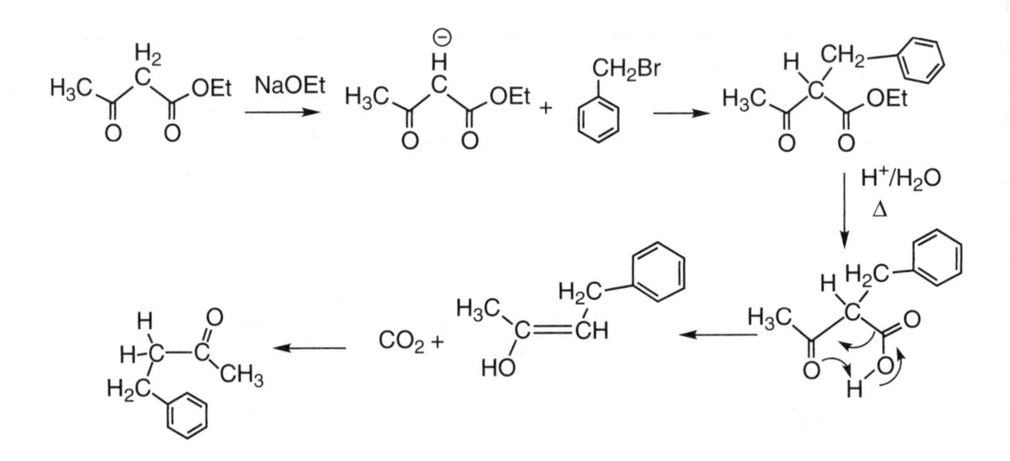

Either CH_3COCH_2R or $CH_3COCHRR'$ methyl ketones are readily prepared in this way. Note that if one tries an alkylation of acetone to prepare $CH_3COCHRR'$, starting with CH_3COCH_2R, one can expect to obtain a mixture of $CH_2R'COCH_2R$ and $CH_3COCHRR'$.

REACTIONS OF SIMPLE ESTER ENOLATES

The Claisen Condensation

When an ester is treated with an alkoxide base, a low concentration of an ester enolate is formed. Normally, an ethyl ester uses NaOEt as a base, a methyl ester uses NaOMe, and so on, so that no interchange of OR groups takes place. The nucleophilic ester enolate reacts with a second, neutral ester molecule in a reaction analogous to the aldol condensation called the **Claisen condensation**. The resulting tetrahedral intermediate ejects RO^-, giving a β-ketoester product. Since the product is much more acidic than the starting ester, the RO^- product deprotonates the β-ketoester. Once the reaction is over, a separate protonation step is needed to obtain the neutral product:

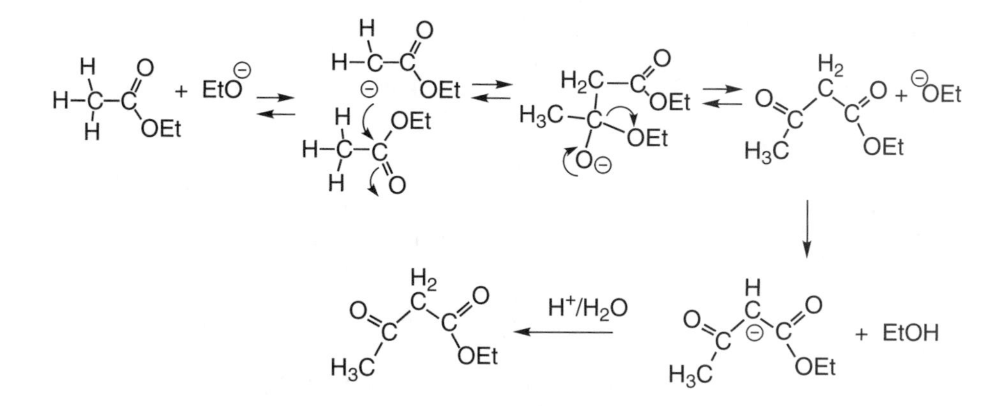

Unlike the aldol condensation, which uses a catalytic amount of base, the Claisen condensation requires one equivalent of base per equivalent of product formed.

All the steps of the Claisen condensation are reversible except for the last deprotonation step of the acidic β-ketoester. Thus, if a nonenolizable β-ketoester is treated with RO^-, a reaction known as the **retro-Claisen condensation** takes place. This reaction results in a breaking apart of the β-ketoester product:

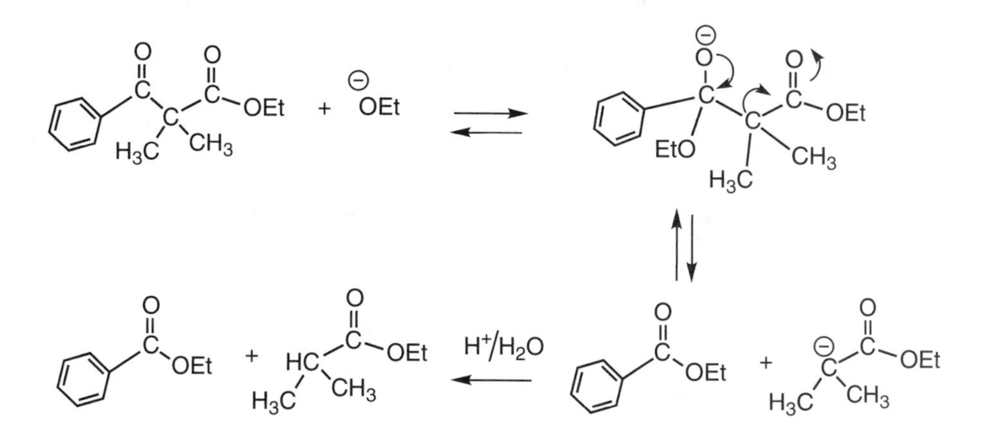

Mixed Claisen Reactions

Just as in the aldol reaction, two different esters can undergo a reaction in a **mixed Claisen condensation**. This reaction is useful only if one of the esters has no enolizable protons so that it cannot form an enolate. Esters of aromatic acids, formic acid, or dialkyl carbonates, RO—CO—OR, are examples of ester derivatives that are nonenolizable:

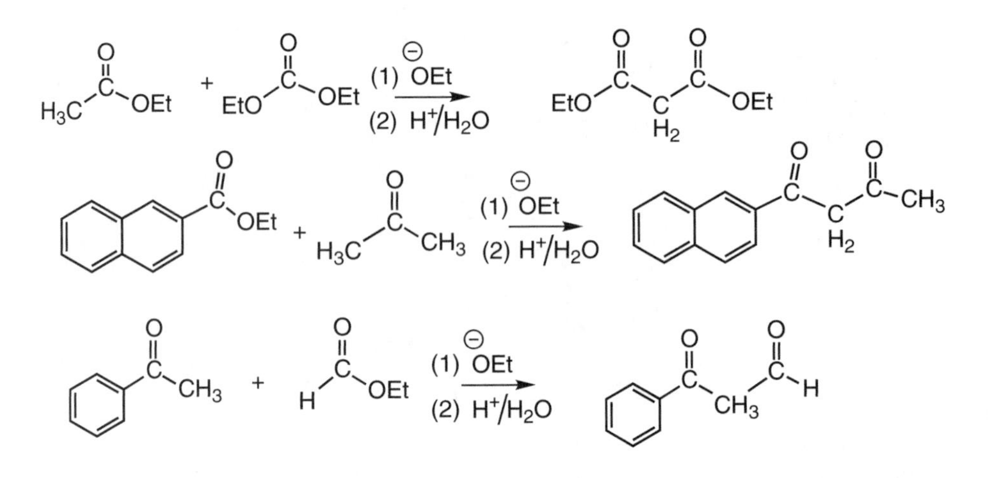

Notice that in the above examples ketone enolates as well as ester enolates can react in mixed Claisen condensations. To summarize, a symmetric Claisen reaction gives a β-ketoester. The reaction of an ester enolate and a dialkyl carbonate gives a 1,3 diester. The reaction of a ketone enolate with a nonenolizable ester gives a β diketone or a β ketoaldehyde.

While Claisen condensations were traditionally carried out in alcohols with alkoxide bases, it is also possible to generate ester enolates with LDA, just as LDA generates ketone enolates. Such enolates need to be prepared at low temperatures to prevent side reactions. Various carbonyl compounds, including ketones and aldehydes, can be added to these preformed ester enolates to give mixed Claisen condensation products:

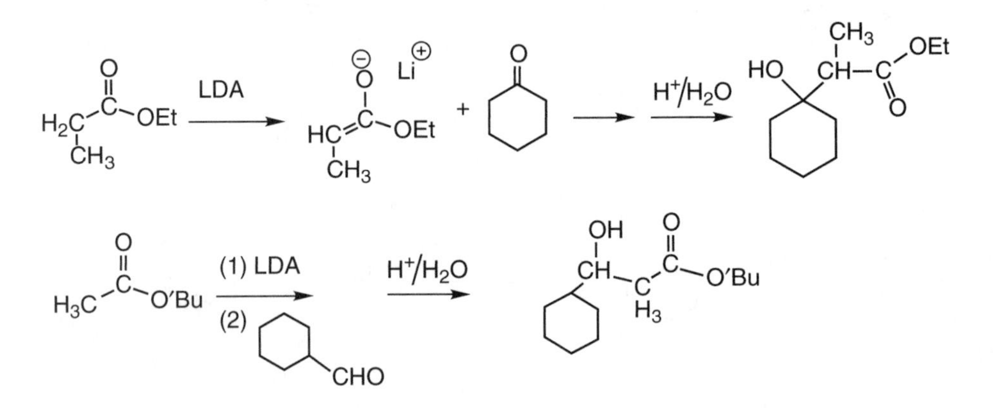

The Dieckmann Condensation

When an intramolecular Claisen condensation takes place to generate a new ring, the reaction is called the **Dieckmann condensation**. Normally five- and six-membered rings can be formed in this way from 1,6 diesters and 1,7 diesters, respectively. The mechanism is the same as in intermolecular Claisen condensation:

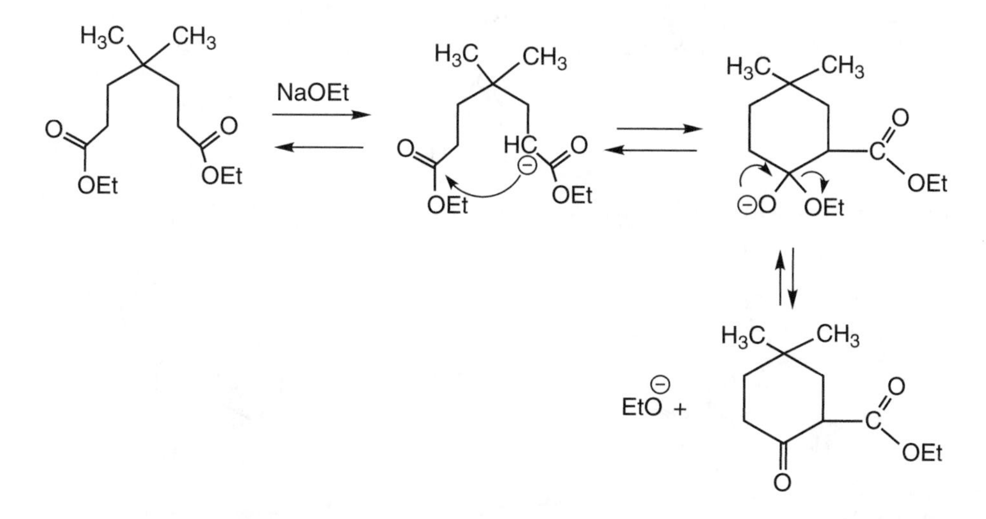

Other ring sizes, such as four- or seven-membered rings do not usually form in a Dieckmann condensation.

EXERCISES

1. Give the expected product from each of the following reactions.

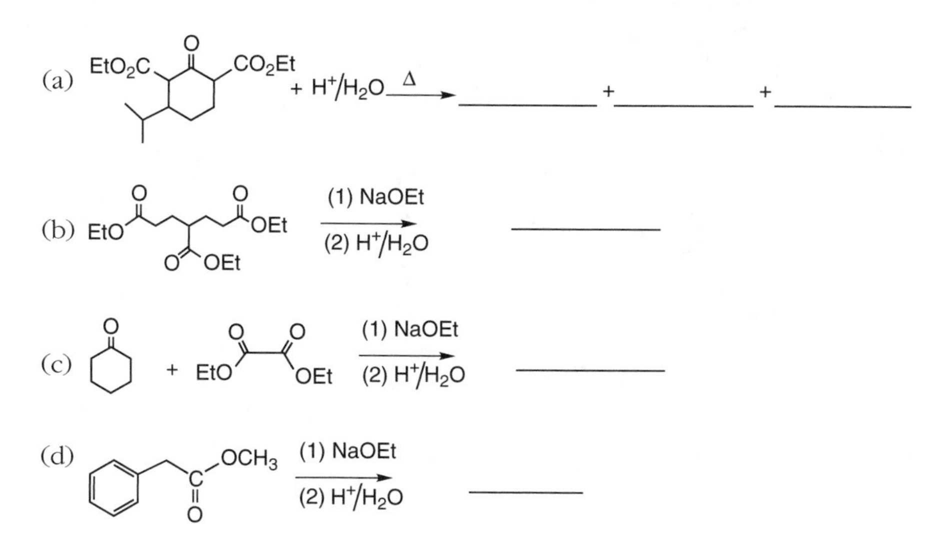

2. Propose a mechanism for the following Claisen condensation.

3. The following reaction takes place. Suggest a mechanism. (*Hint*: What bonds need to break?)

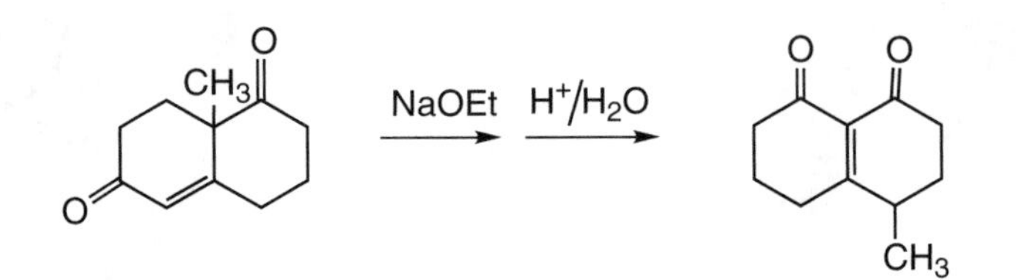

Solutions

1. (a)

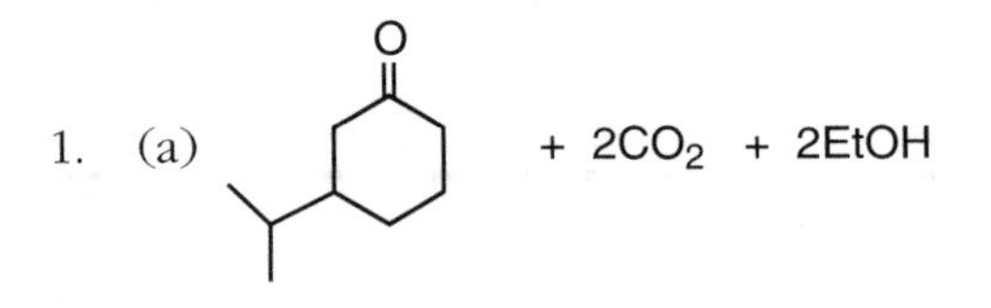

(b)

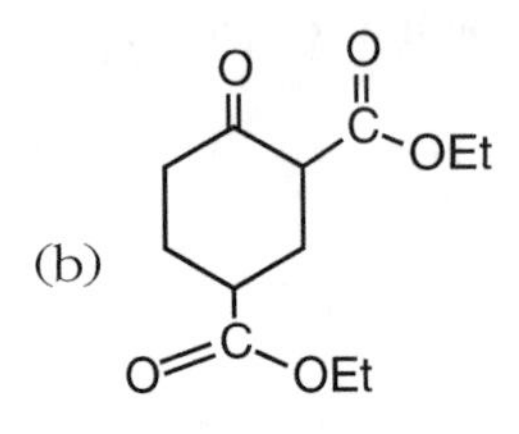

(c)

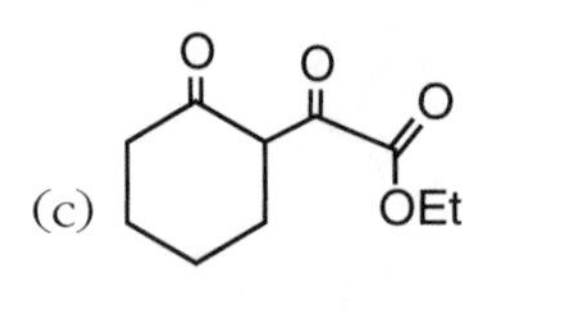

(d)

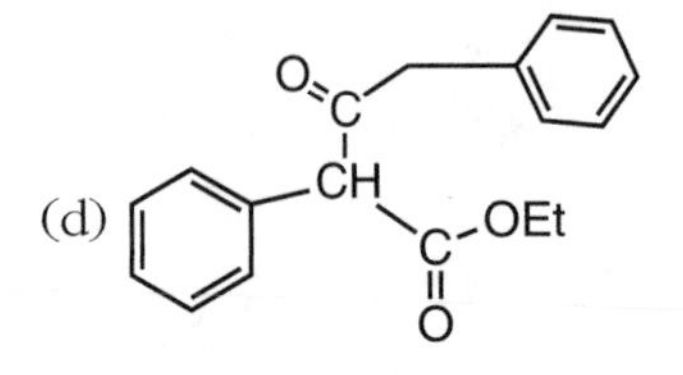

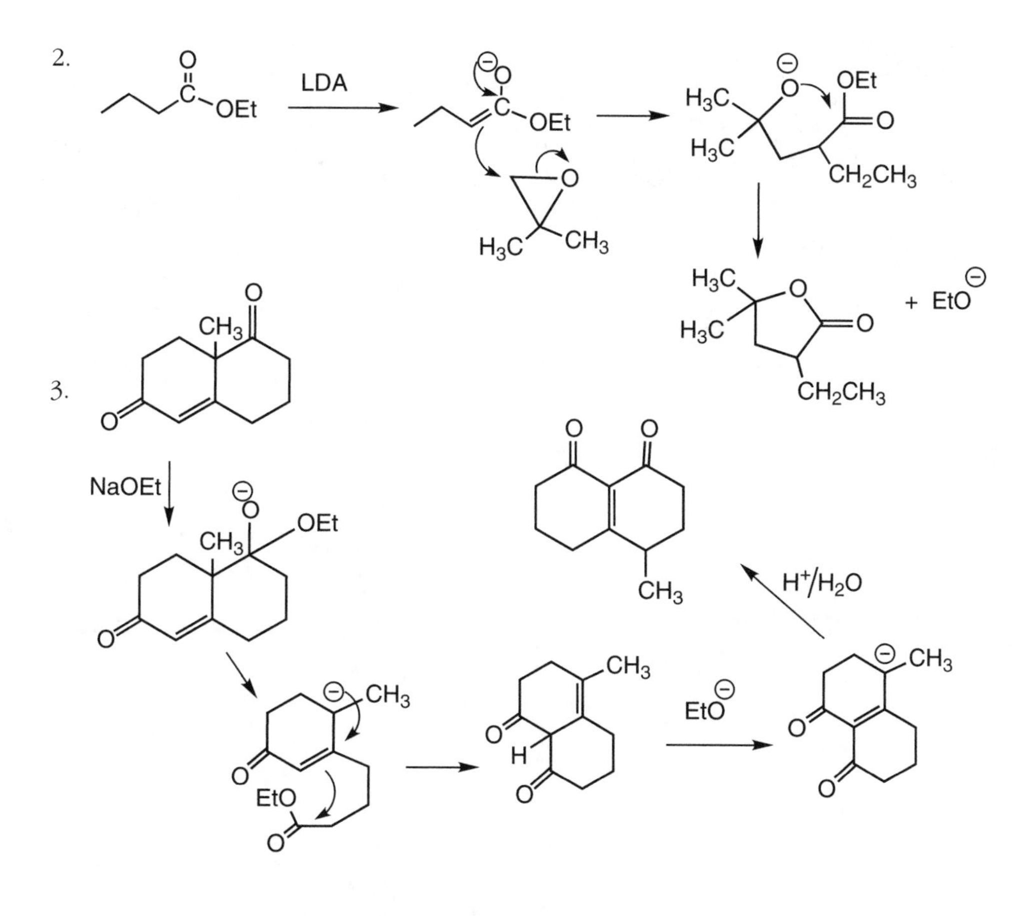

CLAISEN CONDENSATIONS IN BIOCHEMISTRY

In biological systems, thioesters, RCO—SR′, rather than carboxylic acid esters are commonly used. Being more acidic, thioesters form enolates more easily. Given the weaker, longer C—S bond, compared to the C—O bond, thioesters are also more reactive acylating agents. Many biochemical reactions use the thioester acetyl-CoA, whose structure was given in Chapter 21. For example, in the Krebs cycle, also known as the citric acid cycle, mitochondria in essence "burn" acetic acid (in the form of acetyl-CoA), generating energy that results in the production of ATP, the energy currency of cells. Molecules of acetyl-CoA enter the cycle via a Claisen condensation of acetyl-CoA with oxaloacetate:

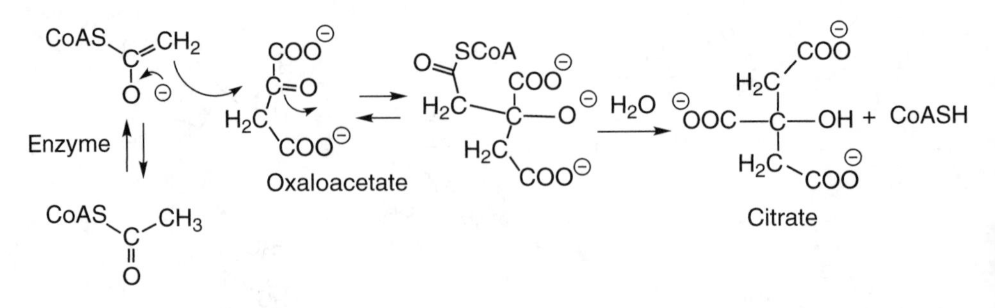

Hydrolysis of the thioester to an acid drives the reaction to completion.

23
AROMATIC MOLECULES I: BASIC PROPERTIES AND REACTIONS OF BENZENE

BENZENE

Benzene is a hydrocarbon that was first isolated by Michael Faraday in 1825 from illuminating gas. Greater quantities can be obtained from coal tar or petroleum. Faraday established that benzene's formula was C_6H_6. This makes benzene an unsaturated compound, with some combination of four rings plus double bonds. A number of derivatives of benzene were isolated from plant sources in which one or more of the hydrogen atoms were replaced by functional groups. For example, the compound toluene, $C_6H_5CH_3$, was found in tolu balsam, a tree resin. Since a number of benzene derivatives either had strong odors or came from plant resins with characteristic odors, benzene and its derivatives came to be known as **aromatic** compounds. A more modern term is **arenes**.

By the middle of the 19th century, it was clear to organic chemists that aromatic compounds showed unexpected reactivity, or rather, lack of reactivity. Most unsaturated compounds, such as ethylene, cyclohexene, and acetylene, reacted with many different reagents to add groups across the double or triple bond in an **addition** reaction:

In contrast, aromatic compounds resisted reactions with reagents such as bromine under mild conditions and reacted to give **aromatic substitution** products under more vigorous conditions.

C_6H_6 + Br_2

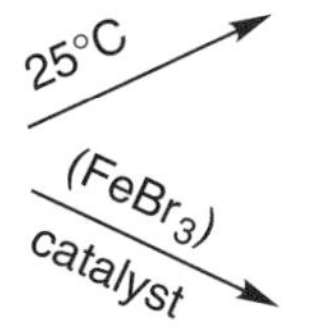

no reaction

C_6H_5Br

The core C_6X_6 unit (where X is some combination of hydrogen, halogen, alkyl, and other groups) was maintained throughout most chemical transformations. In time the word *aromatic* came to mean compounds with unusual stability and resistance to addition reactions.

EXERCISE

From earlier chapters, you probably know that benzene is drawn as cyclic with alternating single and double bonds, as will be discussed in the next section. Try drawing four C_6H_6 isomers containing at least one ring. Remember that triple bonds or two consecutive double bonds in a six-membered ring are unstable.

Solution

Many structures are possible. A few are shown below, all of which are highly reactive but are capable of at least fleeting existence.

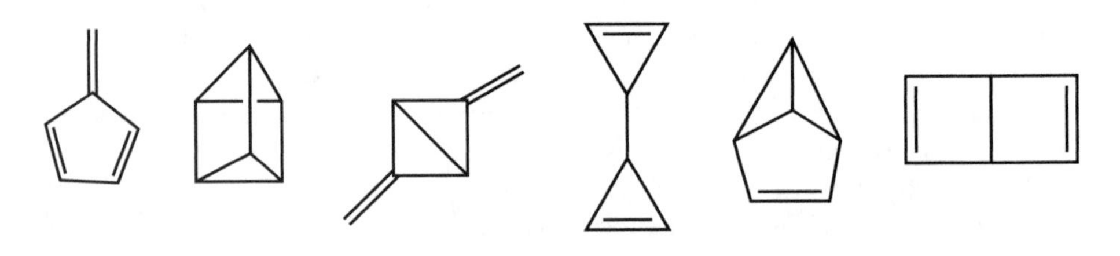

KEKULÉ STRUCTURES AND BONDING IN BENZENE

From our earlier discussion of resonance in organic compounds, we have seen that benzene is a resonance hybrid of two contributing structures:

This formulation of benzene as a cyclic molecule with three double bonds was first proposed by August Kekulé. However, our modern formulation is not exactly the same as the Kekulé structure proposed in 1866. Kekulé thought benzene was a rapidly equilibrating molecule with instantaneous single and double bonds:

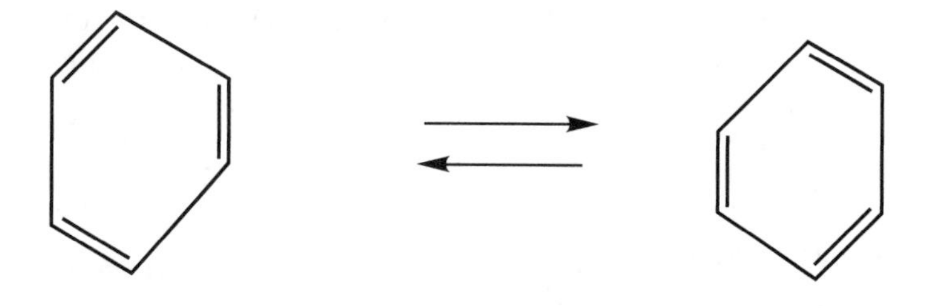

Such a formulation is incorrect. Benzene is a perfectly hexagonal molecule with a carbon-carbon bond length of 1.40 Å, halfway between that of a pure single bond and a pure double bond. The two contributing structures we draw for benzene are a result of our inability to depict benzene's structure accurately using a single line drawing and do not imply any equilibration at all.

Sometimes organic chemists draw benzene as a hexagon with a circle inscribed:

This single drawing tries to convey the idea that all the carbon-carbon bonds in benzene are identical. However, it is easy to become confused by such a drawing and end up with structures with five bonds to carbon, so we will not employ this representation.

Since resonance stabilizes molecules and benzene is a hybrid of two essentially identical contributing structures, we would expect benzene to be an unusually stable organic molecule. The difference in energy between benzene and a hypothetical C_6H_6 compound with alternating single and double bonds is called benzene's **resonance energy** or **aromatic stabilization energy**.

Resonance energy is calculated as shown in Figure 23.1. The conversion of cyclohexene to cyclohexane has a heat of hydrogenation of −28.6 kcal/mol. This means that one mole of cyclohexene and one mole of hydrogen react, with the aid of a platinum catalyst, to give off enough heat to raise the temperature of one kilogram of water 28.6°C. A hypothetical cyclohexatriene, a molecule with no resonance stabilization, would be expected to have a heat of hydrogenation of −85.8 kcal/mol [(3) (−28.6)]. The measured heat of hydrogenation of benzene is only −49.8 kcal/mol. Thus, benzene is more stable than expected by 36 kcal/mol, the resonance energy of benzene. Since a carbon-carbon bond has a bond energy of about 85 kcal/mol, the resonance stabilization energy of benzene is a substantial amount of energy. Whenever benzene reacts by addition reactions instead of substitution reactions, this resonance stabilization is lost. In later sections we will see many other molecules that are stabilized by cyclic resonance structures and are therefore aromatic in the modern definition of the term.

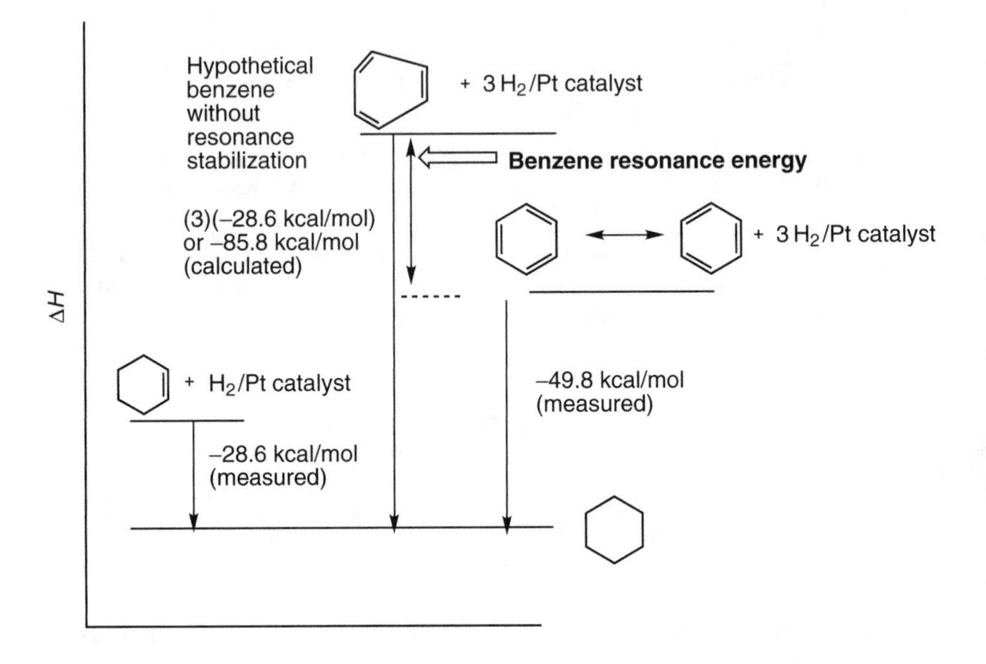

FIGURE 23.1. *Experimental determination of benzene's aromatic stabilization energy.*

EXERCISE

The heat of hydrogenation of (*Z*)-1,3,5-hexatriene is –80.5 kcal/mol. If this molecule is used as a reference instead of cyclohexene, what is the resonance energy of benzene?

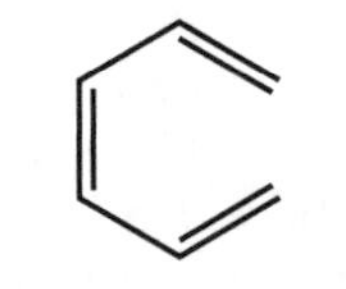

(*Z*)-1,3,5-Cyclohexatriene

Solution

Here the average heat of hydrogenation of a double bond in the reference compound is –26.8 kcal/mol. The difference in energy released on hydrogenation, the resonance energy, is 80.5 − 49.6 or 30.7 kcal/mol. While the exact value for the resonance stabilization energy in benzene depends on the reference compound used, it is always substantial.

Bonding in Benzene: A Hybrid Orbital Description

According to the hybrid orbital description of bonding in benzene, each carbon is sp^2 hybridized and thus has a *p* orbital on each carbon (Figure 23.2).

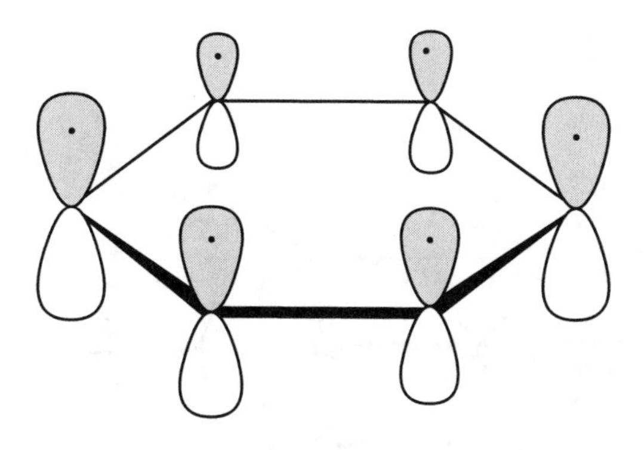

FIGURE 23.2. *The p orbitals in benzene. The electrons in the double bonds are represented as •.*

Since there is one electron in each orbital, benzene possesses a cyclic set of six *p* orbitals and six electrons. From this figure, we see that it is clearly arbitrary where one puts the double bonds, and that the equality of all six C—C bonds in benzene is unsurprising.

From the hybrid orbital picture, one sees a π **framework** above and below the molecular plane and a σ **framework** of σ C—C and C—H bonds. These two classes of bonds interact essentially not at all because they are perpendicular to one another, with the σ bonds in the nodal plane of the *p* orbitals. Most of the chemistry of benzene and its derivatives involves the π-framework bonds.

EXERCISE

Sketch an orbital picture of the benzene-derived carbocation, $C_6H_5^+$, and the benzene-derived anion, $C_6H_5^-$. How does the orbital containing the positive or negative charge interact with the benzene π framework?

Solution

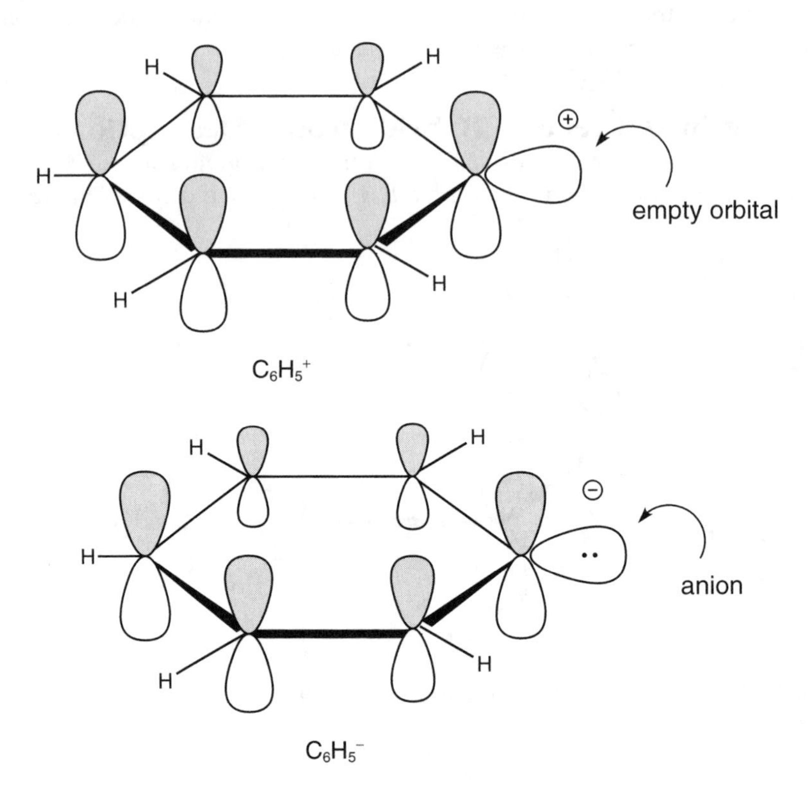

Both the empty orbital of the carbocation and the orbital with an anionic electron pair in the anion are in the plane of the benzene ring and part of the σ framework. They do not interact with the π framework. There is no resonance stabilization of the cation or anion because of the perpendicular orientation of the orbitals.

Molecular Orbital Theory of Bonding in Benzene

The molecule orbital theory of bonding says that electrons exist in molecules within molecular orbitals which, unlike the hybrid orbitals discussed above, can extend over the entire molecular structure. Each molecular orbital can hold no more than two electrons, and each molecular orbital has a definite energy. Some orbitals are bonding orbitals. When electrons are placed in these orbitals, the molecule is stabilized because the shape of the bonding molecular orbitals tends to concentrate electrons between nuclei. Other orbitals are antibonding molecular orbitals. If electrons are placed in these orbitals, electron-electron repulsion destabilizes the molecule. Most stable organic molecules have just enough electrons to fill all the lower-energy bonding molecular orbitals with none left over to put into higher-energy antibonding orbitals.

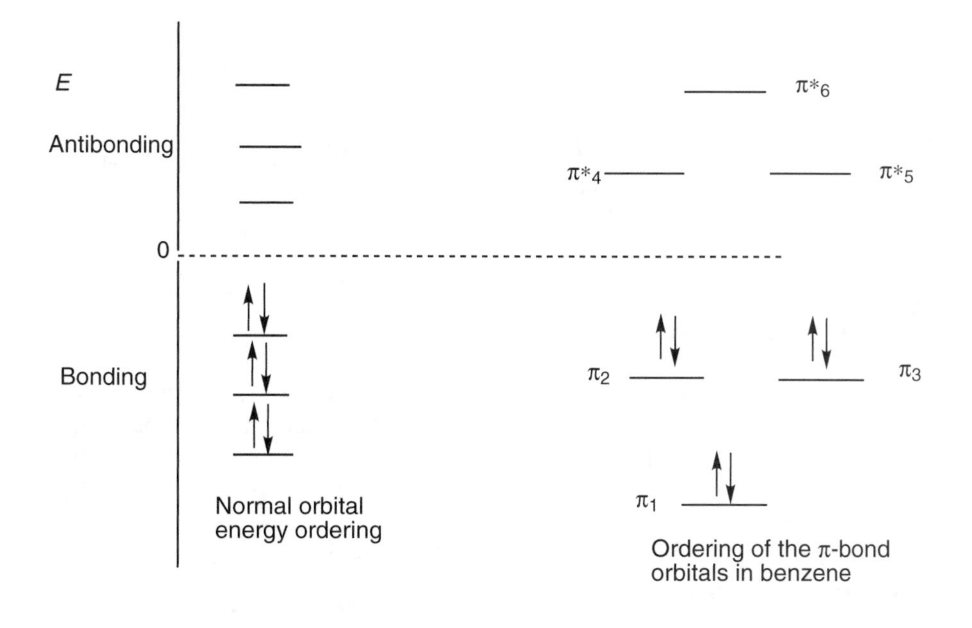

FIGURE 23.3. *Ordering of the molecular orbital energies for a normal organic compound and for benzene. The arrows represent electron pairs.*

For most organic molecules, there is a lowest-energy molecular orbital and each additional molecular orbital has an energy higher than the one before (Figure 23.3). In the case of benzene, the three molecular orbitals holding the six π electrons, orbitals π_1, π_2, and π_3, are ordered so that the two highest-energy occupied orbitals have the same energy. Orbitals with the same energy are said to be **degenerate**. When the energies of electrons in benzene are calculated (using quantum mechanical techniques beyond the scope of introductory organic chemistry), they are found to be lower than the energies of electrons in nonaromatic molecules.

Hückel's Rule

The German chemist Erich Hückel, in an early application of quantum chemistry to organic chemistry in 1931, found a simple rule to predict when molecules will be aromatic and thus have their π molecular orbitals ordered like benzene's. He calculated that any conjugated, closed, planar cyclic molecule with 2, 6, 10, 14, 18, . . . π electrons will be aromatic. The preceding series of numbers can be generated by the formula $4n + 2$, where n is any integer starting with 0. Thus, aromatic compounds are said to obey the **$4n$ + 2 rule**.

The $4n + 2$ rule, or Hückel's rule, expands the class of aromatic compounds beyond benzene and includes carbocations and carbanions. For example, the

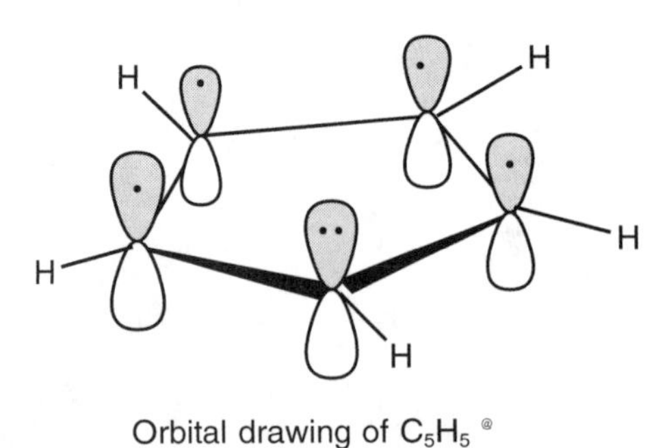

Orbital drawing of $C_5H_5^{\ominus}$

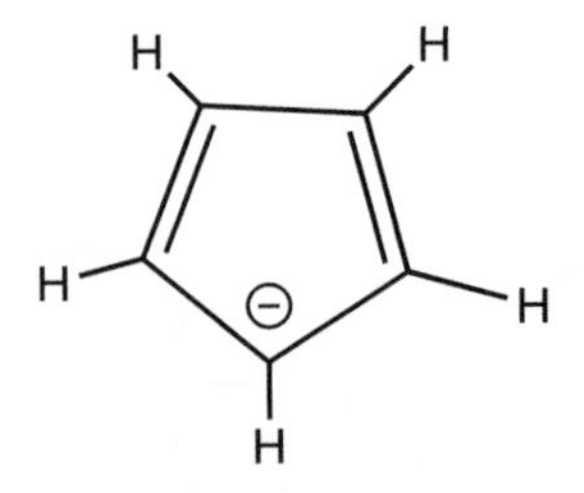

Line drawing of $C_5H_5^{\ominus}$

carbanion $C_5H_5^-$ has a closed planar cycle of six electrons. There are two electron pairs in the two double bonds and an electron pair in the remaining *p* orbital. A closed cycle of electrons means that each atom of the cycle has a *p* orbital that can overlap the others in the ring. For example, the 1,3-cyclohexadienyl anion is not aromatic, even though there are six electrons in the π system, because the cycle of *p* orbitals is not closed. The one CH_2 group has no *p* orbital.

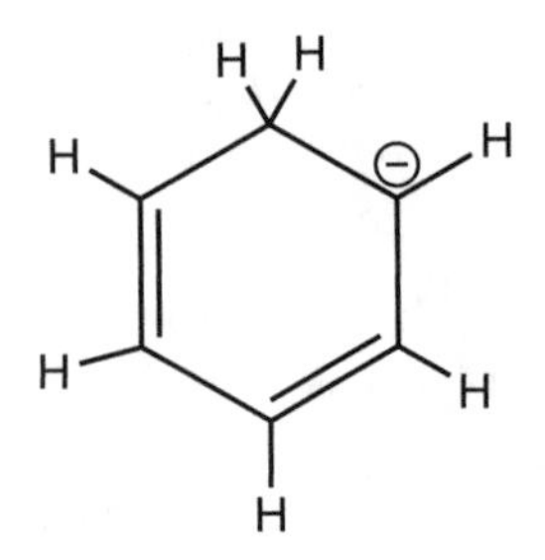

1,3-Cyclohexadienyl anion

An example of an aromatic cation is the 1,3,5-cycloheptatrienyl cation. In this molecule, one empty *p* orbital contributes no electrons to the π system. Nevertheless, the molecule is aromatic because it has a closed cyclic planar system of six electrons distributed over seven *p* orbitals:

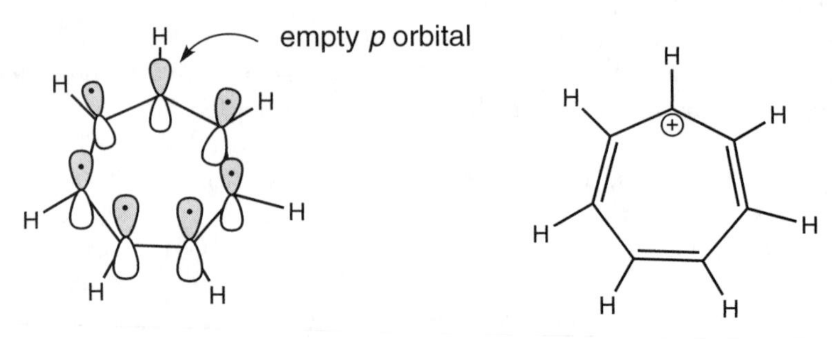

1,3,5-Cycloheptatrienyl cation

For carbanions and carbocations, aromaticity is a relative term. Such species are still reactive compared to neutral organic molecules. However, for charged species they are relatively stable and unusually easy to form. For example, 1,3-cyclopentadiene is only slightly less acidic than water since its deprotonation gives the aromatic $C_5H_5^-$ anion.

Because of the ordering of orbital energies, some closed cyclic planar molecules are unusually unstable. Molecules with $4n$ electrons, that is, with 4, 8, 12, 16, . . . , electrons are said to be **antiaromatic**. The most studied antiaromatic molecule is 1,3-cyclobutadiene. Its arrangement of π-orbital energies is shown in Figure 23.4.

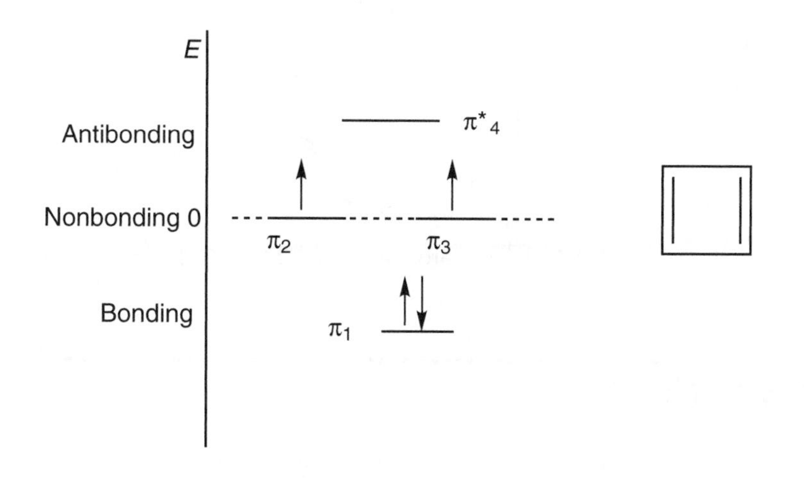

FIGURE 23.4. *Orbital ordering for the π system in 1,3-cyclobutadiene.*

The two orbitals π_2 and π_3 are degenerate and nonbonding. Because of electron-electron repulsion, the two electrons each go into one of the degenerate orbitals. As a result, 1,3-cyclobutadiene behaves as if it has one double bond and two unpaired electrons, instead of two double bonds. This results in an extremely reactive, unstable molecule. Another molecule that could be antiaromatic is cyclooctatetraene. If it were planar, it would have eight π electrons and therefore be antiaromatic ($4n$, where $n = 2$). However, this molecule can avoid antiaromaticity by adopting a nonplanar conformation, the tub form:

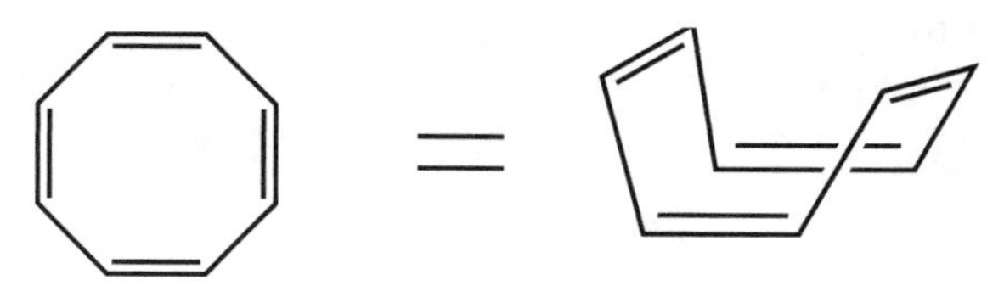

Cyclooctatetraene

Since the molecule is not planar, it lacks one of the properties needed for antiaromaticity and is classified as **nonaromatic**. Cyclooctatetraene reacts with reagents like any other alkene.

There can be a number of compounds with conjugated monocyclic arrays of double bonds. These molecules are called **annulenes**. For example, the compound with nine conjugated double bonds and thus 18 π electrons, named [18]-annulene, has been found to possess aromatic stabilization energy. For annulenes with large rings, some of the double bonds must be trans to avoid angle strain.

[18]-Annulene, aromatic, $4n + 2$ for $n = 4$

EXERCISES

1. For each molecule indicate whether it is aromatic, antiaromatic, or nonaromatic. For molecules that follow the $4n + 2$ or $4n$ rule, indicate how many electrons are in the π system.

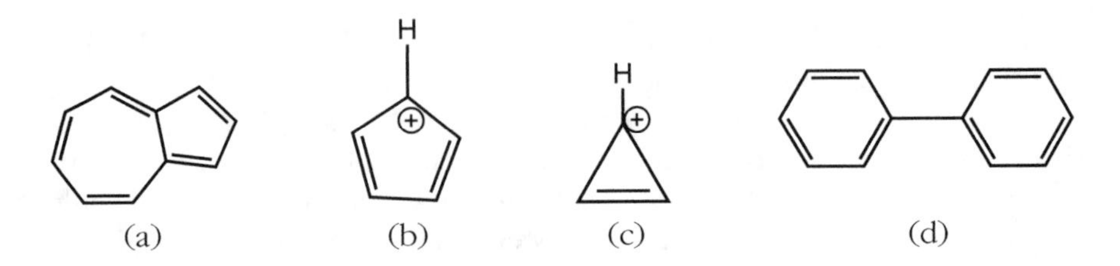

2. Cyclooctatetraene reacts with potassium metal, which is a source of electrons, to form a compound with the formula $K_2C_8H_8$. What is the structure of this salt?
3. If *cis,trans,cis,cis,trans*-[10]-annulene were planar, would it be aromatic? Can you think of a reason why it may not be planar?

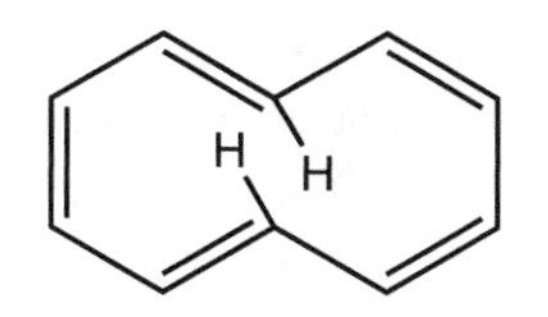

cis,trans,cis,cis,trans-[10]-Annulene

Solutions

1. (a) aromatic, ten electrons; (b) antiaromatic, four electrons; (c) aromatic, two electrons ($4n + 2$, where $n = 0$). (d) aromatic. The two benzene rings form two independent six-electron aromatic systems. There is not a single closed cycle that encompasses the whole molecule.
2. The $C_8H_8^{2-}$ dianion is an aromatic, ten-electron system, having picked up two electrons from the potassium. The advantage of aromaticity makes this dianion planar.

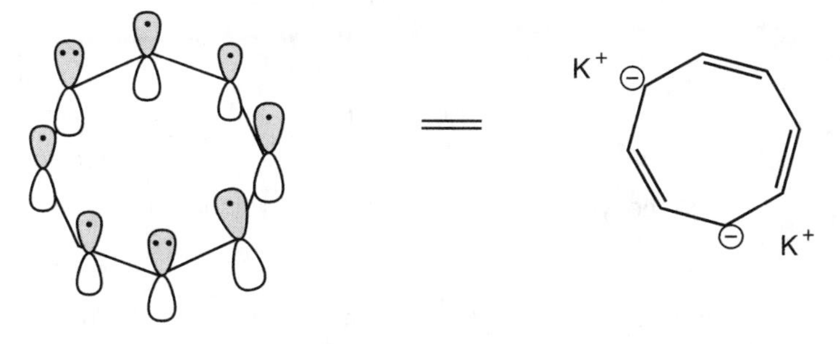

One resonance form for $C_8H_8^{2-}$

3. This annulene would be aromatic because it has ten electrons ($4n + 2$, where $n = 2$). However, the two hydrogens that point into the ring occupy almost the same region of space as if the molecule were planar. So the molecule twists out of planarity to relieve the steric strain. As a result, the molecule is nonaromatic. For most annulenes like this, it is difficult for a student to see if steric strains are present without building models.

TRIVIAL NAMES OF AROMATIC MOLECULES

Most simple benzene derivatives were known long before systematic IUPAC nomenclature rules were developed. Because benzene derivatives are ubiquitous in organic chemistry, they are known by their well-established trivial or common names. Examples of common compounds with carbon, nitrogen, and oxygen substituents are given here.

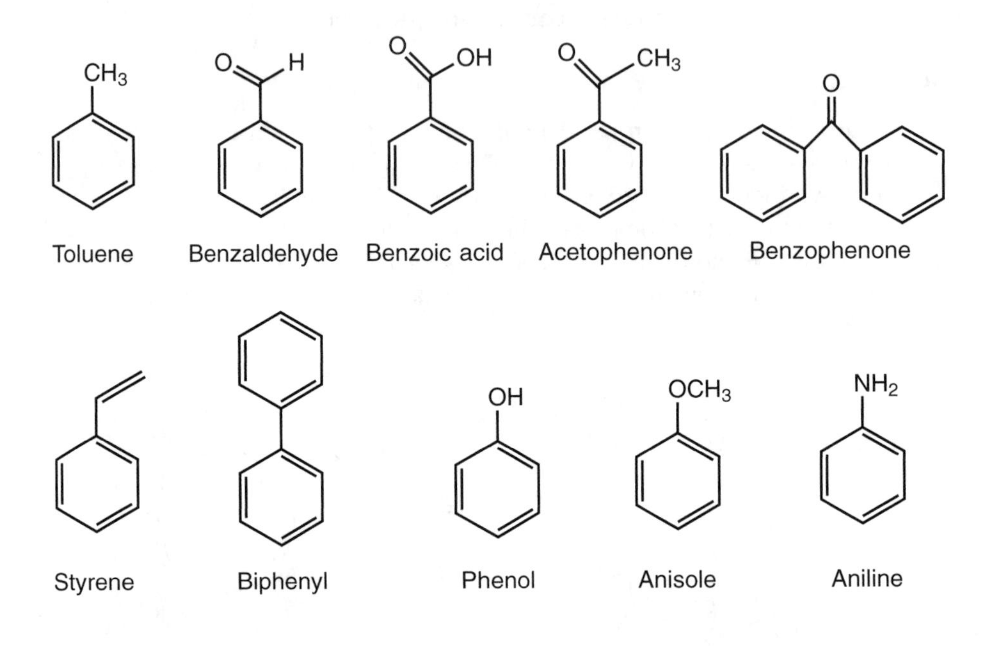

For simple benzene derivatives, the name is produced by taking the name of the functional group and adding the suffix *-benzene*:

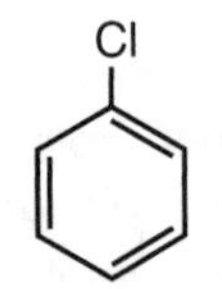

Br

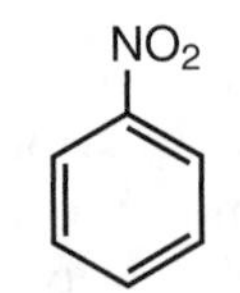

Chlorobenzene Bromobenzene Nitrobenzene

When there are two substituents on the benzene, their relative positions are indicated by the prefixes *ortho-*, *meta-*, and *para-*, which indicate a 1,2, 1,3, or 1,4 relationship, respectively. *Ortho-*, *meta-*, and *para-* are abbreviated *o-*, *m-*, and *p-*.

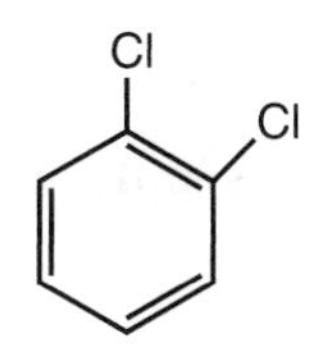

ortho-Dichlorobenzene

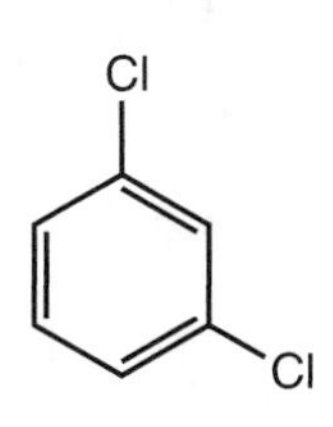

meta-Dichlorobenzene

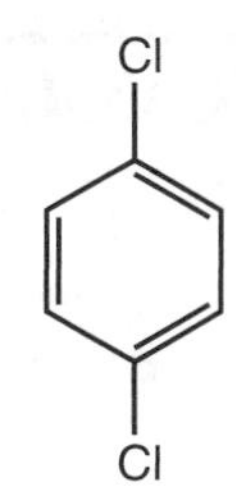

para-Dichlorobenzene

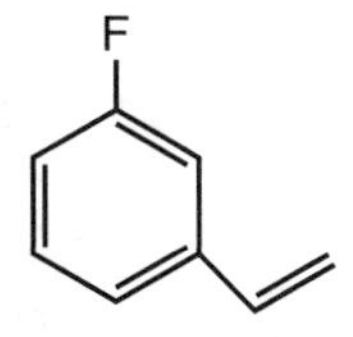

meta-Fluorostyrene

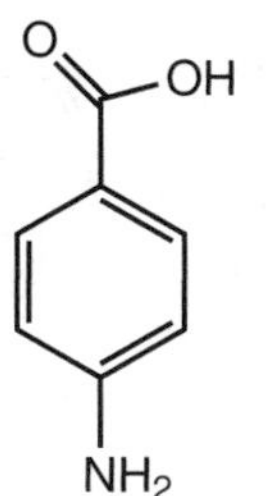

para-Aminobenzoic acid

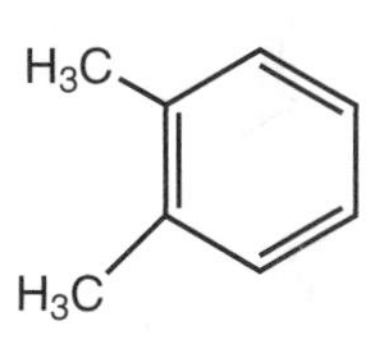

ortho-Xylene (xylene is a dimethylbenzene)

Recall that because of resonance in benzene, there are two ways to draw the ortho isomer:

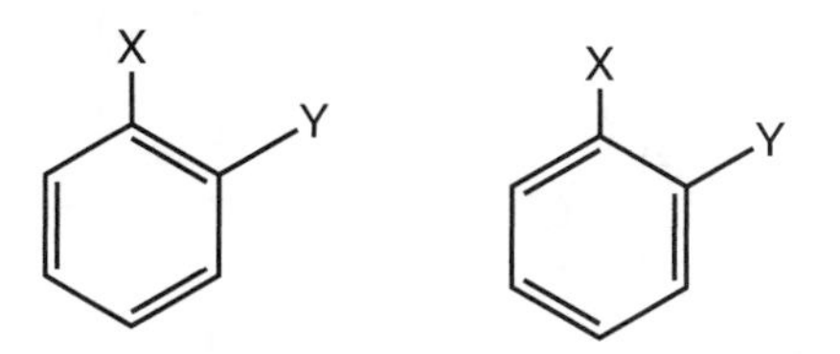

These two representations are exactly the same compound.

When there are three or more groups on a benzene ring, the group with the highest priority provides the naming base. The positions of the other substituents are numbered relative to the group of highest priority. An abbreviated ranking of groups is —COOH > —COOR > —CN > —CHO > —OH > $—NH_2$. Some groups, such as the halogens and the nitro group are named as substituent groups:

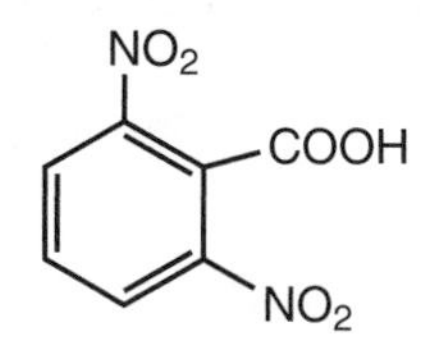

2,6-Dinitrobenzoic acid

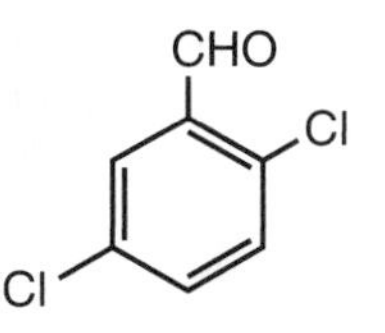

2,5-Dichlorobenzaldehyde

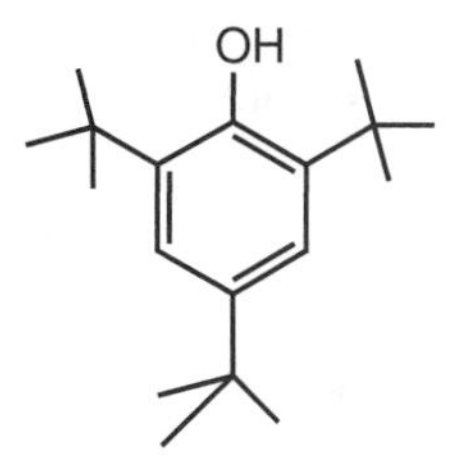

2,4,6-Tri-*tert*-butylphenol

POLYCYCLIC AROMATIC HYDROCARBONS

Aromatic rings that share one or more edges are known as **polycyclic aromatic hydrocarbons**:

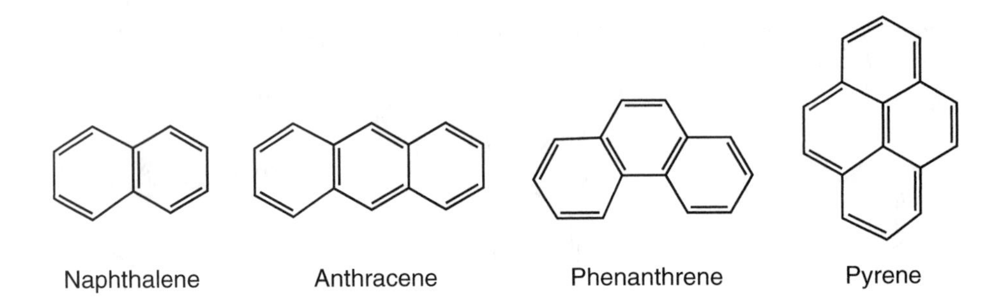

Naphthalene and anthracene are examples of linear fused rings, while phenanthrene is an angularly fused ring. Most fused rings can be drawn as several different Kekulé structures. Phenanthrene, for example, can be represented by five Kekulé structures:

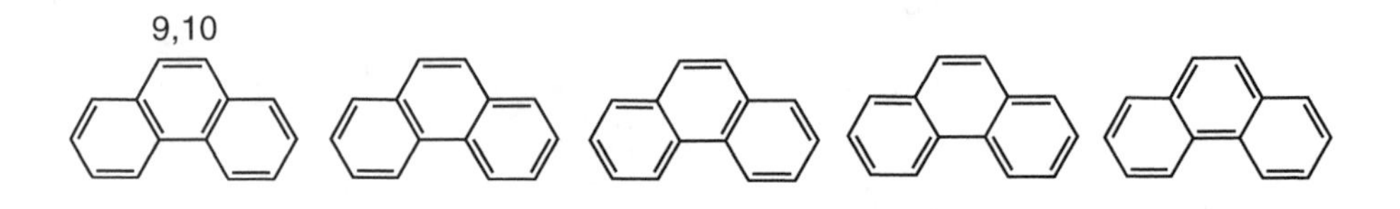

Notice that of the five contributing structures, four have the 9,10 bond as a double bond. In fact, the 9,10 bond essentially reacts as a typical double bond. When several Kekulé structures can be drawn, the most important are those that have as many rings as possible in the benzene form. For example, in the first phenanthrene structure all three rings resemble benzene. In the last representation, only the central ring resembles benzene.

Polycyclic aromatic hydrocarbons exhibit aromatic stabilization energy. However, they possess less stabilization per ring than benzene itself. The measured resonance stabilization energy of naphthalene is 61 kcal/mol, and that of anthracene is 83 kcal/mol. These values are greater than benzene's 36 kcal/mol, but not twice or three times as great.

ELECTROPHILIC AROMATIC SUBSTITUTION OF BENZENE

General Features of Electrophilic Aromatic Substitution

Aromatic molecules are nucleophiles. Thus, they react with electrophiles but only with powerful electrophiles that can temporarily disrupt the aromatic sextet of electrons. When an electrophile adds to benzene, or other aromatic molecules, a transient carbocation is formed. However, the carbocation does not add an anion, leading to an addition product, as seen for alkenes. Instead, the carbocation loses a proton and regenerates the highly stable aromatic molecule. The general free energy diagram for electrophilic aromatic substitution appears in Figure 23.5.

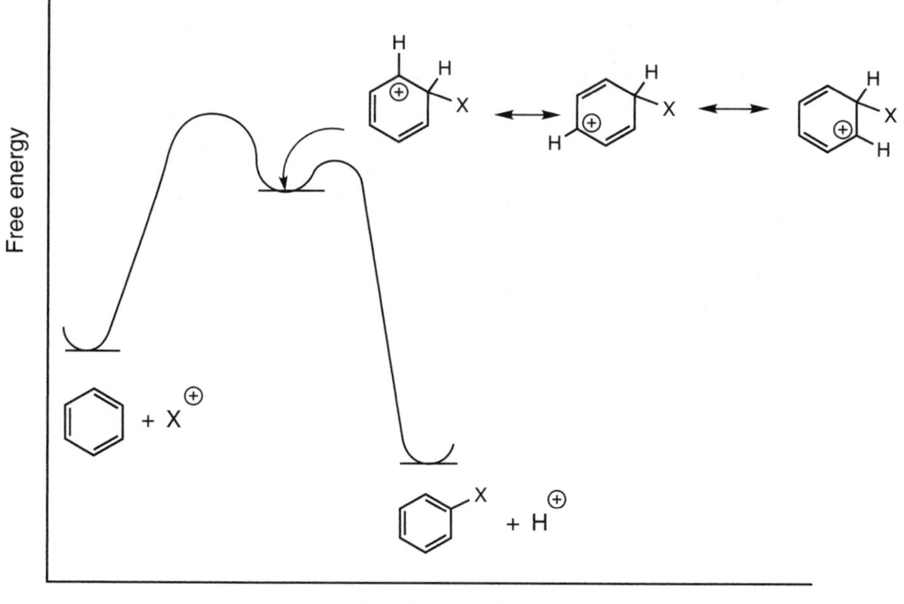

FIGURE 23.5. *Reaction coordinate diagram for a typical electrophilic substitution reaction.*

The first step in electrophilic aromatic substitution is highly endergonic. As we shall see, anything that stabilizes the carbocation intermediate increases the rate of the reaction. The second step, loss of a proton to a base in the reaction mixture, is not rate-limiting. If the carbocation intermediate added an anion to form a cyclohexadiene structure, the 36 kcal/mol of aromatic stabilization energy would be lost, so substitution is favored over addition.

Almost everything that is important about electrophilic aromatic substitution is explained by a careful analysis of the three carbocation resonance structures. Notice that only three carbons of the cationic intermediate support any positive charge. Two of the carbons never hold any charge. The positions 1,2 to the C—X carbon are known as the **ortho** positions, the position 1,4 to the C—X carbon is the **para** position, and the 1,3 positions are designated the **meta** positions. No charge is localized at the meta positions in this resonance structure. We will often use the ortho, meta, and para terms to denote relative positions (1,2 or 1,3 or 1,4) on a benzene ring.

Halogenation of Benzene

Bromine and chlorine are not sufficiently electrophilic to react directly with benzene. One must add a Lewis acid, usually $AlCl_3$ or $FeCl_3$ for Cl_2, and $FeBr_3$ for Br_2. The Lewis acid reacts with the halogen. For example, $FeBr_3$ reacts with Br_2 to give $Br^+FeBr_4^-$. Elemental fluorine is so reactive that it essentially burns the benzene molecule. Special methods are used to introduce both F and I onto aromatic rings. The overall reaction for the chlorination of benzene is

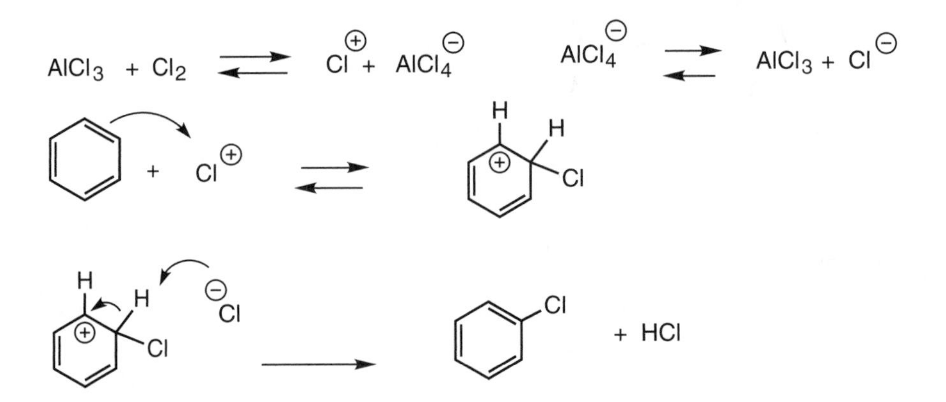

Nitration of Benzene Rings

The nitro group is a very versatile group in aromatic chemistry. It can be converted to numerous other groups by multistep reactions that will be covered in Chapter 24. The species that reacts with aromatic rings is NO_2^+, an extremely powerful electrophile as a consideration of its resonance forms suggests. Normally, NO_2^+ is made by the reaction of nitric acid, HNO_3, with sulfuric acid. One can also buy the salt nitronium tetrafluoroborate, NO_2BF_4.

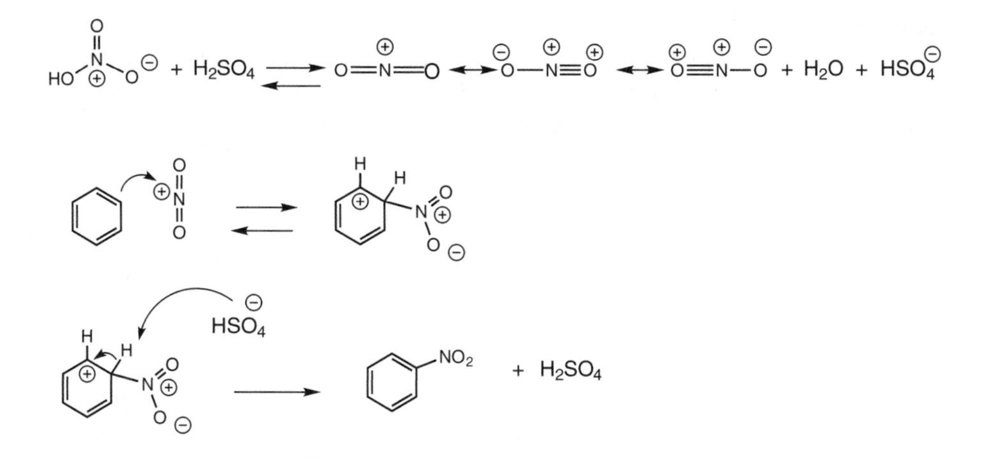

Aromatic Sulfonation

Unlike halogens or nitric acid, sulfur trioxide, SO_3, can react directly with arenes to give sulfonic acids. Because sulfur is a third-row element, one can draw structures that formally violate the octet rule. These structures have a formal positive charge on sulfur, which helps to explain its electrophilicity. Normally, sulfur trioxide is used as a solution in H_2SO_4 which is called oleum.

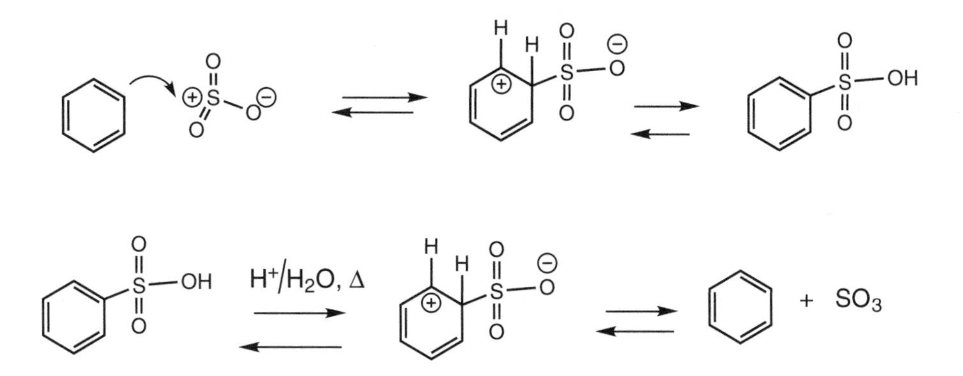

Aromatic sulfonation is unusual in that it is readily reversible. Heating an aromatic sulfonic acid in acid results in protonation of the aromatic ring and the ultimate loss of SO_3.

EXERCISE

What product would result from heating benzene in aqueous D_2O/D_2SO_4?

Solution

A proton or deuteron, D^+, is an active electrophile. It would be expected to reversibly add to a benzene ring. Ultimately, if the concentration of D_2O were high enough, one would obtain C_6D_6.

24
AROMATIC MOLECULES II: REACTIONS OF SUBSTITUTED BENZENES, PHENOLS, AND ANILINES

When one studies the electrophilic aromatic substitution of substituted molecules, such as chlorobenzene, aniline, or nitrobenzene, one finds a range of reactivity patterns. Some groups, such as —OH, cause the ring to undergo a reaction much faster than benzene itself. We will call these groups donating groups, designated D. Other groups, such as the nitro group, cause the aromatic ring to react much more slowly than benzene. These groups will be designated electron-withdrawing groups, W. Furthermore, different isomers are formed from benzene substituted with D groups or W groups. Electron-donating groups direct the new substituent to the ortho and para positions. Electron-withdrawing groups direct the new substituent to the meta position. The only important exceptions to this pattern are halogens. They are weakly W groups, but they direct substituents to the ortho and para positions.

One can measure the rates of reaction of substituted benzenes with a standard electrophile and order the substituents from highly activating (reacts faster than benzene) to highly deactivating (reacts more slowly than benzene). Table 24.1 summarizes the effects of common groups. As will shortly become clear, these reactivity patterns are explained by the resonance properties of carbocation intermediates in electrophilic aromatic substitution.

FRIEDEL-CRAFTS ALKYLATION AND ACYLATION

Friedel-Crafts Alkylation

Before looking at aromatic substitution in general, we will consider a specific case where alkyl groups are placed on benzene rings. When benzene is treated with an alkyl chloride, such as CH_3Cl, in the presence of the Lewis acid $AlCl_3$, a mixture of products is obtained, including toluene,

TABLE 24.1. ORDER OF SUBSTITUENTS FROM HIGHLY ACTIVATING TO HIGHLY DEACTIVATING IN ELECTROPHILIC AROMATIC SUBSTITUTION

Substituent	Effect	Directing	Trend
$-NH_2$, $-NHR$, $-NR_2$; $-OH$, $-OR$	Strongly activating	D Groups, ortho/para-directing	More activating ↑
$-NH-C(=O)-R$		D Groups, ortho/para-directing	
$-R$, $-Ar$, $-HC=CH_2$		D Groups, ortho/para-directing	
$-H$			
$-F$, $-Cl$, $-Br$, $-I$	Weakly deactivating	Ortho/para-directing	
$-C(=O)-R$; $-C(=O)-OR$	Deactivating	W groups, meta-directing	
$-C{\equiv}N$; $-SO_3H$; $-NH_3^{\oplus}$, $-NR_3^{\oplus}$; $-NO_2$, $-CF_3$	Strongly deactivating	W groups, meta-directing	More deactivating ↓

ortho-dimethylbenzene (or *o*-xylene), *para*-dimethylenebenzene (or *p*-xylene), and 1,2,4-trimethylenebenzene. If the reaction continues for a long time, some hexamethylenebenzene can be isolated. Very little *meta*-dimethylbenzene (or *m*-xylene) is produced. From our classification above, this means that a methyl group is a D group.

The alkylation of aromatic molecules with a mixture of an alkyl halide and a Lewis acid is known as **Friedel-Crafts alkylation**. The mechanism for this reaction involves the formation of a carbocation by $RCl + AlCl_3 \rightarrow R^+ + AlCl_4^-$. The carbocation is not a free alkyl cation since CH_3Cl can be used. It is partly stabilized by association with the $AlCl_4^-$. The mechanism for the addition of R^+ to an aromatic ring is analogous to the other aromatic substitution reactions covered in Chapter 23:

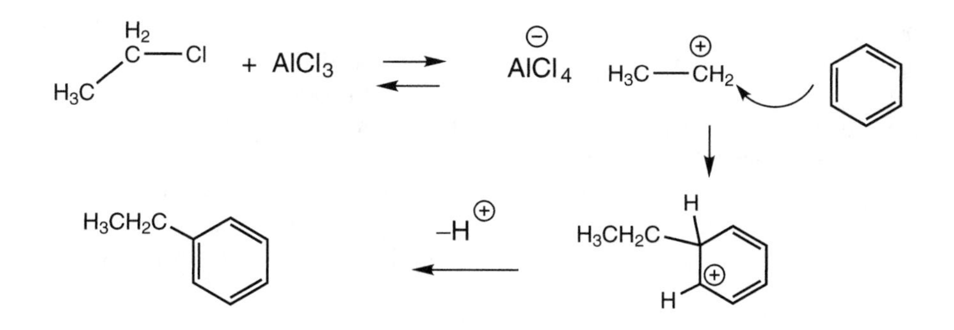

According to Table 24.1, an alkyl group is treated as an activating group and directs new substituents to the ortho and para positions. This directing effect is due to the fact that alkyl groups stabilize carbocations:

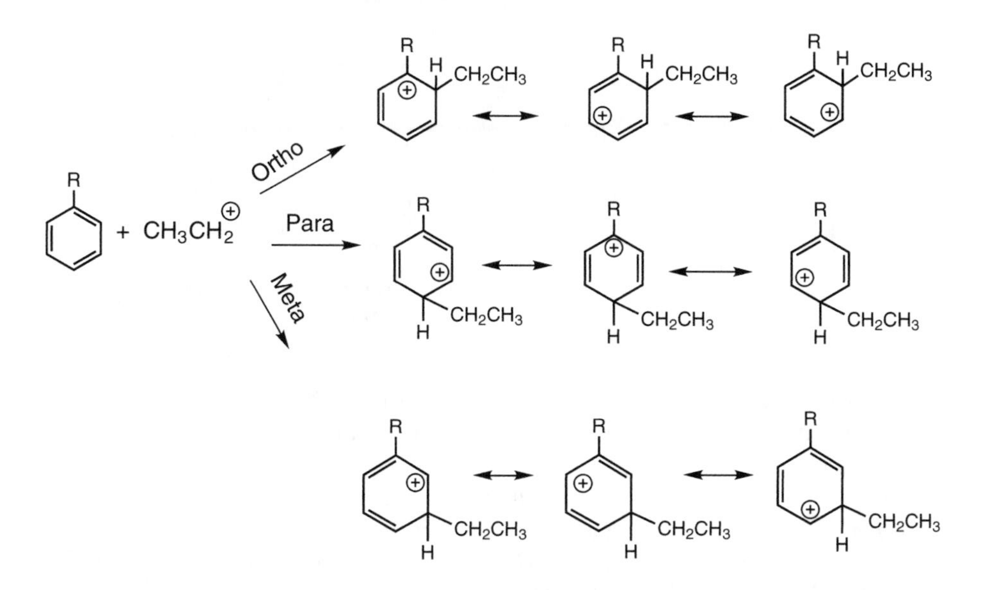

The carbocation intermediates that lead to the ortho or para dialkyl isomer have one of the resonance structures with a carbocation located on an alkyl-substituted carbon. In contrast, the carbocation that leads to the meta dialkyl isomer has the positive charge only on CH-substituted carbons. Thus, the ortho and para carbocation intermediates are more stable and form in preference to the meta isomer.

Because alkyl groups are inductively electron-donating compared to hydrogen, alkyl-substituted benzene rings are more electron-rich than benzene itself. In other words, they are better nucleophiles in electrophilic aromatic substitution reactions than benzene, and they react faster. This reactivity causes a problem in Friedel-Crafts alkylations since the product is more reactive than the starting material. As soon as some of the monoalkylated product builds up, it reacts in preference to benzene, giving the dialkylated product which can react further. Eventually one has a mixture of products that are monoalkylated, dialkylated, trialkylated, and so on. Normally, an excess of benzene is used to overcome this problem.

A second problem with Friedel-Crafts alkylation is possible rearrangement of the carbocation electrophile. A primary carbocation undergoes hydride transfer to form a secondary carbocation, leading to a mixture of alkyl products:

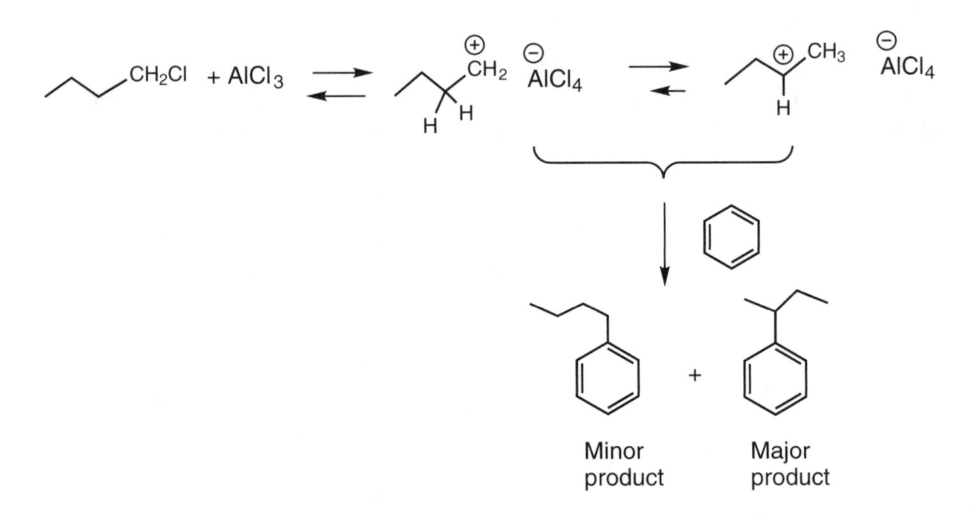

Secondary halides give minor amounts of rearrangement products, but Friedel-Crafts alkylation with primary halides is a poor reaction because of the mixture of products produced.

Carbocations for Friedel-Crafts alkylations can be obtained in a number of ways in addition to the reaction of RCl with $AlCl_3$. For example, the addition of a proton to an alkene generates a carbocation. Alternatively, protonation of a 2° or 3° alcohol, followed by the loss of water, produces a carbocation. Either of these routes works well in intramolecular Friedel-Crafts alkylations:

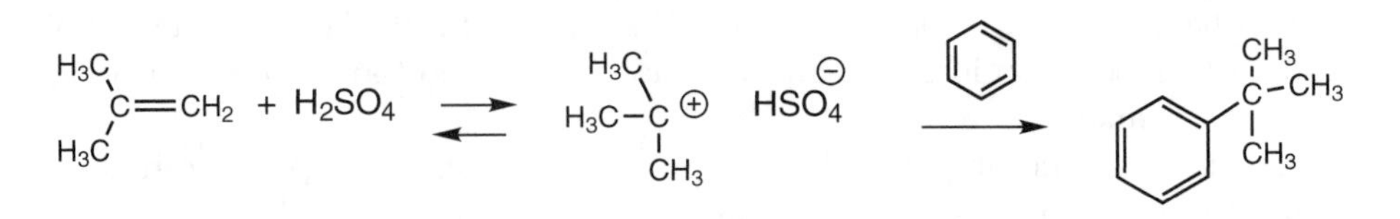

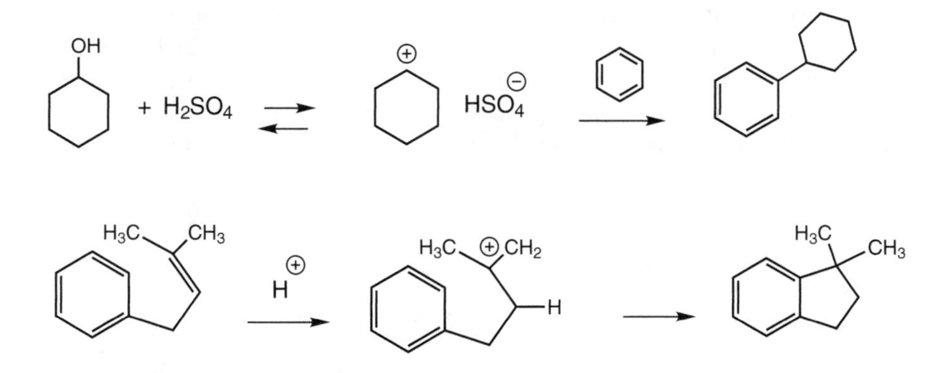

Friedel-Crafts Acylations

A reaction that is generally more useful than Friedel-Crafts alkylation is **Friedel-Crafts acylation**. When an acid chloride and $AlCl_3$ are mixed, an **acylium ion** is formed:

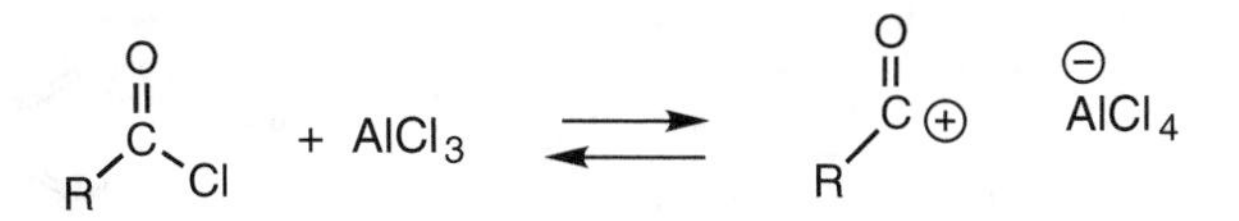

Like the carbocations formed in Friedel-Crafts alkylations, acylium ions are not free carbocations but strongly associated ion pairs. They are relatively stable and do not undergo any rearrangement reactions. Acylium ions react with aromatic rings as long as there are no deactivating groups on the ring. An acylium ion is only a moderately reactive electrophile. It does not react with deactivated molecules such as nitrobenzene or aryl ketones. Thus, Friedel-Crafts acylation is free of two of the problems of Friedel-Crafts alkylation. There are no rearrangement products, and the reaction stops after one acyl group adds onto the ring:

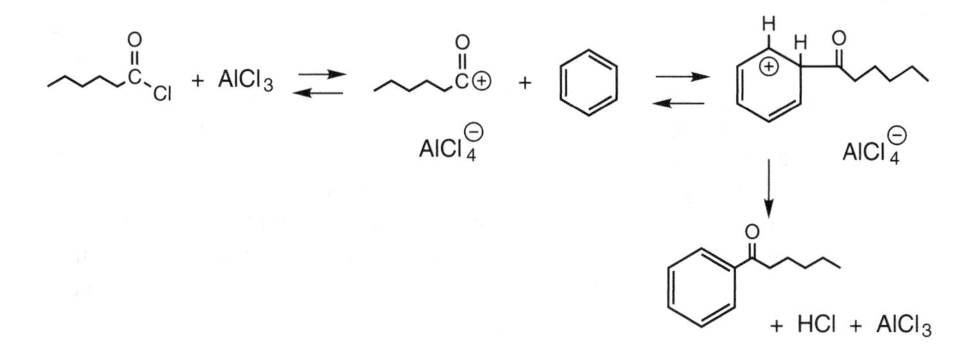

The Friedel-Crafts acylation reaction is even more useful than it initially seems because aryl ketones are readily deoxygenated to give hydrocarbons. Three synthetic methods are commonly used for this deoxygenation: the Clemmensen reaction (zinc amalgamated with mercury and HCl), the Wolff-Kishner reaction (hydrazine, H_2NNH_2, and NaOH), and reduction with Pd and H_2:

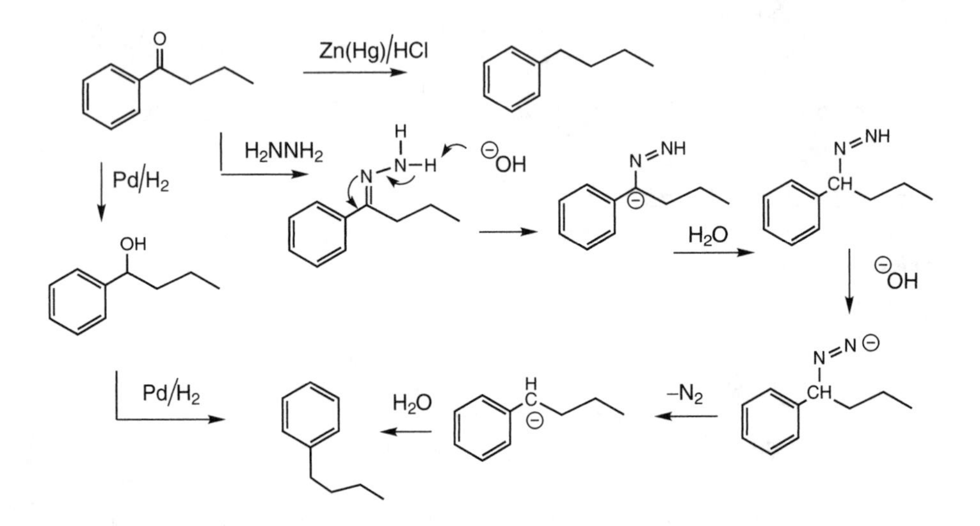

The Wolff-Kishner reaction proceeds with initial formation of a hydrazine, which undergoes base-promoted decomposition to a hydrocarbon. Hydrogenation with Pd initially gives a benzylic alcohol that undergoes **hydrogenolysis**, replacement of a benzylic group (here an —OH) with an H. While the Wolff-Kishner reaction deoxygenates alkyl or aryl ketones, only ketones with at least one aryl group undergo Clemmensen or Pd/H_2 deoxygenation.

Example

A common type of problem requires one to: start with benzene and any other reagents and prepare *n*-hexylbenzene. Using a Friedel-Crafts alkylation reaction (benzene plus 1-chlorohexane and $AlCl_3$) is always wrong. The result will be a mixture of products arising from carbocation rearrangements. Instead, always use a two-step procedure to make a linear alkyl substituent: Friedel-Crafts acylation followed by deoxygenation. The reagent used for deoxygenation depends on what other functional groups are present in the final product. For example, if there is a double bond elsewhere in the molecule, using Pd/H_2 would hydrogenate it as well as the ketone group.

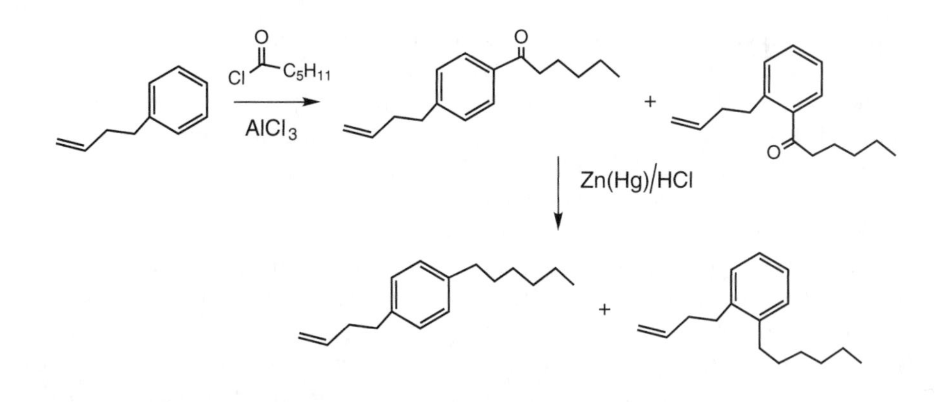

SUBSTITUTION REACTIONS ON ACTIVATED AND DEACTIVATED BENZENES

Activated Benzene Derivatives

The property that unites all the donor groups in Table 24.1 is that each is able to stabilize a carbocation. An alkyl group stabilizes a carbocation by hyperconjugation. The oxygen and nitrogen groups have lone electron pairs that can stabilize a positive charge through resonance:

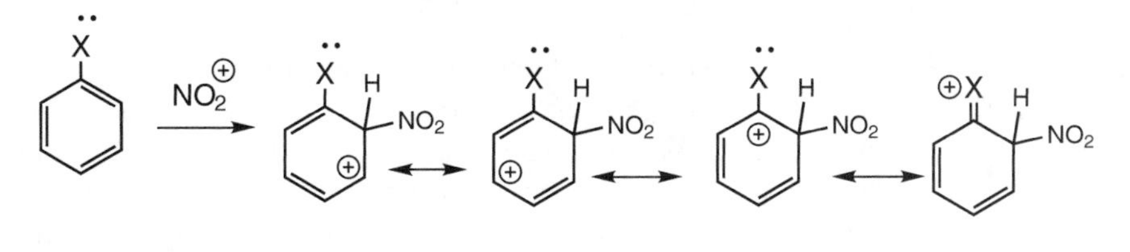

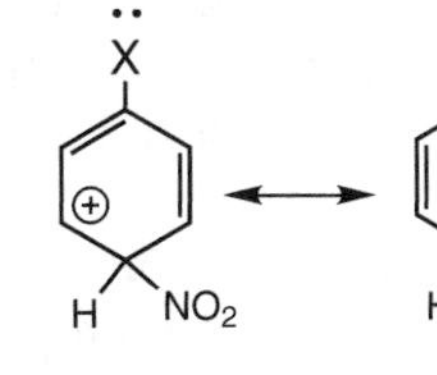
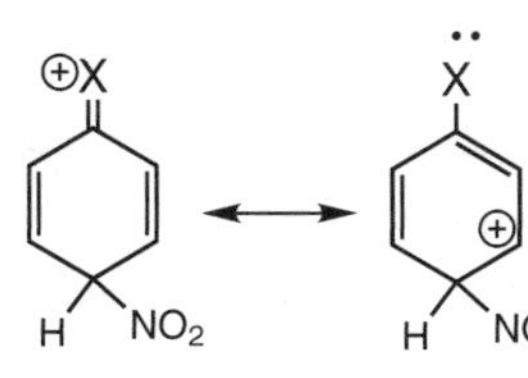

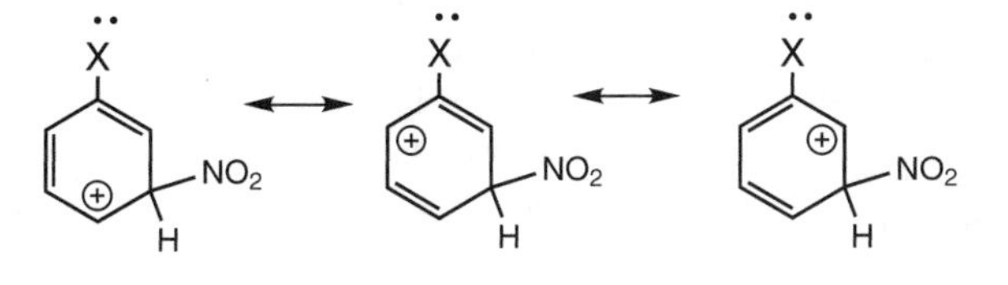

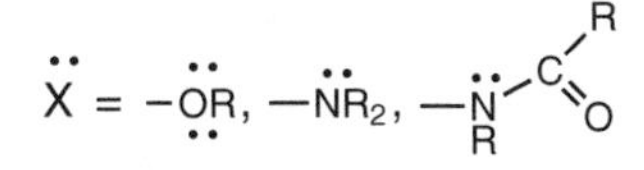

Only in the meta isomer does the positive charge not interact with the nonbonding electron pair.

The Hammond postulate explains why the relatively stable ortho and para intermediates are formed more rapidly in electrophilic aromatic substitution on activated benzenes. The initial formation of the carbocation intermediate is endergonic. Thus, the rate-determining transition state resembles the stable carbocations, the ortho and para isomers.

Amine and oxygen substituted groups react with electrophiles more rapidly than benzene itself. This observation implies that substituted benzenes are more electron-rich than benzene. However, nitrogen and oxygen are inductively electron-withdrawing. This inductive effect is overwhelmed by the electron-donating nature of the lone pair electrons in the π system. One can draw resonance structures for the ground state of the substituted benzenes that emphasize this electron donation:

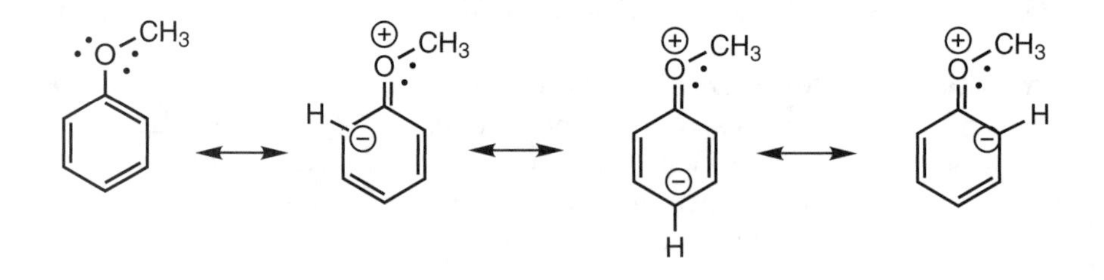

In general, resonance effects trump inductive effects in organic chemistry.

Because activating, or D, groups, promote both ortho and para substituents, one usually obtains a mixture of both products. Although there are two ortho positions and one para position, the para product is often the predominant product. This is because steric factors favor it. Predicting which isomer, ortho or para, will be the major isomer formed is not really possible in introductory organic chemistry. Given a synthetic problem, the student should assume that a mixture of products will result and that they can always be separated to obtain the desired isomer.

A problem with electrophilic aromatic substitution reactions of anilines (amine-substituted benzenes) is that the amine group is basic and can be readily protonated. While the $—NR_2$ group is ortho-para-directing, the amine cation, $—NHR_2^+$, is meta-directing. One way around this problem is to acylate the amine, giving an amide, $—NHCOCH_3$. The amide nitrogen is not basic, so under acidic conditions it directs substitution to the ortho and para positions, and in a separate step the amide can be hydrolyzed back to the amine group.

EXERCISE

Write a mechanism that explains why heating *t*-butylbenzene in aqueous acid gives benzene.

Solution

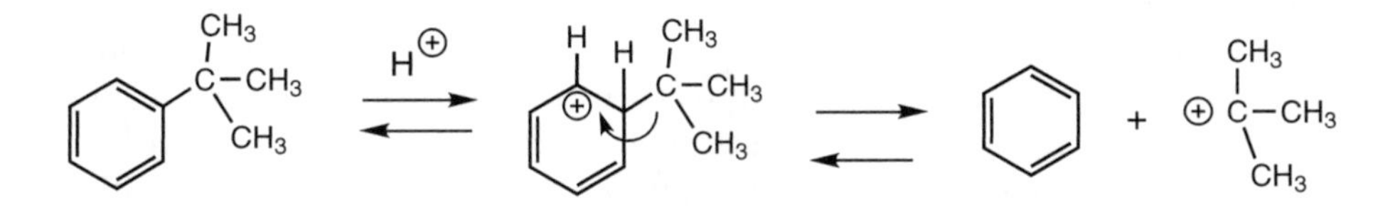

The stability of the *t*-butyl carbocation allows it to compete with H^+ as a leaving group in electrophilic aromatic substitution.

Substitution Reactions on Deactivated Benzene Derivatives

The W substituents listed in Table 24.1 all direct substitution to the meta position. This directing effect arises because all the substituents destabilize an adjacent positive charge:

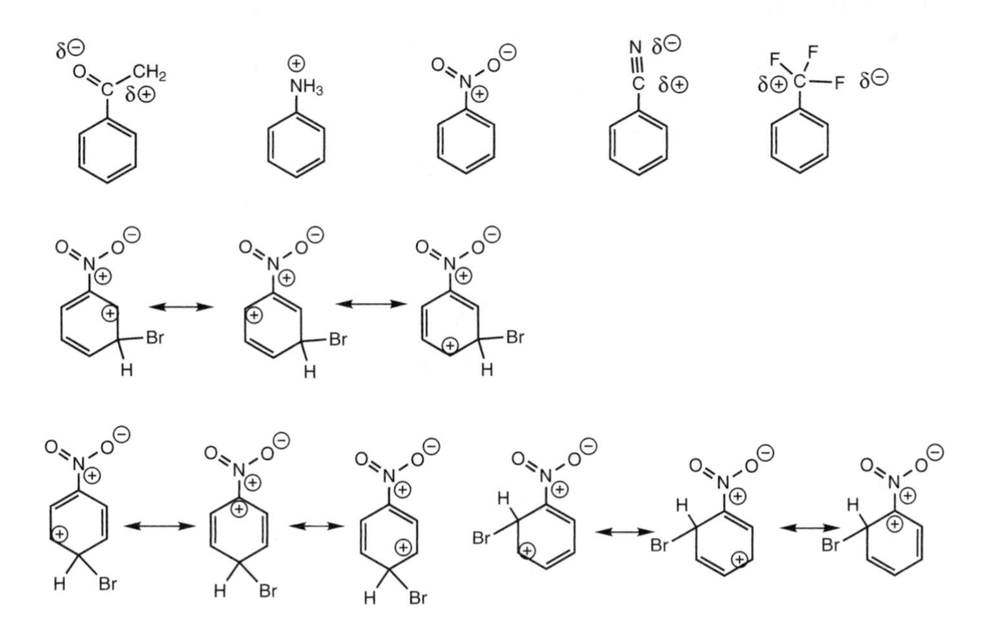

All W groups have a full positive charge, a formal positive charge, or a partial positive charge next to the aromatic ring. When a new group enters the ortho or para position, one of the resonance structures places the two positive charges next to one another. This unfavorable coulombic interaction is avoided only when a meta isomer is formed. Meta isomers form because they arise from the least unfavorable resonance structures.

Because all W groups are electron-withdrawing, they make the benzene ring a poorer nucleophile compared to benzene. In other words, they deactivate the benzene ring. Some less reactive electrophiles, especially the acylium ion in Friedel-Crafts acylation, do not react with W-substituted aromatic rings.

Halogens are a special case in electrophilic aromatic substitution because they are deactivating, relative to benzene, but direct substitution to the ortho and para positions. Since halogens have nonbonding electron pairs, they are similar to oxygen and nitrogen and are capable of stabilizing an adjacent positive charge through resonance. However, the large size of the orbitals on halogens (at least —Cl through —I) means they overlap poorly with carbon orbitals. Halogens are inductively electron-withdrawing relative to —H but do not donate electron density to the ground state of the ring through resonance as —OR or —NR_2 groups can. The net result is overall deactivation because of inductive effects, but small stabilization of ortho- and para-substituted carbocation intermediates, relative to the meta isomer, due to resonance.

Reactions of Multiply Substituted Benzene Rings

When a benzene ring has several substituents, aromatic substitution takes place ortho and para to the most activating group. Since D groups promote reactivity and W groups retard reactivity, when a reaction takes place, it is dominated by the D group. The effect of a W group is to make the reaction slower than it would be if the W group were absent:

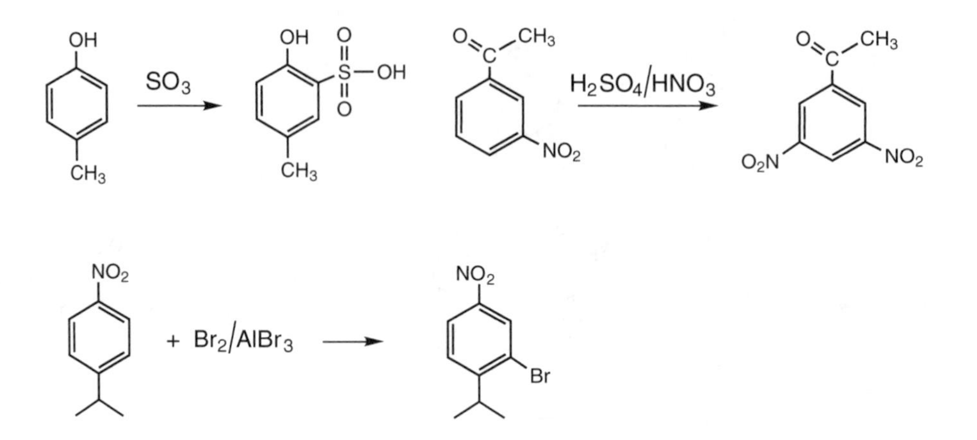

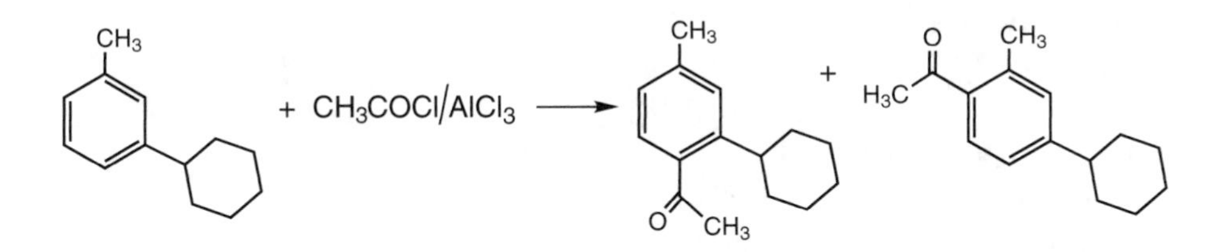

In the third example above, substitution at the position between the two substituents (ortho to them both) is difficult because of steric effects. Usually 1,2,3-trisubstituted isomers are minor products because of steric factors.

EXERCISES

Give the expected product or products for the following reactions.

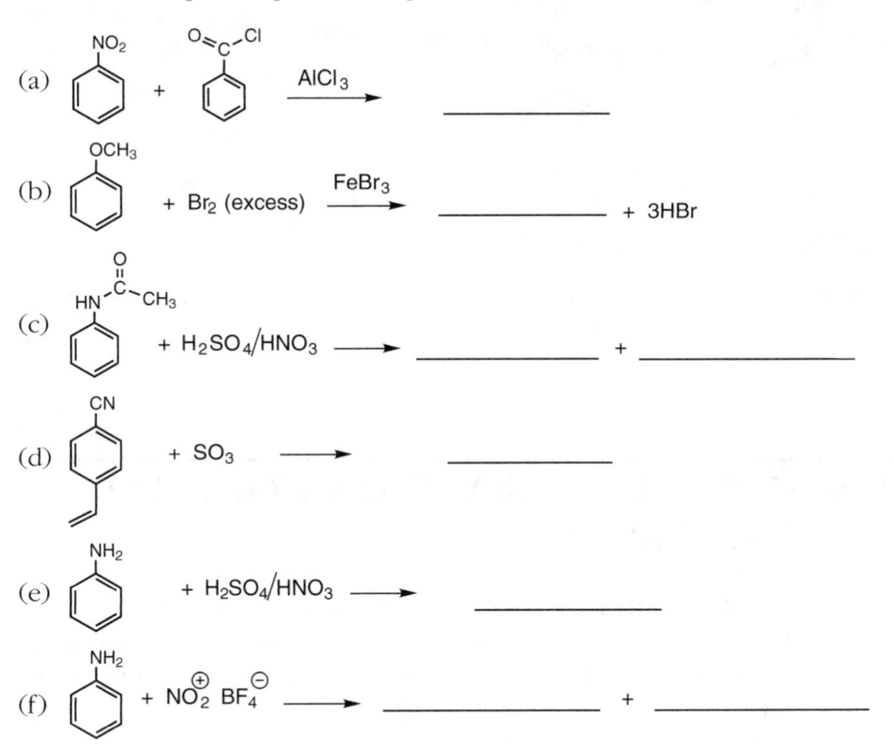

Solutions

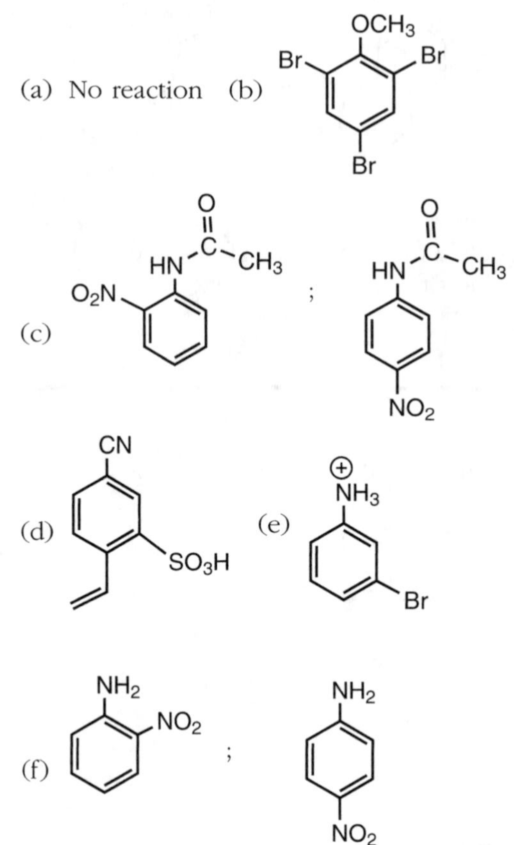

SYNTHESIS AND REACTIONS OF ARYL DIAZONIUM SALTS

Reduction of Aromatic Nitro Compounds to Amines

Scores of reagents have been reported to reduce aromatic nitro groups to amines. Metallic tin in the presence of HCl (which generates $SnCl_2$) is one common reagent. A second useful procedure, which avoids acid, is to reduce the nitro group with Pd/C and hydrogen:

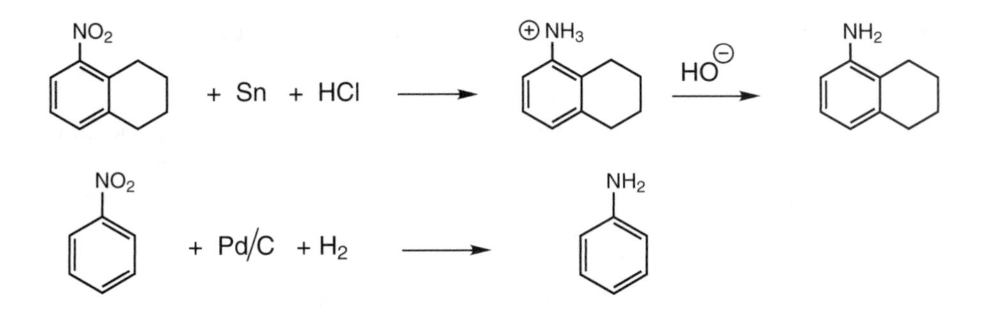

Several intermediates are formed during the reduction, including aryl nitroso compounds, Ar—NO, and aryl hydroxylamines, ArNH—OH. These intermediates are cleanly reduced to the final amine product.

Conversion of Anilines to Aryl Diazonium Salts

Aromatic amines (such as **aniline**, $H_2N{-}C_6H_5$) react with nitrous acid, HONO, to give **aryl diazonium salts**, ArN_2^+. Nitrous acid is prepared by the reaction of sodium nitrite, $NaNO_2$, with an acid such as HCl:

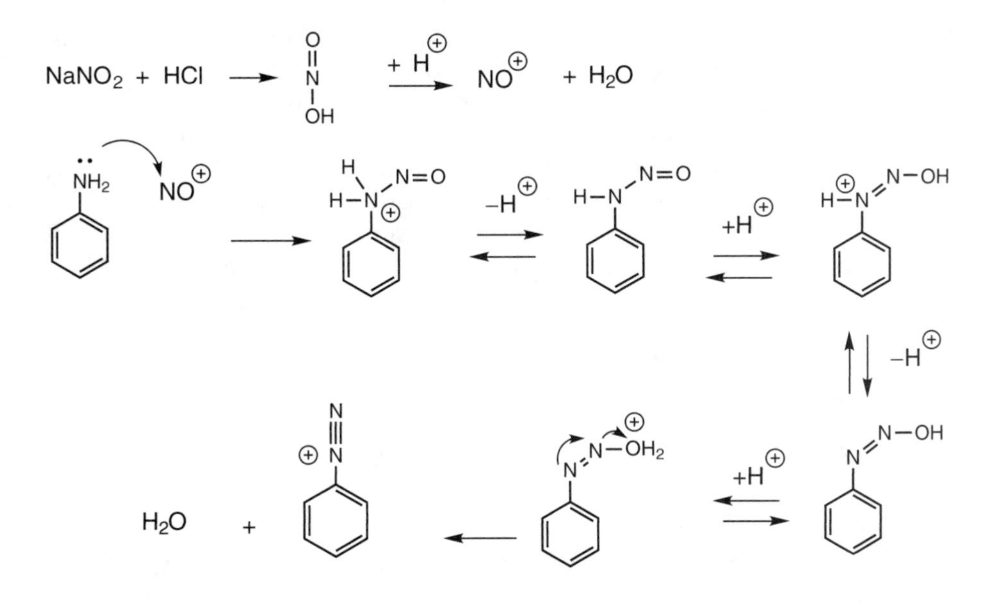

Aryl diazonium salts are potentially explosive when isolated in crystalline form. However, they are useful synthetic intermediates and normally are used as solutions. When HCl is used to generate HONO, the aryl diazonium salt has Cl^- as the counterion. One can also use H_2SO_4, which gives ArN_2^+ HSO_4^-.

Substitution Reactions of Aryl Diazonium Salts

Although N_2 is a superb leaving group, an aryl carbocation is such a high-energy species that aryl diazonium salts do not spontaneously dissociate. However, ArN_2^+ intermediates react with many nucleophiles to give substitution products. These reactions are not simple nucleophilic displacements and may involve radical intermediates. Whatever the mechanism, the nucleophile always ends up on the carbon that was bonded to the N_2 group:

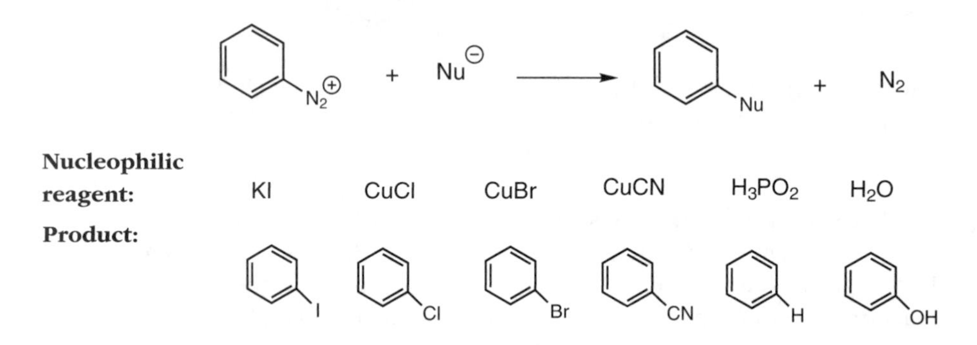

The use of copper salts to introduce Cl, Br, and CN substituents is known as the **Sandmeyer reaction**. Heating of the diazonium salt in water is one of the few general ways to synthesize phenols. The ArN_2^+ HSO_4^- salt must be used in water (rather than the ArN_2^+ Cl^- salt) so that water is the only reactive nucleophile. The reaction with **hypophosphorus acid**, $HP(OH)_2$, at first may not seem to be useful. However, recall that diazonium salts are prepared from the strongly activating $—NH_2$ group. The amine group can direct groups onto a ring and then, in a subsequent series of steps, be removed:

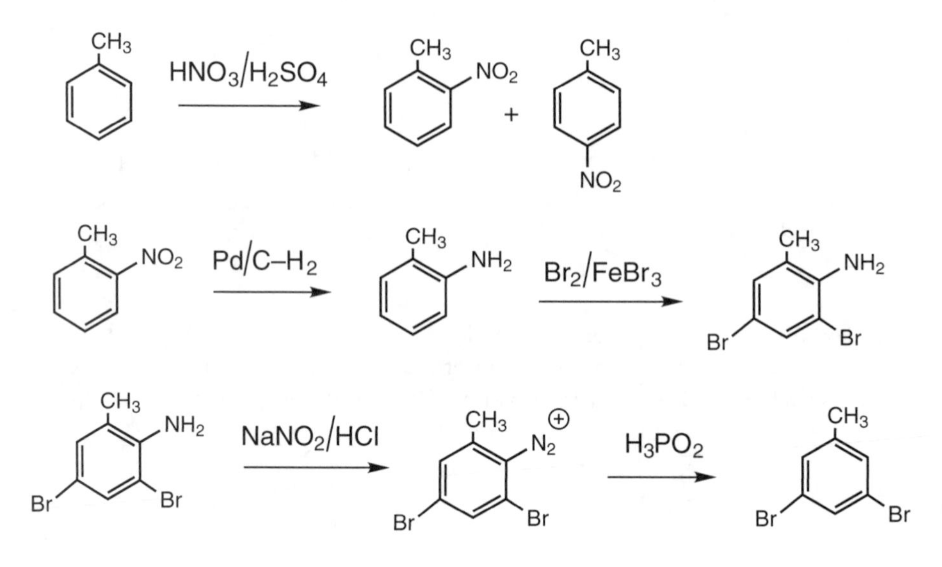

The product, 3,5-dibromotoluene, looks odd because both groups are ortho-para-directing but are all meta to one another. However, they are ortho and para to the departed diazonium group. This same strategy can be used to place groups ortho or para to a meta-directing group:

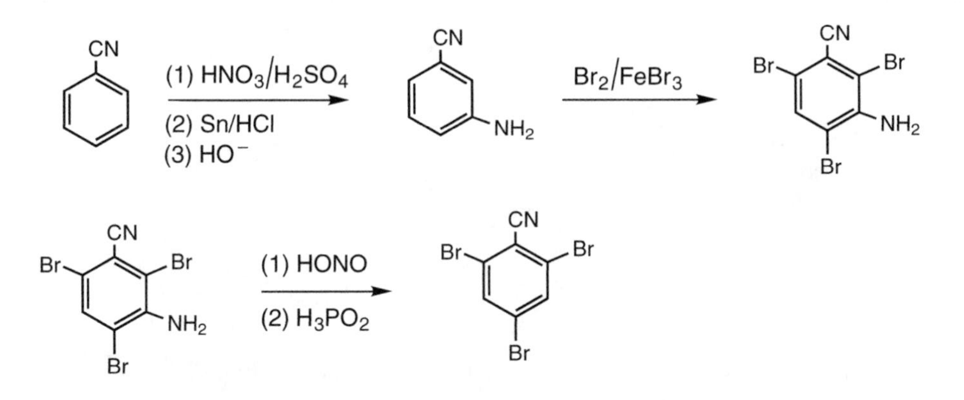

Very electron-rich aromatic rings, such as phenols or anilines, can react with aryl diazonium salts on nitrogen to give **azobenzenes**, Ar—N=N—Ar′. This class of compounds is normally brightly colored, and many dyes are based on this functional group:

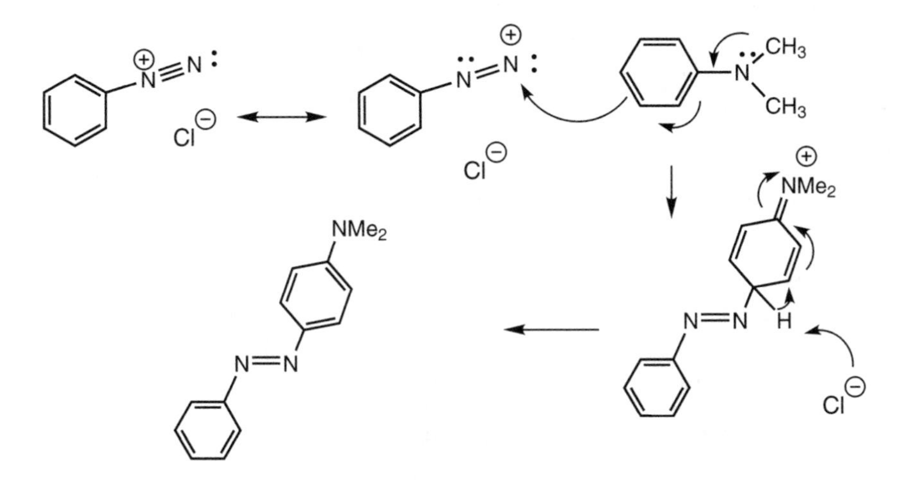

Side-Chain Reactions of Alkyl Benzenes

A benzene ring is resistant to most reduction and oxidation reagents. However, alkyl groups attached to arenes are readily oxidized. We have already covered NBS oxidations (Chapter 13). These reactions can be used to make benzylic bromides.

Alkyl groups are oxidized all the way to carboxylic groups with either H_2CrO_4 (chromic acid) or alkaline $KMnO_4$. Methyl groups, linear alkyl groups, and branched alkyl groups are all oxidized to the —COOH group. The mechanisms of these oxidations involve an initial free radical abstraction of a benzylic C—H bond by the Cr or Mn reagent. The only unreactive groups are tertiary alkyl groups because they have no benzylic C—H bonds:

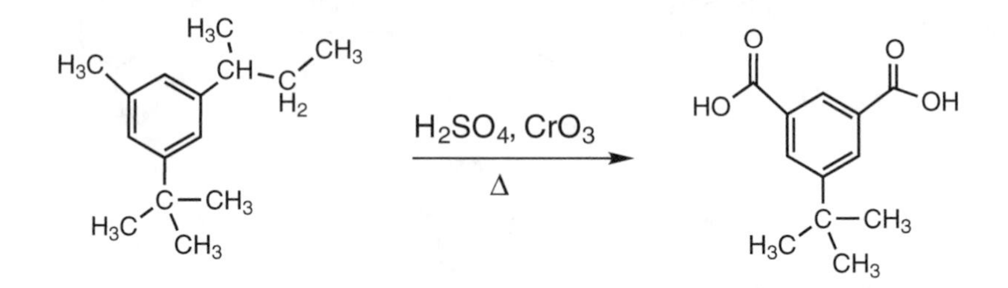

Functional groups that can be oxidized, such as alkenes, alkynes, alcohols, and aldehydes, react first under these very vigorous reaction conditions.

Multistep Syntheses of Arenes

At this point, the student should have a command of reactions that form carbon-carbon bonds and reactions that change one functional group to another. The following examples of multistep syntheses use reactions not only from this chapter but from earlier chapters as well.

Examples

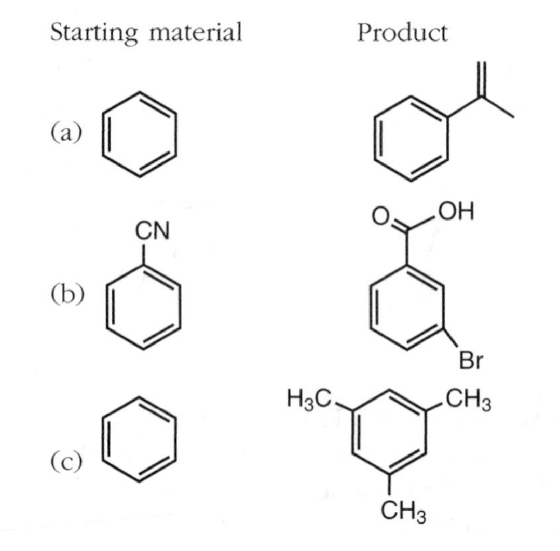

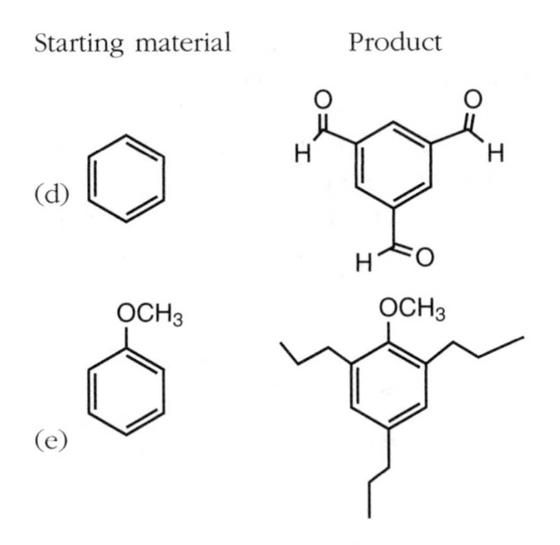

Solutions: In Example (a), one needs to attach an sp^2 carbon to a benzene ring. A Friedel-Crafts reaction with $CH_2{=}CHBr{-}CH_3$ will not work because one cannot generate a carbocation from an sp^2 carbon. Instead, a three-step reaction is needed involving Friedel-Crafts acylation, alkylation of the ketone, and elimination. Alternatively, bromination followed by cuprate alkylation of an aryl bromide is an excellent procedure.

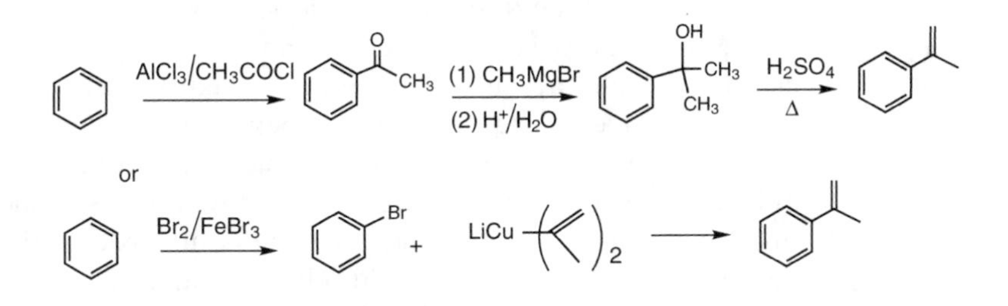

To solve Example (b), one must recall that nitriles are hydrolyzed to carboxylic acids.

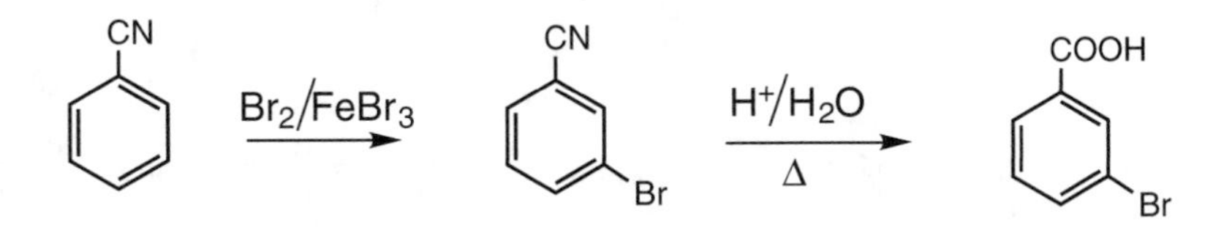

Example (c) is another case of ortho-para-directing groups all meta to one another on a ring. This kind of problem uses the "disappearing —NH_2" trick where all the groups are ortho or para to an —NH_2 that used to be on the ring. Because amines are bases, one needs to protect the amine as an amide during the alkylation. Otherwise, the basic amine group will form a complex with the $AlCl_3$ Lewis acid and not react.

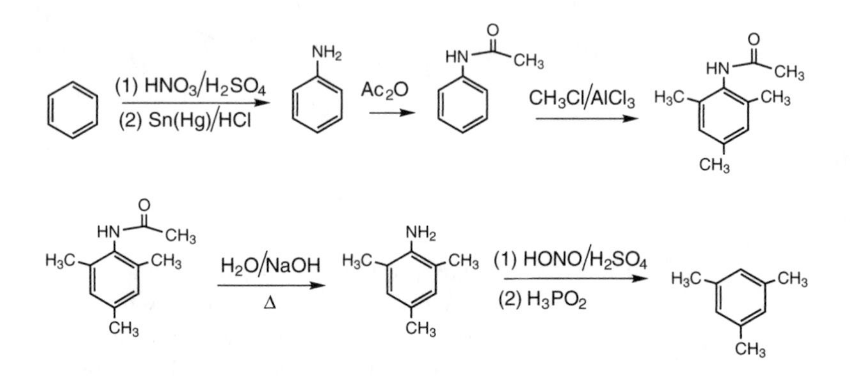

Example (d) has three meta-directing groups meta to each other. However, we know that Friedel-Crafts acylations can be used only to introduce one acyl group. A further complication in this case is that the desired acid chloride, formyl chloride, ClCHO, is unstable and decomposes to HCl and CO. It cannot be used in a Friedel-Crafts acylation. An indirect method must be used to introduce aldehyde groups onto a benzene ring. Since Example (c) provides a way to make 1,3,5-trimethylbenzene, we can use this as a precursor and oxidize the methyl groups to aldehydes. In addition to the method shown, one could oxidize the methyl groups to —COOH groups with chromic acid or permanganate, reduce the acids to alcohols with $LiAlH_4$, and oxidize the alcohols to aldehydes with PCC.

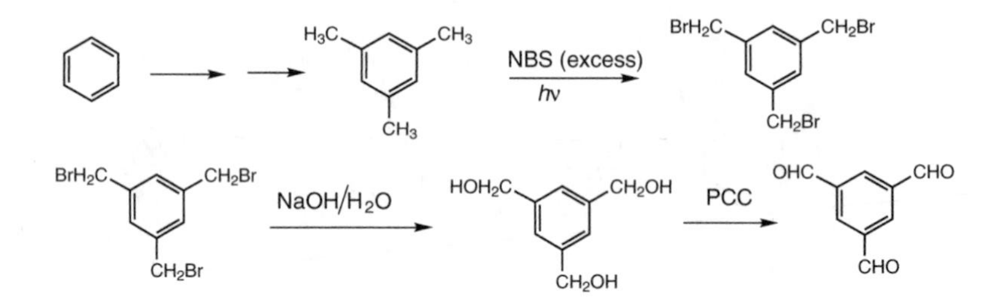

Example (e) raises a question of alkylation. Since Friedel-Crafts alkylation cannot be used to introduce *n*-propyl groups, because of rearrangements, an acylation-reduction route is suggested. However, Friedel-Crafts acylation cannot be used on deactivated benzene rings. Acylation with CH_3CH_2COCl should occur twice on the starting material since $—OCH_3$ is a strongly activating group. However, acylation may or may not take place on an intermediate with two deactivating acyl groups and one $—OCH_3$ group. A safer synthesis would involve multiple bromination of the starting material and replacement of the —Br groups with $LiCu(n\text{-}Pr)_2$.

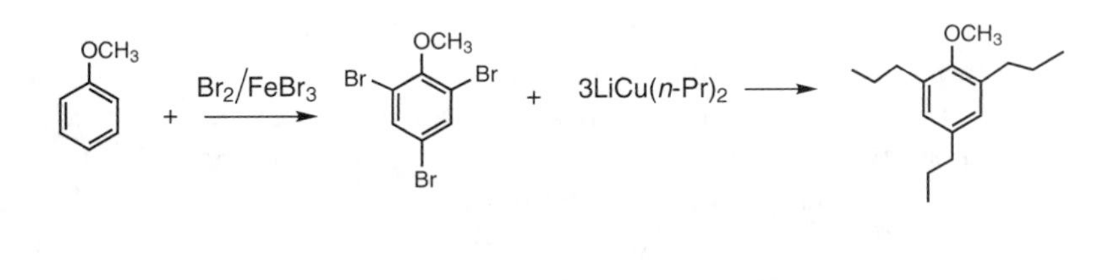

PHENOLS

Acidity of Phenols

Unlike carboxylic acids, where only the inductive effects of substituents can affect acidity, both resonance and inductive effects can change the acidity of phenols. Except for strongly electron-withdrawing groups such as the nitro group, most substituents have a minor influence on acidity, as summarized in Table 24.2.

TABLE 24.2. ACIDITY OF SUBSTITUTED PHENOLS

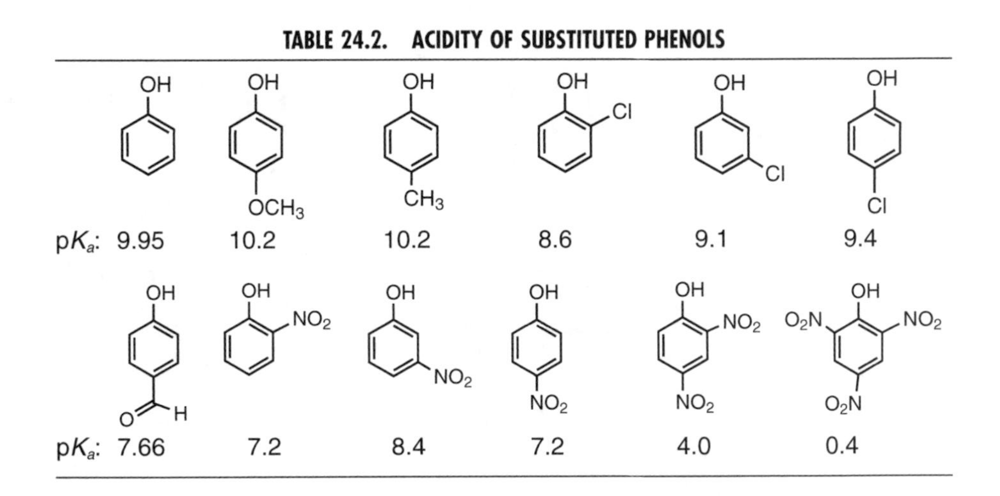

	OH	OH, OCH3	OH, CH3	OH, Cl	OH, Cl	OH, Cl
pK_a:	9.95	10.2	10.2	8.6	9.1	9.4

	OH, O, H	OH, NO2	OH, NO2	OH, NO2	OH, NO2, NO2	OH, O2N, NO2, O2N
pK_a:	7.66	7.2	8.4	7.2	4.0	0.4

A halogen substituent is able to stabilize the negative charge of a phenoxide anion by inductive effects. Since inductive effects fall off with distance, *ortho*-chlorophenol is the most acidic and *para*-chlorophenol is the least acidic. A nitro group can stabilize a negative charge through inductive effects or resonance:

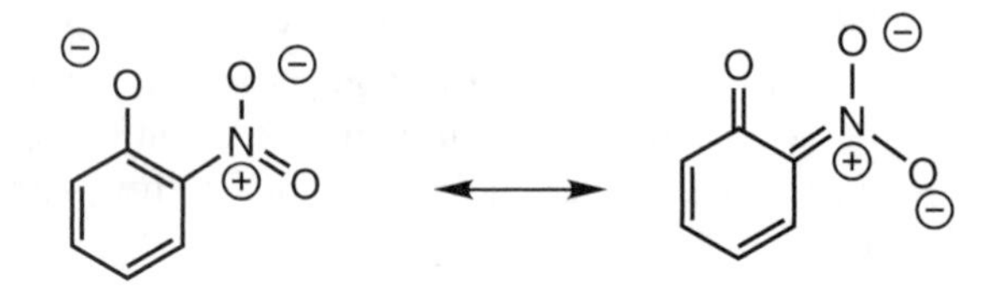

In the ortho isomer, there is some anion-anion repulsion as well as favorable inductive and resonance stabilization of the phenoxide anion. In the meta nitro isomer, only inductive effects stabilize the conjugate base. In the para isomer, resonance stabilization dominates. As a result, the ortho and para nitrophenols are equally acidic, while the meta isomer is less acidic. The resonance and inductive effects are additive, so phenols substituted with multiple nitro groups are strong acids. Picric acid, 2,4,6-trinitrophenol, in addition to being an explosive, is a stronger acid than HF.

Reactions of Phenols

One of the most important phenol derivatives is aspirin, **o-acetylsalicylic acid**. Salicylic acid derivatives are found in willow tree bark and were used in folk medicine for centuries, but the bark is so bitter that many pain sufferers found pain preferable to chewing the bark. Salicylic acid is *o*-hydroxybenzoic acid. It is prepared by the **Kolbe-Schmitt reaction**, which involves heating sodium phenoxide with CO_2 and acidification to give the acid:

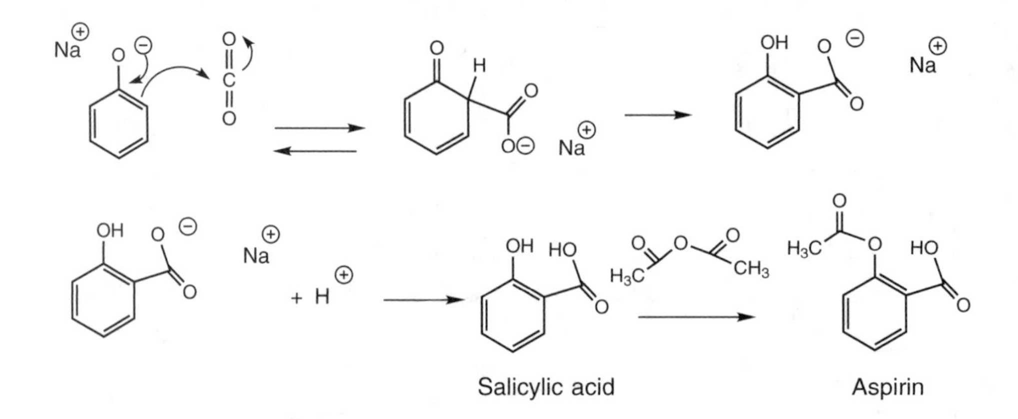

The ortho isomer is the thermodynamic isomer in the Kolbe-Schmidt reaction because there is an intramolecular hydrogen bond between the phenolic OH and the carboxylic acid group.

Allyl aryl ethers undergo a thermal rearrangement reaction known as **Claisen rearrangement**. The movement of electrons is exactly the same as that seen in the Diels-Alder reaction except that an intramolecular reaction occurs rather than an intermolecular reaction:

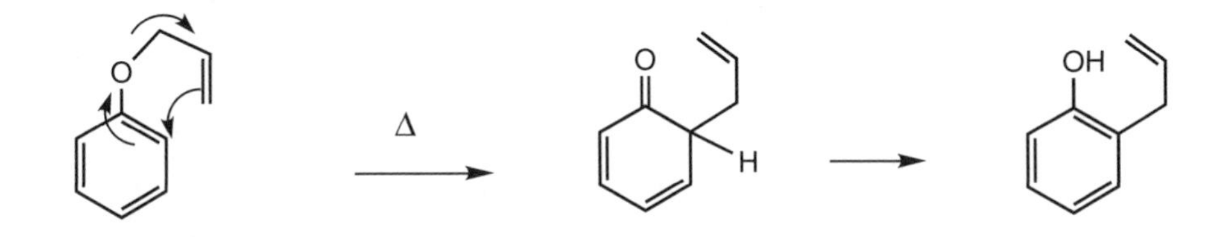

The ether is made in a standard Williamson ether synthesis reaction of sodium phenoxide and allyl halide. The mechanism for formation of the phenol is the same as that for a ketone-enol rearrangement except that, because of aromatic resonance stabilization, the enol form is much more stable.

1,4 Diphenols are known as ***para*-hydroquinones**. Numerous oxidizing agents convert them to ***para*-quinones**, including Cr^{+6} reagents. Reducing agents convert quinones to hydroquinones:

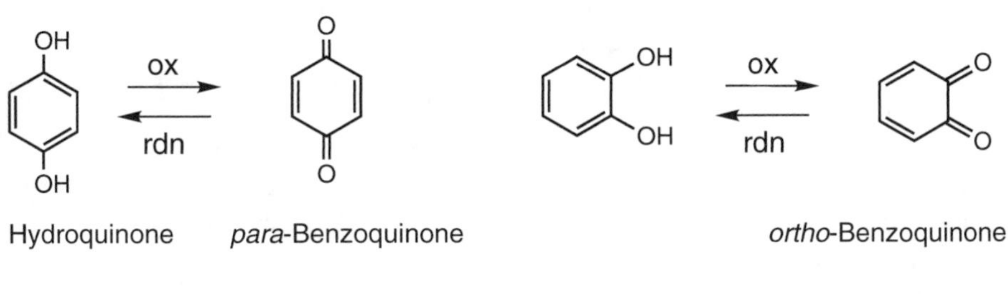

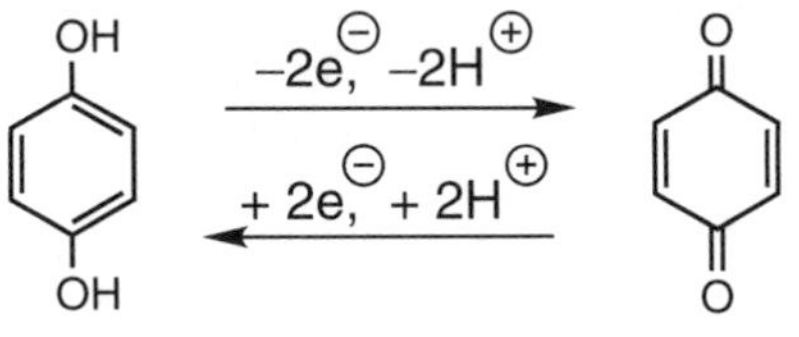

1,2 Diphenols are oxidized to ortho quinones, which are normally less stable than para quinones. The quinone-hydroquinone equilibrium is important in biochemical electron transfer. Substituted versions of these molecules act essentially as molecular batteries to store and release electrons.

There is an important difference in how oxidation and reduction reactions are depicted in organic chemistry and in biochemistry. Most organic reactions are shown as two-electron atom transfer reactions. For example, when a ketone is reduced to an alcohol, a hydride reagent, a source of H^-, is depicted. We use curved arrows to show the flow of electron pairs. In biochemistry, electrons are normally depicted as being added or removed one electron at a time. Thus, *para*-benzoquinone is reduced to hydroquinone by adding an electron, then a proton, then an electron, and then another proton. In fact, such single-electron transfer processes are how many substrates are reduced or oxidized in biochemistry.

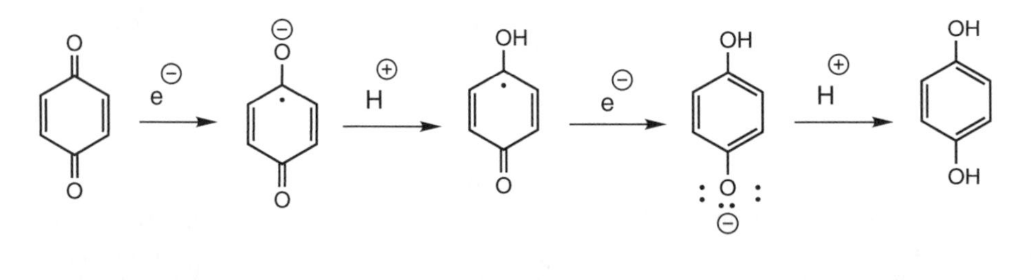

ANILINES

Basicity of Aromatic Amines

Amines are organic bases, and anilines are weak bases. To the extent that the nitrogen lone electron pair is delocalized into the aromatic ring, an aromatic amine is a weaker base than an amine with alkyl substituents. As in phenols, groups that stabilize negative charge affect the acid-base properties of aniline derivatives.

There are two ways to measure the basicity of aniline derivatives. Since a strong acid has a weak conjugate base, one can measure the pK_a of the amine's conjugate acid. The more acidic the RNH_3^+, the weaker the free base RNH_2. Alternatively, one can define the parameter pK_b, where $K_b = [RNH_3^+][HO^-]/[RNH_2]$ and $pK_b = -\log K_b$. We can use the pK_a of the amine's conjugate acid to measure amine basicity, or pK_b, since for an amine, $pK_a + pK_b = 14$. Table 24.3 summarizes amine basicity. By comparison, the pK_a of the protonated alkyl amine $CH_3NH_3^+$ is 10.6. Because protonation of the nitrogen lone electron pair prevents resonance interactions with the aromatic ring, aromatic amines are less basic (more difficult to protonate) than alkyl amines. A substituent that is electron-withdrawing, because of either resonance or inductive effects, decreases the basicity of an aromatic amine, and electron-donating groups (such as alkyl groups) slightly increase the amine's basicity.

TABLE 24.3. BASICITY OF AROMATIC AMINES

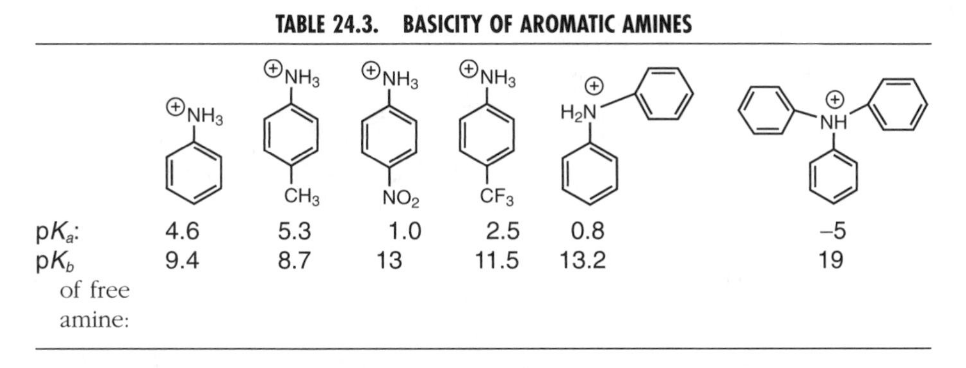

	Anilinium ($C_6H_5NH_3^+$)	p-CH_3	p-NO_2	p-CF_3	$(C_6H_5)_2NH_2^+$	$(C_6H_5)_3NH^+$
pK_a:	4.6	5.3	1.0	2.5	0.8	–5
pK_b of free amine:	9.4	8.7	13	11.5	13.2	19

Synthesis of Aromatic Amines Through Benzyne Intermediates

Very strong bases, under vigorous conditions, can cause aryl halides to undergo an elimination reaction to generate a **benzyne**, C_6H_4, intermediate. For example, heating chlorobenzene in sodium amide/ammonia gives aniline via benzyne formation:

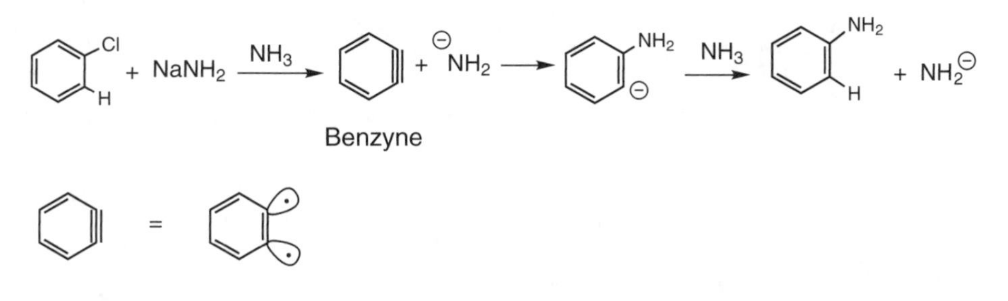

Benzyne is an extremely strained alkyne. A π-bond needs to have its *p* orbitals parallel for good overlap. In benzyne, the *p* orbitals in the plane of the ring are not parallel because of the geometry of the σ-bond framework. Consequently, this weak bond undergoes addition reactions with almost any reagent to give substitution products.

EXERCISES

1. Order the following compounds from most acidic to least acidic.

(a)

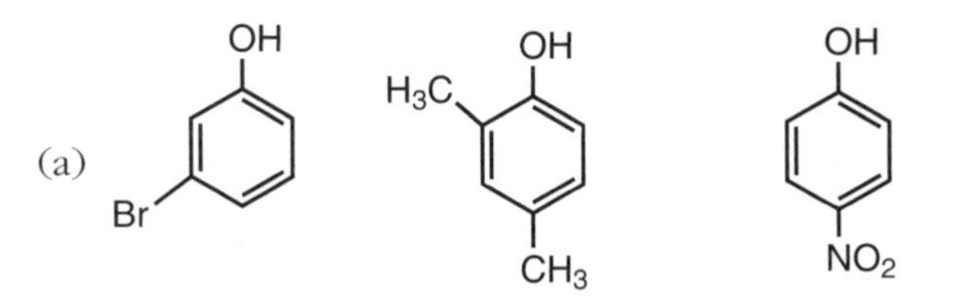

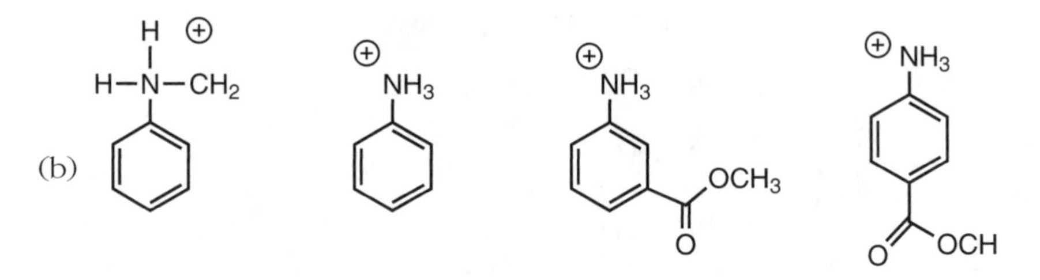

2. Give a mechanism for the following reaction.

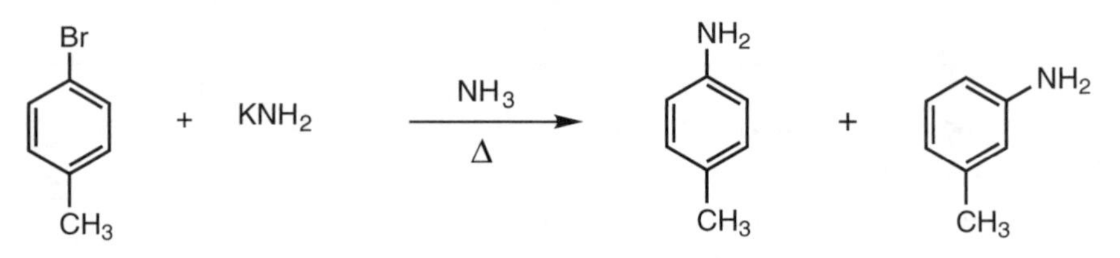

Solutions

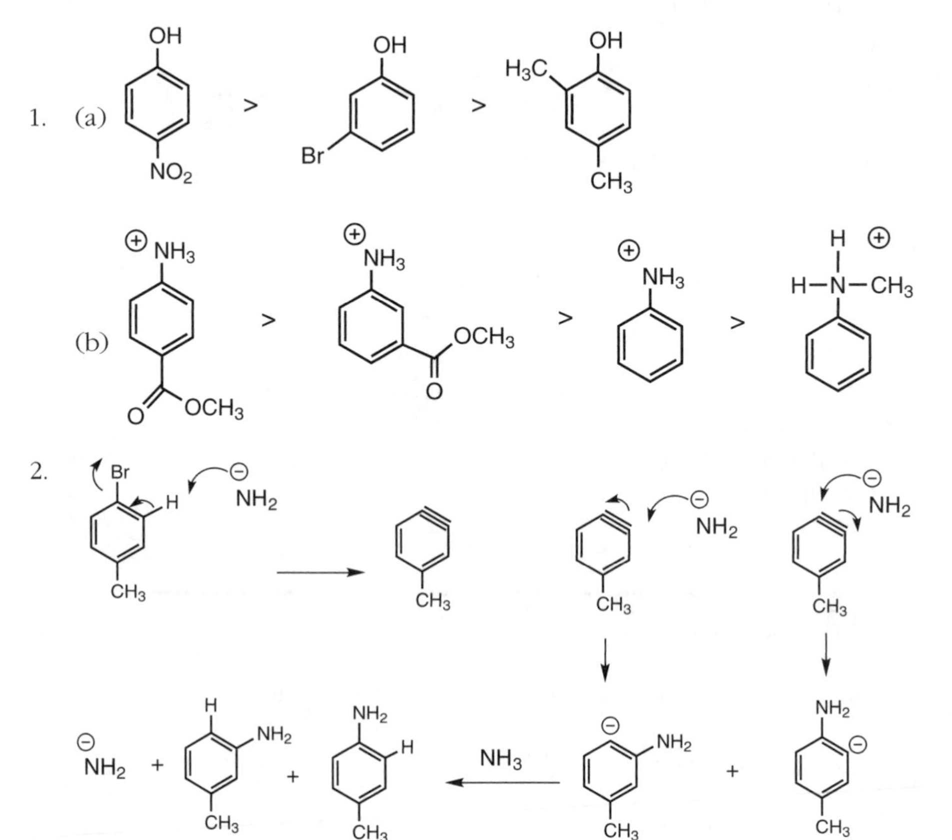

25
SYNTHESIS AND REACTIONS OF AMINES

The synthesis and properties of numerous nitrogen-containing functional groups have already been discussed. Amides, imines, enamines, nitriles, diazonium salts, and nitro compounds have been covered in detail. In this chapter, the properties and reactions of organic derivatives of ammonia—amines—will be outlined. Amines are extremely important in organic chemistry and medicinal chemistry. Most weak to moderate bases in organic chemistry and biochemistry are amines. Many of the structures of drugs used in human medicine are found to be amines, in part because most amines are protonated at physiological pH and are thus water-soluble as salts.

STRUCTURE AND BASICITY OF AMINES

Structure of Alkyl Amines

Ammonia and alkyl amines (**primary**, RNH_2; **secondary**, R_2NH; and **tertiary**, R_3N) are **pyramidal**, and the nitrogen can be considered sp^3 hybridized. Unlike sp^3 hybridized carbons, however, an amine undergoes **inversion**, with the lone electron pair oscillating between one side of the molecule and the other:

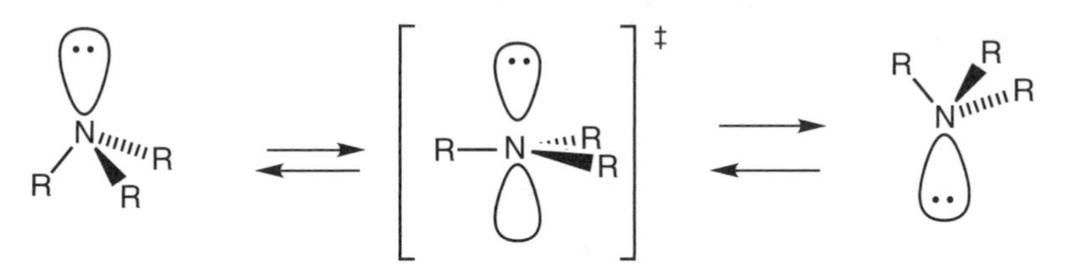

The transition state for this inversion has the nitrogen lone pair in an unhybridized *p* orbital. Since the energy barrier to nitrogen inversion is very low, an amine with three different R groups is not optically active. Although for one instant such a compound is chiral, because there are four different groups around the tetrahedral nitrogen, inversion converts one enantiomer to the other. This racemization takes place even at very low temperatures. **Quaternary** ammonium salts, R_4N^+, have four N—C bonds and no lone electron pair.

As a result, these compounds are potentially chiral and can be resolved into stable enantiomers.

Placing groups on the amine nitrogen that can interact through resonance with the lone electron pair stabilizes the nitrogen inversion transition state and flattens out the nitrogen. When one of the R groups is aromatic, resonance causes a flattening of the bonds around the amine by approximately 20°. When one of the R groups is an acyl group, the resulting amide nitrogen is planar.

EXERCISE

What is the geometry of the amine nitrogen in *p*-nitroaniline?

Solution

Because of the strong resonance interaction between the amine lone electron pair and the nitro group, the $—NH_2$ group should be planar.

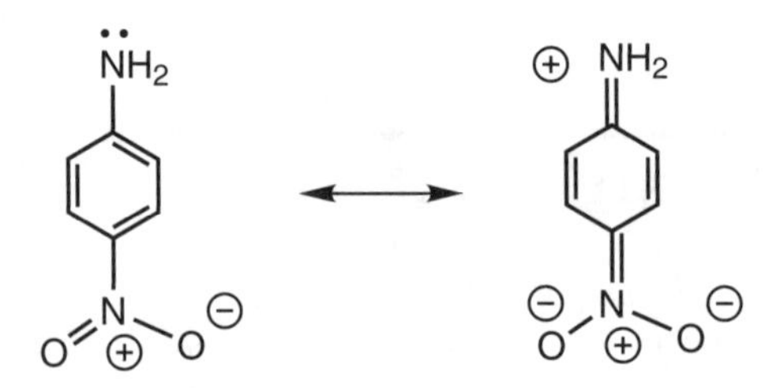

Imines, compounds with a carbon-nitrogen double bond, can be considered to have an sp^2 hybridized nitrogen. In this case the nitrogen is planar, with the lone pair in an sp^2 hybrid orbital. Pyridine can be considered an aromatic version of an imine, and it is a weak base (pK_a of its conjugate acid is 5.25).

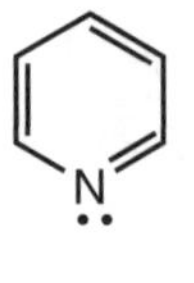

Pyridine

Basicity of Amines

A summary of amine base strength is given in Table 25.1, in the form of the pK_a values of the amine's conjugate acids. Recall that a strong base has a weak conjugate acid, so the higher the pK_a value, the stronger the base.

TABLE 25.1 BASE STRENGTHS (MEASURED AS CONJUGATE ACID pK_a VALUES) FOR AMINES

Amine	pK_a of Conjugate Acid
NH_3	9.25
H_2NCH_3	10.65
$HN(CH_3)_2$	10.73
$N(CH_3)_3$	9.81
$N(CH_2CH_3)_3$	10.75
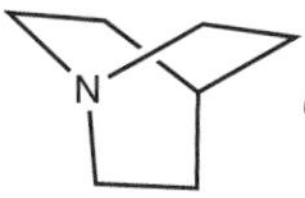 Quinuclidine	10.60
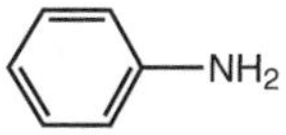 C_6H_5—NH_2	4.6
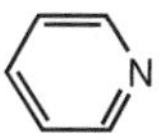 Pyridine	5.25
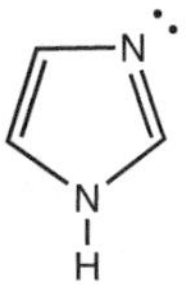 Imidazole	6.95

Several features are involved in determining the basicity of amines. First, alkyl groups are electron-donating relative to hydrogen and serve to destabilize the nitrogen lone electron pair. Thus, methylamine is more basic than ammonia, and dimethyamine is more basic still. However, tertiary alkyl amines are less basic than one would expect because of steric crowding in the tertiary amine salt. The importance of steric crowding is seen in quinuclidine, where the alkyl groups are tied back by the bicyclic rings, leading to increased basicity compared to that of trimethylamine. Electron-withdrawing groups act to decrease amine basicity. For example, in ethylenediamine, $H_2NCH_2CH_2NH_2$, protonation of one amine decreases the basicity of the other —NH_2 group because of inductive and field effects.

The nitrogen lone electron pairs in pyridine and imidazole are in sp^2 orbitals, with 33 percent *s* character. They are lower in energy and thus less basic than the electron pair in the amine sp^3 orbital, with 25 percent *s* character. Of the two nitrogens in imidazole, the lone pair on the NH nitrogen is needed to complete an aromatic sextet and is not the site of protonation. Instead, the sp^2 nitrogen is protonated.

The strongest neutral amine bases are derivatives of **guanidine**:

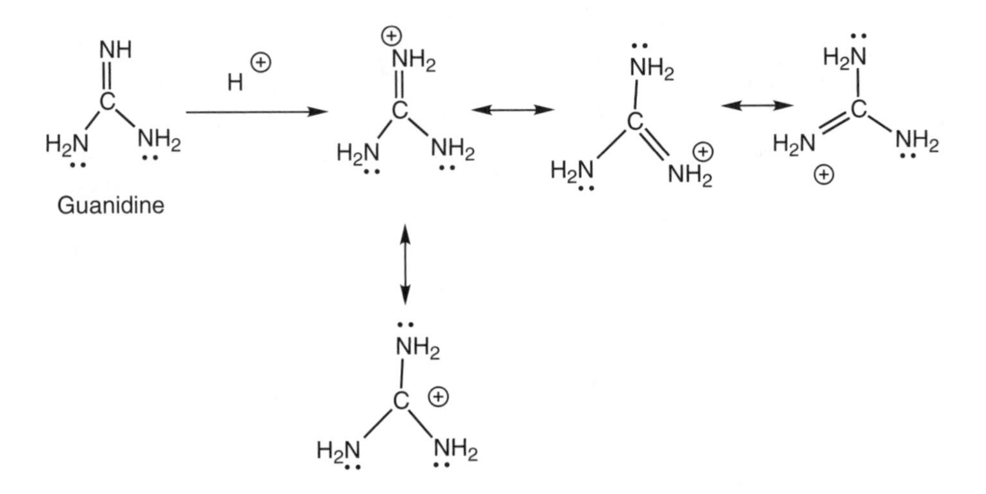

Protonation of the =NH nitrogen creates an extensively delocalized (and thus stable) cation. This conjugate acid has a pK_a of 13.6.

Example

4-Dimethylaminopyridine is a strong organic base often used in organic synthesis. Why is it a strong base?

Solution: The answer is that the resonance donation of electron density to the pyridine ring:

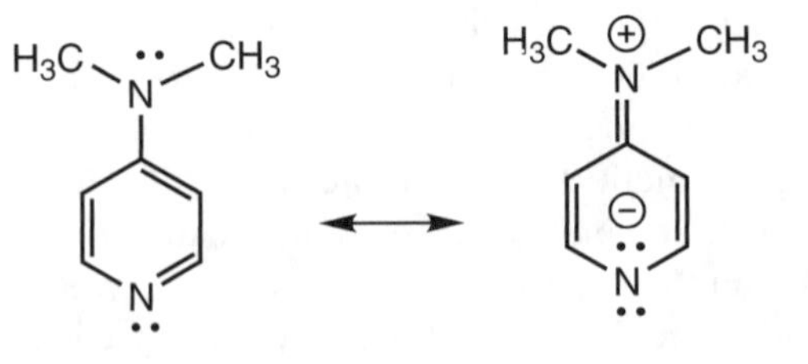

The nonbonding electrons on the aromatic nitrogen are perpendicular to the aromatic π system. However, resonance donates electron density to that nitrogen from the —NMe_2 group, generating electron-electron repulsion and making the pyridine nitrogen a stronger base.

SYNTHESIS OF AMINES

Nucleophilic Reactions With Alkyl Halides or Epoxides

Since ammonia and amines are nucleophiles, they react with electrophiles such as alkyl halides to form new C—N bonds. The initial product is a protonated amine that must be neutralized with base to give the neutral amine. This kind of reaction is of limited synthetic value because the product can react further with more RX to give mixtures of products:

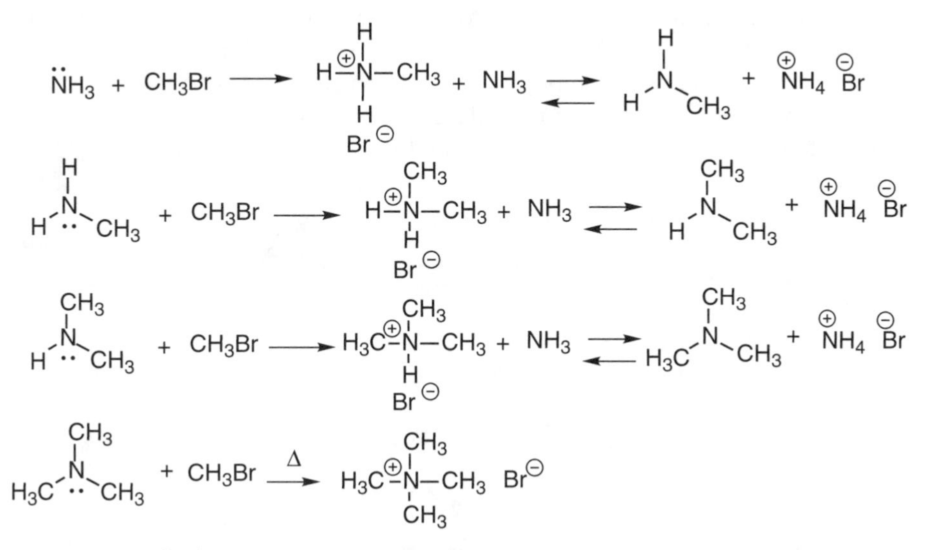

Because alkyl groups are electron-donating, a 1° amine is a better nucleophile than ammonia. Thus, as soon as its concentration reaches measurable levels, it competes with ammonia for the electrophile. Ultimately, if the reaction takes place at a higher temperature, the quaternary ammonium salt can be synthesized. Only if NH_3 is present in great excess can the 1° amine be isolated in good yield.

A better way to synthesize primary amines is the **Gabriel synthesis**, which uses potassium phthalimide as the source of nucleophilic nitrogen:

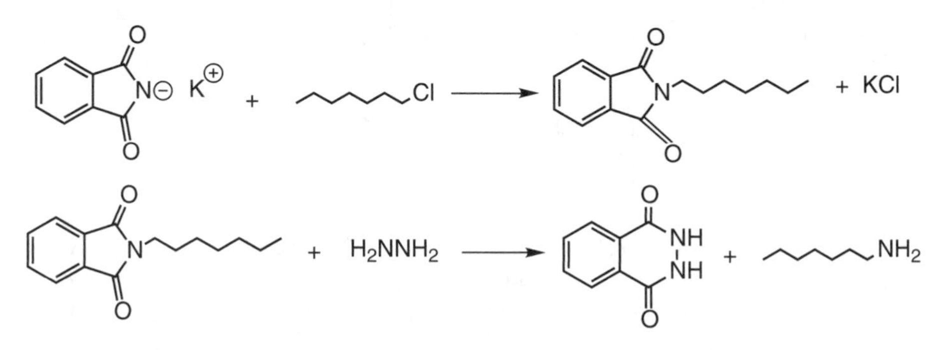

The initial product, an *N*-alkyl phthalimide, is the product of an S_N2 reaction. Primary halides or tosylates make the best electrophiles, and dipolar aprotic solvents such as DMSO are the best solvents. Imides, like amides, can be hydrolyzed in an aqueous base, but a reaction with hydrazine (which is mechanistically similar to the transesterification reaction) liberates the amine product in excellent yield. Imides, like amides, are not bases, so the *N*-alkyl phthalimide product does not react further with alkyl halides.

The reaction of amines with epoxides takes place when **epoxide resins** are used as adhesives or coatings. When one buys epoxy glue, there are often two tubes of material that are mixed to form the glue that eventually sets into a rigid resin. One of the tubes contains a diamine, such as $NH_2CH_2CH_2NH_2$, and the other contains a compound with two or more epoxy groups. As the epoxy cures, the amine groups undergo S_N2 reactions with the epoxides, leading to polar polymers that stick to and bond with surfaces:

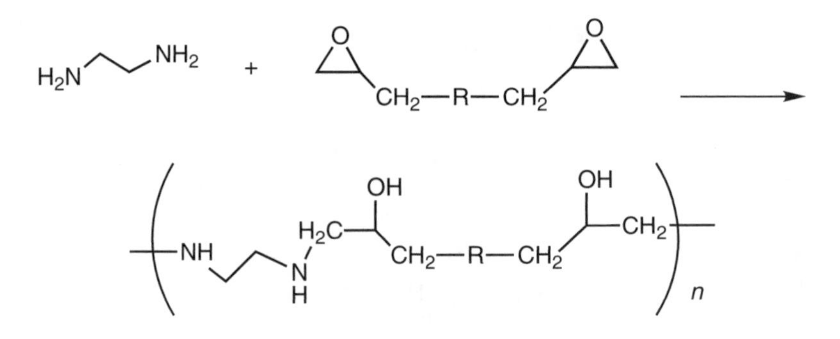

Amines From the Reduction of Azides and Other Functional Groups

The azide anion, N_3^-, is an excellent nucleophile. Since alkyl azides undergo facile reduction to primary amines, this two-step process provides another valuable route to primary amines:

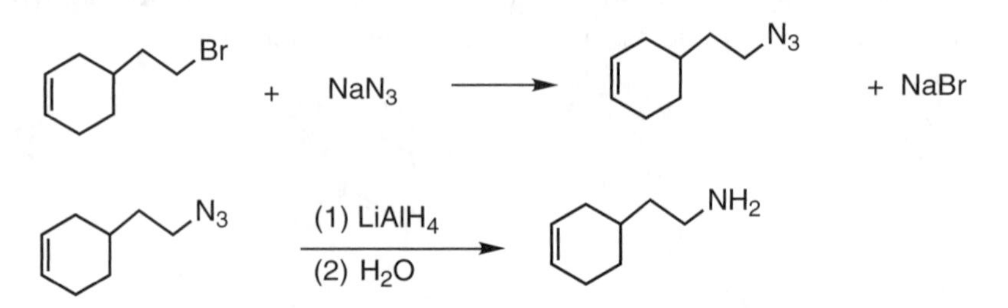

The azide group is reduced by $LiAlH_4$ or by using catalytic hydrogenation with Pt and H_2. Amines are also the product when $LiAlH_4$ reduces nitriles or amides, as described in Chapter 20. Primary, secondary, or tertiary amines can be made by reducing amides with $LiAlH_4$, depending on whether $RCONH_2$, $RCONHR'$, or $RCONR'_2$ is the substrate.

Reductive Amination

A general method for synthesizing amines, which is similar to the way in which amines are synthesized in biochemistry, is **reductive amination**. In this reaction, an aldehyde or ketone reacts with an amine to form an imine. In a second step, the C=N bond is reduced to a single bond:

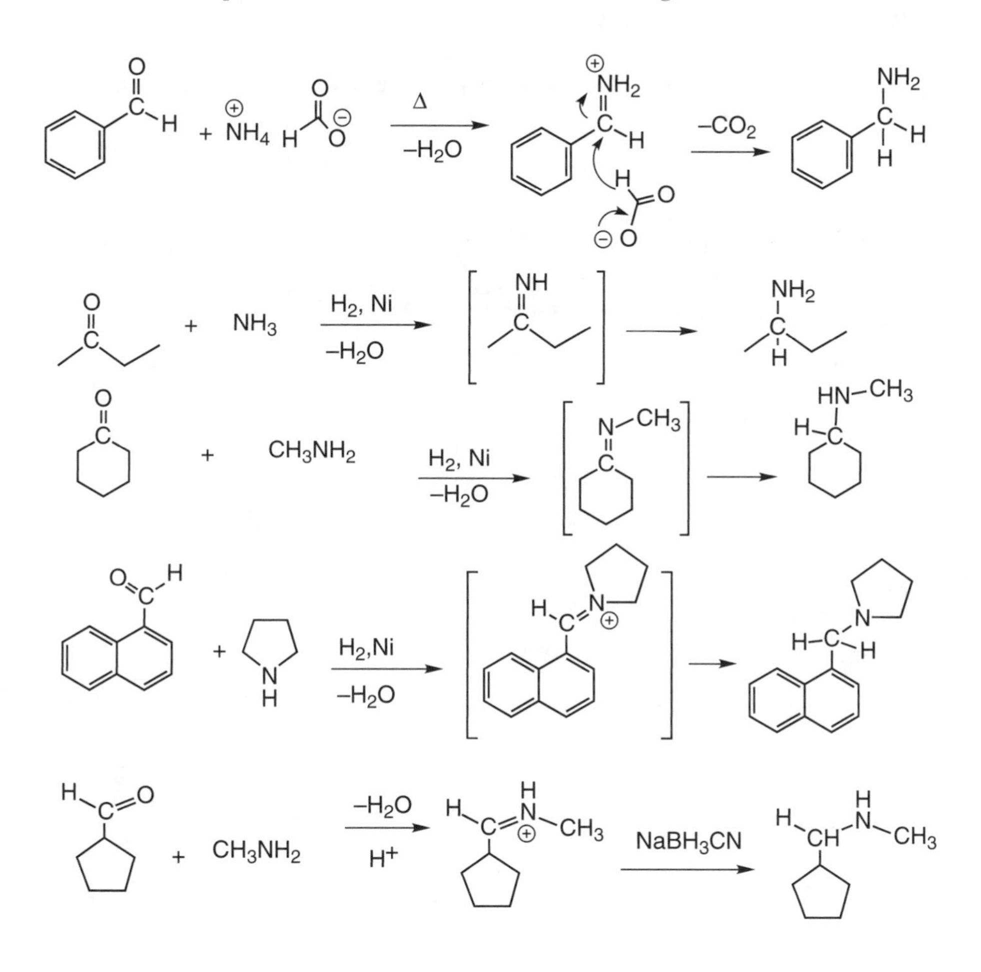

Numerous ways to reduce the imine intermediate are available. Heating an aldehyde or ketone with ammonium formate generates an imine that can be reduced by the formate anion, which decomposes to CO_2 and H^-. A nickel hydrogenation catalyst is frequently used to reduce imines to amines without the reduction of carbonyl compounds. Alternatively, the protonated imine can be reduced with **sodium cyanoborohydride**, $NaBH_3CN$, a derivative of

$NaBH_4$. The electron-withdrawing cyano group deactivates the reducing agent so that it does not reduce aldehydes or ketones. It does, however, reduce iminium salts to amines. Finally, secondary amines can react with aldehydes and ketones to give iminium salts that are reduced to amines. In fact, virtually any amine of the type $R_2CH—NR'_2$ can be prepared by reductive amination with the appropriate choice of carbonyl compound and amine or ammonia.

The Hofmann Rearrangement

A reaction that is particularly valuable for preparing 1° amines bound to tertiary carbons is the **Hofmann rearrangement**. It involves treating a primary amide with Br_2 and NaOH:

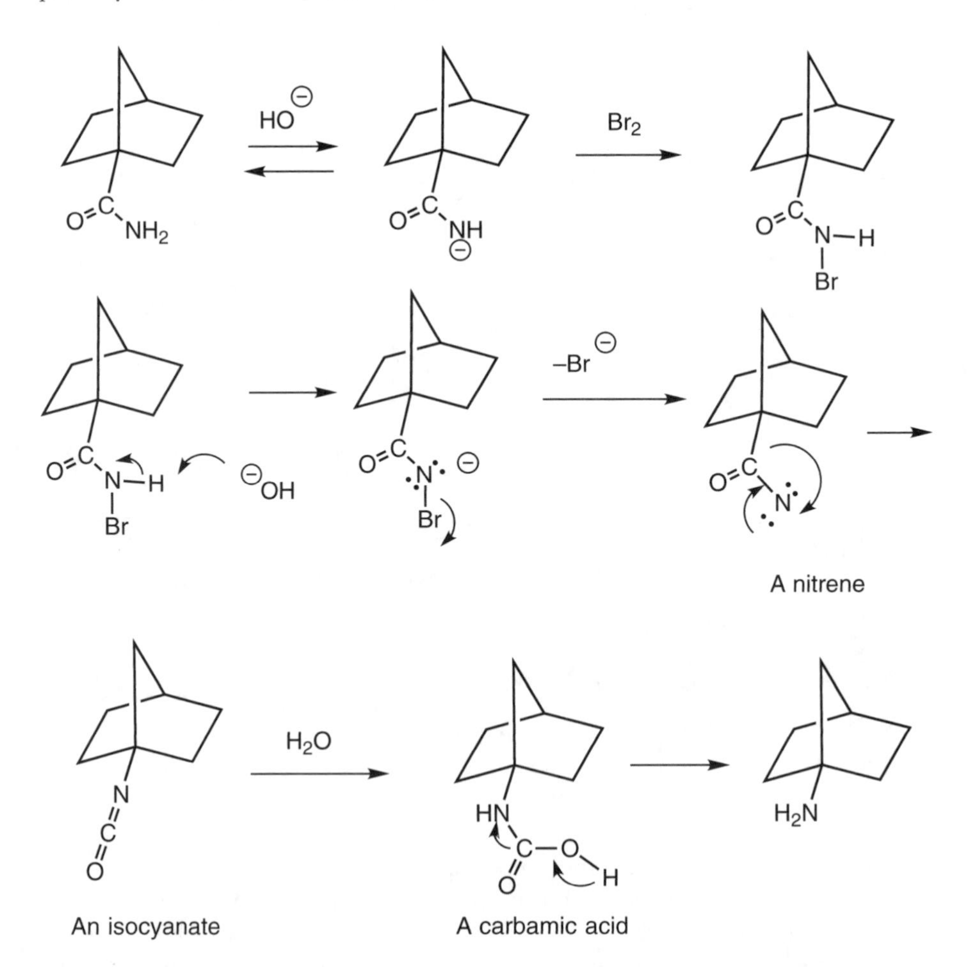

The mechanism for the Hofmann rearrangement is complicated. An *N*-bromoamide is generated and deprotonated. This anion undergoes a reaction called α **elimination** in which an anion, in this case Br^-, departs. This elimination leaves behind a neutral, electron-deficient reactive intermediate. The **nitrene** intermediate is a neutral nitrogen with only six valence electrons. It is **isoelectronic** with a carbocation; that is, it has the same number of valence electrons as a carbocation. Isoelectronic species have similar reactivity. Both a nitrene and a carbocation have a vacant *p* orbital, and rearrangements can produce a more stable species. In the case of a nitrene, rearrangement of the carbon substituent from CO to N produces an **isocyanate**, —NCO, functional group. This adds water to give a **carbamic acid**. Carbamic acids, —NHCOOH, can be thought of as adducts of amines and CO_2 and are unstable. They decompose to an amine and CO_2, which produces the final product of the Hofmann rearrangement, an amine. The stereochemistry of the migrating carbon is maintained in the rearrangement.

EXERCISES

1. Give syntheses for the following amines.

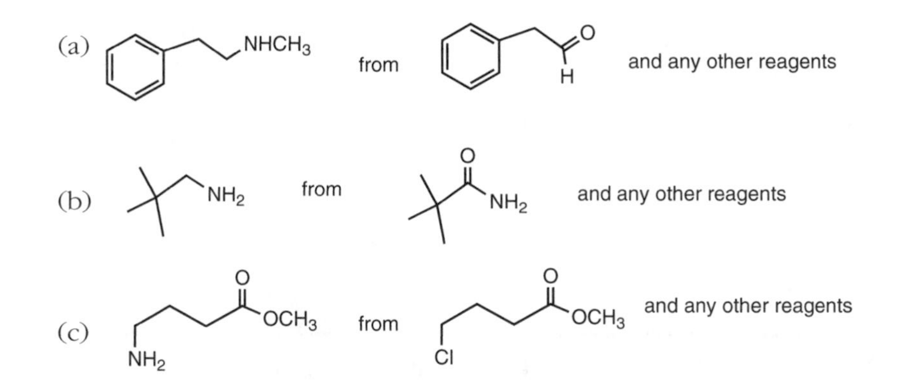

2. Heating a primary amine, formaldehyde, and a reducing agent such as sodium formate with an acid catalyst gives the tertiary *N,N*-dimethylamine. Give a mechanism for this reaction.

Solutions

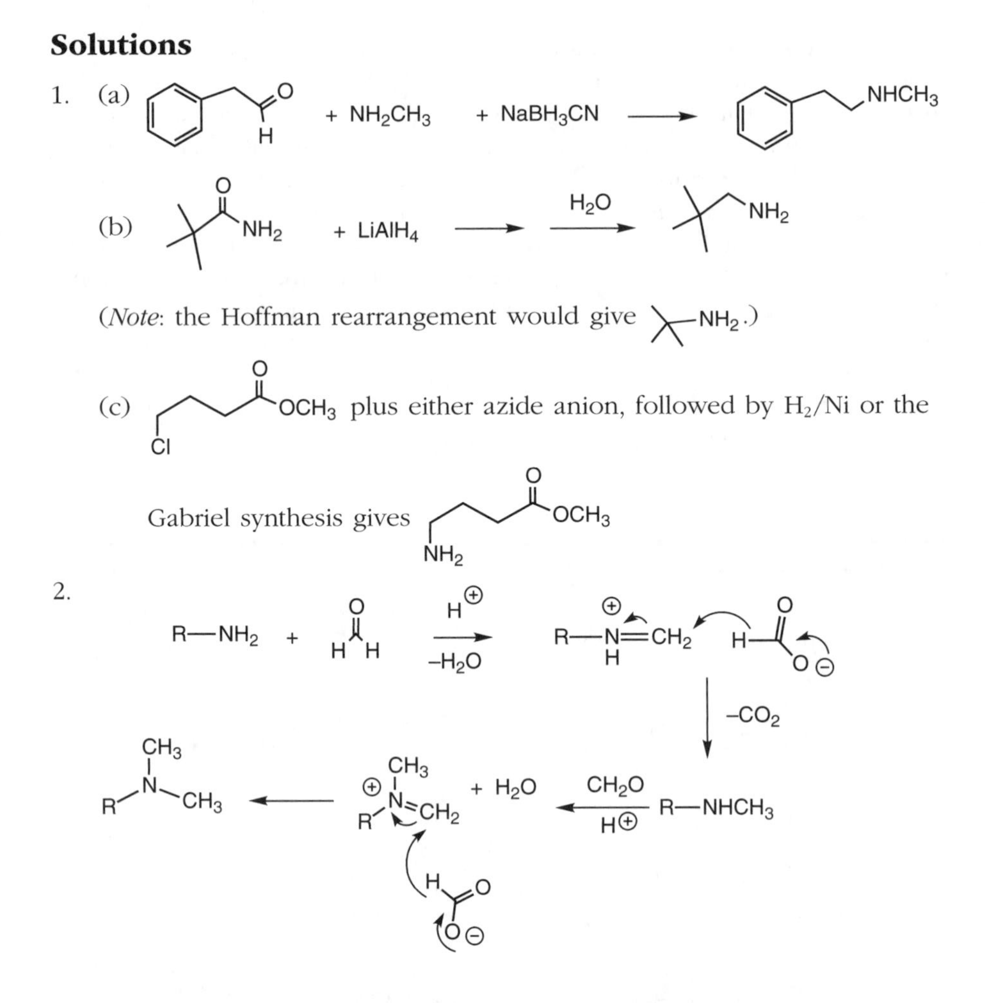

REACTIONS OF AMINES

The Hofmann and Cope Eliminations

The **Hofmann elimination** was, for many years, one of the major routes for forming alkenes. When a quaternary ammonium halide reacts with Ag_2O in water, it forms a quaternary ammonium hydroxide, $R_4N^+ HO^-$. Heating this salt causes it to decompose into an alkene, a tertiary amine, and water:

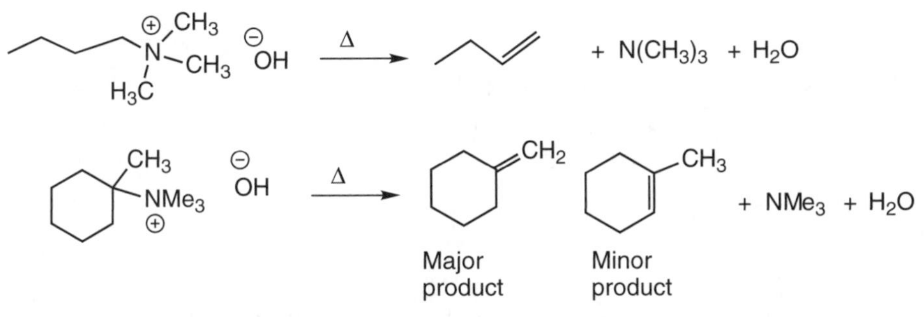

The mechanism for this reaction is similar to an E2 elimination in that the proton and the tertiary amine leaving group must be anti. However, the major product follows **Hofmann's rule**, which states that the elimination gives the less substituted alkene. The isomer formed is the regioisomer opposite that formed from other eliminations that follow Saytzeff's rule. Saytzeff products are alkenes with the most substituted (most stable) double bond (Chapter 9). Normally Hofmann eliminations are carried out with $RNMe_3^+$ salts so that there is no ambiguity about which R group undergoes elimination.

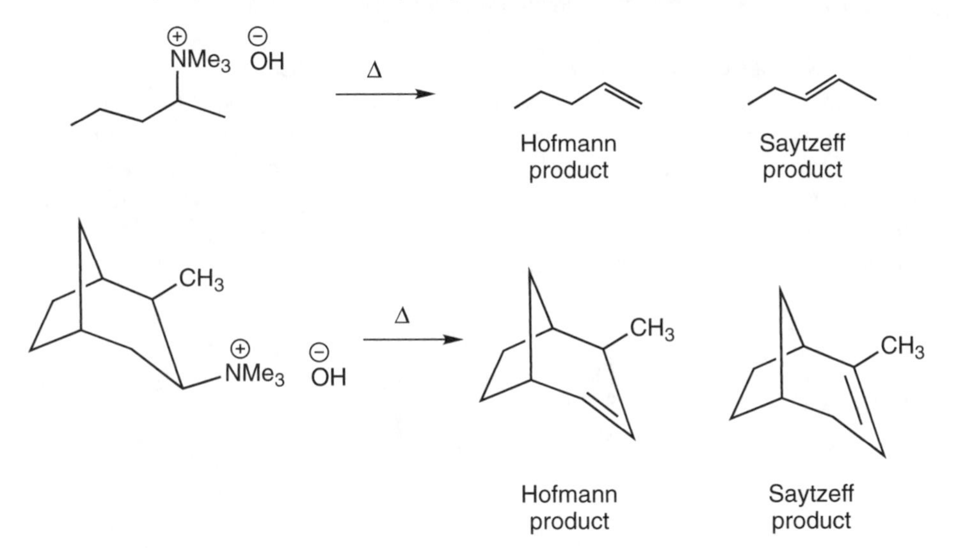

The **Cope elimination** is a variant of the Hofmann elimination. In this reaction, a tertiary amine reacts with hydrogen peroxide to give an **amine oxide**:

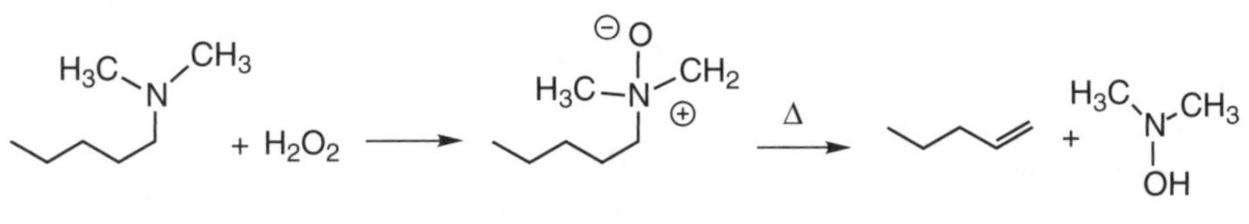

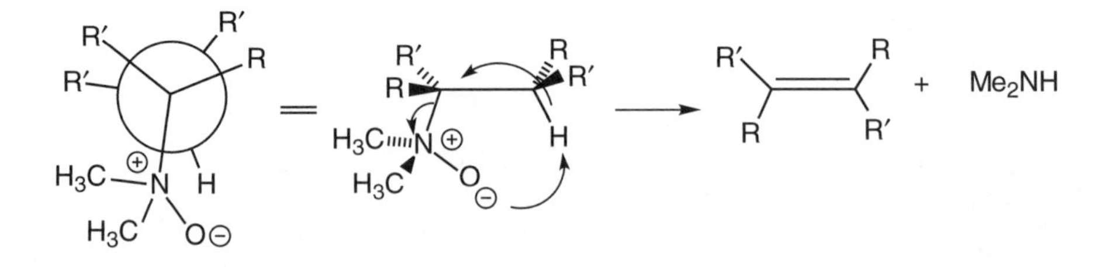

The Cope elimination proceeds through a cyclic transition state. The amine oxide O^- and the C—H bond undergoing elimination must be syn to one another. Just like the E2 elimination, the Cope elimination is under stereoelectronic control. However, the E2 reaction requires an anti transition state and the Cope elimination requires a syn transition state. When there are several syn C—H bonds beta to the amine oxide, each reacts, giving a mixture of alkenes. Neither the Hofmann rule nor the Saytzeff rule is followed. An advantage of the Cope elimination is that it takes place at a lower temperature (100°–130°C) than the Hofmann elimination (above 150°C).

Synthesis and Reactions of Mannich Bases

An important reaction for introducing a tertiary amine beta to a carbonyl group is the **Mannich reaction**. A secondary amine (normally dimethylamine), formaldehyde, and a ketone are mixed with an acidic catalyst. First an iminium salt is formed, and this reactive electrophile alkylates the enol form of the ketone. The ultimate product is the protonated form of a β-dimethylamino ketone, a **Mannich base**:

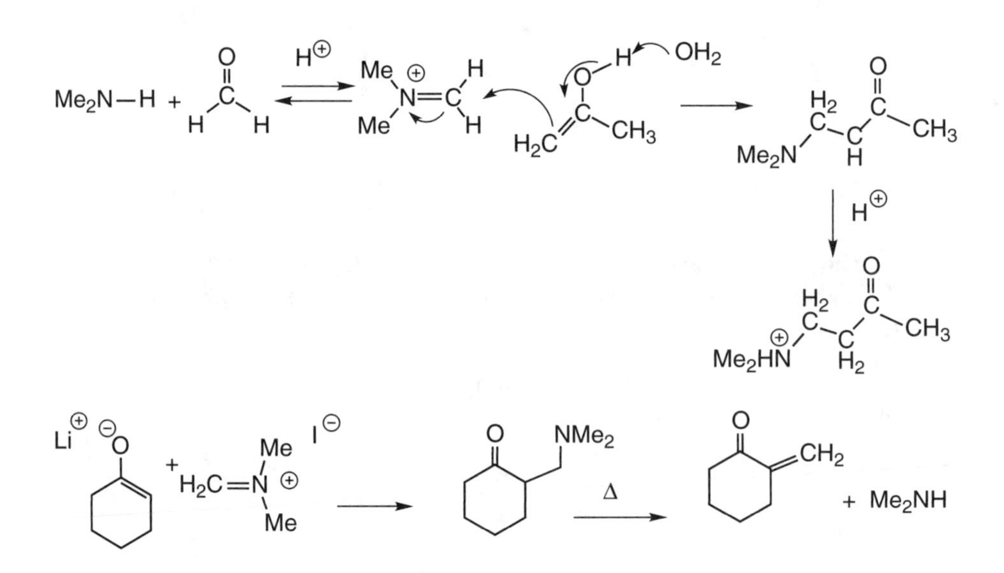

The iodide salt of $Me_2N{=}CH_2^+$ is a reasonably stable solid that can be used to make Mannich bases from enolates. Mannich bases or their protonated forms undergo elimination when heated to give α,β-unsaturated ketones. This elimination takes place because, unlike that in simple amines, the C—H bond beta to the amine is acidified by the carbonyl group and the product is stabilized by conjugation.

Formation of *N*-Nitrosoamines

If nitrous acid, HONO, is treated with a primary amine, an unstable alkyl diazonium salt, RN_2^+, is formed in a mechanism identical to the formation of aryl diazonium salts. However, because N_2 is such a good leaving group, these alkyl diazonium salts give a complex mixture of E1 and S_N1 products. Secondary amines react with nitrous acid to give stable *N*-nitrosoamines. These compounds are potent carcinogens:

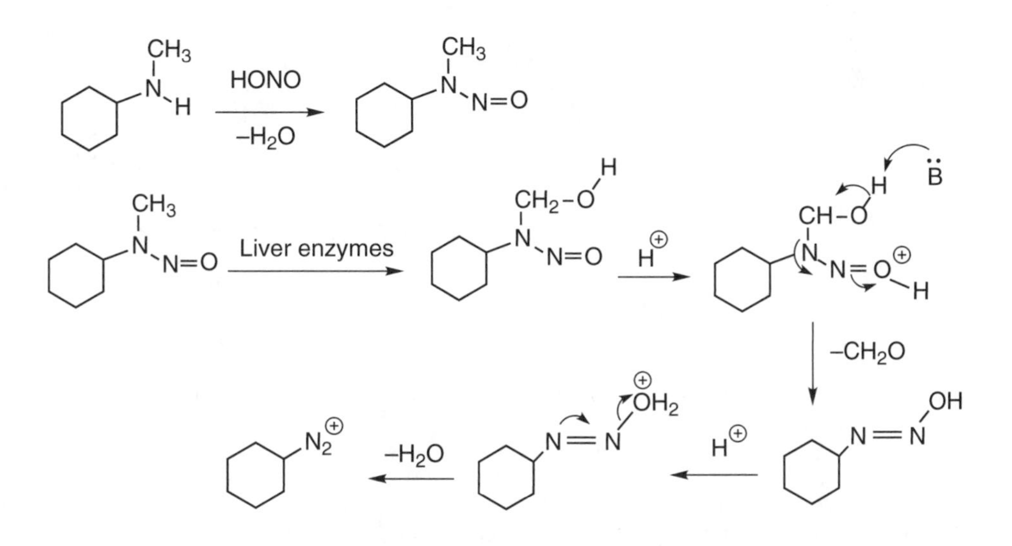

The liver contains enzymes that oxidize C—H bonds to C—OH bonds so that hydrophobic molecules become hydrophilic and can be excreted by the kidneys. However, in the case of *N*-nitrosoamines, this reaction leads to alkyl diazonium ions. These potent electrophiles react with many of the body's nucleophiles, including groups on DNA. When DNA is alkylated, it can lead to the affected cells becoming cancerous, in a complex series of steps. Interestingly, the mouth contains enzymes that reduce nitrates to nitrites, which means that the body is constantly exposed to traces of *N*-nitrosoamines produced by food and saliva in the acidic environment of the stomach. The body contains several repair systems that repair damage to DNA because DNA is continually exposed to damage from naturally occurring carcinogens.

The Beckmann Rearrangement

Oximes rearrange under acidic conditions to give amides via the **Beckmann rearrangement**. The amides can undergo hydrolysis to give an amine and an acid, if desired:

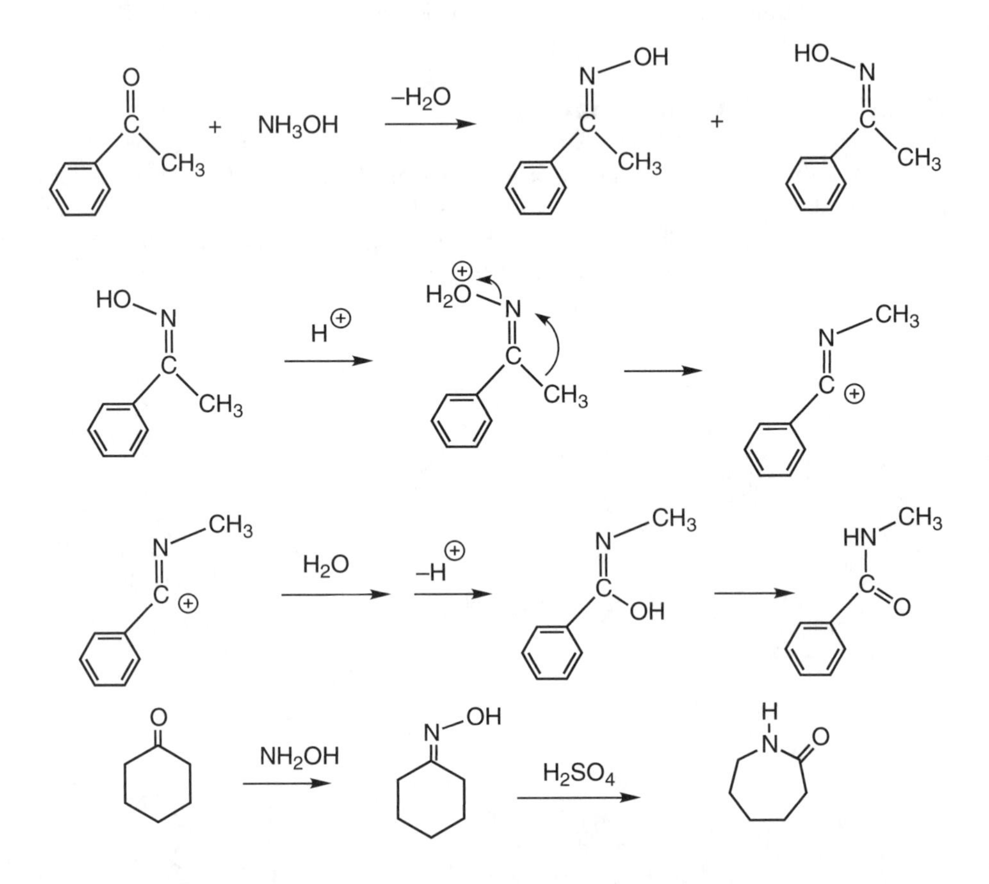

Ketones and aldehydes react with hydroxylamine to give oximes. These compounds can exist in two isomers, which are able to invert. Protonation of the oxime takes place on oxygen, and a loss of water is concerted with migration of the group trans to the water. The resulting carbocation (which is stabilized by the lone pair of electrons on nitrogen) adds water, leading to the enol form of an amide, which tautomerizes to an amide. A cyclic oxime undergoes ring expansion to a lactam.

EXERCISE

Prior to the invention of spectroscopic methods, the structure of unknown cyclic amines was determined, in part, by repeated Hofmann eliminations until all C—N bonds were broken. What alkene would result from repeated Hofmann eliminations on the following amine?

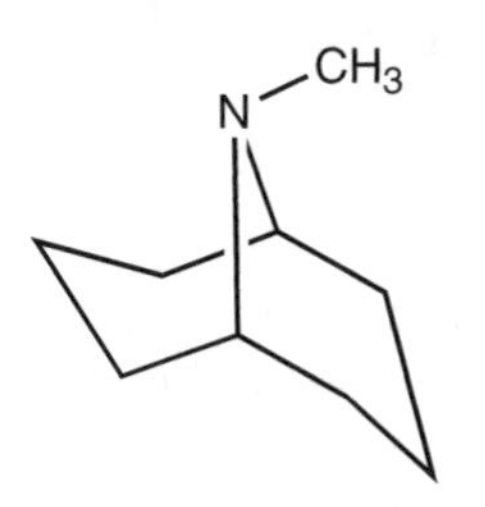

Solution

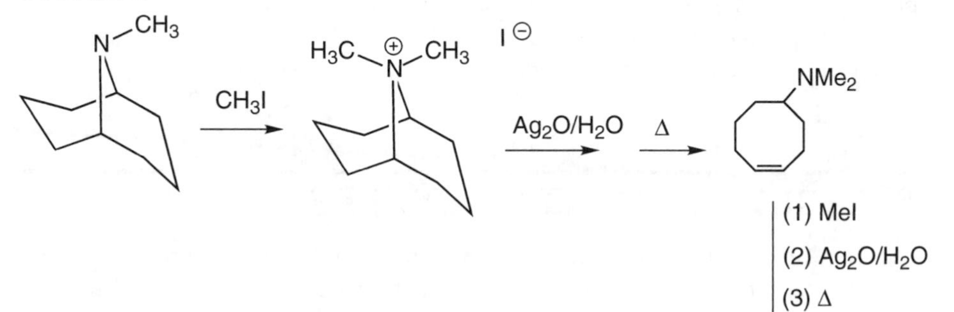

26
HETEROCYCLIC CHEMISTRY

Heterocycles are cyclic molecules with one or more noncarbon atom in the ring. Commonly, the heteroatom is oxygen, nitrogen, or sulfur, but essentially any element can be part of a heterocycle. Heterocycles are found in nature, in DNA and RNA, as parts of amino acids, and as vitamins and enzyme cofactors. Many of the drugs used in human medicine are heterocyclic compounds. This chapter provides only the briefest outline of heterocyclic chemistry, but any study of biochemistry or pharmacology at the molecular level will repeatedly encounter heterocycles.

SATURATED HETEROCYCLES

When an oxygen, nitrogen, or sulfur atom is part of a saturated ring, its chemistry is similar to that of an acyclic functional group. The only important exceptions are three-membered rings, where ring strain enhances their reactivity. Systematic names of saturated heterocycles use **oxa**, **aza**, and **thia** to denote the presence of oxygen, nitrogen, and sulfur, respectively. When substituents are placed on a saturated heterocyclic ring, the heteroatom is given the relative position 1.

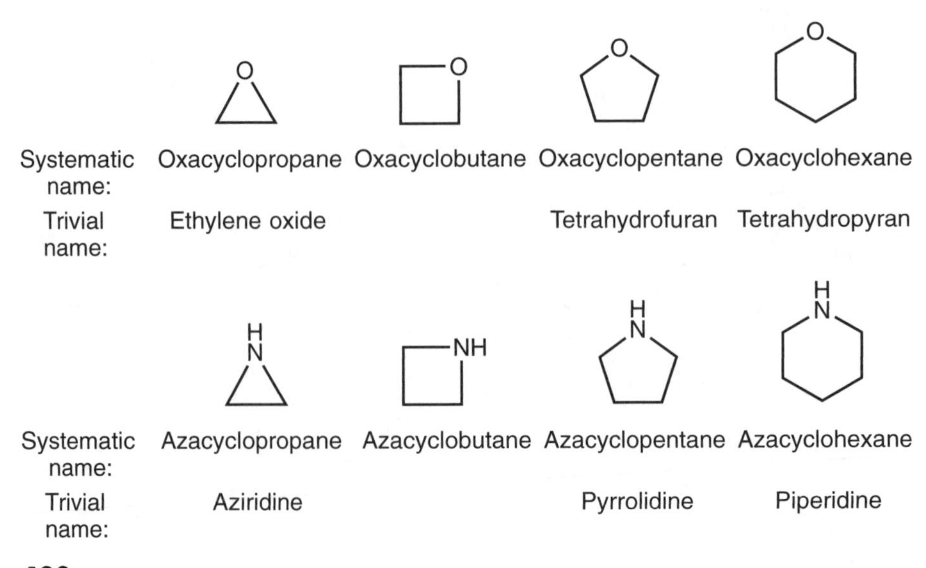

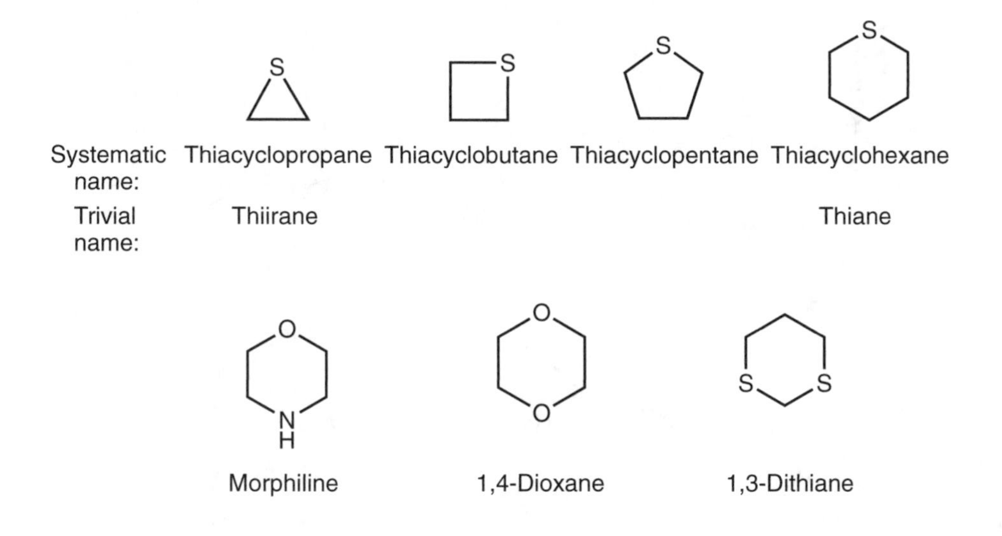

Tetrahydrofuran (THF) is primarily used as a solvent, particularly for Grignard reactions. 1,4-Dioxane is also a commonly used solvent. The secondary amines pyrrolidine, piperidine, and morphiline are used as organic bases and for the preparation of enamines from aldehydes and ketones. The inductive effect of oxygen makes morphiline a weaker base (pK_a of its conjugate acid is 9.3) than piperidine or pyrrolidine (whose conjugate acids have pK_a values of 11.1 and 11.3, respectively).

The sulfur compound 1,3-dithiane is useful for preparing aldehydes and ketones. The CH_2 between the sulfurs is slightly acidic because the polarizability of sulfur can stabilize a negative charge. The resulting anion can react in an S_N2 reaction with the usual electrophiles to give a new C—C bond. Hydrolysis with the aid of a thiophilic (sulfur-loving) mercury salt replaces the C—S bonds with C—O bonds, generating a new carbonyl group:

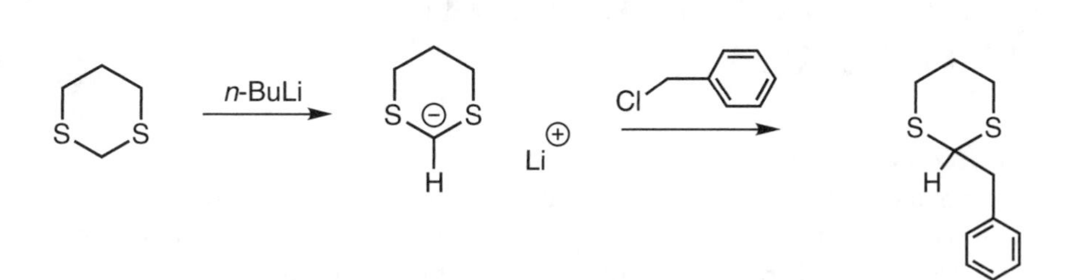

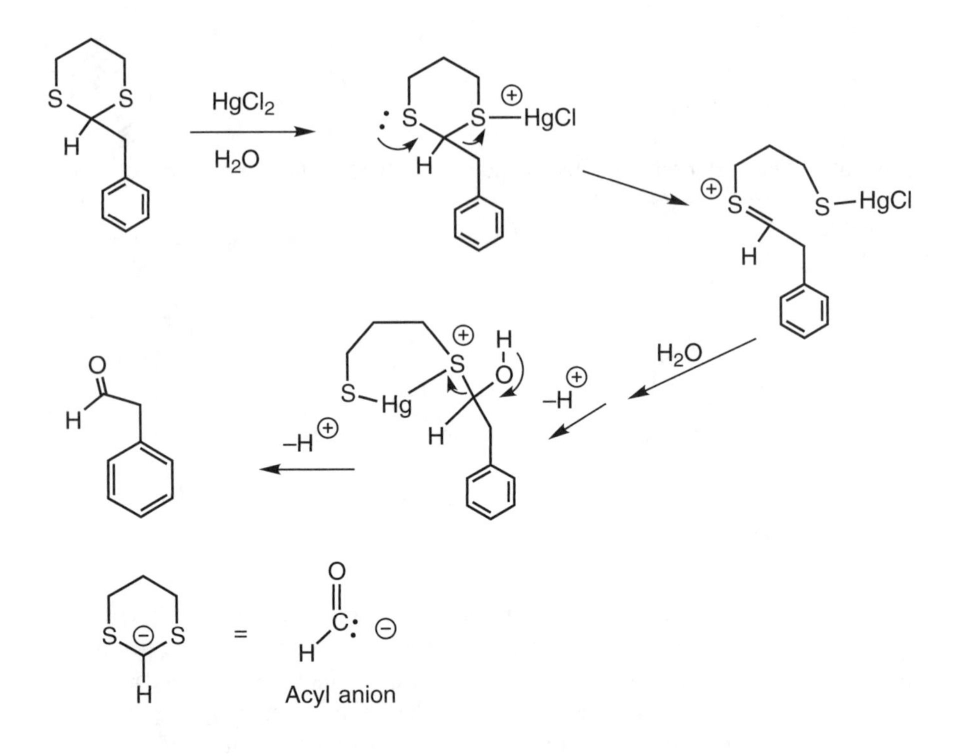

The anion of 1,3-dithiane gives the same product (after hydrolysis) as an **acyl anion** would. Acyl anions are themselves too unstable to use in organic synthesis.

AROMATIC FIVE-MEMBERED RING HETEROCYCLES WITH ONE HETEROATOM

Aromaticity

An aromatic sextet of electrons can arise from three double bonds, as in benzene, or from two double bonds and a nonbonding pair of electrons on a heteroatom. Thus, **furan**, **pyrrole**, and **thiophene** are aromatic:

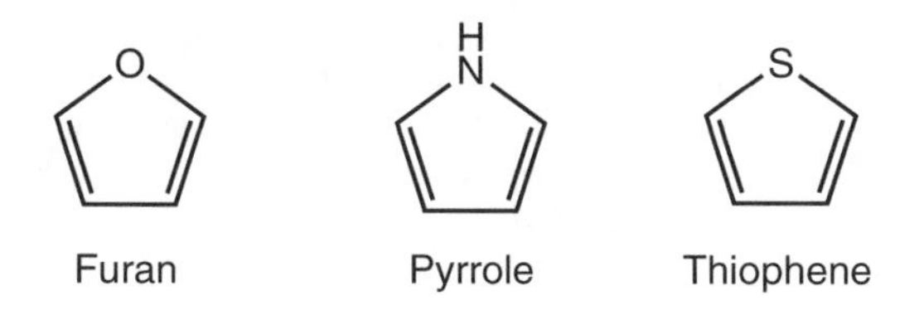

Furan Pyrrole Thiophene

In furan and thiophene, one lone electron pair overlaps with the π system, and the other lone pair is in the plane of the ring, perpendicular to the aromatic sextet. With pyrrole, the N—H bond is in the plane of the ring and the lone pair overlaps with the π system to give six electrons:

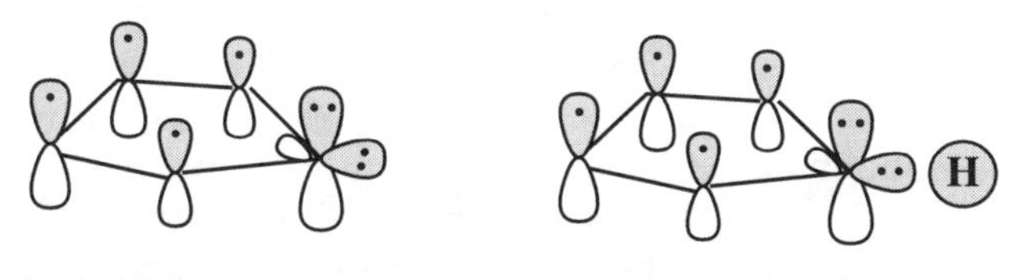

Orbitals of thiophene and furan

Orbitals in pyrrole

The resonance stabilization energy of benzene is greater than that of these **heteroaromatic** compounds. The order of aromaticity is benzene > thiophene > pyrrole > furan. In fact, in the presence of strong acids, furan undergoes reactions characteristic of vinyl ethers (Chapter 18):

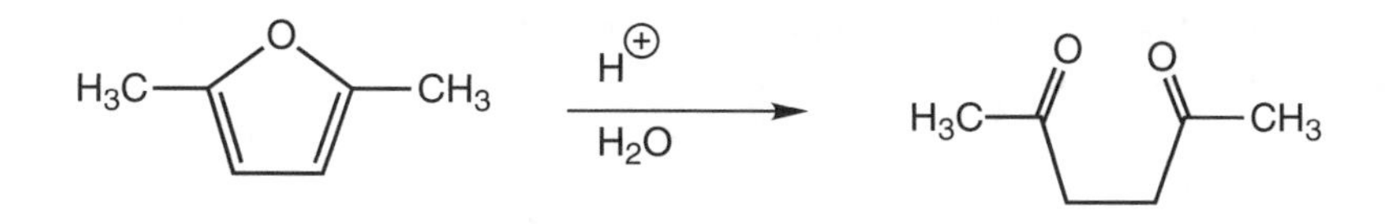

All three of these ring systems undergo electrophilic aromatic substitution and are much more reactive than benzene. In part, this reactivity difference arises because the rate-determining step in electrophilic aromatic substitution is the first step, which breaks up the aromatic π system. Since thiophene, pyrrole, and furan have less stabilization to lose than benzene, the intermediate is lower in energy and the overall reaction proceeds more rapidly.

Reactions of Thiophene, Pyrrole, and Furan

All three of these heteroaromatic rings undergo electrophilic aromatic substitution, preferentially at C-2. The reactivity order is pyrrole > furan > thiophene because of several factors, including the electronegativity of the heteroatom and the resonance stabilization of the aromatic ring. The reason substitution takes place at C-2 in preference to C-3 is the greater number of resonance structures that can be drawn for the C-2 intermediate:

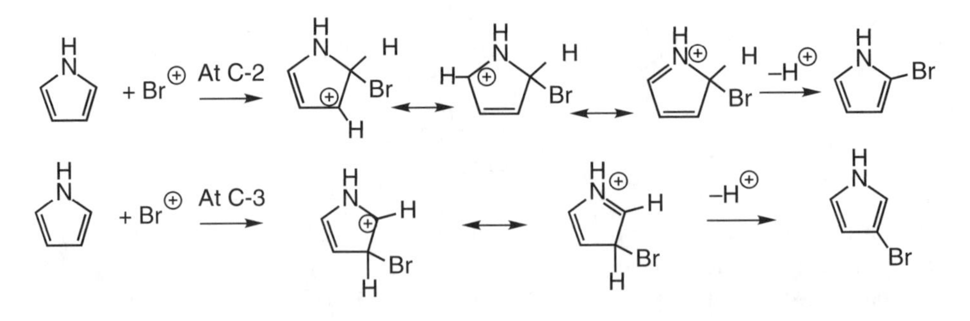

The different reactivities of the three five-membered aromatic rings are seen in the conditions needed for Friedel-Crafts acylation. Thiophene reacts with CH_3COCl using the Lewis acid $SnCl_4$. Furan reacts with CH_3COCl catalyzed by the milder Lewis acid BF_3. Pyrrole reacts directly with acetic anhydride to give 2-acetylpyrrole with no need for a catalyst. Because pyrrole is so reactive, ordinary nitration conditions lead to polymerization. Nitropyrroles have to be made using acetyl nitrate formed by reacting acetic anhydride and nitric acid.

PYRIDINE AND RELATED HETEROCYCLES

Aromaticity and Reactivity of Pyridine and Its Derivatives

Pyridine is the nitrogen analog of benzene where one CH group has been replaced by N. Other nitrogen analogs of benzene include the diazines **pyridazine**, **pyrimidine**, and **pyrazine**:

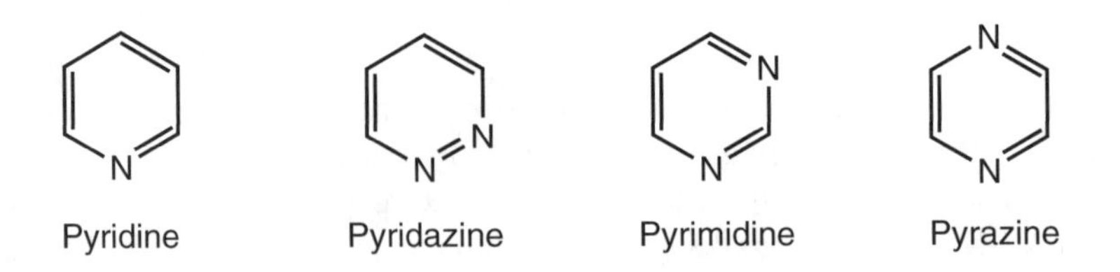

All these heterocycles are aromatic, and all are weak bases. The nitrogen lone pair electrons lie in the plane of the ring and do not interact with the π system. The aromatic stabilization energy for pyridine is 32 kcal/mol, only slightly less than that for benzene. In the case of pyrimidine, the aromatic stabilization energy is reduced to 26 kcal/mol, still a substantial stabilization.

Pyridine and diazines are much less reactive in electrophilic aromatic substitution reactions than benzene. Since nitrogen is more electronegative than carbon, the pyridine ring is less electron-rich than benzene. Furthermore, any electrophile tends to react with the most basic site in pyridine, which is the nitrogen lone electron pair, rather than the π system. Diazines are more-or-less unreactive in electrophilic aromatic substitution reactions.

When pyridine reacts in an electrophilic aromatic substitution reaction, the electrophile is directed to the 3 position. This isomer is favored because it avoids the resonance structures with a positive charge on the nitrogen atom that forms when substitution takes place at the 2 or 4 position:

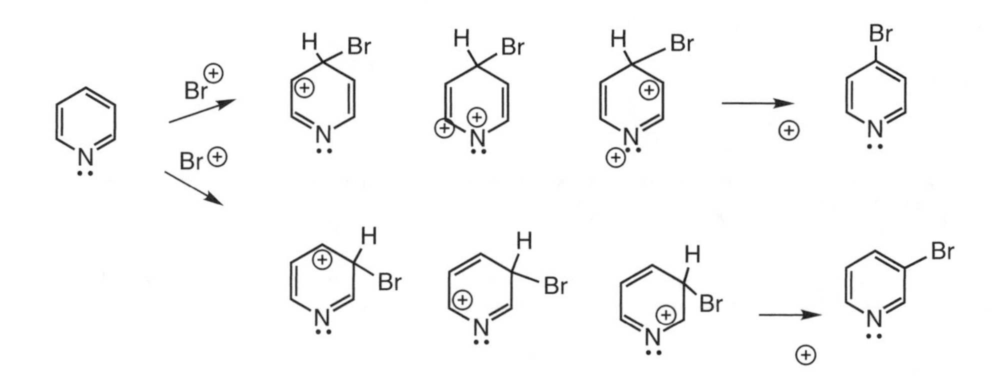

Pyridine is similar in reactivity to nitrobenzene, but if the pyridine is protonated, it is 10^{12} times less reactive. Therefore, very vigorous conditions are needed to carry out electrophilic substitutions. Pyridine can be nitrated to 3-nitropyridine only with H_2SO_4 and HNO_3 at temperatures above 300°C.

Because pyridine is electron-deficient, it reacts with nucleophiles. For example, pyridine reacts with CH_3Li at the 2 position:

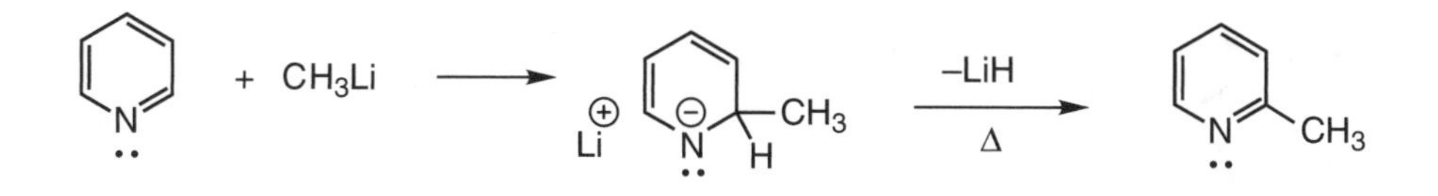

Heating the anion or treatment with a mild oxidizing agent rearomatizes the ring. When the pyridine ring has a leaving group at the 2 or 4 position, nucleophiles replace the leaving group in an addition-elimination reaction. Mechanistically, this substitution resembles what happens when a nucleophile reacts with an acid chloride:

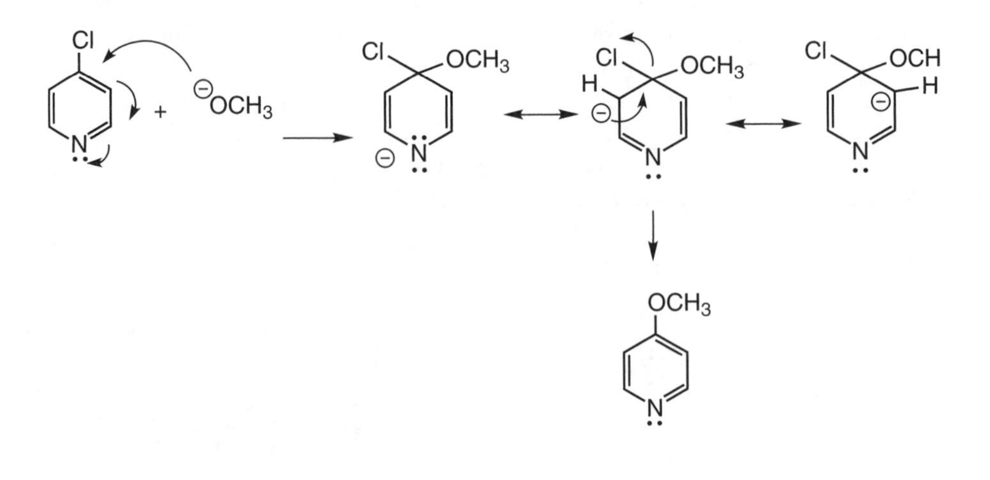

EXERCISES

1. Draw an orbital diagram for pyrimidine showing all the π bonds and lone electron pairs.

2. Give the expected product for the following reactions.

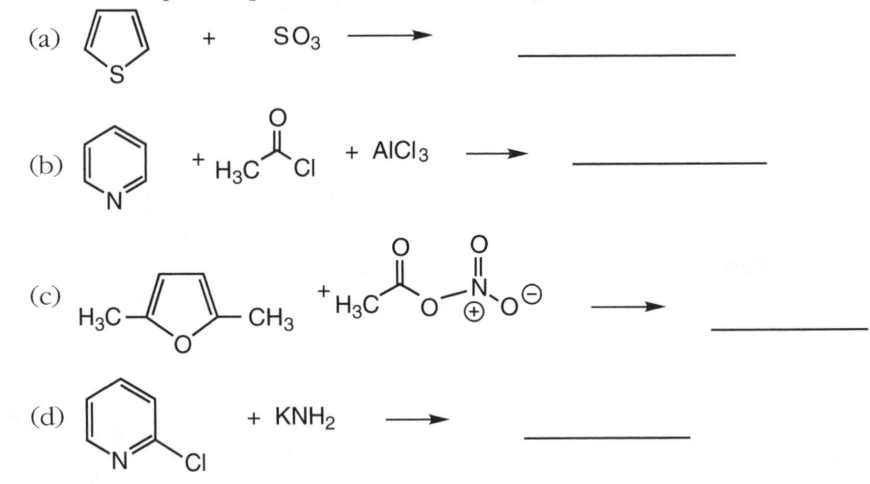

Solutions

1.

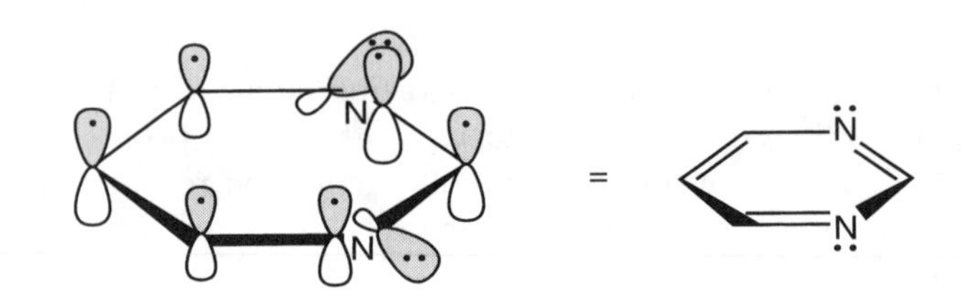

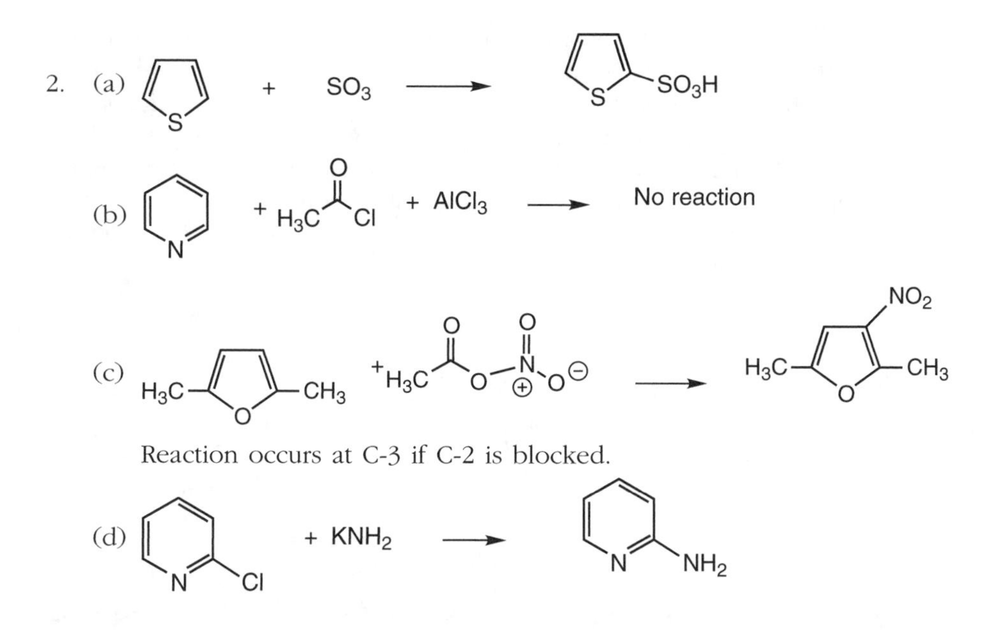

Reaction occurs at C-3 if C-2 is blocked.

Heterocycles With Two Rings

There are several important heterocycles with two aromatic rings, including **quinoline**, **isoquinoline**, and **indole**:

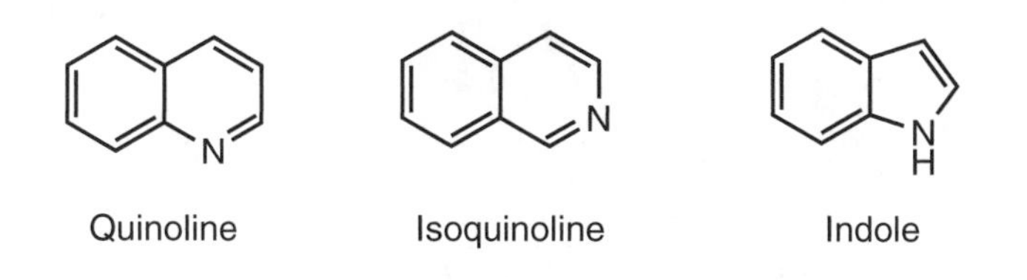

Electrophilic aromatic substitution takes place readily on the C_6 ring in quinoline and isoquinoline, giving both the C-5 and C-8 isomers:

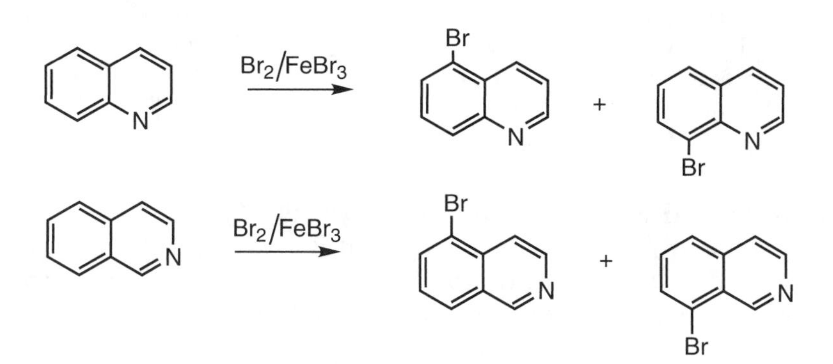

The synthesis of most nitrogen-containing heterocycles involves either Michael additions or imine formation to create C—N bonds, followed by at least one dehydration step to generate a double bond. An example of this kind of reaction is **Skraup quinoline synthesis**, which involves the Michael reaction of an aniline derivative with a conjugated aldehyde or ketone under acidic conditions. The initial product is a dihydroquinoline that must be oxidized to the final quinoline product with a mild oxidizing agent such as nitrobenzene or HgO:

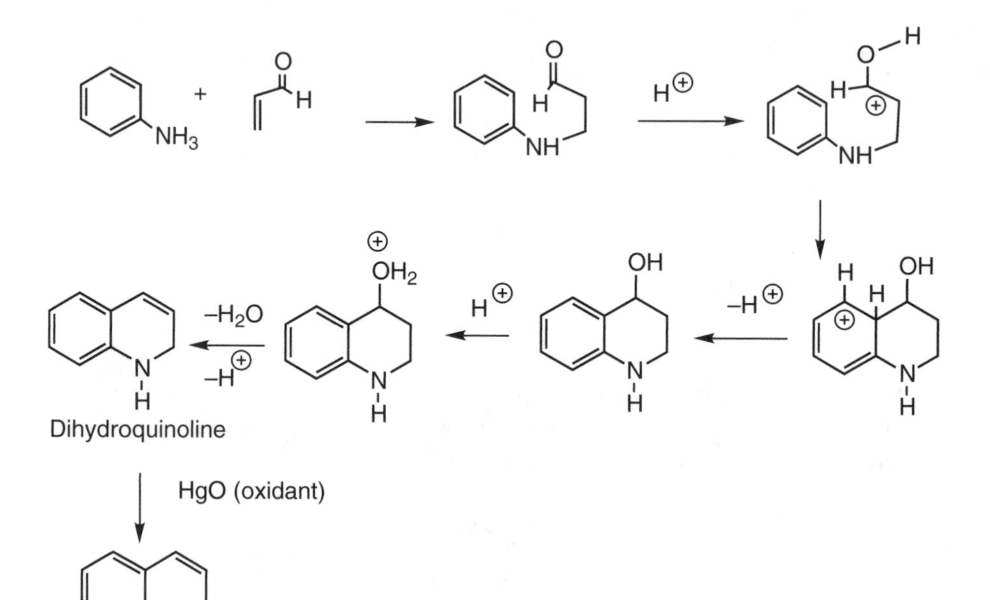

Indoles are similar to pyrroles in that they are very reactive in electrophilic aromatic substitution reactions. However, the C_6 ring permits other resonance structures to exist, leading to preferential substitution at C-3:

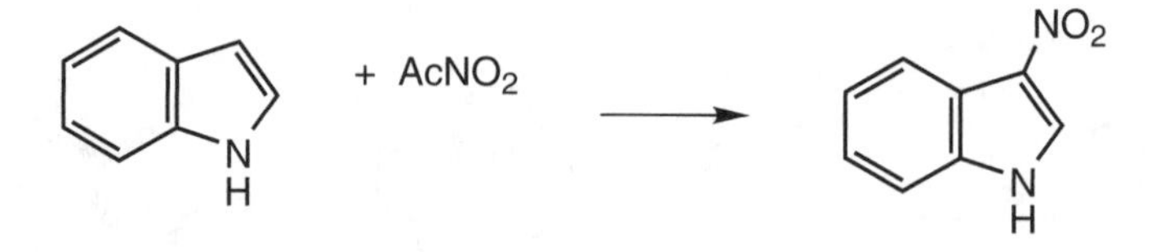

Structure and Synthesis of Porphyrins

Porphyrins are tetrapyrrole heterocycles that have a central cavity capable of binding metal ions. Heme, or iron protoporphyrin IX, the iron-containing pigment in hemoglobin that binds oxygen, is a porphyrin:

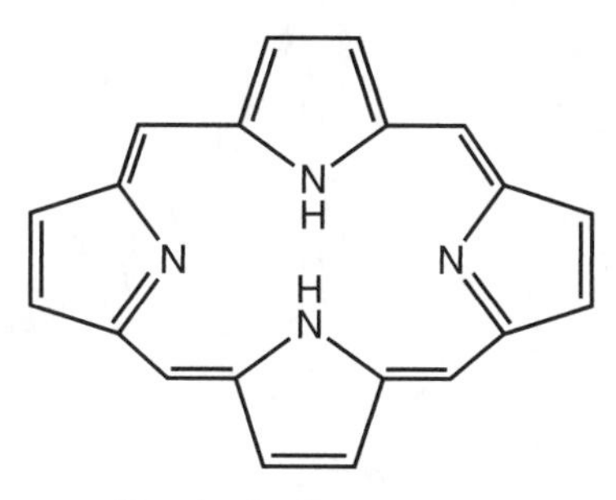

Porphyrin ring system

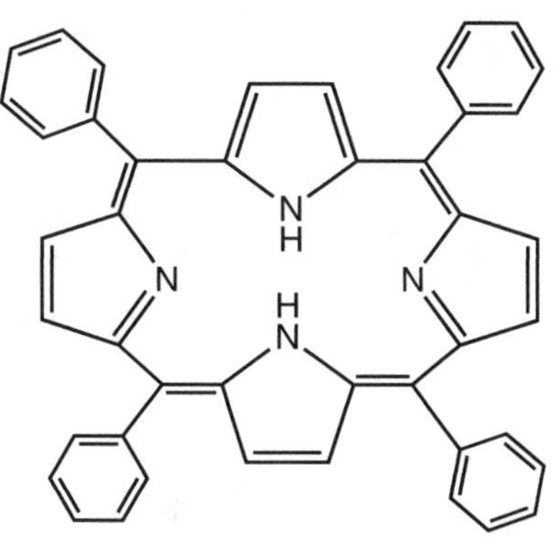

Tetraphenylporphyrin

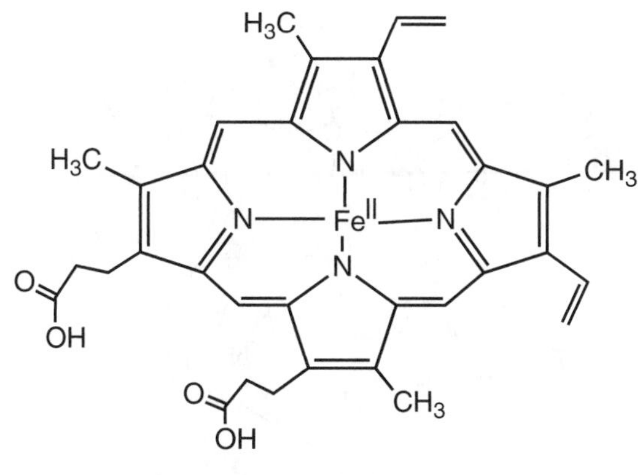

Iron protoporphyrin IX

Porphyrins are highly colored and very stable. Although synthetic routes to porphyrins often produce low yields, the properties of porphyrins make them easy to isolate. The synthesis of tetraphenylporphyrin from pyrrole and benzaldehyde with acid catalysis illustrates the general mechanism for porphyrin synthesis:

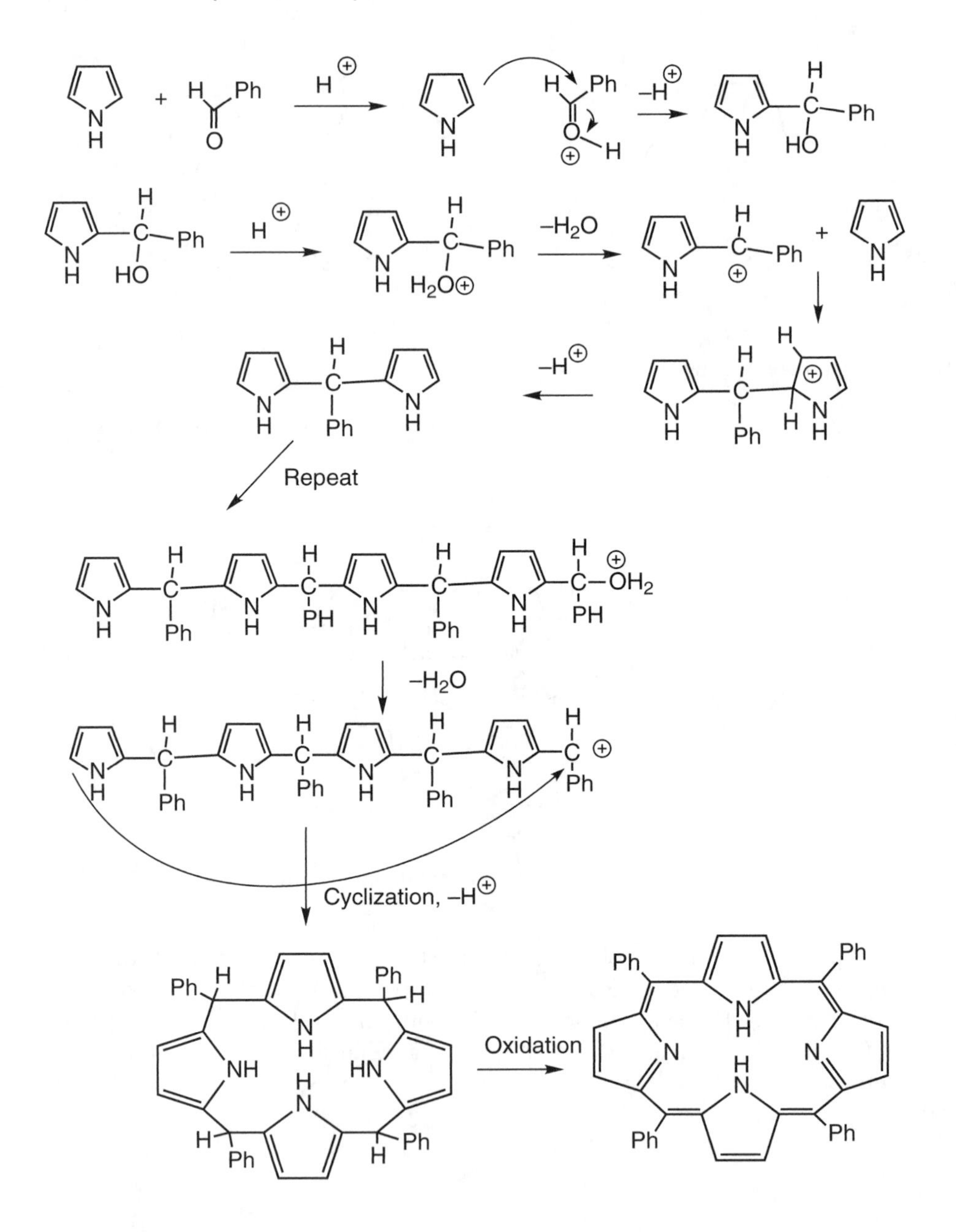

The liver breaks down porphyrins to linear tetrapyrroles, which are then broken down further. If a person has liver problems, linear tetrapyrroles can accumulate. The yellow color of patients with jaundice arises from these yellow porphyrin derivatives in the body.

27
CARBOHYDRATES

The classic definition of **carbohydrates** is compounds that have the formula $(CH_2O)_n$. Frequently they are known as sugars or **saccharides**. Carbohydrates play numerous roles in biochemistry. Everyone knows that sugars are used in metabolism as a source of quick energy. However, polymers of sugars, **polysaccharides**, make up starch and cellulose. Furthermore, sugars are a central component of biological recognition systems. The ABO blood group system depends on the types of carbohydrate structure on individual red blood cells. DNA and RNA are based on the sugars ribose and deoxyribose. One of the most active areas of organic and biochemical research is glycobiology, which studies the synthesis and properties of sugars and their adducts with lipids, proteins, and other molecules.

CARBOHYDRATE STEREOCHEMISTRY

R,S and D,L Nomenclature

The simplest carbohydrate is the three-carbon sugar **glyceraldehyde**, CHO—CH(OH)—CH_2OH. It possesses a single stereogenic center, so that it can exist as an R or S isomer (see Chapter 5 for a review of R and S).

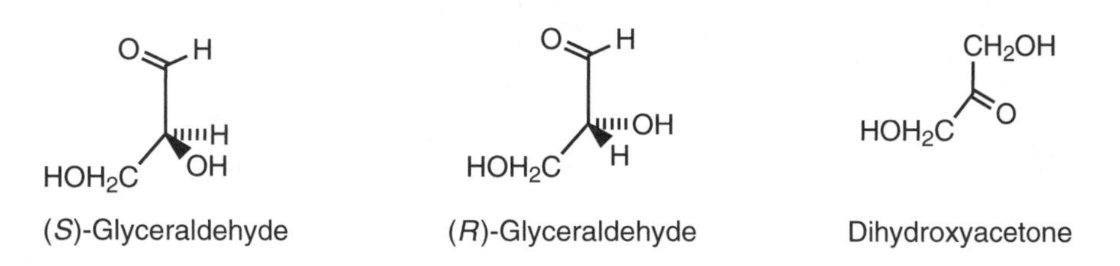

(*S*)-Glyceraldehyde (*R*)-Glyceraldehyde Dihydroxyacetone

Glyceraldehyde is an example of an **aldose**, a carbohydrate whose most oxidized functional group is an aldehyde. The isomeric dihydroxyacetone is a **ketose**, a carbohydrate whose most oxidized functional group is a ketone. The suffix *-ose* designates this compound as a carbohydrate. Each glyceraldehyde enantiomer is an **aldotriose** (*-triose* because it has three carbons, and *aldo-* for aldehyde). Four-carbon sugars are **tetroses**, five-carbon sugars are **pentoses**, and so on.

The R,S system of nomenclature was invented after World War II. However, sugars have been studied since before 1900. Chemists in the 1900s knew glyceraldehyde was chiral, and they had ways of converting glyceraldehyde to essentially all other carbohydrates. However, they had no way of knowing the exact three-dimensional structure of glyceraldehyde or any other molecule. The organic chemist Emil Fischer invented a way to describe the *relative* stereochemistry of any sugar based on glyceraldehyde.

One enantiomer of glyceraldehyde rotates the plane of polarized light to the right; so it is the (+) isomer. We now know that (+)-glyceraldehyde is (*R*)-glyceraldehyde. Fischer guessed that (+)-glyceraldehyde has what we now call the R configuration (he had a 50/50 chance of getting it right, and he did). He called this form of (+)-glyceraldehyde D-glyceraldehyde. He called the enantiomer L-glyceraldehyde, which is (–)-glyceraldehyde.

Fischer had a specific way of drawing carbohydrates known as a **Fischer projection**. He placed the most oxidized group at the top of the structure and drew the carbon chain straight down from this carbon. Substituents come off the chain at right angles and point out of the paper. The intersection of the horizontal and vertical lines is the location of one of the carbons in the main carbon chain. Some examples are

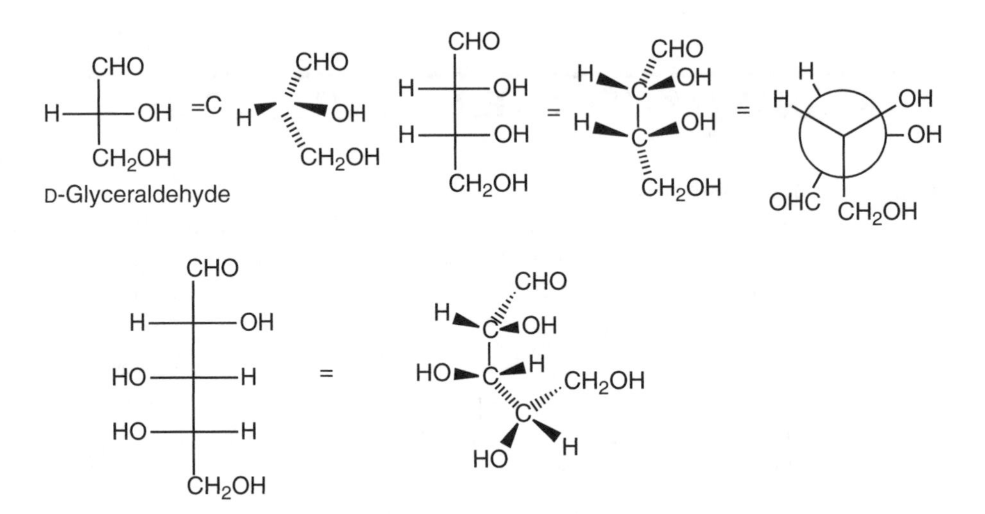

In a Fischer projection, the carbon chain folds back around on itself and the substituents (usually —H and —OH) reach out like welcoming arms. The carbohydrate molecule does not actually adopt this eclipsed conformation in solution; it is just a way to depict a given enantiomer or diastereomer.

It is very easy to look at a Fischer projection and decide if the carbohydrate is D or L. Correctly drawn with the most oxidized group at the top of the chain, look at the bottom stereogenic center. If the —OH group is on the right, the sugar is a D sugar. If the —OH group is on the left, the sugar is an L sugar. All the other sugar positions are irrelevant. This rule comes from the fact that the D and L system is a relative system based on the glyceraldehyde molecule. Any carbohydrate can potentially be made from glyceraldehyde by carrying out reactions on the —CHO group. From this point of view, it is important only to know if one starts from D-glyceraldehyde (leading to a D sugar) or L-glyceraldehyde (leading to an L sugar).

There is no fundamental correlation between R and S, D and L, and (+) and (−). Some D sugars have the relevant stereogenic center R, others S. Also, some D sugars rotate the plane of polarized light to the right, (+), and others to the left, (−):

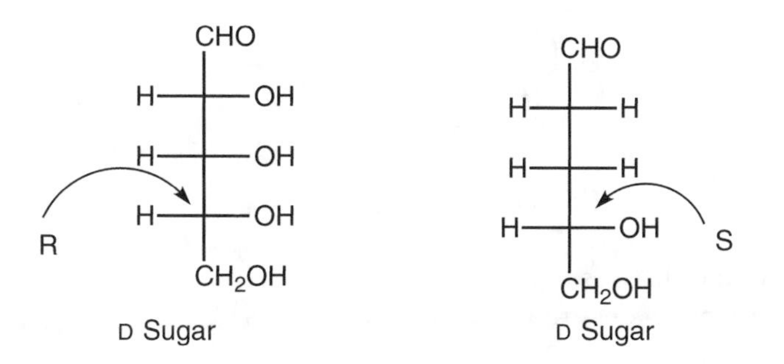

Enantiomers and Diastereomers in Carbohydrates

Except for trioses, all carbohydrates can exist as diastereomers. For example, there are two D-aldotetroses, D-erythrose and D-threose:

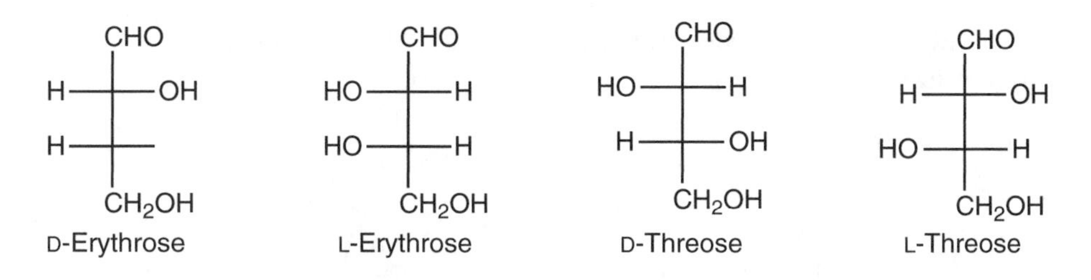

It is very important to recognize what enantiomers and diastereomers look like in carbohydrate chemistry. Notice that for a molecule with more than one stereogenic center, one must invert *all* the stereocenters to obtain the enantiomer. Thus, D-erythrose and L-erythrose are enantiomers. However, if one inverts just the lower H—C—OH center in D-erythrose, one obtains not L-erythrose, but L-threose. This is because pairs of diastereomers have some stereogenic centers that are the same and others that are inverted. As diastereomers, D-erythrose and D-threose have different physical properties and different reactivities. As enantiomers, D-erythrose and L-erythrose have identical physical properties and reactivities toward achiral reagents.

The carbohydrates erythrose and threose have given their names to two general kinds of diastereomer isomers, **erythro** isomers and **threo** isomers. If one takes diastereomers with two adjacent chiral centers and draws the molecules in an eclipsed conformation, with the carbon chain groups eclipsing (as one does in a Fischer projection), the isomer with similar groups eclipsed is the erythro isomer, and the isomer with dissimilar groups eclipsed is the threo isomer:

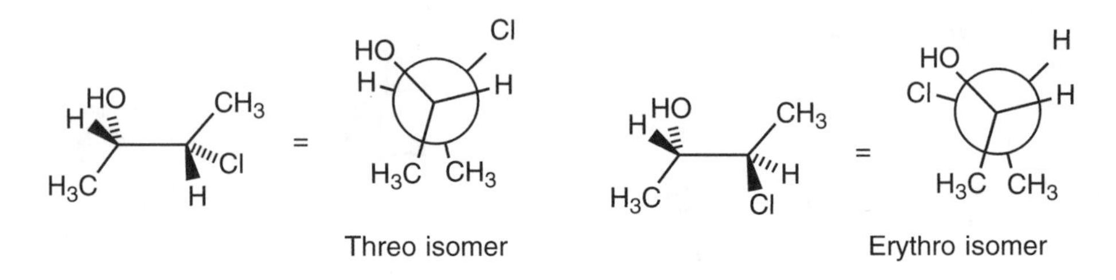

The erythro and threo isomer nomenclature applies only when the groups on the stereogenic centers are reasonably analogous.

EXERCISES

1. Draw the two enantiomeric erythro isomers of 2,3-dihydroxypentane and assign each stereogenic center as R or S.
2. Draw Fischer projections of the following carbohydrates and indicate if the molecule is D or L.

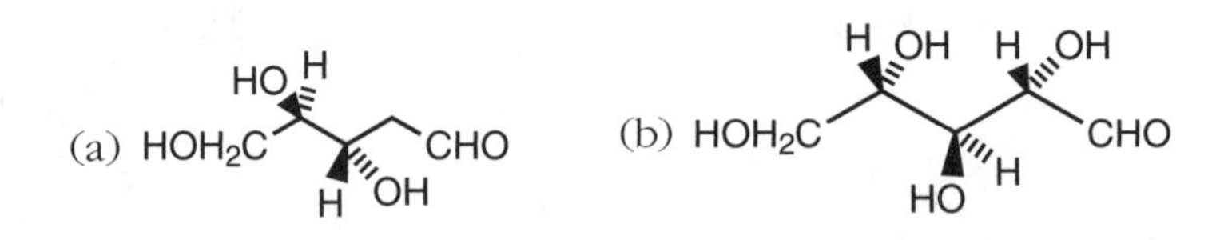

3. Draw the Fischer projection of (2*R*,3*R*)-2,3,4-trihydroxybutyraldehyde. Is this carbohydrate D or L?

Solutions

1.

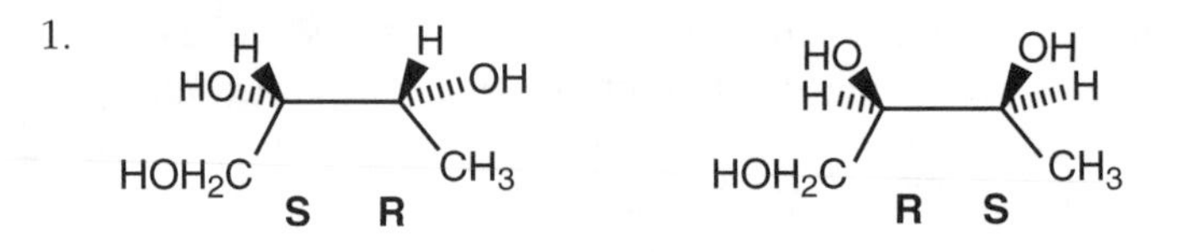

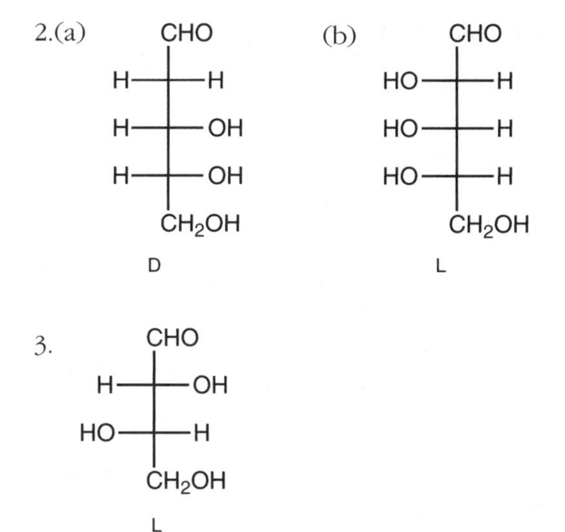

STRUCTURES OF THE 15 D-ALDOSES WITH THREE TO SIX CARBONS

Most of the carbohydrates found in nature are D sugars. Furthermore, although seven or eight carbon sugars are occasionally encountered, most naturally occurring carbohydrates have six or fewer carbons. The names and Fischer projections of all the aldoses with six or fewer carbons are shown in Table 27.1.

TABLE 27.1. D-ALDOSES WITH THREE TO SIX CARBONS

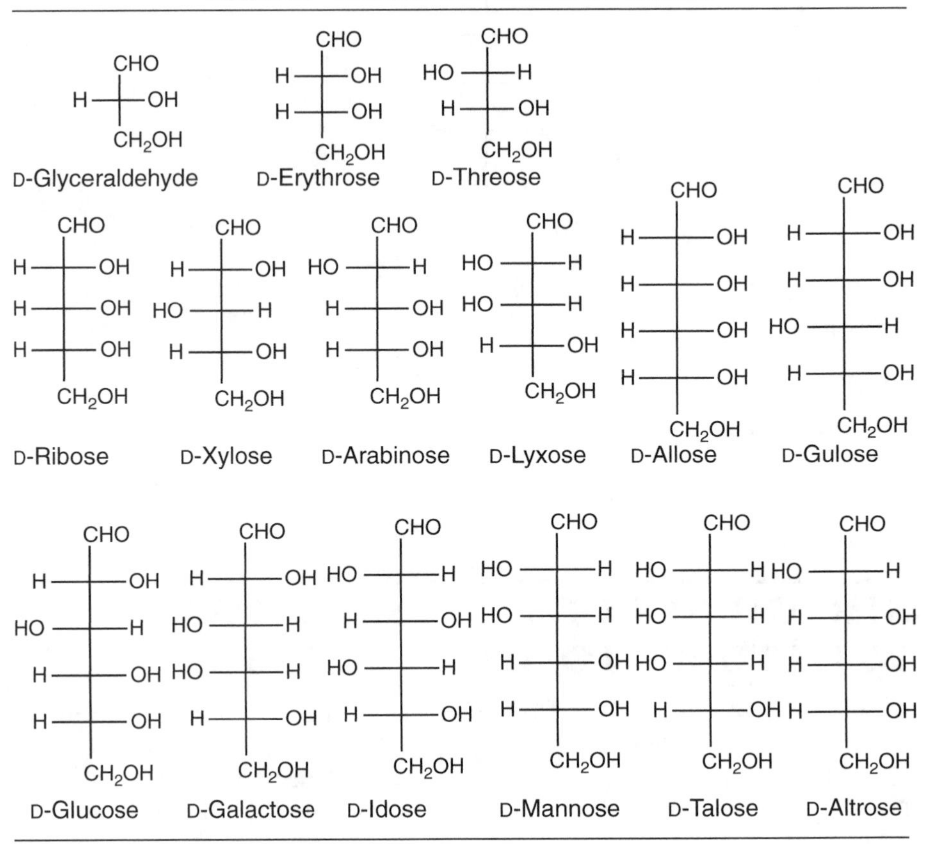

Each of the sugars in Table 27.1 is a **monosaccharide**, a single sugar. Some carbohydrates are composed of two sugars linked together. These are called **disaccharides**. Some sugars have been reduced and have an —H in place of an —OH group. These are called **deoxy sugars**. The most common deoxy sugar is 2-deoxy-D-ribose, which is the sugar component of DNA. Some sugars have one of the —OH groups oxidized to a ketone, while the initial —CHO group has been reduced to a —CH_2OH group. **Fructose** is an example of this kind of ketose. If the terminal —CHO group is reduced to an alcohol, so that every carbon is attached to an —OH group, the molecule is an **alditol**. The alditol of D-glucose is D-glucitol. If the terminal —CHO group is oxidized to a carboxylic acid group, the product is called an **aldonic acid**. The aldehyde oxidation product of D-glucose is D-gluconic acid:

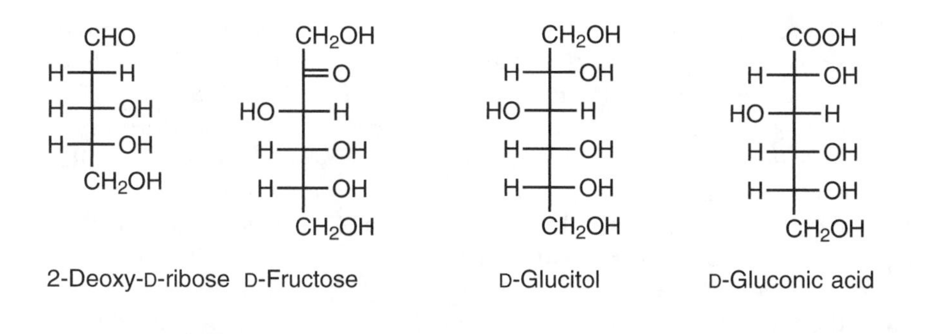

REACTIONS OF MONOSACCHARIDES

Cyclization

Carbohydrates are polyfunctional compounds with an aldehyde and two or more hydroxyl groups. Recalling the discussion of hemiacetal formation in Chapter 18, it should not be surprising that a carbohydrate can form an internal hemiacetal by having one of its —OH groups react with the —CHO group. Actually, only 0.02 percent of the sugar is the acyclic aldehyde form in water. If the hydroxy group on C-5 of glucose reacts with the —CHO group to form a hemiacetal, a six-membered ring is formed. The —CHO carbon that was sp^2 hybridized becomes an sp^3 hybridized carbon and a new stereogenic center:

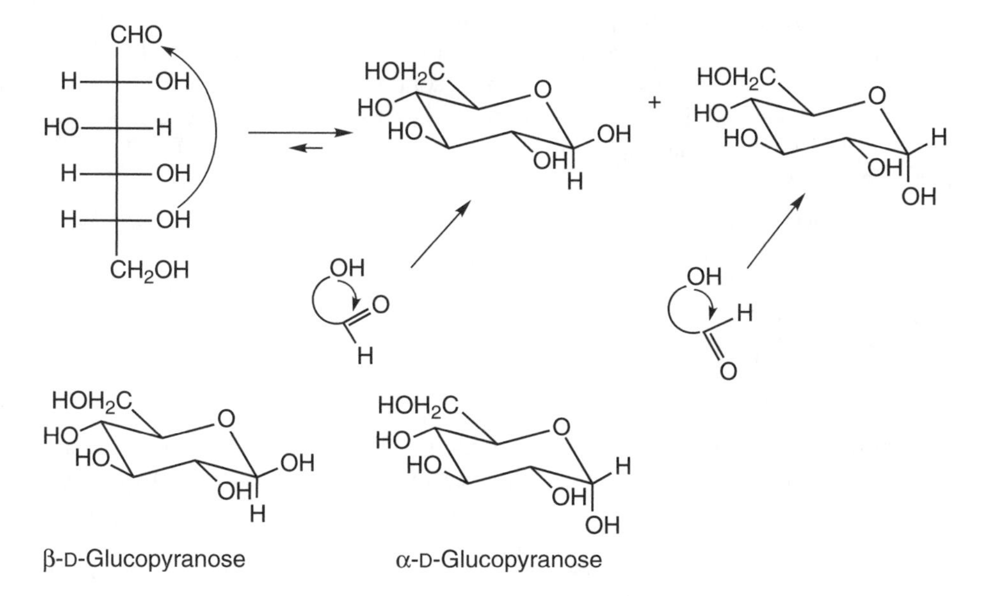

Depending on the orientation of the carbonyl group, the new stereocenter can have an axial —OH group or an equatorial —OH group. The stereochemistry of all the other centers is determined by the structure of the acyclic sugar. The ring carbon with two C—O bonds (the former —CHO carbon) is called the **anomeric center**. As will be shown shortly, the anomeric center has enhanced reactivity compared to the other carbon in the ring. The diastereomer with the anomeric —OH group equatorial is called the β **anomer**. The diastereomer with the anomeric —OH group axial is called the α **anomer**. When sugars cyclize to six-membered rings, they are called **pyranoses**, after the six-membered oxygen heterocycle pyran. Glucose can also cyclize to a five-membered ring form if the hydroxyl group at C-4 adds to the aldehyde group. These are called **furanose** forms. The diastereomer with the —OH group on the side of the ring opposite the —CH_2OH group is the α anomer, and the diastereomer with the —OH group on the same side of the ring as the —CH_2OH group is the β anomer:

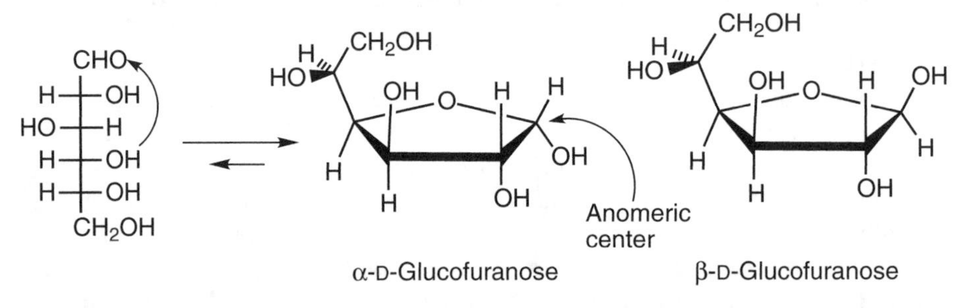

Drawing chair pyranose forms of carbohydrates is easier than it seems if one remembers that D-glucose has all its substituents (except for the anomeric carbon) in the most stable equatorial position. For example, to draw α-D-idose, note from Table 27.1 that the configuration of D-idose differs from that of D-glucose at C-2, C-3, and C-4. Thus, three centers must be axial in the pyranose form. Likewise, β-D-mannose differs from D-glucose only at C-2, so the C-2 —OH group is axial and C-3 and C-4 —OH groups are equatorial.

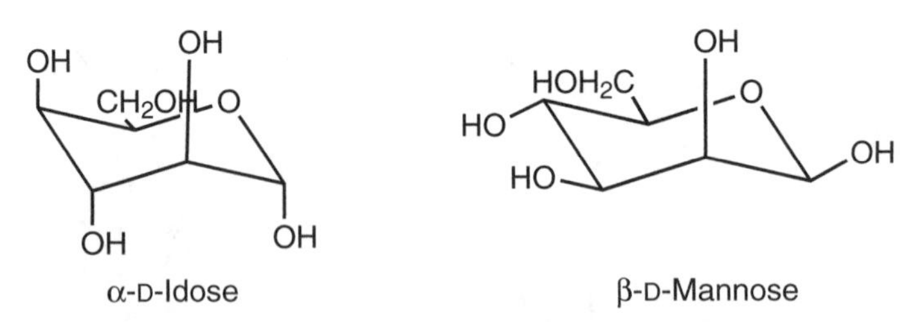

There is one subtle point about drawing the chair forms of pyranose sugars. The alternative chair form, with the left carbon pointing down and the right carbon pointing up, corresponds to the L sugar:

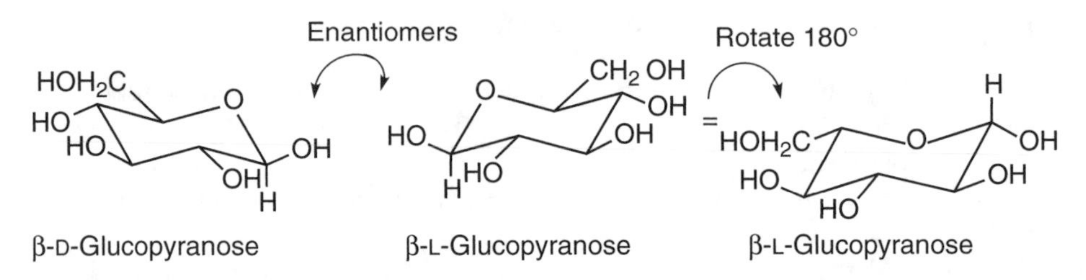

In addition to the usual chair forms for drawing six-membered rings, sugar chemists have an alternative way to draw six-membered rings (the student will by now be expecting that sugar chemists have an alternative way to do everything). As shown below for β-D-glucose, a **Hayworth projection** depicts a cyclic sugar as a flat ring with substituents pointing straight up or straight down. The ring is viewed along the edge, and the oxygen is at the top (five-membered ring) or top right (six-membered ring). The anomeric carbon is on the right side, and the $—CH_2OH$ group points up. The anomeric —OH is up in the β anomer and down in the α anomer. Groups on the right side of a Fisher projection point down, and groups on the left side of a Fisher projection point up.

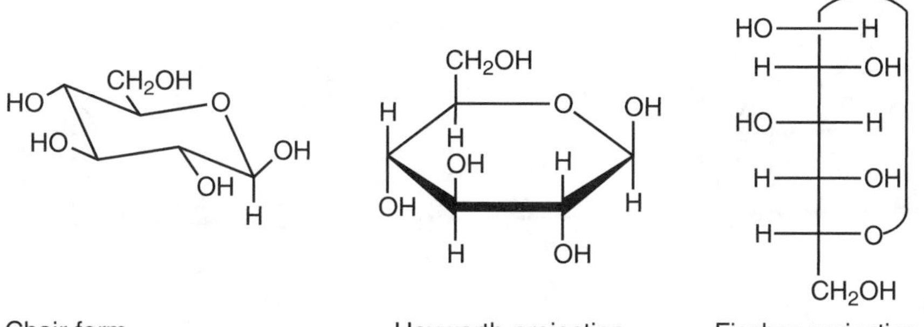

Mutarotation

The α and β forms of cyclic sugars are diastereomers. Diastereomers that differ at only one stereogenic center are called **epimers**, and the two molecules are said to be **epimeric**. If a pure sample of, for example, either β-D-glucose or α-D-glucose is added to water, in time an equilibrium mixture of β-D-glucose (64 percent) and α-D-glucose (36 percent) forms. This equilibration of anomeric —OH epimers is called **mutarotation**. Mutarotation can take place by reversible ring opening, rotation, and ring closing of the cyclic hemiacetal:

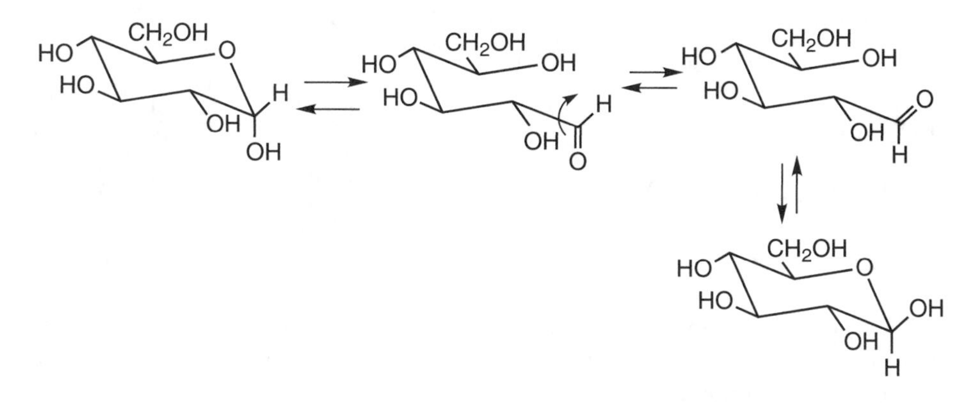

Reactions at the Anomeric Carbon

Under acidic conditions, substitution reactions can take place at the anomeric carbon by what is essentially an S_N1 reaction, assisted by the ring oxygen. For example, in acidic methanol the anomeric —OH group is replaced by an $—OCH_3$ group, and all the other —OH groups are unreactive:

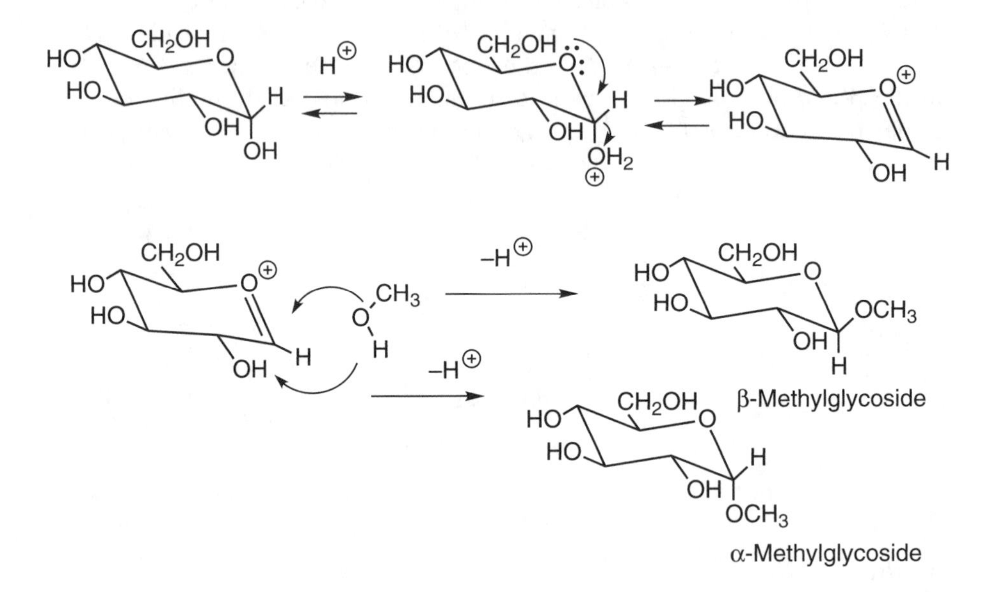

The cationic sugar intermediate is stabilized by the ring oxygen. All the other —OH groups would generate only a 1° or a 2° carbocation, which is not nearly as stable. Since the nucleophile can approach the ring from the top or bottom face, a mixture of anomers is formed. The product is a carbohydrate acetal, which is known as a **glycoside**. The ether bond exocyclic to the ring from the anomeric center is known as a **glycosidic bond**. A glycoside is named by replacing the *-e* ending of a sugar's name by *-ide* and ends with *-pyranoside* or *-furanoside* depending on whether the ring is five- or six-membered:

Ethyl β-D-mannopyranoside

Methyl α-D-fructofuranoside

Nucleophiles other than oxygen can react to form bonds at the anomeric carbon. For example, aromatic amines or heterocycles with N—H bonds can react to form an anomeric mixture of *N*-glycosides:

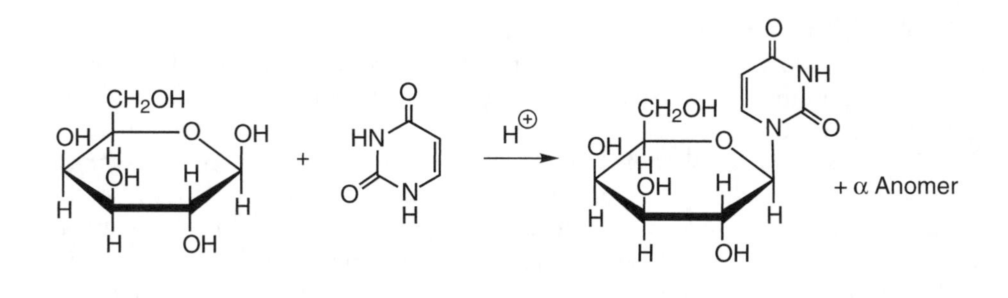

EXERCISES

1. Draw Hayworth projections for β-D-fructofuranose and α-D-fructopyranose.
2. What products would you expect from the reaction of β-D-glucose with anhydrous HCl?

Solutions

1.

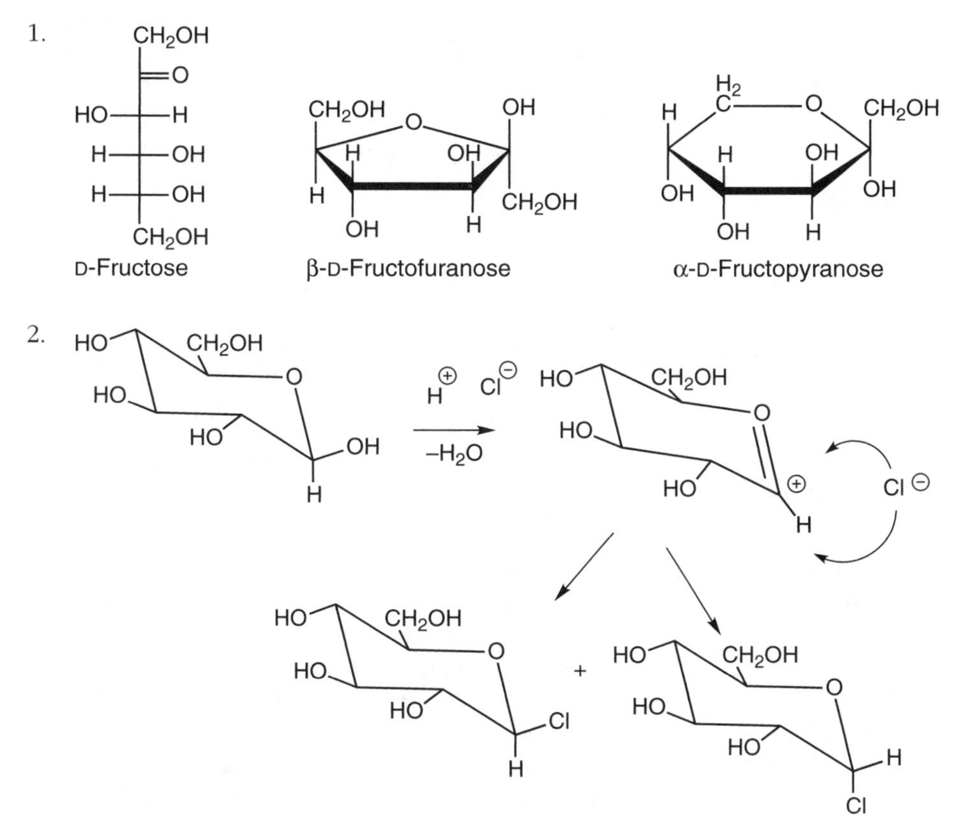

The Anomeric Effect

When glucose or other sugars react with an alcohol under an acid catalyst, the equilibrium mixture consists of more of the α anomer (axial) than of the β anomer (equatorial). Similarly, when other electronegative groups such as halogens, amines, or phenols bond to the anomeric carbon, more of the axial α anomer is produced at equilibrium than one would expect from steric factors. This preference of electronegative groups for the axial position at the anomeric site is called the **anomeric effect**. The source of this preference is thought to be in the interaction of the lone pair electron orbitals on oxygen with the electronegative group.

Oxidation and Reduction of Aldoses

Because a carbohydrate is in equilibrium with open-chain aldehyde and cyclic hemiacetal forms, reagents that react with aldehydes react with carbohydrates. Although the equilibrium concentration of the open-chain form is very low, it is constantly replenished in a rapid equilibrium. $NaBH_4$ or catalytic hydrogenation reduces aldoses to alditols. The structure of D-glucitol is shown on p. 417; this compound is also known as D-sorbitol. It is used as a sugar substitute because it is approximately 60 percent as sweet as cane sugar (sucrose). Xylitol, the alditol of the five-carbon sugar xylose, is another sucrose substitute found in chewing gum and other food products.

Several reagents are selective for the oxidation of aldoses to aldonic acids without oxidizing the primary and secondary alcohol groups in a carbohydrate. The best known is **Tollen's reagent**, an ammonical silver solution, $Ag(NH_3)_2^+$:

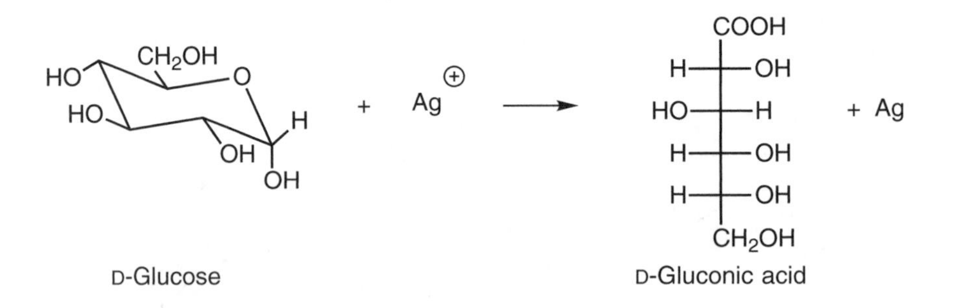

The by-product of this reaction is metallic silver. If it takes place in a clean glass vessel, the silver deposits as a silver mirror, and this reaction is still used to deposit a silver coating on surfaces. Carbohydrates that reduce Ag^+ (or Cu^+) to the metal are known as **reducing sugars**. Reducing sugars have a free —OH at the anomeric carbon since these compounds can ring-open to aldehydes. Methyl glycosides or other sugar acetals are not reducing sugars.

An aldose can undergo what is in effect an internal oxidation-reduction reaction when treated with dilute base. This reaction involves an **enediol** intermediate, an alkene with —OH groups on each alkene carbon. The carbonyl group can continue moving up and down the carbon chain with extended exposure to base.

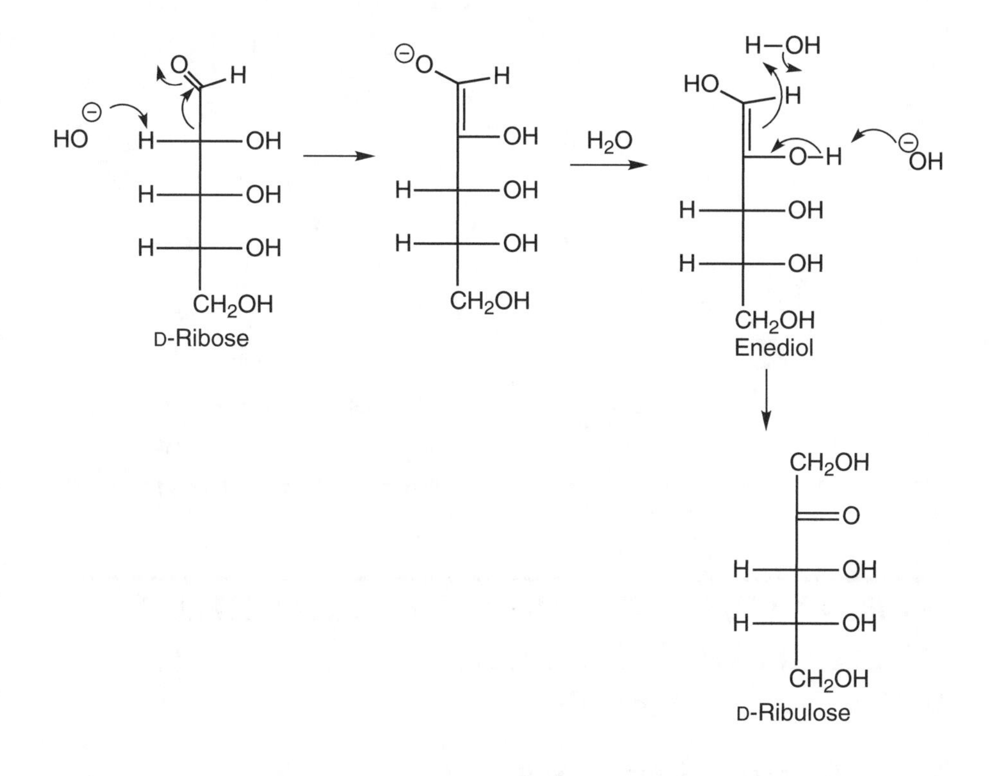

The Kiliani-Fisher Synthesis

In our discussion of D and L nomenclature and glyceraldehyde, we noted that one could convert glyceraldehyde to any other sugar (or at least easily to any other aldose). This transformation is performed by the **Kiliani-Fisher synthesis**. In this chain elongation process, the aldose is treated with HCN, which converts the aldehyde to a cyanohydrin. A pair of diastereomers is formed which can be separated. Then the —CN group is reduced to an aldehyde group. In the original synthesis, this reduction was carried out first by hydrolysis of the —CN to —COOH and then by reduction of the acid to an aldehyde. A modern variant reduces the —CN group to an imine by catalytic hydrogenation with $Pd/BaSO_4$, and the imine undergoes rapid hydrolysis to the aldehyde:

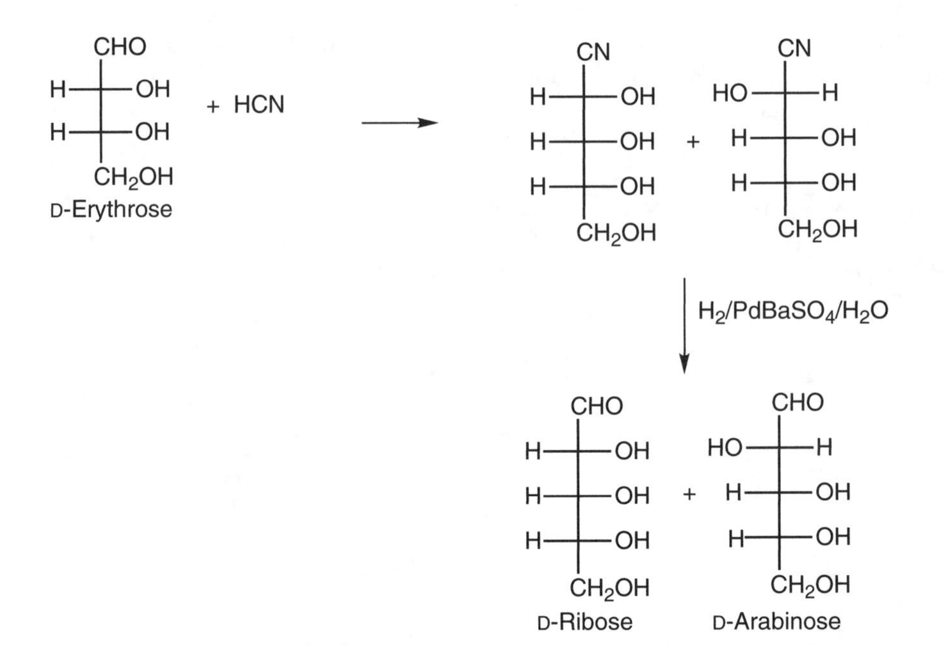

DERIVATIVES OF MONOSACCHARIDES, DISACCHARIDES, AND POLYSACCHARIDES

Monosaccharide Derivatives

Some carbohydrates have had one of the —OH groups along the chain replaced by an amine group or an amide group. The most common **amino sugar** is *N*-acetyl-D-glucosamine. This sugar makes up the polysaccharide **chitin**, which makes up the exoskeleton of insects and lobsters. Other amino sugars are found in numerous antibiotics.

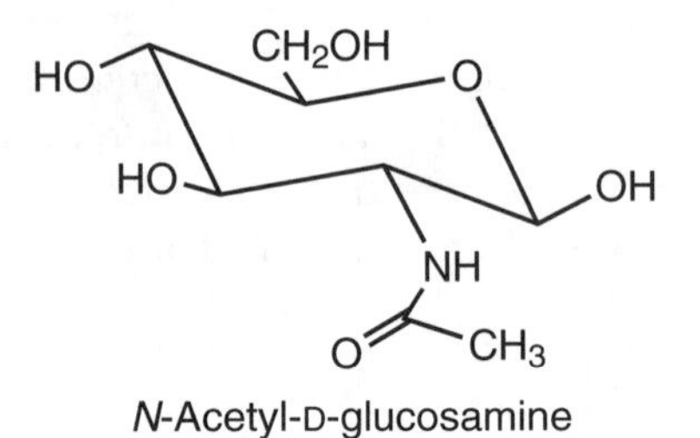

N-Acetyl-D-glucosamine

Ascorbic acid, vitamin C, is a modified carbohydrate. Most animals synthesize their own vitamin C; however, guinea pigs, chimpanzees, fruit bats, humans, and a few other organisms have lost the ability to biosynthesize this antioxidant. Although the word *vitamin* implies a life-giving amine, ascorbic acid contains no nitrogen. It is biosynthesized from D-glucose. Notice that because of the rules governing Fischer projections, when one reduces the aldehyde of D-glucose to $—CH_2OH$ and oxidizes the bottom $—CH_2OH$ to a —COOH group, one has to rotate the sugar, which by definition becomes an L sugar:

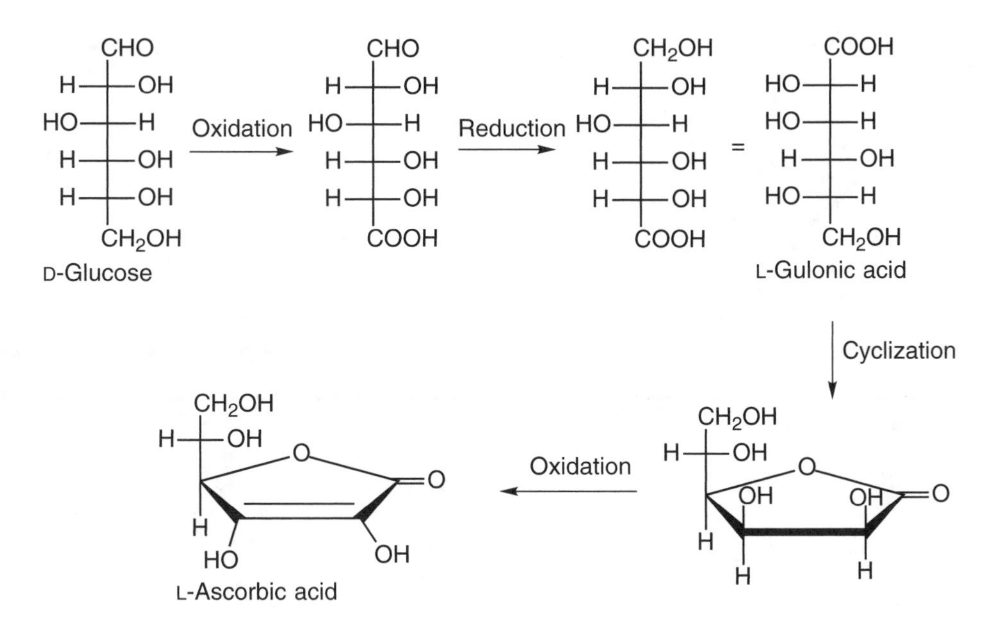

Disaccharides

Two sugars joined together form a **disaccharide**. Normally, disaccharides are glycosides (sugar acetals) where the anomeric carbon of one sugar is bonded to the —OH group of a second sugar. The most usual kind of disaccharide linkage is between the anomeric carbon, C-1, of one sugar (denoted α-1 or β-1 depending on whether the second sugar is α or β) with the hydroxyl at C-4 of the second sugar (denoted 4′). Maltose, for example, is a disaccharide of two glucose sugars joined by an α-1,4′-glycosidic linkage, and cellobiose is a disaccharide with a β-1,4′-glycosidic linkage:

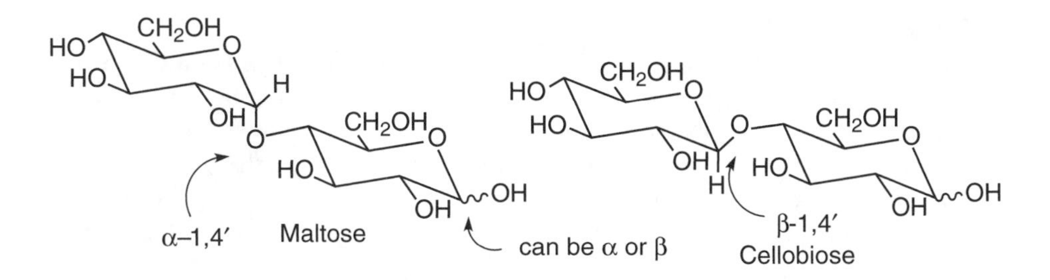

Notice that specifying the second sugar as glucose establishes the stereochemistry at C-4′. Only in the case where two sugars are joined 1,1′ would one have to specify each linkage as α or β.

The most important disaccharide is cane sugar, or table sugar, sucrose. Sucrose is an α-1,β-2′-D-glucopyranosyl-D-fructofuranoside:

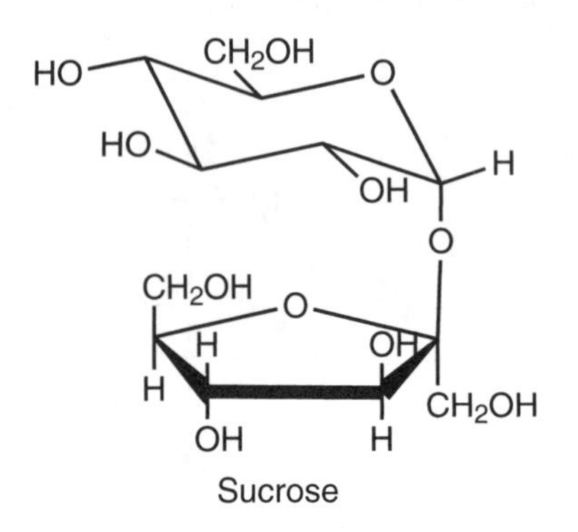

Sucrose

EXERCISE

Milk sugar, lactose, is a disaccharide of D-galactopyranose bonded at its anomeric carbon β-1,4′ to D-glucopyranose. Draw its structure using sugar chair diagrams.

Solution

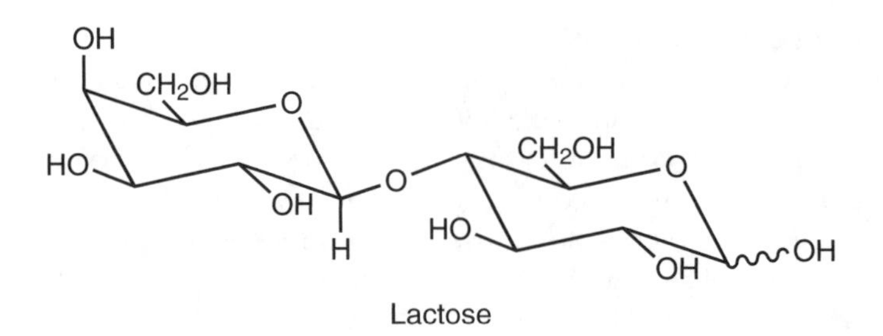

Lactose

Polysaccharides

Polymers of monosaccharides joined together are called **polysaccharides**. They are some of the most common and important biological molecules. **Cellulose** is a linear polysaccharide of β-1,4′-D-glucopyranose units. Hydrogen bonding between polymer chains produces the strength found in plant fibers. Although hydrolysis of the glycoside units would product glucose, most animals cannot obtain nutrition from cellulose because they lack the enzymes that degrade cellulose. Animals that eat grass have microorganisms that colonize their digestive systems and provide the enzymes that break down cellulose:

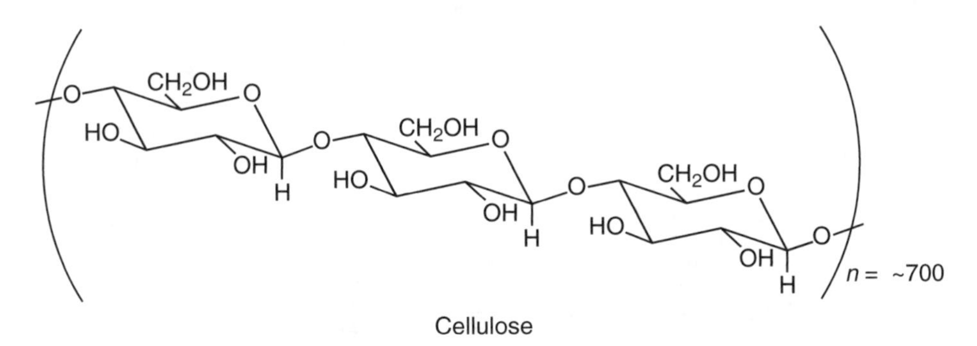

Cellulose

Starch is an energy storage polysaccharide found in potatoes and other plants. It is composed of a **branched polymer** where, in addition to a linear chair, branches extend off the main chain. Starch contains both α-1,4- and α-1,6-glycosidic linkages. The unbranched polymer is known as **amylose**, and the branched polymer is called **amylopectin**. Because the linkage is α (axial) at the anomeric center, it is less stable and easier to hydrolyze than the β (equatorial) linkage found in cellulose. Animals, including humans, store glucose as a polymer similar to amylopectin, called **glycogen**, in the liver and muscles. When a runner "hits the wall," his or her glycogen stores are used up.

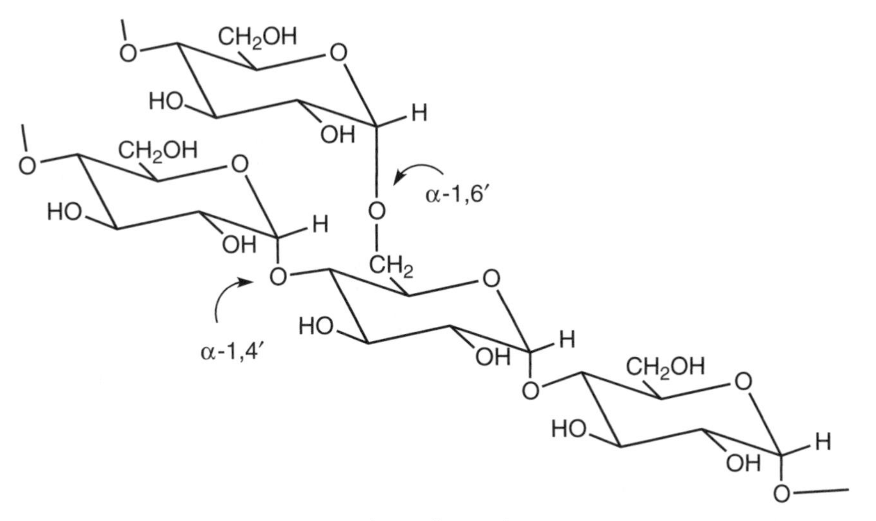

Amylopectin or glycogen

Some polysaccharides have more than one carbohydrate in the repeating unit. **Heparin**, the polysaccharide that is released during injury and inhibits blood clot formation, is made up of a complex repeating unit of glucosamine, glucuronic acid, and iduronic acid. Some glucosamines are *N*-acetylated, *N*-sulfonated, or sulfonated at C-6, and the iduronic acid residues have a sulfate group at the C-2 hydroxyl:

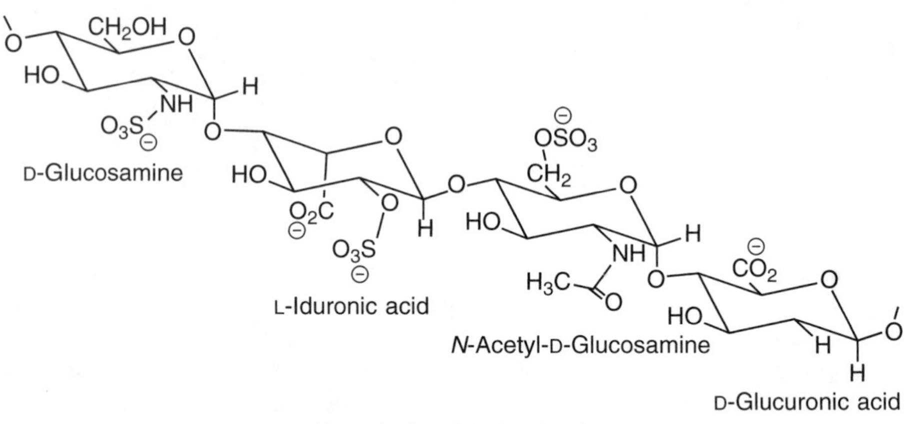

Heparin (partial structure)

28
THE ORGANIC CHEMISTRY OF PROTEINS AND NUCLEIC ACIDS

PROTEINS

Protein Building Blocks: α-Amino Acids

Proteins are polymers of α-amino acids, bifunctional molecules of the type $^{-}OOCCHRNH_3^{+}$. Because these molecules possess both an acidic group and a basic group, they exist as **zwitterions**, or **internal salts**, where the base is protonated by the internal acid. The neutral form, $HOOCCHRNH_2$, is present to an insignificant degree in solution or in the solid state. The protein polymer is built up by the formation of an amide bond, known as a **peptide bond** where amino acids are concerned, between the $—COO^-$ group of one α-amino acid and the $-NH_3^+$ group of a second α-amino acid. A molecule made up of two α-amino acids connected through a peptide bond is called a **dipeptide**. Three α-amino acids connected through two peptide bonds, $^{+}NH_3CHRCO—NHCHR'CO—NHCHR''COO^-$, is called a **tripeptide**, and so on. Polymers of α-amino acids with a few to about 50 α-amino acid units are defined as **polypeptides**, while longer polymers are called proteins. Proteins are central to understanding biochemistry. Some are structural, such as collagen and silk. Proteins make up the molecular machines that create movement in muscles and the channels that allow ions to flow in and out of cells to carry nerve impulses. Essentially all the **enzymes** that catalyze biochemical reactions are proteins.

In biology information flows from DNA to RNA to protein. The sequence of particular α-amino acids in proteins is encoded in the sequence of DNA bases known as genes. Only 20 α-amino acids are encoded by DNA. Although hundreds of amino acids are found in biology, all but these 20 are prepared by secondary biochemical reactions. The structures of these genetically encoded α-amino acids are shown in Table 28.1. Each α-amino acid has a standard three-letter abbreviation and a one-letter abbreviation, which are also listed.

TABLE 28.1. THE 20 GENETICALLY ENCODED α-AMINO ACIDS

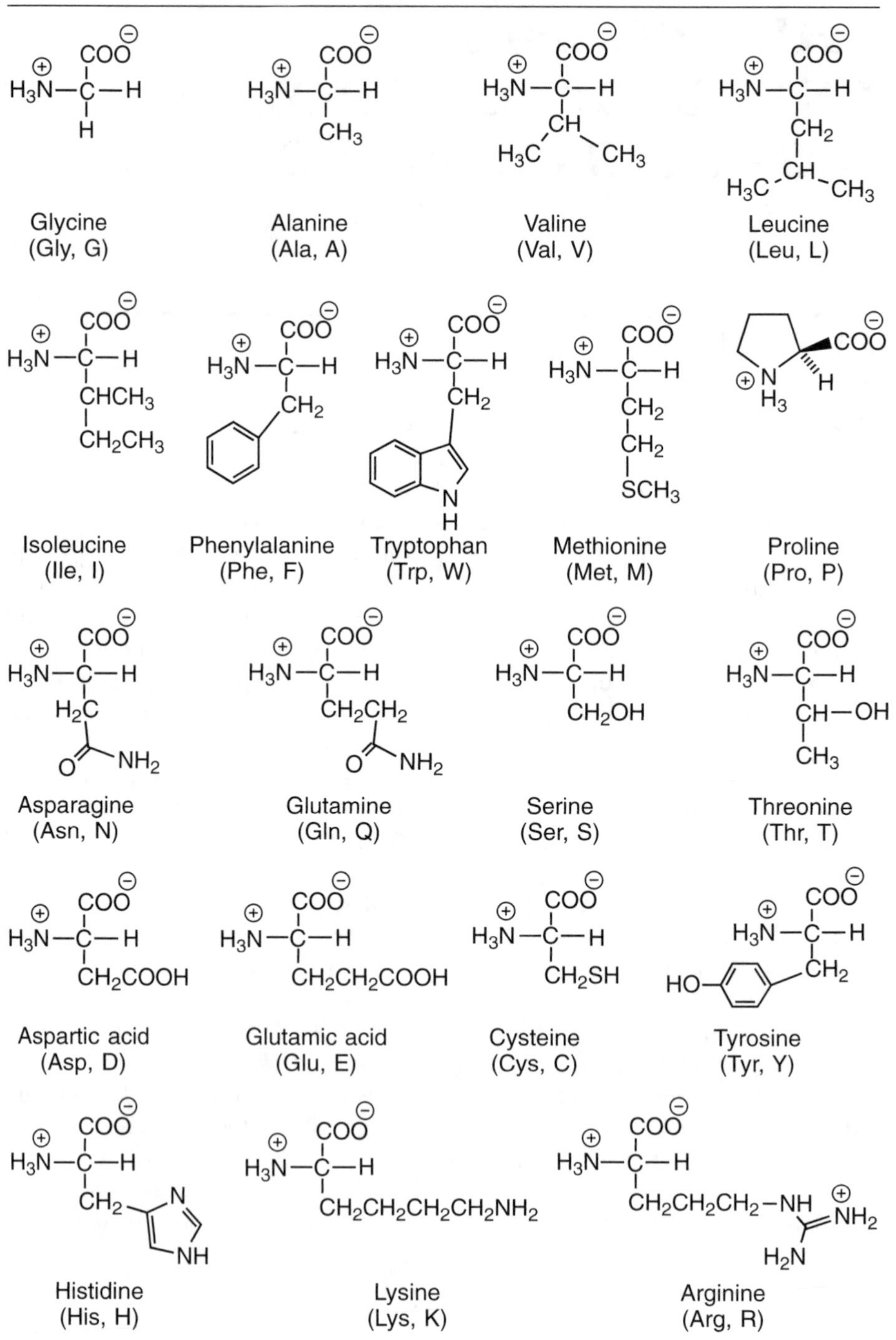

Except for glycine, all these amino acids are chiral at the **α carbon**. All the amino acids in Table 28.1 (except for the cyclic amino acid proline) are drawn as Fischer projections, making the point that all the genetically encoded amino acids are L enantiomers. Amino acids are normally grouped into four classes. The first nine amino acids in Table 28.1, from glycine to proline, are said to be nonpolar because their side chains are hydrophobic. The next four in the table, from asparagine to threonine, are polar amino acids since their side-chain groups are good at hydrogen bonding. There are four acidic amino acids, from aspartic acid to tyrosine, because each has a side-chain group that is deprotonated at pH 7. Finally, the three amino acids histidine, lysine, and arginine are basic amino acids since their side-chain groups are protonated at pH 7.

EXERCISES

1. Which α-amino acids have a stereogenic center in their side chain?
2. All the chiral α-amino acids have the S configuration at the α carbon, except one. Which one has the R configuration?

Solutions

1. Isoleucine (which is R at the side-chain stereogenic center) and threonine (which is also R).
2. Cysteine since the CH_2SH group has a higher priority than COOH.

Synthesis and Biosynthesis of α-Amino Acids

Numerous reactions can be used to prepare α-amino acids. One of the oldest is the **Strecker** synthesis, which involves the reaction of an aldehyde with NH_3 and NaCN, to give an α-amino nitrile, $RCH(NH_3)CN$. The mechanism of this process involves first the reaction of the aldehyde with ammonia to form an imine and then the addition of NC^- to the C=N double bond. In the second step, the nitrile is hydrolyzed under acidic conditions to give the carboxylic acid group of the α-amino acid. Some scientists have suggested that the Strecker synthesis might be an example of a **prebiotic** reaction, one that synthesizes the building blocks of life in the absence of life and took place on the ancient earth or in interstellar space. Of course, the product of the Strecker synthesis is a racemic mixture.

Several different biochemical pathways synthesize α-amino acids. One of the most important uses pyridoxamine phosphate, a derivative of vitamin B_6, and an α-ketoacid. In this reaction, the vitamin and the ketone group of the α-ketoacid form an imine, this imine undergoes allylic rearrangement, and hydrolysis produces an α-amino acid and the aldehyde form of the vitamin. The only difference between this biochemical mechanism and ordinary organic reactions is that it takes place within the interior of an enzyme, a chiral environment due to the chiral α-amino acids that make up the enzyme. Thus, only a single enantiomer of the α-amino acid is made:

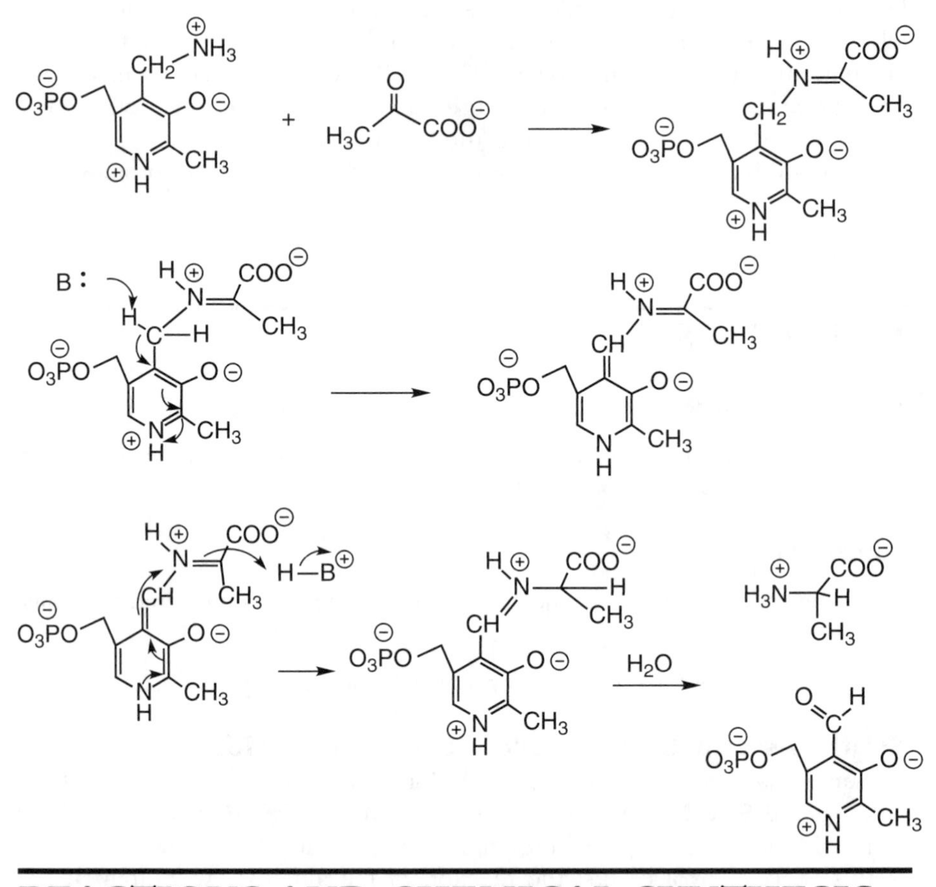

REACTIONS AND CHEMICAL SYNTHESIS OF PEPTIDES

Reaction of Amino Acids with Ninhydrin

Because proteins are polymers held together by peptide (amide) bonds, they slowly hydrolyze and release traces of α-amino acids in the presence of water and heat. Thus, there are small amounts of α-amino acids in fingerprints and bodily fluids. The reagent ninhydrin reacts rapidly with α-amino acids at room temperature to give a product with a deep-purple color. This reaction allows one to observe fingerprints and is very useful in forensic chemistry:

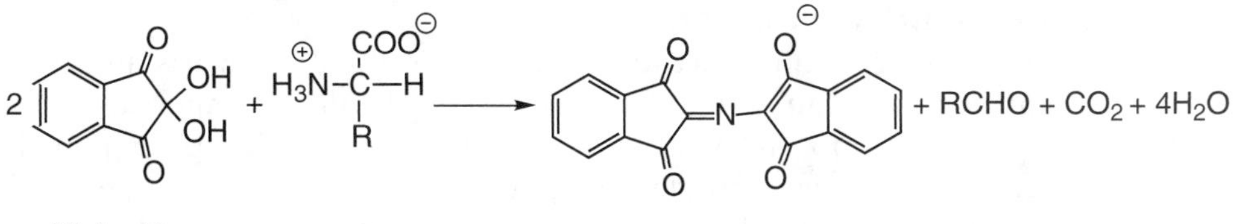

Ninhydrin

EXERCISE

Outline a mechanism for the reaction of ninhydrin with an α-amino acid.

Solution

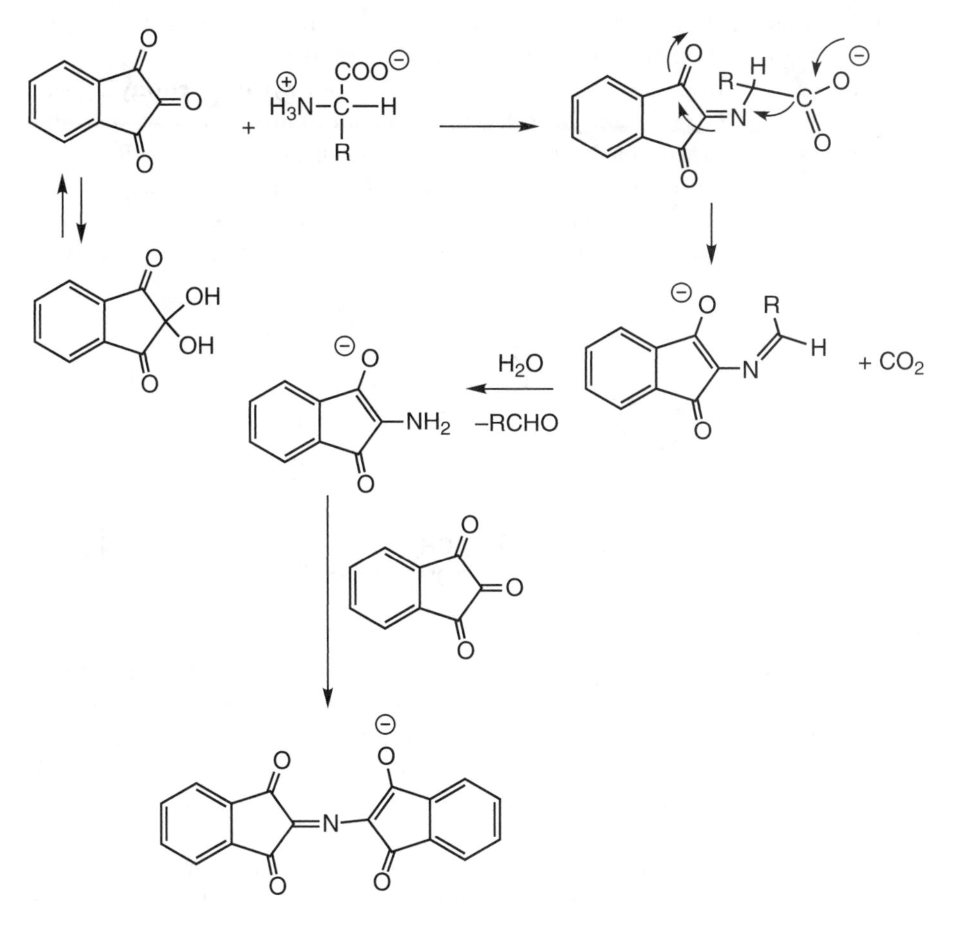

Peptide Bond Formation

When two amino acids react to give a peptide bond, each has an amino group and a carboxylate group. One has to control which α-amino acid supplies the $—NH_2$ group and which supplies the —COOH group for peptide bond formation. The same problem arises in the synthesis of longer peptides. The general solution to this problem is to use α-amino acid derivatives with some of their functional groups differentiated with protecting groups. For example, to synthesize the dipeptide $^{+}NH_3$-Gly-Ala-COO^-, one starts with $XNHCH_2COOH$ (where X is a protecting group for an amine) and $NH_2CH(CH_3)COY$ (where Y is a protecting group for an acid). Next the amine and acid react to form an amide bond, and finally the protecting groups are removed to give the dipeptide. If a tripeptide is desired, one can take the protected dipeptide and remove only one of the protecting groups, say the acid one, to obtain $XNHCH_2CONHCHCH_3COOH$. Then a third protected amino acid, such as $NH_2CH(CH_2Ph)COY$, can be added. After coupling and deprotection, the tripeptide $^{+}NH_3$-Gly-Ala-Phe-COO^- would result.

The $—NH_2$ group can be protected by the **benzyloxycarbonyl** group (abbreviated **Z**) or by the ***tert*-butoxycarbonyl** (Boc) group. In each case the amine is converted to a **carbamate** functional group, RO—CO—NHR, which, like an amide, is nonbasic. The benzyloxycarbonyl group is put on using $PhCH_2OCOCl$, and the *tert*-butoxycarbonyl group is added using (*tert*-Bu-O-$CO)_2O$. The benzyloxycarbonyl group is removed by hydrogenolysis with Pd/C—H_2, while the Boc group is removed with HBr:

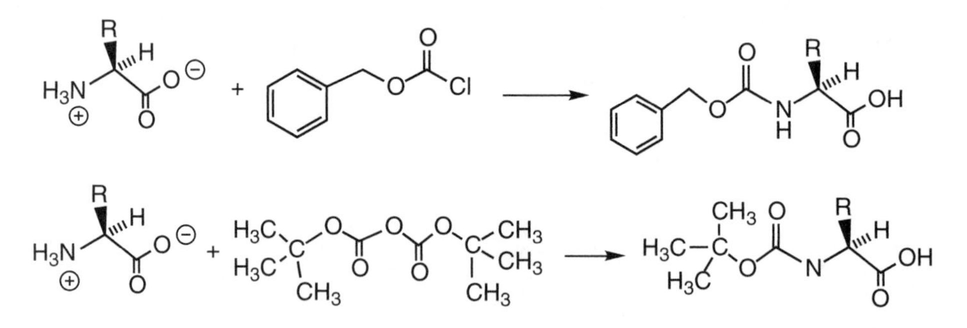

The carboxyl groups of amino acids are usually protected as esters. Methyl esters can be used, which are deprotected by basic hydrolysis. Alternatively, benzyl esters can be used, which are deprotected by hydrogenolysis. Many side-chain functional groups, such as the —OH group of serine or the —SH group of cysteine, need to be protected as well, and various protecting groups are available for each α-amino acid.

Once the α-amino acid derivatives are properly protected, one needs a way to form the amine bond. While conversion of an acid group to an acid chloride and reaction with the free amine portion of a protected α-amino acid to give an amide bond are possible, this reaction can result in racemization of the chiral α carbon. Since proteins contain more than 50 amino acids residues, racemization can produce more than 2^{50} diastereomers. Thus, mild amide-forming reactions are required in protein synthesis. A useful reagent is DCC, which was discussed in Chapter 21. Adding an amine, carboxylic acid, and DCC produces the amide bond without significant racemization.

EXERCISE

Describe the synthesis of the tripeptide $^{+}NH_3$-Phe-Ile-Phe-COO^- starting from the appropriate protected α-amino acids.

Solution

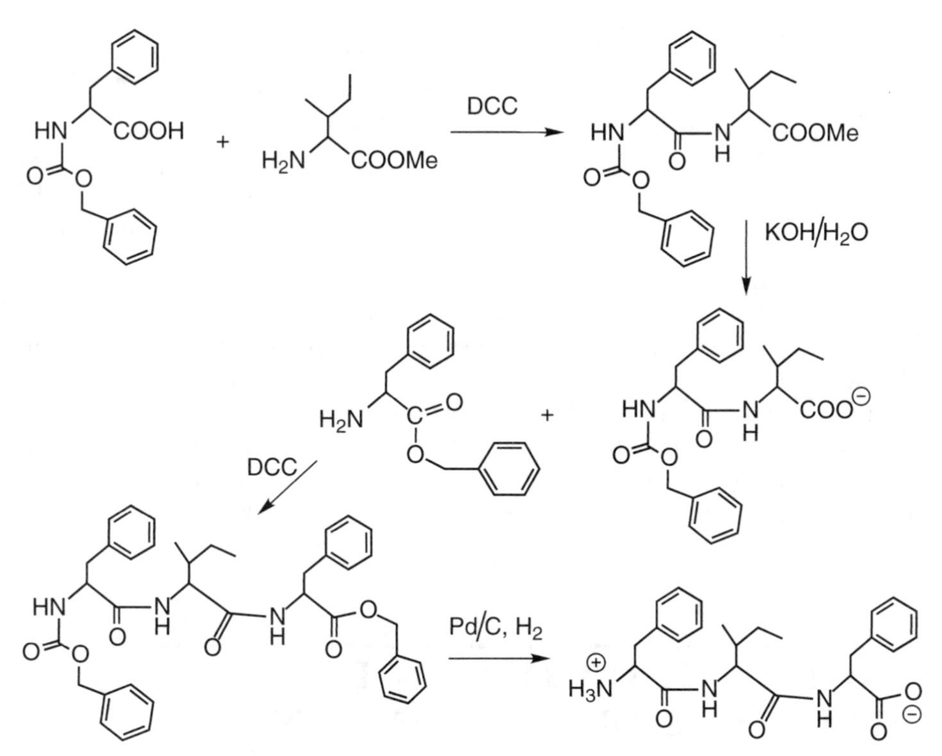

Solid Phase Peptide Synthesis

The most effective way to make peptides and proteins (as well as other types of biological molecules) is by **solid phase synthesis**. In this method, the first residue of the protein chain is chemically attached to a polymer bead. Reagents and protected α-amino acids then flow over this bead, leading to extension by one residue. By-products can be washed from the growing chain on the bead, and the cycle is repeated as many times as is necessary to give the correct number of residues. Finally, the chain is chemically cleaved from the bead and any remaining protecting groups are removed.

One polymer, or **resin**, used as a support in solid phase protein synthesis is **Merrifield resin**, named after R. Bruce Merrifield who invented solid phase synthesis in 1962. This support is a derivative of polystyrene, containing 2 percent divinylbenzene added to form links between the polystyrene chains and rigidify it. A few of the benzene rings undergo Friedel-Crafts alkylation with $ClCH_2OCH_3$, which introduces chloromethyl groups. These benzylic chloride groups react with an N-protected α-amino acid at the O of the carboxylate group to form an ester:

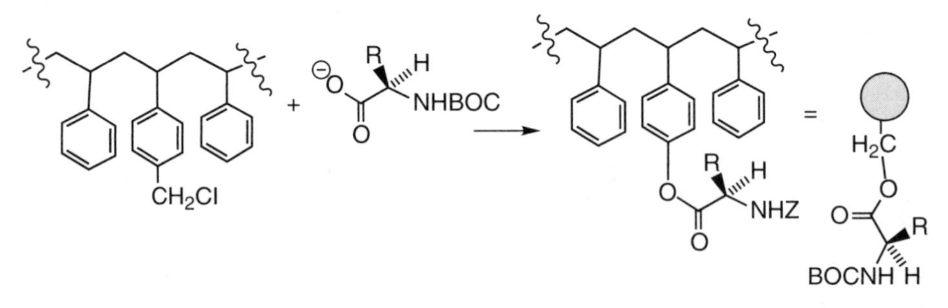

A major advantage of solid phase synthesis is that a large excess of reagents can be added at each coupling step (the step that makes the peptide bond), promoting a maximum yield. All the unreacted reagents can be washed from the polymer bead, making purification very simple. High yields in individual coupling steps are vital to obtaining an overall good yield of product. For example, if one carries out eight coupling steps, each with an average yield of 90 percent, the yield of the final product is $(0.9)^8$, or 42 percent. However, if the average yield is 99 percent, over eight steps the yield is $(0.99)^8$, or 92 percent.

PROTEIN STRUCTURE AND STRUCTURE DETERMINATION

Primary, Secondary, Tertiary, and Quaternary Structure

Proteins are large molecules, yet they have a definite structure. Once formed, they spontaneously fold into a compact shape. As yet, it is impossible to predict this shape, but techniques such as X-ray crystallography can determine the location of each heavy (nonhydrogen) atom in the molecule if the protein forms single crystals. The structure of a protein is usually normally divided into four levels. The order of amino acids along the protein chain is known as the **primary structure**. Normally, the primary structure is listed starting at the NH_3^+ end, the **N terminus**, and ends at the last COO^-, the **C terminus**.

The peptide bond is planar, but there is rotation around the single bonds attached to the α carbon:

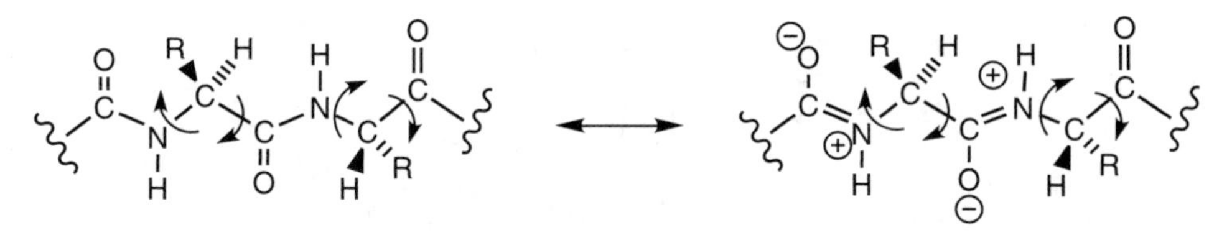

Thus, parts of the peptide backbone are rigid and other parts have some rotational freedom. The chain can fold into two important conformation families stabilized by hydrogen bonds between a backbone N—H hydrogen bond donor and a backbone C=O hydrogen bond acceptor. In one of these conformations, called an **α-helix**, the backbone coils into a right-handed helical form. In the second conformation, called a **β-sheet**, the same kinds of C=O · · · HN hydrogen bonds form between portions of two or more backbone chains passing by one another. The small protein crambrin (a 46 amino acid residue protein) is shown below to illustrate the backbone shape of an α-helix and a β-sheet:

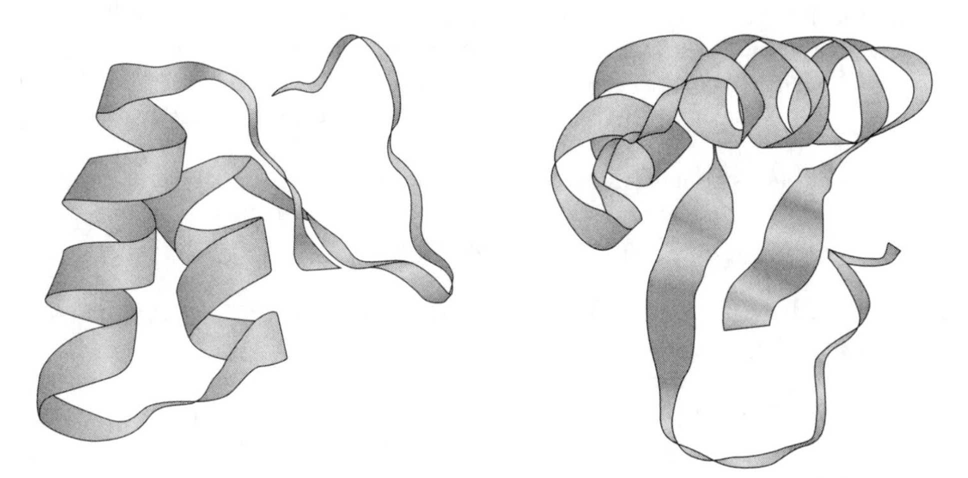

The regions of a protein that contain these ordered α-helix and β-sheet conformations are considered the **secondary structure**. Not all parts of a protein exist in these highly ordered structures. Some regions have a chain turn or loop. The overall three-dimensional structure of a protein is called its **tertiary structure**. The tertiary structure of crambrin is shown following:

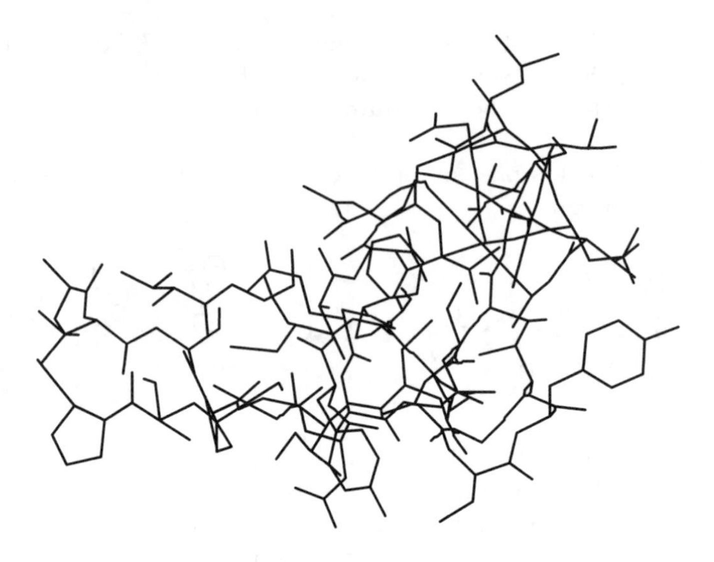

One can see some of the residues that make up crambrin from this diagram (which omits hydrogen atoms). At the lower right, one can recognize a tyrosine side-chain group, and at the lower left, a proline residue.

Almost 20,000 protein tertiary structures are known, and more are being solved each day. The National Institutes of Health of the United States support a database, the protein database, of all these structures. It is freely accessible from the Internet, and the current address is http://www.rcsb.org/pdb. Programs that allow one to view these structures are also freely available, including RasMol, its successor Protein Explorer, and Swiss PDB Viewer. Using these terms in a web search provides links to sites where these programs can be downloaded.

Some proteins contain more than one polypeptide chain. Hemoglobin, for example, is made up of four independent chains. The packing of these chains is considered a protein's **quaternary structure**.

Chemical Means for Determining Protein Primary Structure

The most powerful chemical method for determining a protein's primary sequence is **Edman degradation**, which removes residues from the N terminus one amino acid at a time. The reaction involves treating the protein with phenyl isothiocyanate, followed by acid treatment (usually HF):

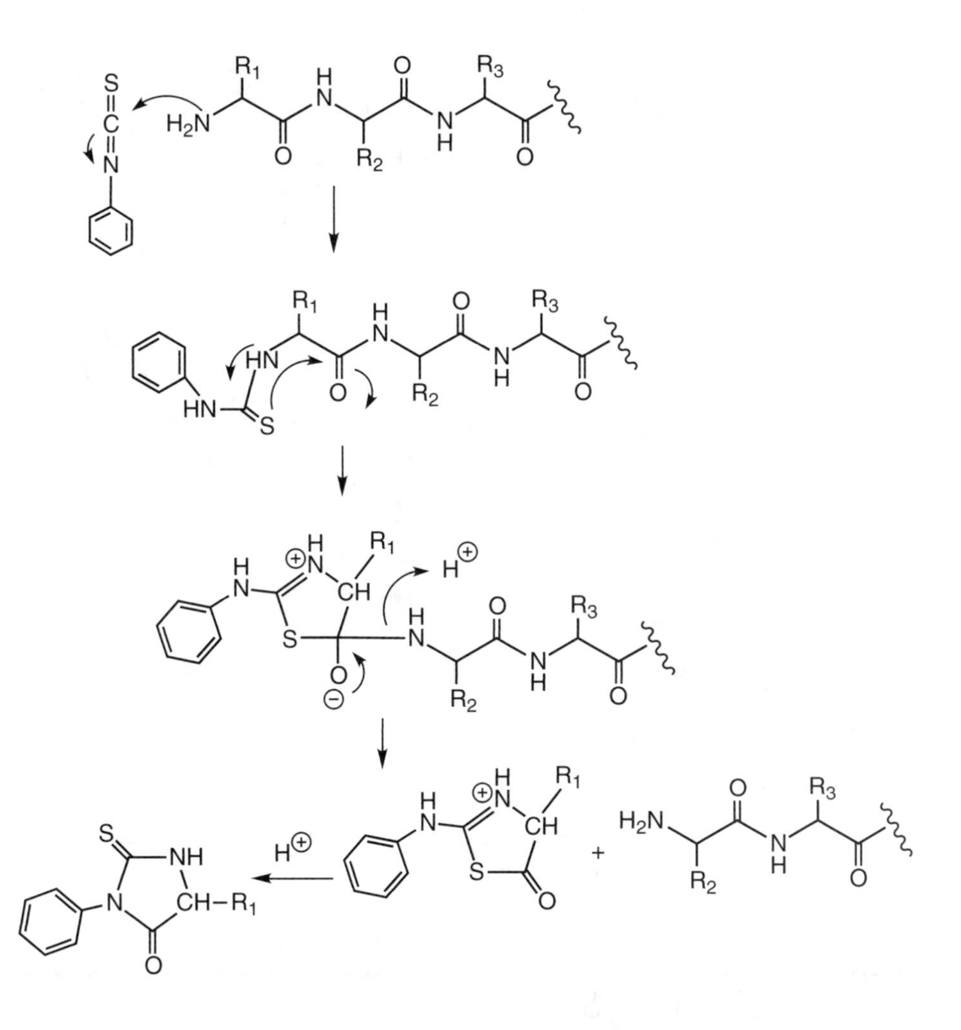

The final products are a heterocycle, a **phenylthiohydantoin** (PTH), and the peptide chain with a new N terminus, corresponding to the second amino acid residue. The nature of the R_1 group in the PTH, which can be compared with an authentic sample, determines the first amino acid at the N terminus. Automated machines can, under favorable circumstances, carry out 50 cycles of Edman degradation.

Several enzymes are available that hydrolyze the peptide bonds of proteins at restricted positions. A chemical reagent that exhibits selectivity is **cyanogen bromide**, CNBr, which cleaves the protein backbone to the C-terminal side of methionine:

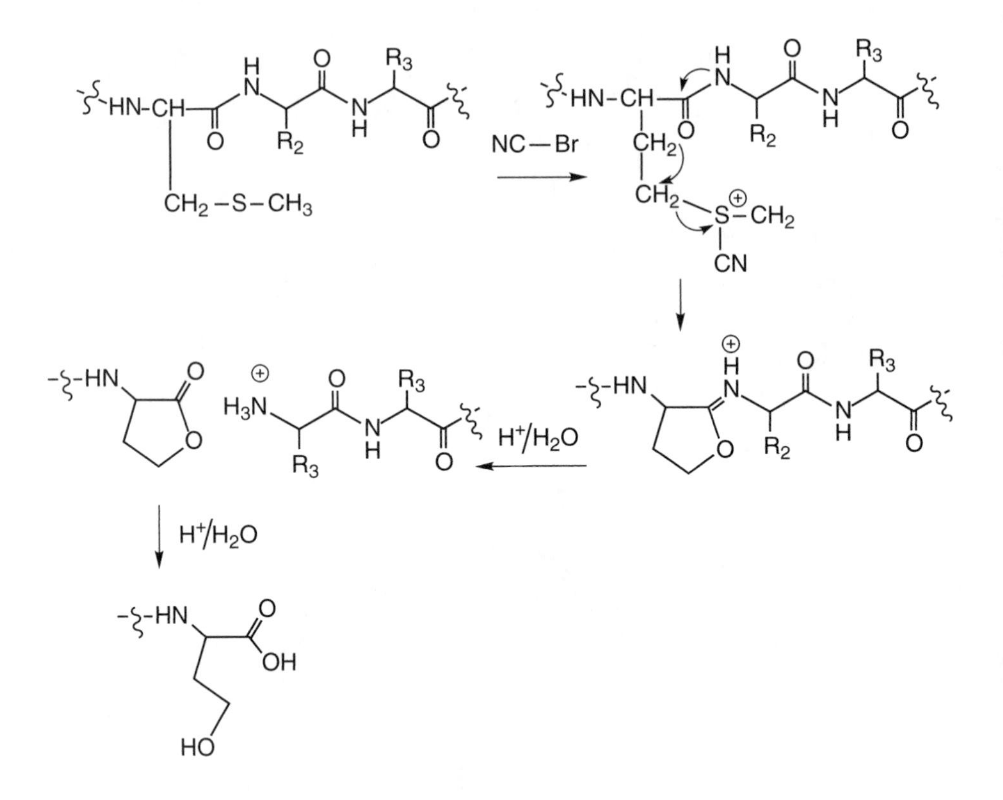

Since methionine is not a particularly common amino acid, only a few cleavage sites are present in most proteins.

NUCLEIC ACIDS

Nucleic acids, DNA and RNA, are polymers of sugars held together by phosphodiester bonds. The sugar groups have heterocyclic bases attached at the anomeric carbon. This section will focus on the structure of these biological molecules.

Properties of Phosphodiesters

A **phosphodiester** is a derivative of phosphoric acid, H_3PO_4, with two OR groups in place of two of the OH groups. In **deoxyribonucleic acid (DNA)** the R groups are derivatives of the sugar 2′-deoxyribose, and in **ribonucleic acid (RNA)** the R groups are derivatives of ribose.

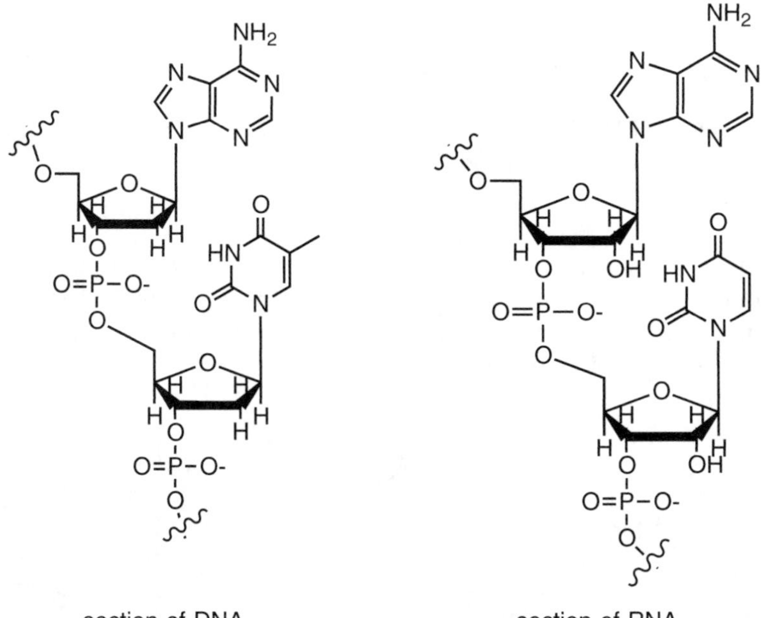

section of DNA section of RNA

The properties of a phosphodiester are similar to those of a carboxylic acid ester except that the phosphodiester has a negative charge at pH 7 because of the acidity of phosphoric acid. Thus, it is resistant to attack by nucleophiles. DNA can be heated in NH_4OH solution for hours without decomposition. In nature, enzymes that hydrolyze the phosphodiester backbone of DNA or RNA use Mg^{2+} to neutralize the charge on the phosphate group to facilitate a nucleophilic attack on the P atom. A nucleophilic attack on tetrahedral phosphodiesters follows the same geometry (known in phosphorus chemistry as the **in-line geometry**) seen for a S_N2 reaction at a tetrahedral carbon:

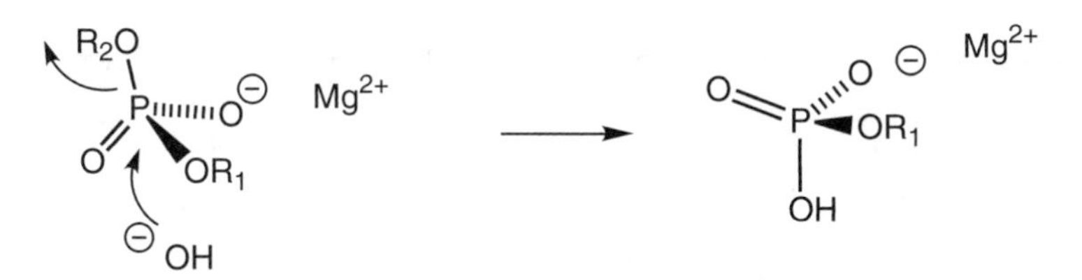

Structure of Nucleosides and Nucleotides

The individual sugar-heterocycle unit in DNA and RNA is called a **nucleoside**. All DNA nucleosides have the same five-carbon sugar, 2′-deoxyribose, and one of four heterocyclic bases, **adenine**, **guanine**, **cytosine**, or **thymine**. In RNA, the sugar is ribose, and the base thymine is replaced by **uracil**. The other three bases are the same:

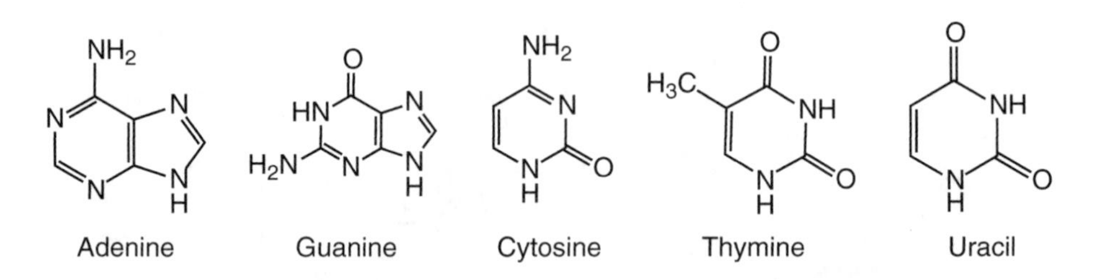

Both adenine and guanine are heterocycles that belong to the **purine** family. The monocyclic bases all belong to the **pyrimidine** family. The structure of the four nucleosides that make up RNA and the four deoxynucleosides that make up DNA are shown here.

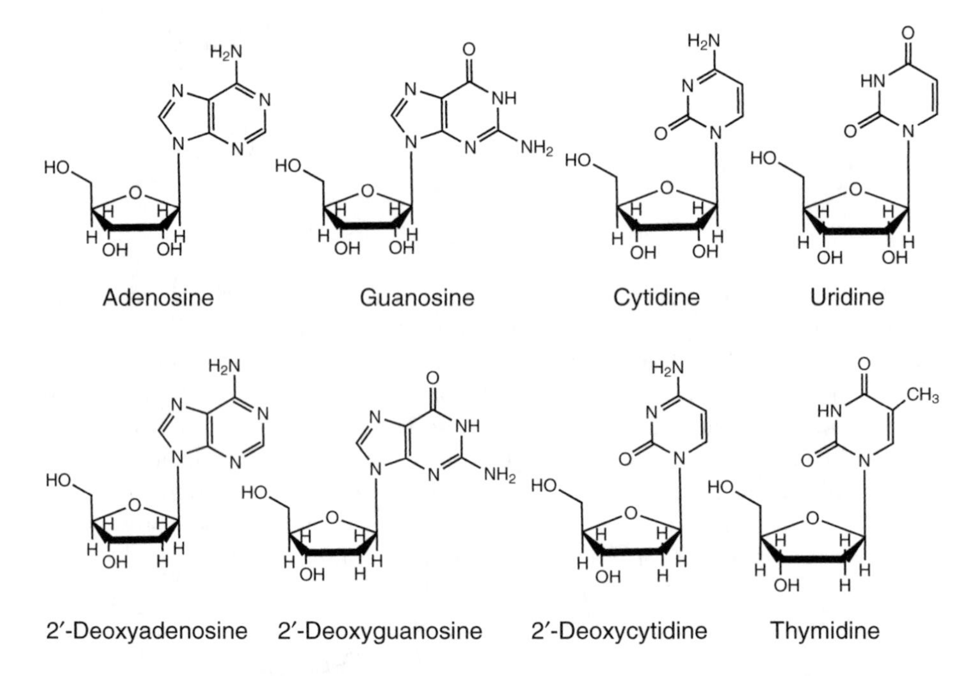

The numbering of the sugar carbons and the atoms in the purine and pyrimidine bases is as follows.

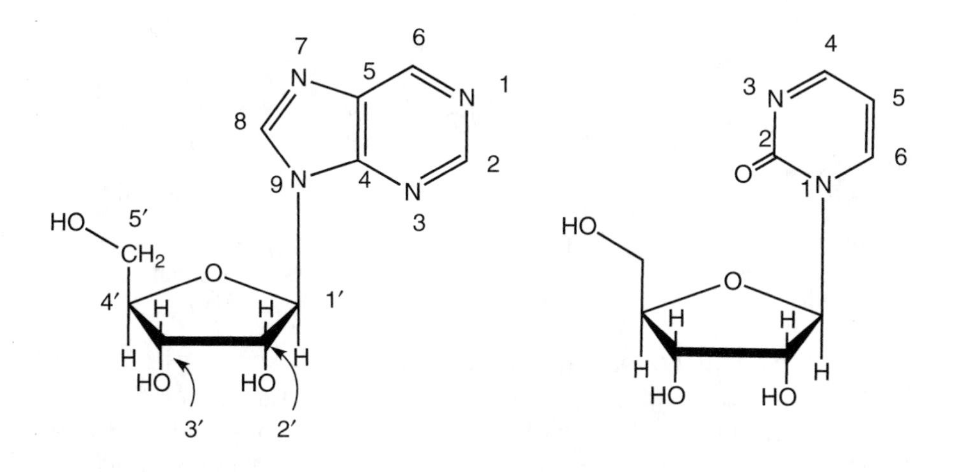

EXERCISE

The antiviral and anti-AIDS drug with the trivial name azidothymidine (AZT) is a derivative of thymidine with an azide group at the 3′ carbon, replacing the hydroxyl group. Draw its structure.

Solution

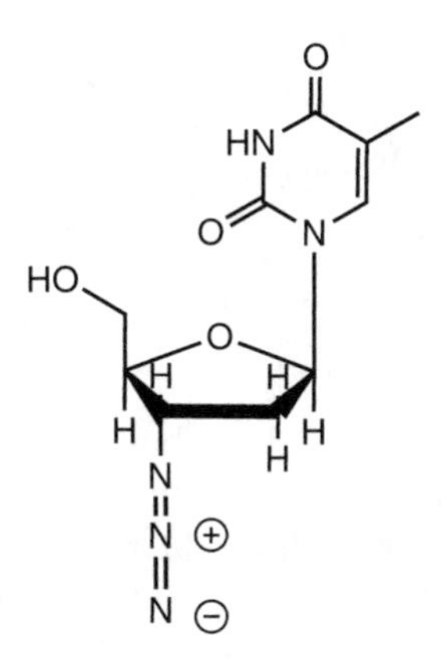

A **nucleotide** is a nucleoside (either DNA- or RNA-derived) with a phosphate group attached to the 5′-OH or 3′-OH group. A nucleoside can be a **monophosphate**, a **diphosphate** (also called a pyrophosphate), or a **triposphate**. The well-known carrier of biochemical energy is a nucleotide, **adenosine 5′-triposphate (ATP)**:

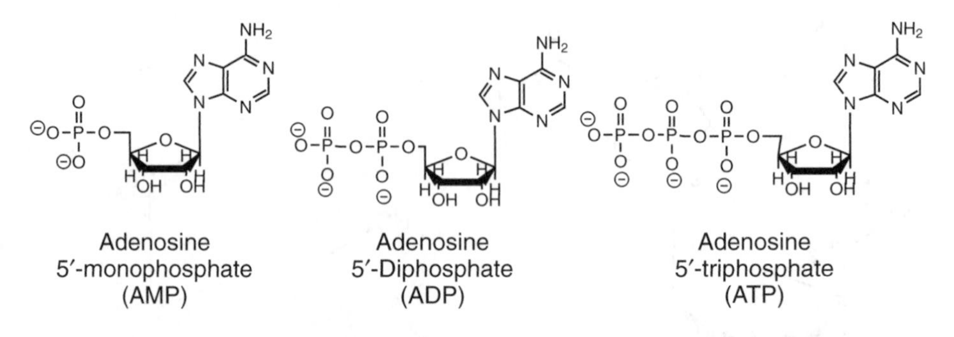

Structure of the DNA Double Helix

DNA is the carrier of genetic information. The sequence of bases, A (adenine), G (guanine), C (cytosine), and T (thymine), is converted to an RNA sequence, and this sequence is read by the cell's protein synthesis machine, the **ribosome**, to make proteins. When a cell divides, each daughter cell needs a complete copy of the DNA carried by the parent cell or it will not be viable. Nature has solved this copying problem by storing the DNA information in a DNA duplex in which two DNA stands wrap around one another. Each DNA base-pairs reliably with its partner, forming two kinds of **DNA base pairs**, the adenine-thymine (A-T) base pair and the guanine-cytosine (G-C) base pair:

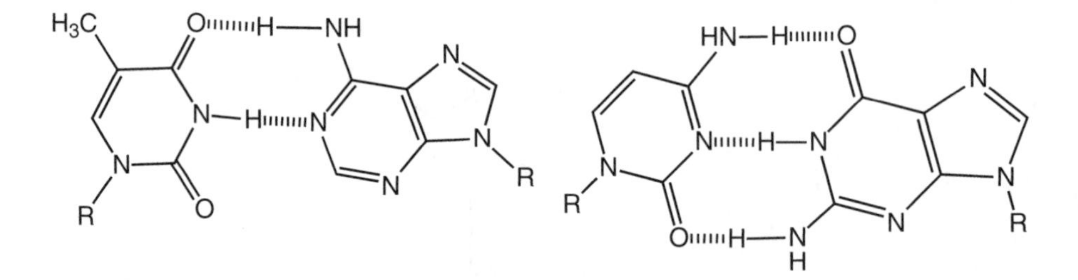

Each A base-pairs with T, and each G base-pairs with C. If you know the sequence of one DNA strand, you know the sequence of the other strand from the base-pairing rules. DNA is duplicated (or **replicated**) by using one strand as the template for the other strand. Therefore, when a cell divides, each daughter cell receives one original DNA strand and one copy strand in a process known as **semiconservative replication**.

Another very important property of duplex DNA, besides the base-pairing pattern, is that the two strands are **antiparallel**. One strand has a 5′ carbon at the top, and the opposite, or **complementary** strand terminates at the 3′ carbon:

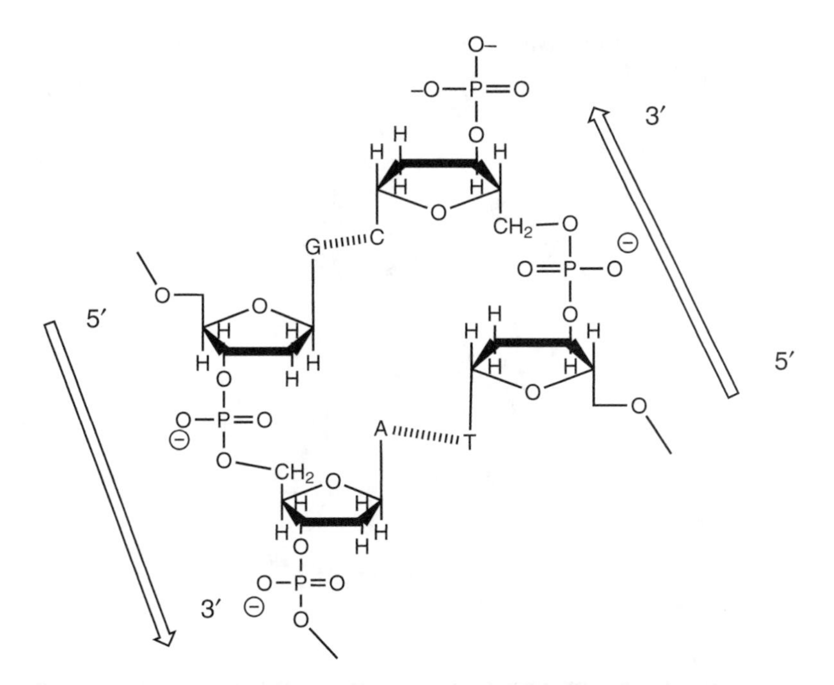

A representation of a duplex with 12 base pairs is shown below.

A representation of the same DNA strands as ribbons, emphasizing the way the duplex wraps around itself, is given below.

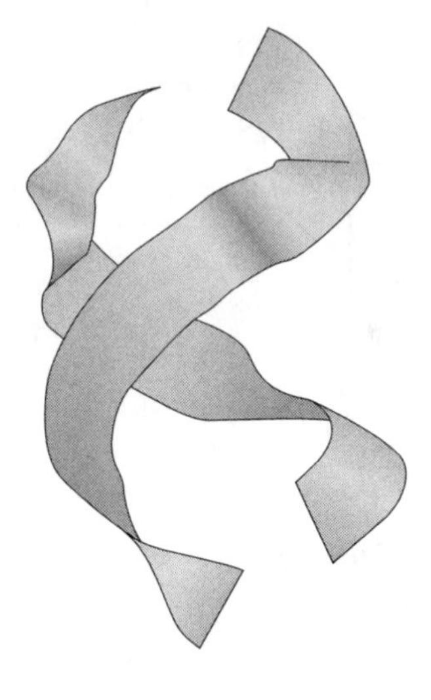

EXERCISE

What is the complementary strand to 5′-AATGCCGACT-3′?

Solution

The answer can be written as 3′-TTACGGCTGA-5′ or as 5′-AGTCGGCATT-3′. Nucleotide sequences are normally written in the 5′-to-3′ direction, so writing TTACGGCTGA would be incorrect since it would be assumed to be 5′-TTACGGCTGA-3′. The antiparallel nature of the DNA duplex is responsible for much of the confusion in questions concerning DNA sequences.

Relative Stabilities of DNA and RNA

Unlike DNA, which is quite stable in basic aqueous solution, RNA is rapidly broken down. The reason for this difference is the 2′-OH group in RNA, which can serve as an intramolecular nucleophile for attack on the 3′-phosphate group:

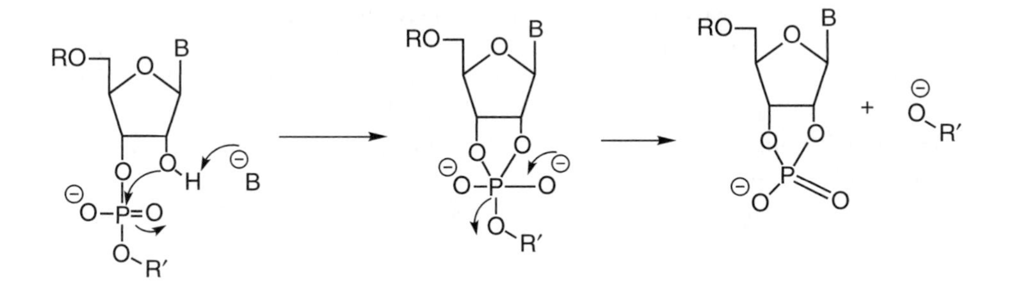

The 2′-OH group forms a cyclic intermediate which breaks down to a cyclic phosphodiester, leading to a break in the RNA chain. DNA is at least 3 billion times less reactive than RNA to hydrolytic chain scission since in DNA an intramolecular mechanism is impossible.

Both DNA and RNA are broken down by treatment with strong acids or alkylating agents. Several different weakly basic or nucleophilic sites are present in DNA and RNA, including the anionic oxygen of the phosphodiester group and various nitrogen atoms on the bases. For example, in DNA, alkylating agents such as CH_3I can react with guanine at N-7, leading to loss of the purine base, a reaction called **depurination**:

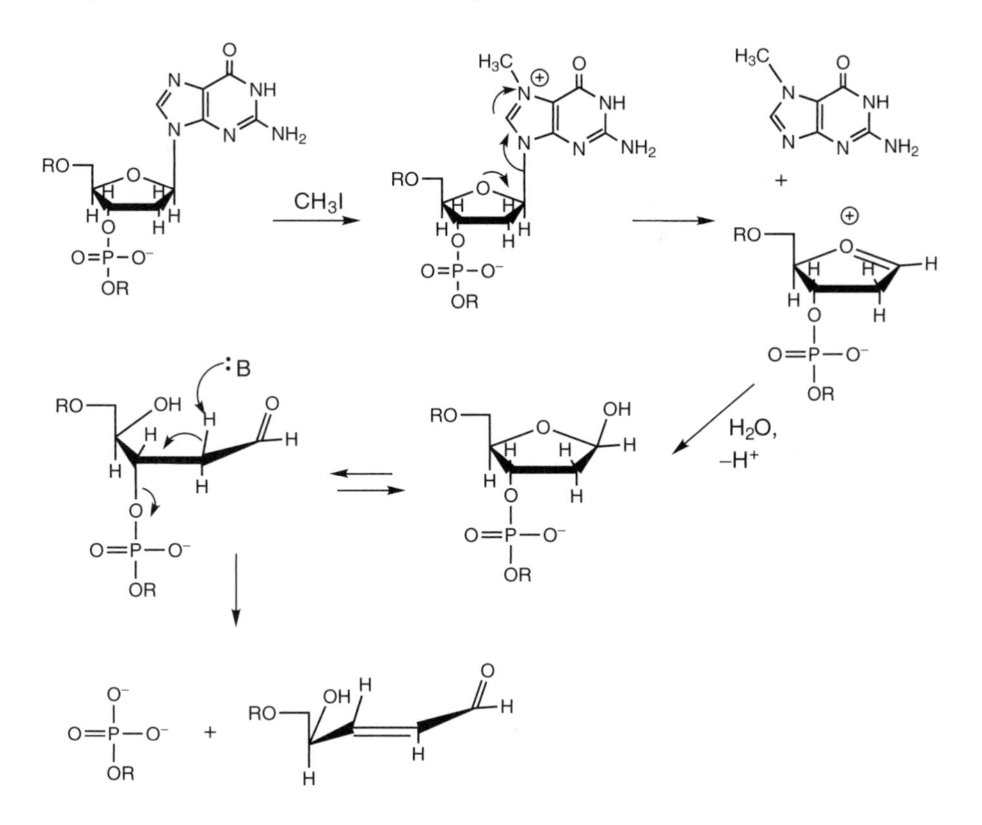

Once the purine is alkylated, it becomes a leaving group and the C—N bond at the anomeric center breaks. The sugar carbocation reacts with water, giving a hemiacetal that is in equilibrium with the open-chain sugar aldehyde form. A weak base can initiate an E2 elimination, which results in a chain break. This is why alkylating agents are **mutagens** (compounds that cause mutations). Alkylating agents cause DNA chain breaks, and the body's DNA repair systems that heal the breaks are error-prone, leading to changes in the order of DNA bases, or mutations.

29
LIPIDS AND RELATED NATURAL PRODUCTS

Natural products are molecules produced by living organisms. Normally the term is restricted to smaller monomeric compounds such as cholesterol rather than larger biological polymers such as DNA or proteins. This chapter will present some examples of largely nonpolar natural products and introduce their structures and properties. In some cases the pathways by which they are made in nature, their **biosynthesis**, will be discussed.

Fatty Acids and Triglycerides

Fatty acids were briefly mentioned in Chapter 20. They are carboxylic acids with long, unbranched carbon chains. Fatty acids are classified into two broad groups, **saturated fatty acids**, which have no carbon-carbon double bonds in the chain, and **unsaturated fatty acids**, which have one or more double bonds in the chain. As we will see below, almost all the double bonds in naturally occurring fatty acids are cis. Some important fatty acids are shown here.

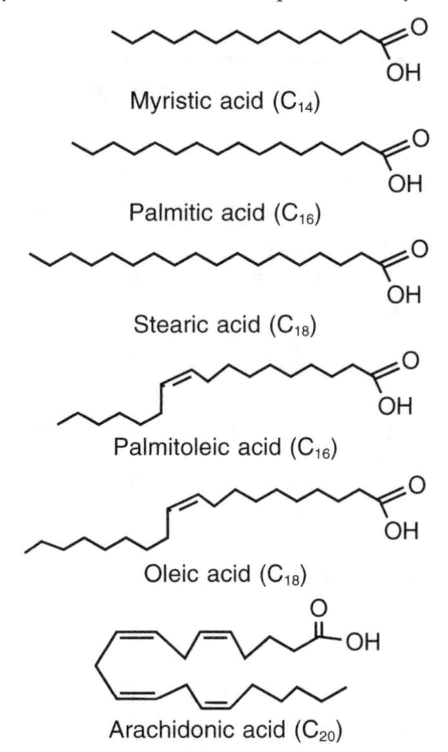

Myristic acid (C_{14})

Palmitic acid (C_{16})

Stearic acid (C_{18})

Palmitoleic acid (C_{16})

Oleic acid (C_{18})

Arachidonic acid (C_{20})

Palmitoleic acid and oleic acid are examples of **monounsaturated fatty acids**. Oleic acid is found in olive oil. Arachidonic acid is an example of a **polyunsaturated fatty acid**. Because fatty acids are built up of acetate units (reacting through Claisen condensations to form carbon-carbon bonds), almost all fatty acids have an even number of carbon atoms.

In nature, fatty acids are found as esters, often of the triol glycerin, $HOCH_2CHOHCH_2OH$. These triesters of glycerin are known as **triacylglycerols** or **triglycerides**. Some triglycerides have the same fatty acid esterified to each —OH group, and others have two or three different fatty acids attached to glycerol. Triglycerides that are solid at room temperature are called **fats**, and those that are liquid at room temperature are called **oils**.

The melting temperature and fluidity of a fat or an oil depend on the number of double bonds in the fatty acid. Saturated fats have their aliphatic chains packed tightly together, so they are relatively high-melting. Cis double bonds (but not trans double bonds) cause kinks in a hydrocarbon chain. These kinks prevent the chains of unsaturated fatty acids from packing together in an orderly manner, which leads to lower melting temperatures for oils. Vegetable oils are processed by partly (but not totally) hydrogenating their double bonds to produce margarine. In general, margarines sold in tubs have more unsaturated groups than those sold as sticks, since sticks have to be harder in order to maintain their shape.

PHOSPOLIPIDS AND LIPID BILAYERS

One of the primary alcohol groups of glycerol can be esterified with a phosphate ester, and the other two —OH groups can form esters with fatty acids. This kind of glycerol derivative is known as a **phosphoglyceride**. These lipids are the building blocks of animal cell membranes. Different phosphoglycerides are known, and they are named according to the nature of the phosphate group.

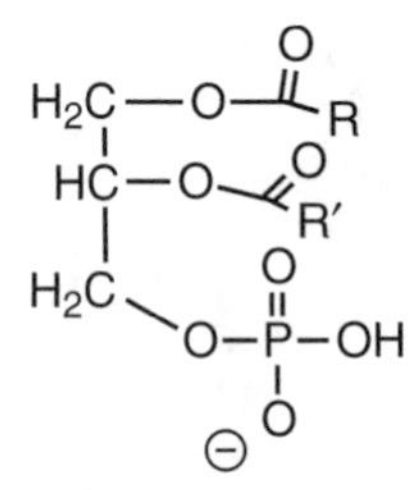

A phosphitidic acid

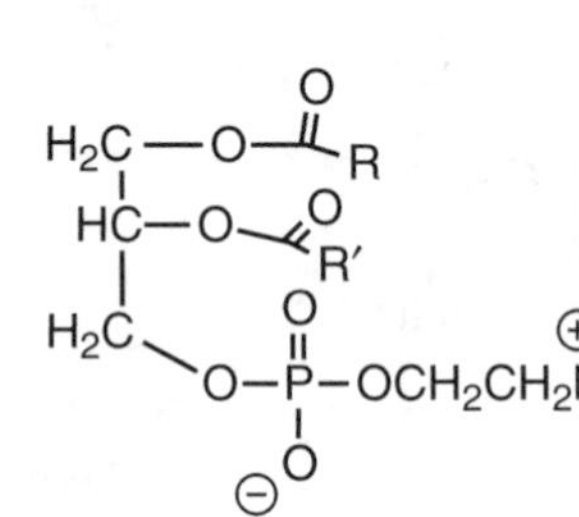

A posphatidylethanolamine

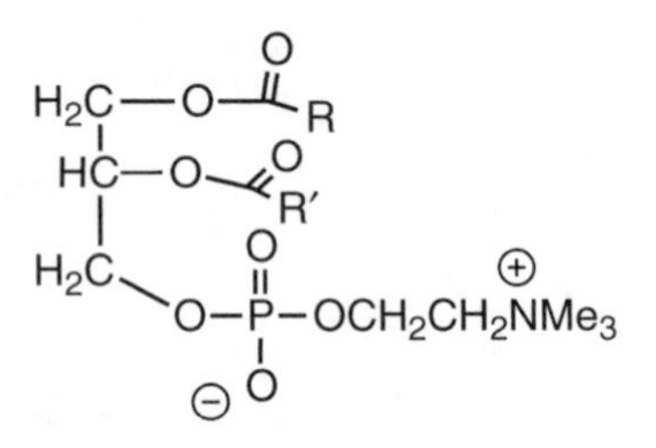

A phosphitidylcholine or lecithin

Other groups can bond to the phosphate group in addition to the ones shown. Real cell membranes are a complex mixture of different phosphoglycerides, **membrane-bound proteins**, cholesterol, and other compounds.

The polar phosphate group of a phosphoglyceride is known as the **polar head group**, and the hydrocarbon chains of fatty acids are referred to as **hydrocarbon tails**. When phosphoglycerides are added to water, they form a **lipid bilayer**, a cross section of which is shown here. In a cell membrane, which is composed of a lipid bilayer, one set of polar head groups faces the outside of the cell and the other interacts with the inside, or cytoplasm, of the cell. The hydrocarbon chains pack in the middle of the bilayer to avoid water. The lipid chains prevent polar molecules or ions from freely diffusing into or out of the cell.

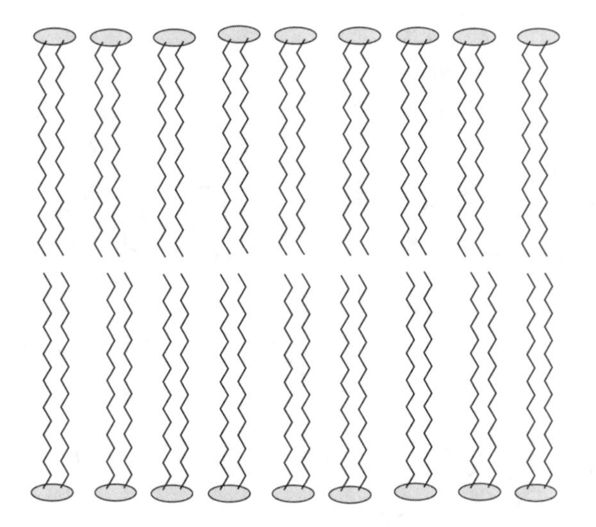

Recall that pure fatty acids form micelles in water, which have a much different structure than a lipid bilayer. The reason for this difference is the shape of a fatty acid vs. that of a phosphoglyceride. Fatty acids are approximately cone-shaped, and cones pack in three dimensions to form spheres, which approximate the shape of a micelle. Phosphoglycerides are approximately cylindrical. Cylinders pack parallel to one another, leading to the bilayer structure.

PROSTAGLANDINS

Prostaglandins are metabolites of the C_{20} polyunsaturated fatty acid arachidonic acid (see above for its structure). They participate in numerous important biological processes, including the regulation of blood platelet aggregation (involved in blood clotting and strokes), inflammation, blood pressure, and the induction of labor in pregnancy. Prostaglandins are required for life, but the human body cannot make arachidonic acid from C_2 precursors. One of several

C_{18} or C_{20} polyunsaturated fatty acids (so-called **essential fatty acids**) must be present in the diet so that it can be converted to arachidonic acid.

An iron-containing enzyme, of a class known as **cyclooxygenases**, creates the basic prostaglandin structure by adding two molecules of O_2 to arachidonic acid:

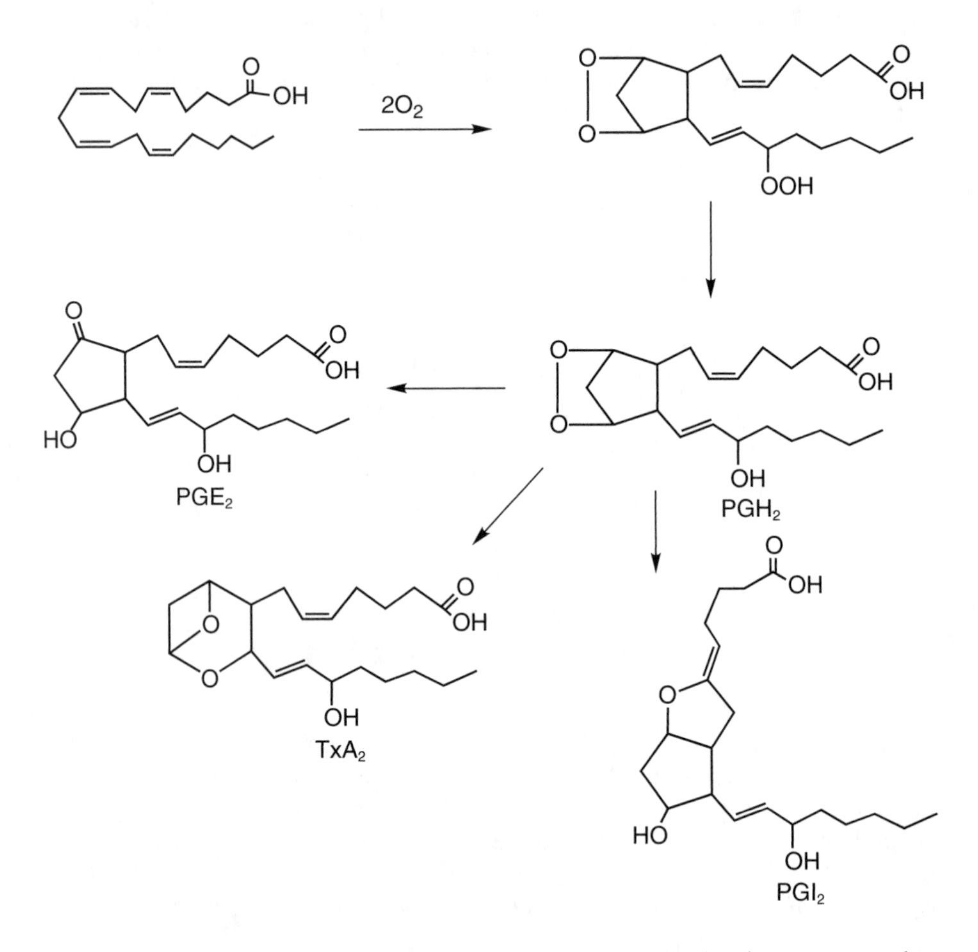

The cyclooxygenase enzyme first generates the prostaglandin **PGH_2**. Other enzymes then catalyze the rearrangement of this **dialkyl peroxide** (a molecule of type ROOR) to the thromboxane **TxA_2** and the prostacyclin **PGI_2**. Both types of molecules are extremely unstable. Thromboxanes constrict blood vessels and aggregate blood platelets, while prostacyclins dilate blood vessels and interfere with blood platelet aggregation. Prostaglandin **PGE_2** is one more stable example of the general prostaglandin family. This family has a five-membered ring with one C_7 chain, ending in —COOH and an adjacent C_8 chain with one allylic alcohol group.

Aspirin works by inhibiting the first step of prostaglandin biosynthesis, the cyclooxygenase reaction, by chemically modifying a serine —OH group of the enzyme:

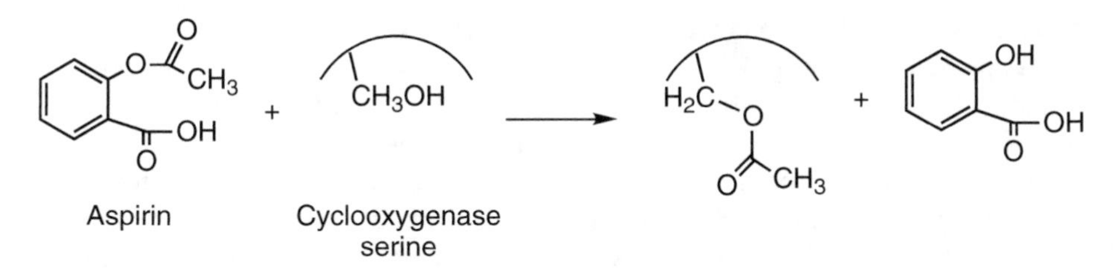

The acyl enzyme product is sterically unable to carry out the oxidation of arachidonic acid, leading to a reduction in the amount of prostaglandin synthesis. In time, the enzyme ester hydrolyzes back to the active serine form. Only one small dose of aspirin every other day is needed to tie up enough of the enzyme to lessen the rate of prostaglandin-promoted blood clotting, which can lead to heart attacks and strokes.

EXERCISE

Give a mechanism for the following synthesis of this prostaglandin analog.

Solution

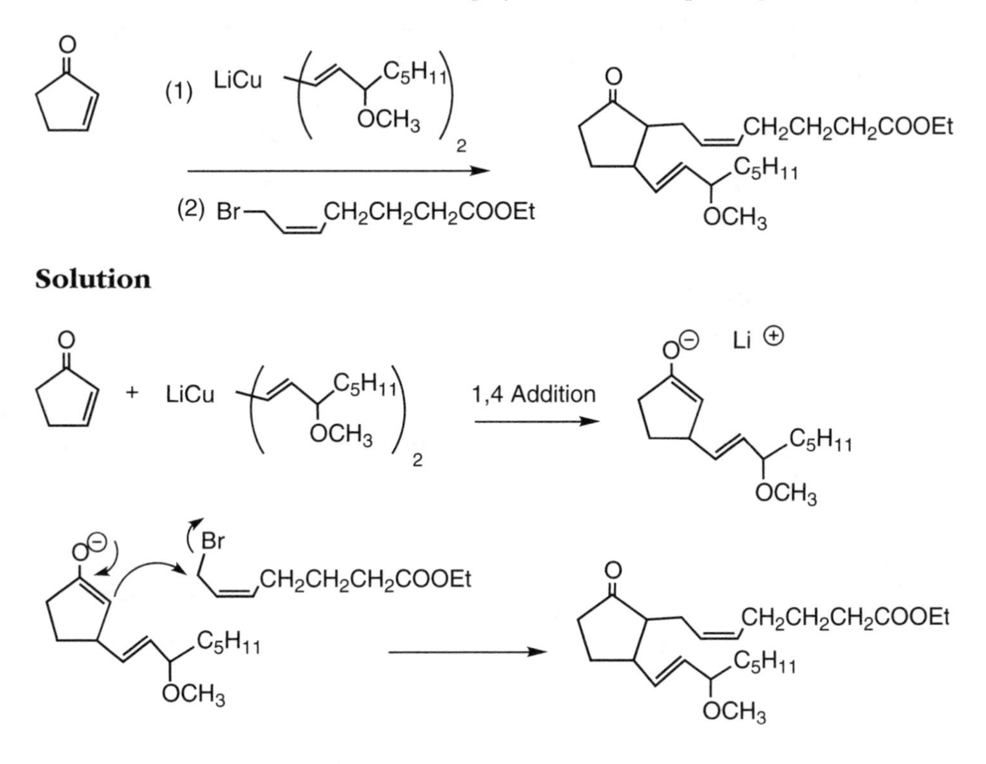

TERPENES AND STEROIDS

Structure and Biosynthesis of Terpenes

Terpenes are natural lipid products derived from common 5-carbon precursors. Many examples are known with 10, 15, 20, 25, 30, and 40 carbon atoms. C_{10} terpenes are known as **monoterpenes**, C_{15} compounds as **sesquiterpenes**, C_{20} compounds as **diterpenes**, and C_{25}, C_{30}, and C_{40} terpenes as **sesterpenes**, **triterpenes**, and **tetraterpenes**, respectively. Terpenes are responsible for many of the flavors, fragrances, and colors of plants. Some are plant hormones, pheromones, poisons, or drugs. While the original definition of a terpene covered only hydrocarbons, it is currently extended to related functionalized $(C_5)_n$ compounds.

Some structures of terpenes are shown here.

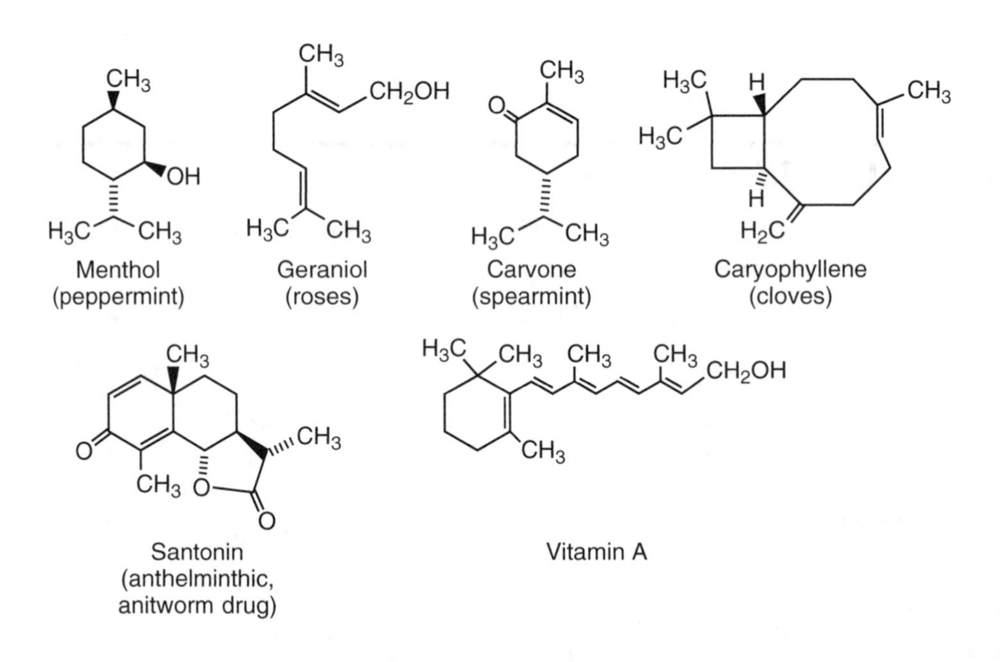

All terpenes are built up of **isoprene units**, normally by a head-to-tail joining:

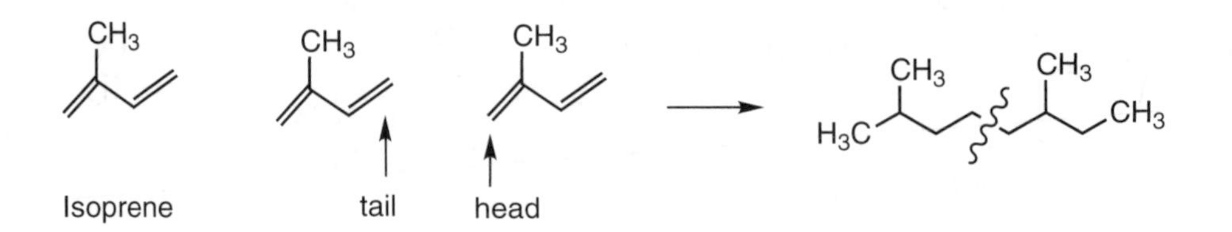

An isoprene unit is just a five-carbon unit with the connectivity of the molecule isoprene. Often terpenes (also called **isoprenoids**) have some cyclized units, forming one or more rings.

Examples One should be able to locate the isoprene units [$(C)_2C—C—C$] that make up any terpene. At times, there is more than one way to localize these units. When possible, favor the pattern that corresponds to head-to-tail joining.

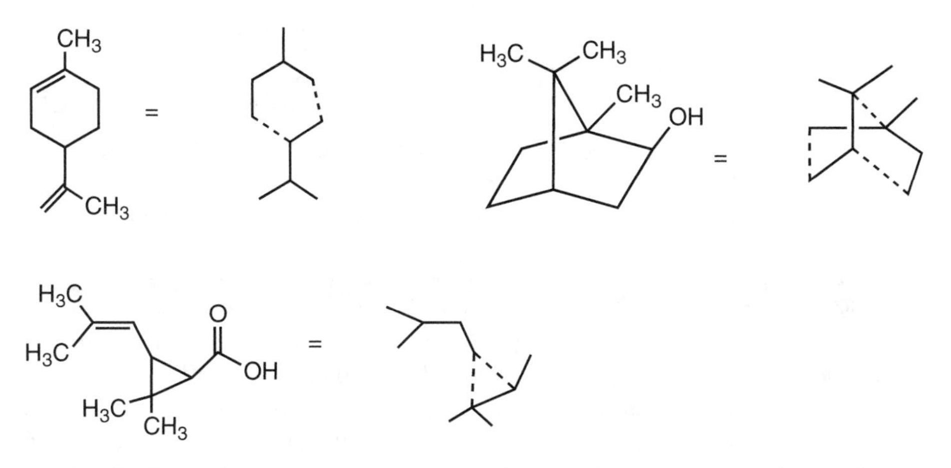

The biological isoprene unit is not the 1,3-diene isoprene but rather a C_5 alcohol derivative, **isopentenyl pyrophosphate** (3-methyl-3-butenyl pyrophosphate). This compound is enzymatically reversibly isomerized to the allylic alcohol derivative **dimethylallyl pyrophosphate** (3-methyl-2-butenyl pyrophosphate). The pyrophosphate group is an anhydride of phosphoric acid. It is a good leaving group and is the leaving group nature uses for many S_N1 and S_N2 reactions.

In a carbon-carbon bond-forming reaction, leading to the C_{10} terpene **geranyl pyrophosphate**, the double bond of isopentenyl pyrophosphate reacts as a nucleophile with the electrophilic carbon bonded to the allylic pyrophosphate group of dimethallyl pyrophosphate:

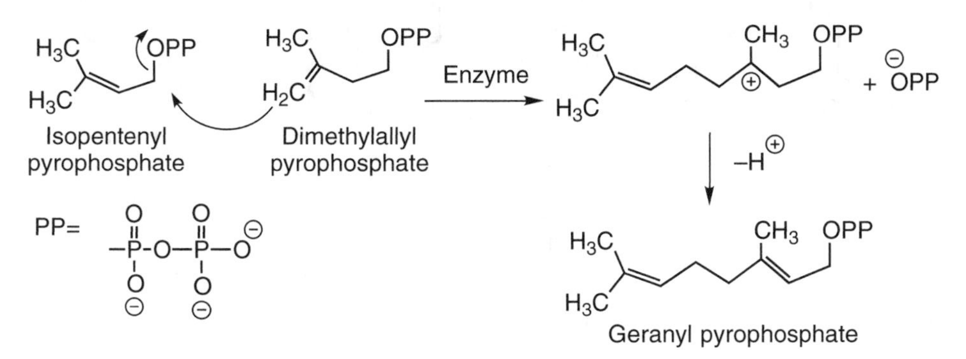

The intermediate carbocation in the addition reaction loses a proton to give geranyl pyrophosphate. A further addition reaction between geranyl pyrophosphate and dimethallyl pyrophosphate gives the sesquiterpene **farnesyl pyrophosphate**:

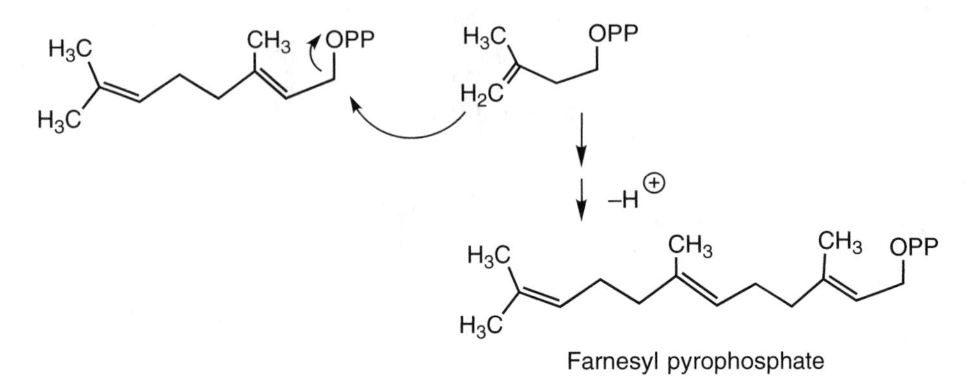

Farnesyl pyrophosphate

Further addition reactions can give diterpenes and larger terpenes.

Most cyclic terpenes arise from intramolecular reactions between a nucleophilic double bond in a linear terpene and one of the electrophilic carbons of the allylic pyrophosphate unit. The initial carbon-carbon bond-forming reaction generates a carbocation, and rearrangements or deprotonations lead to the natural product. In some cases further oxidations or other functional group transformations lead to the final product:

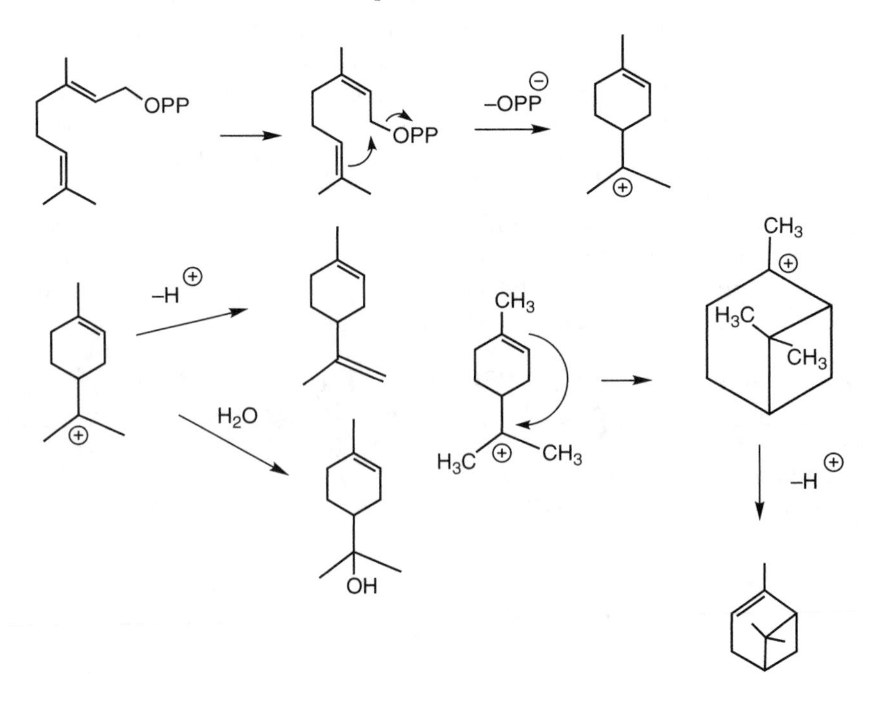

In a specific organism, a single enzyme produces one or a small number of closely related terpenes. However, the same terpene pyrophosphate can give a variety of different products in different organisms, depending on how a given enzyme folds up the substrate.

Structure and Biosynthesis of Steroids

Steroids are tetracyclic terpene-derived natural products. Some commonly encountered steroids are cholesterol (present in most animal cell membranes), the anti-inflammatory drug hydrocortisone, the sex hormones estrogen and testosterone, and a precursor of vitamin D, 7-dehydrocholesterol. The four rings that make up the steroid nucleus are designated A, B, C, and D:

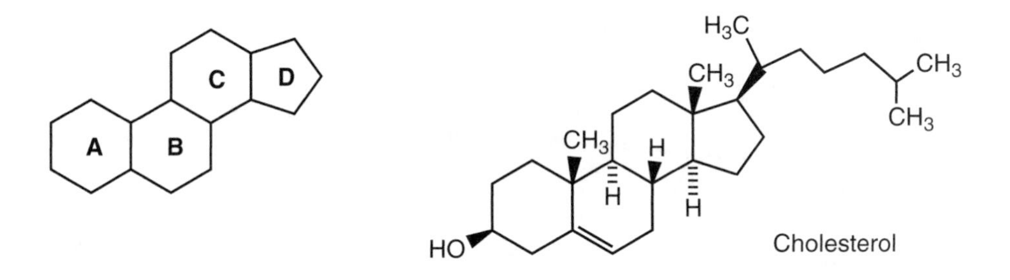

Most steroids have a carbon chain, sometimes aliphatic as in cholesterol and sometimes functionalized, attached to ring D.

Steroids are biosynthesized by one of the most remarkable reactions in all of chemistry and biochemistry. The acyclic triterpene squalene is the immediate precursor of the steroid ring system (squalene is itself unusual in that it consists of two farnesyl units joined tail to tail). Squalene is first selectively enzymatically oxidized to squalene 2,3-epoxide. Then an enzyme folds the squalene into a very precise conformation and initiates cyclization by protonating the epoxide oxygen. A series of carbocation-alkene additions take place sequentially, producing all four rings from one acyclic molecule:

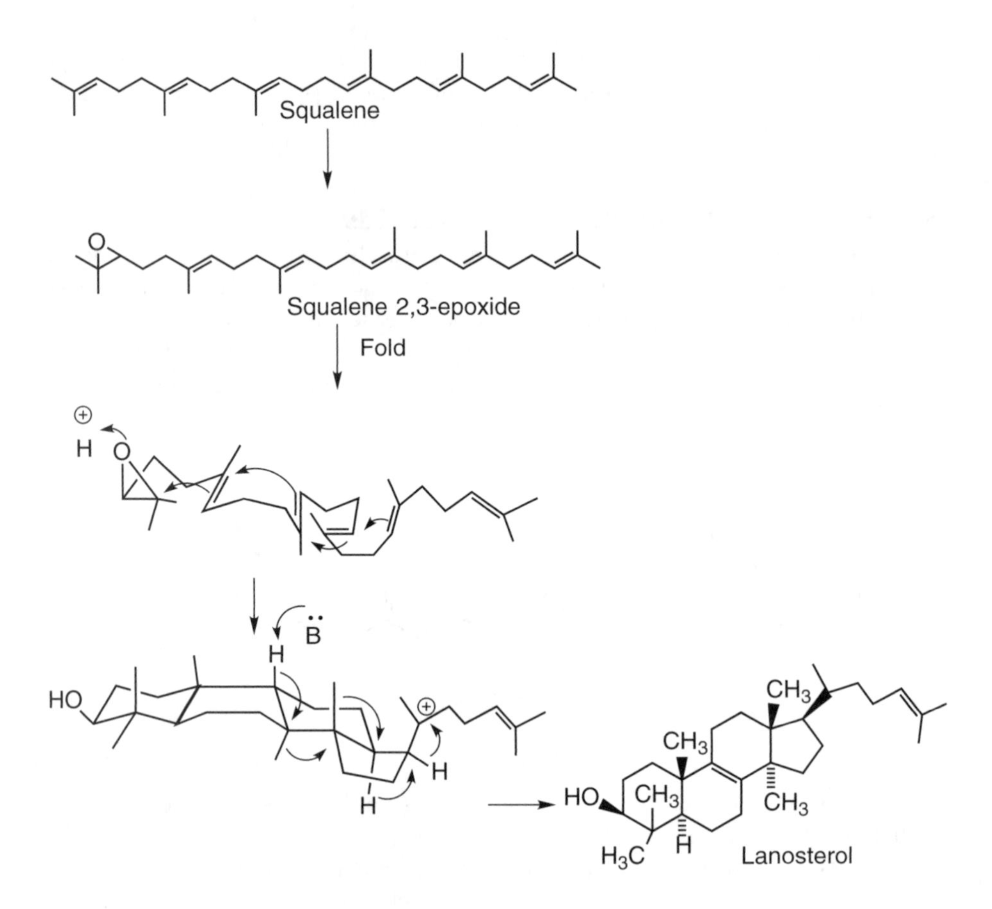

The carbocation first formed on cyclization undergoes a complex series of elimination rearrangements at the enzyme active site to give lanosterol. If one counts the number of carbons in cholesterol, one finds that there are 27, not the 30 expected for a triterpene. In a further complex series of reactions requiring many additional enzymes, lanosterol is converted to cholesterol. It is thought that lanosterol is further converted to cholesterol because cholesterol, but not lanosterol, causes lipid bilayer cell membranes to have biologically favorable fluidic properties.

Unlike cholesterol, sex hormones such as androsterone (an **androgen** or male sex hormone) and progesterone (an **estrogen** or female sex hormone) do not have long alkyl chains on ring D. Both cortisone (a **glucocorticoid** steroid hormone that decreases inflammation and is synthesized in the adrenal cortex) and aldosterone (a **mineralocorticoid** steroid hormone that regulates the absorption of ions by the kidneys, thereby regulating blood pressure) have two-carbon oxygenated side chains on ring D:

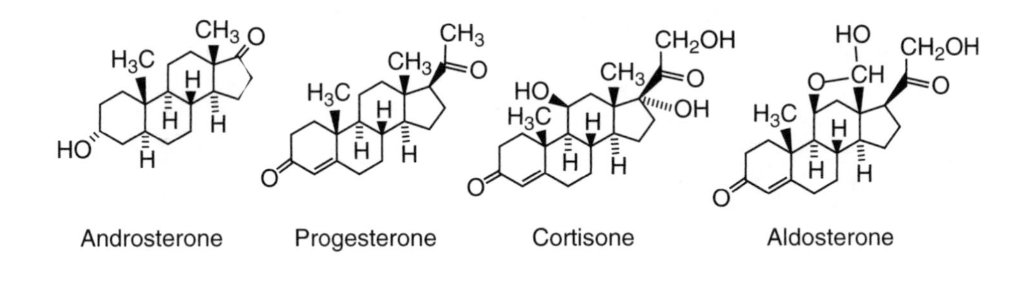

EXERCISE

Draw the three-dimensional (chair form) structure of androsterone.

Solution

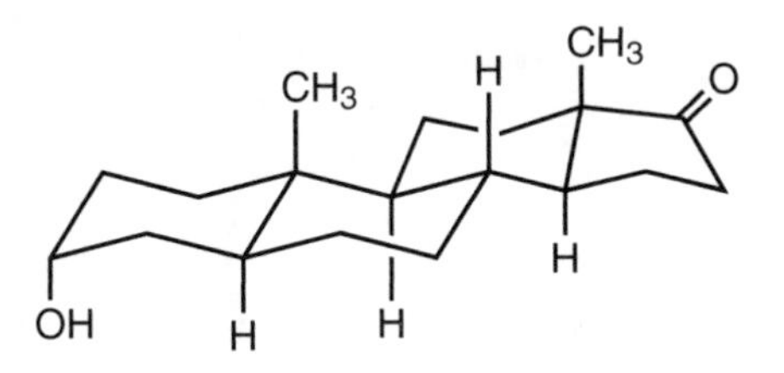

INDEX

B

C

D

E

F

O

P